MATERIALS SCIENCE AND TECHNOLOGIES

MAGNETIC THIN FILMS: PROPERTIES, PERFORMANCE AND APPLICATIONS

Materials Science and Technologies

Additional books in this series can be found on Nova's website under the Series tab.

Additional E-books in this series can be found on Nova's website under the E-books tab.

Condensed Matter Research and Technology

Additional books in this series can be found on Nova's website under the Series tab.

Additional E-books in this series can be found on Nova's website under the E-books tab.

MATERIALS SCIENCE AND TECHNOLOGIES

MAGNETIC THIN FILMS: PROPERTIES, PERFORMANCE AND APPLICATIONS

JOHN P. VOLKERTS
EDITOR

Nova Science Publishers, Inc.
New York

For permission to use material from this book please contact us:
Telephone 631-231-7269; Fax 631-231-8175
Web Site: http://www.novapublishers.com

Library of Congress Cataloging-in-Publication Data

Magnetic thin films : properties, performance, and applications / editor,
John P. Volkerts.
p. cm.
Includes index.
ISBN 978-1-61209-302-4 (hardcover)
1. Magnetic films. 2. Thin films--Magnetic properties. I. Volkerts, John
P.
TK7871.15.M3M34 2010
621.3815'2--dc22
2010051741

Published by Nova Science Publishers, Inc. † New York

CONTENTS

PREFACE

This new book presents topical research in the study of magnetic thin films, including developments in giant magnetoresistance and tunneling magnetoresistance based spintronic devices with perpendicular anisotropy; magnetic properties and domain wall propagation of FePt thin films; preparation of thin ferrite films on silicon substrates and dependence of texture of magnetic thin films on different substrates and orientations. Also discussed in this compilation are the properties, preparation and MEMS applications of rare earth magnetic thin films.

Chapter 1 - This chapter reviews rare-earth magnetic thin films, specifically samarium cobalt and neodymium iron boron, with a brief introduction to their similarities and differences. Understanding the preparation and principles behind magnetic thin films is crucial in evaluating their magnetic performances. In addition to the widely used pulsed laser deposition and magnetron sputtering techniques, new fabrication routes such as tape casting and screen printing that have emerged in the last decade as well as other potential alternatives will also be discussed.

Intense research over recent decades has improved the properties of these magnetic thin films to suit various industrial applications from magnetic recording to micro-electro-mechanical system (MEMS) devices. This review critically assesses these research activities and highlights the differences in properties and performance of the magnetic thin films produced by each of these techniques.

The practicality of these rare-earth magnetic thin films as an alternative to permanent magnetic thin films has been debated over the years. Therefore, the requirements for each application in terms of magnetic properties, adaptability and environmental stability will be evaluated and rationalized with respect to rare-earth magnetic thin films. The ongoing challenges of preparing Sm-Co and Nd-Fe-B magnetic thin films as well as the inherent limitations are also addressed.

Chapter 2 - Doped manganites i.e. $RE_{1-x}A_xMnO_3$ (RE is a rare earth ion, A is a bivalent cation e.g. Sr, Ca, Pb etc) are known to exhibit a large change in the resistivity on application of magnetic field. This phenomenon is known as magnetoresistance (MR). These manganites also exhibit a change in resistivity as a function of electrical field, known as electroresistance (ER). Both MR and ER are known to depend on the microstructure of the samples. In this chapter the authors present their results on the microstructure dependence of MR and ER in $La_{0.6}Pb_{0.4}MnO_3$ (LPMO) thin films with different cyrstallinity namely, single crystalline (SC), nano-crystalline (NC) and polycrystalline (PC) deposited using pulsed laser deposition technique. The inter-granular and intra-granular magnetic and transport properties of these films were investigated using ferromagnetic resonance (FMR) technique, temperature dependent magnetization, and low temperature charge transport measurements. The SC, NC and PC films exhibits the maximum MR values of 60%, 30% and 100% respectively at 1 Tesla magnetic field. Using FMR technique, we demonstrated that large numbers of spin glassy grain boundaries present in the NC films contributes to high field MR (HFMR), however the antiferomagnetic ordering of Mn spins present at grain boundaries in PC films contribute to low field MR (LFMR). To further delineate the role of grain boundaries from the bulk single crystalline regions, we prepared a bi-crystal grain boundary junction of LPMO. The current-voltage characteristics of LPMO grain boundary junction show that transport is dominated by multistep inelastic tunneling along with a small contribution of elastic tunneling, which is responsible for LFMR. It has been observed that only NC and PC films exhibits the ER property, it indicates that ER is an extrinsic properties of CMR materials and arises only due to magnetic disorder present at grain boundaries.

Chapter 3 - Thin films exhibit interesting magnetic properties including unusual anisotropy, and surface ferromagnetism. Recently, dilute concentrations of transition metal cation doped in semiconducting oxide thin films showed interesting ferromagnetic properties which has been proposed as a promising candidate for applications including spintronics and magneto-optoelectronics. Several different transition metal doped oxide thin films have shown range of magnetic properties from paramagnetism to spin clusters to weak or strong ferromagnetism. The broad range of properties in these oxide thin films is related to the defects. The defects include oxygen and metal cation vacancies, coherently intergrown secondary phase segregations within the oxide lattice, and surface and interface defects. In this chapter we will discuss the magnetic properties of transition metal doped wide bandgap semiconducting oxide thin films. After including a brief introduction to the research on the oxide thin films, the authors will discuss the work from their studies with focus on the role of defects, surface and interface disorder, and clusters on the magnetic properties of nanostructured and crystalline thin films. The role of oxygen vacancies in triggering the ferromagnetic properties is proposed. In the case of Co doped ZnO introduction of oxygen defects induces the ferromagnetism. Further the ferromagnetic to nonmagnetic state and vice versa can be switched by introducing and removing oxygen vacancy defects. To unambiguously establish the Co^{2+}/O^{2-}defects which significantly influence the magnetic property, the possible role played by the secondary impurity phase and the surface and interface disorder are studied. The authors show by the structural and magnetic property correlations of Co3-yZnyO thin films that the introduction of oxygen vacancy defects in these films lead to a crystalline Zn:CoO core with a disordered oxygen deficient amorphous Co:ZnO surface. The disordered surface layers are shown to be ferromagnetic which couples with the antiferromagnetic core therefore exhibiting an exchange bias coupling. The oxygen vacancy defects are also shown to induce weak ferromagnetism in Cr doped and a pure In2O3 thin film which is intrinsic to the material with the evidence of spin polarized charge carriers present in these thin films. Such materials exhibit O and In vacancies with associated In-In clustering which is predominantly found in the vicinity of the surfaces. With the nonmagnetic doping such as Cu in ZnO thin films ferromagnetic properties are observed with the film magnetization decreasing with the increasing concentration of Cu. Structurally at high concentrations of Cu (>3 at.%) doping, CuO nanophase inclusions within the basal planes of ZnO are evidenced. However the origin of ferromagnetism with large magnetic moment (1.6

μB/Cu) at low concentrations (<1 at.%) remains unclear with possible contributions from the coherently intergrown nanoclusters or due to the doping of Cu at Zn site with associated oxygen defects. Systematic investigations on the magnetic properties of CuO-ZnO heterostructures reveal the observed magnetization cannot be accounted for solely by spins localized near the CuO–ZnO interface or in the CuO layer. The possible role by the oxygen defects in the transition metal doped semiconducting oxides and even in undoped semiconducting oxides such as In2O3 and TiO2 will be discussed in this chapter.

Chapter 4 - In recent years, there has been a dramatic increase in the interest towards application of giant magnetoresistance (GMR) spin-valves and magnetic tunneling junctions (MTJs) with perpendicular anisotropy in spintronics, such as a spin transfer switching (STS) magnetic random access memory (MRAM), ultra high density magnetic information devices, and low field detection spin oscillators. This interest is driven by the fact that spin-valves and MTJs with perpendicular anisotropy are expected to provide technical promises such as high thermal and magnetic stabilities that will allow the realization of extremely low dimensional and high reliability devices in more advanced spintronic applications [1-2]. In this chapter, the recent developments in GMR and tunneling magnetoresistance (TMR) spin valves and devices with perpendicular anisotropy will be reviewed and presented with distinct seven sections to understand their technical roles in advanced spintronics applications. The first section will deal with the physical origins of perpendicular anisotropy of the magnetic materials used for GMR spin-valves studied up to now. The fabrication of the magnetic thin films with perpendicular anisotropy including the optimization of film deposition conditions and the fabrication process of nano-meter sized devices will be included in this section. The GMR performance in various spin-valve structures with perpendicular anisotropy and their magnetic and thermal stabilities will also be disused in this section. The second section will focus on the physical nature of GMR and its correlation with interlayer coupling in different kinds of spin valves with perpendicular anisotropy. A newly proposed physical model of the GMR behavior interpreted in terms of the physical correlation between perpendicular anisotropy and magentostatic energy and its extension to the understanding of underlying physics of perpendicular interlayer coupling including RKKY oscillation and Néel coupling types of indirect exchange coupling will be dealt with in this section. The third section will discuss on a physical model of exchange bias and the effects of nanopatterning on the exchange bias characteristics in perpendicularly magnetized ferrimagentic/anti-ferromagentic thin films with perpendicular anisotropy for optimizing exchange biased GMR spin valves with perpendicular anisotropy. The fourth section looks at anomalous peak behavior observed in Hall effect measurements of exchange biased spin valves with perpendicular anisotropy. The fifth section will focus on magnetic tunnel junctions (MTJs) with perpendicular anisotropy. The basic theories of tunneling magnetoresistance (TMR), the initial and the recent research achievements of MTJs with perpendicular anisotropy in current spintronics will be reviewed and presented in this section. The sixth section will look at the current and potential applications of GMR and TMR devices with perpendicular anisotropy including spin-transfer switched MRAM and spin oscillator devices. The physical mechanisms and the research into the optimization of perpendicular anisotropy materials for these applications will be discussed. Finally, this chapter will be concluded with the survey on the advantages of GMR and TMR devices with perpendicular anisotropy and the future challenges targeting for the further developments in advanced spintronics applications.

Chapter 5 - In this chapter, a variation of magnetic domain and correlated magneto-impedance of a micro-fabricated rectangular-shaped magnetic thin film is discussed. In a certain case of the magnetic element, which is made from an amorphous film with in-plane uniaxial anisotropy and low magnetostriction, the Landau-Lifshitz-like magnetic domain is observed. A change in structure of magnetic domain occurs as a function of external magnetic field. The Landau-Lifshitz magnetic domain consists of contiguous domains with anti-parallel magnetic-momentum with existence of closure domains in the edge area of the striped element. The dimensions of the rectangular strip discussed in this chapter, especially the width, must be almost in the same order of the width of a domain area. The high-frequency impedance of the element is determined by the high-frequency permeability, and the permeability is determined by the domain structure, the direction of magnetic moment in each domain and fixed force of the magnetic moment. The variation of the domain structure and correlated magneto-impedance of the element is characterized by a direction of magnetic easy axis. In this chapter, a magnetic phenomenon and an analytical explanation is shown for the rectangular magnetic element with the width of some tens of microns, therefore these element has the Landau-Lifshitz-like magnetic domain. The phenomenon is based on the variation of magnetic domain and correlated magneto-impedance induced by external magnetic field as a parameter of the in-plane direction of uniaxial magnetic easy axis.

Chapter 6 - Spintronics belongs to one of the most quickly developing areas of science and technology; it is based on the control of the processes of transfer of spin current between the elements of electronic devices. Researches of a spin dependent transport and spin relaxation in solids, search for new materials with the high degree of an electron spin polarization and development of methods for active control of spin states in solid-state circuits constitute the main directions of the spintronics.

Chapter 7 - The ferrites are ceramic magnetic materials consisting of the entire family of iron oxides including spinels, garnets, hexaferrites, and orthoferrites. These materials have significant potential in applications ranging from millimeter wave integrated circuitry to transformer cores and magnetic recording.

The ferrite thin films have application in microwave devices, micro-inductors, micro-transformers, magnetic recording, gas sensors, catalysts and as shield for the electromagnetic interference. The RF– magnetron sputtering is a useful technique for depositing dense and homogeneous thin films of insulating ferrite compounds. The Authors have used RF-magnetron sputtering to deposit Cu , Zn and CuZn spinel ferrite thin films on glass substrates at room temperature in oxygen (O_2), argon (Ar) and mixture of ($Ar+O_2$) environment.

A detailed study of the effect of process gas environment and pressure on the crystal structure, magnetic and optical properties of these ferrite thin films is undertaken. The XRD and AFM studies confirm the nanocrystalline nature of the as deposited films. The observed changes in the thin film crystal structure, magnetization and optical properties as a function of deposition conditions are attributed to the random distribution of the cations among the tetrahedral A-sites and octahedral B-sites during the deposition process.

The presence of multivalent cations with differing sites preferences gives rise to structural and magnetic disorder. The change in disorder leads to the change in the properties of thin films.

Chapter 8 - Chemically ordered $L1_0$ (CuAu-I type structure) FePt thin films with a FCT (face centered tetragonal) structure possess a high K_u (magnetic anisotropy constant) value ($>10^7$ erg/cm^3) that allows them to have very small thermally stable magnetic grains (~ 2.6

nm) and makes them the most promising candidates for recording media having an ultra-high areal density. Usually, as-deposited FePt alloy thin films are a disordered FCC (face centered cubic) phase and tend to show a (111) texture.

A high temperature (above 550 °C) post annealing is needed to form an ordered $L1_0$ phase. In order to develop FePt films as perpendicular magnetic recording media, $L1_0$ FePt ordering temperature has to be reduced and the easy axis of the films should be perpendicular to the film surfaces. Moreover, the grain size and exchange coupling between grains should be decreased. Thus, this chapter reviews two main issues for $L1_0$ FePt thin film to be successful perpendicular magnetic recording (PMR) media: (1) how to lower $L1_0$ FePt ordering temperature while getting crystallographic texture controlled at the same time, and (2) how to reduce grain size and inter-granular exchange coupling. The first issue is related to the study of the intrinsic relationships between the chemical ordering, lattice constant, and anisotropy energy of FePt films. It is demonstrated that residual stress can assist to expand the a-axis and shrink the c-axis of FePt films and thus favors the chemical ordering of the films at relatively low temperatures. The second issue covers the study of the structural and magnetic properties of FePt films grown on a CrX alloy underlayer (X=Mo, Ru, Ti, or W). It is demonstrated that the texture of FePt films strongly depends on the texture of CrX underlayers, for which Cr (200) texture and substrate temperature (T_s) are the key parameters for the growth of the FePt (001) films. To decrease the grain size and maintain the (001) preferred orientation in FePt films, the effect of carbon additive on the structural and magnetic properties of FePt films is also discussed. A good FePt (001) texture can be maintained even with C content up to 20 vol.% in FePt films by a fine control of sputtering deposition process.

Chapter 9 - 10 years ago, Dietl theoretically predicted that the ferromagnetism (FM) at high temperature could be obtained in many semiconductors such as ZnO, GaAs, GaN, etc., if the authors dope Mn along with a certain concentration of holes. Reports of Curie temperatures well above room temperature for wide-band gap oxides doped with a few percent of transition-metals have triggered intense interest in these materials as potential magnetic materials for spintronics.

The origin of the magnetism is debated; in some systems, the FM can be attributed to nanoparticles of a ferromagnetic secondary phase, but in others, properties are found which are incompatible with any secondary phase, and an intrinsic origin related to structural defects is implicated: Their experimental results on TiO_2, HfO_2, In_2O_3, ZnO, and SnO_2 thin films have confirmed that magnetism is certainly possible in pristine semiconducting oxides, and the observed FM is most probably due to oxygen vacancies and/or defects. The assumption for FM due to oxygen vacancies/defects in TiO_2 thin films is strongly confirmed by X-ray magnetic circular dichroism measurements (XMCD): There is a presence of XMCD signals at both O K and Ti L2,3 edges. It shows that the FM in TiO_2 films stems from both O-2p and Ti-3d electrons. The authors theoretical model also suggests that confinement effects must play a key role in shaping up magnetic properties of thin films, or more generally, of low dimension structured oxides.

A new picture of defect-based magnetism is emerging. There is a need for a type of dilute magnetic thin films— a system that is easy to prepare and reproduce. Device applications can follow by design. Once the mechanism is better understood, the next challenge will be to generate stable and controllable defect structures where the authors can get the benefit of this unusual high-temperature FM.

Chapter 10 - In this chapter, the authors present the magnetic properties of thin epitaxial FePt layers (5-40 nm) deposited by molecular beam epitaxy on MgO(001) and Pt(001) samples. The tuning of growth conditions, as well as the use of ion irradiation, can be used to control the ordering of the FePt alloy within the $L1_0$ phase, and thus the uniaxial magnetic anisotropy of these samples.

The authors then focus on samples with high anisotropy, which possess ultra-thin domain walls. The authors present the magnetic properties of these samples, as well as some of their transport properties. Using microstructural characterizations, domain observations and micromagnetic calculations, The authors show that hysteresis properties are linked to the interaction of domain walls with structural defects. In these layers, the magnetization reversal can be analyzed as an invasion percolation process without trapping, leading to fractal geometries of the reversed domain. This thermo-activated growth process involves domain wall motion by avalanches, where one single depinning event controls the propagation dynamics over large distances.

The authors show that this property can be used in nanostructures to observe the interaction of a domain wall with a single defect. This allows realizing current-induced depinning in nanostructures, and measuring the spin-transfer efficiency in FePt samples.

Chapter 11 - The magnetoimpedance (MI) effect is defined as the change of the impedance experienced by an ac current flowing through magnetic materials when an external dc magnetic field is applied. This effect is promising in the application of micromagnetic field sensors with high sensitivity and quick response.

If the film is magnetically soft and has a well-defined anisotropy axis and large saturation magnetization, it will help enhance the MI effect due to the increased interaction with the external magnetic field. It is well known that FeCo based alloy thin films exhibit very large saturation magnetization.

However, FeCo based thin films also have very large saturation magnetostriction. Such large saturation magnetostriction causes the degradation of soft magnetic properties and no-detectable magnetoimpedance effect. This paper used a soft magnetic layer or nonmagnetic layer as interlayer to improve the soft magnetic properties of FeCo based films effectively, and investigated the effect of laminating layer on the microstructure and magnetoimpedance effects of FeCo based single or sandwiched thin films. Results show that the MI ration of single or sandwiched thin films with laminating Permalloy layer is enhanced obviously. The improvement of MI ratio of single or sandwiched films with Permalloy laminating layer can be explained by exchange induced ripple reduction mechanism.

Chapter 12 - Thin-films of $(Ni,Zn)Fe_2O_4$ were grown by means of RF sputtering on Si (100) and (111) substrates. Films with a thickness up to 100 nm were prepared for analysis purposes, enabling the optimization of the sputter process. The purpose of these ferrite thin films is the preparation of MFM cantilever-coatings for use with a high-frequency magnetic force microscope (HF-MFM). As a basis for these probes, the authors employ commercial, micromachined silicon cantilevers which exhibit (100)-oriented Si surfaces on the shank, and (111)-oriented surfaces on the pyramid-like tip end. The substrates were not additionally heated during the evaporation.

Chapter 13 - The knowledge about microstructure and grain morphology of magnetite materials is very important in order to understand the partially puzzling magnetic properties. The authors have, therefore, investigated a variety of magnetite samples concerning details of their respective microstructures. (001)- and (111)-oriented magnetite thin films were grown on MgO substrates (film thickness 200 nm) by means of oxygen-plasma-assisted molecular beam epitaxy and by laser-ablation.

Chapter 14 - Measurements on electrical resistivity of thermally evaporated $Mn_{100-x}Ni_x$ films (with x = 0.5, 1.5 and 2.5 at. %) have been carried out over the temperature range from 300 to 1.4 K using the van der Pauw four probe technique. The films were grown on a glass substrate held at a temperature of 300 K in an ambient pressure of 2 x 10^{-6} torr. All the films show the usual resistance minimum, a notable characteristic of α-Mn but their low temperature behavior reveal a tendency towards saturation of the resistivity as the temperature approaches zero. This saturation is reminiscent of the Kondo effect.

In: Magnetic Thin Films
Editor: John P. Volkerts
ISBN: 978-1-61209-302-4

Chapter 1

A REVIEW OF PREPARATION, PROPERTIES AND APPLICATIONS OF RARE EARTH MAGNETIC THIN FILMS

James Wang [1]***, Jo Ann Gan*** [1]***, Yat Choy Wong*** [2] ***and Christopher C. Berndt*** [1]

[1] Industrial Research Institute Swinburne, Faculty of Engineering and Industrial Sciences, Swinburne University of Technology, Hawthorn 3122, Australia

[2] Faculty of Engineering and Industrial Sciences, Swinburne University of Technology, Hawthorn 3122, Australia

ABSTRACT

This chapter reviews rare-earth magnetic thin films, specifically samarium cobalt and neodymium iron boron, with a brief introduction to their similarities and differences. Understanding the preparation and principles behind magnetic thin films is crucial in evaluating their magnetic performances. In addition to the widely used pulsed laser deposition and magnetron sputtering techniques, new fabrication routes such as tape casting and screen printing that have emerged in the last decade as well as other potential alternatives will also be discussed.

Intense research over recent decades has improved the properties of these magnetic thin films to suit various industrial applications from magnetic recording to micro-electro-mechanical system (MEMS) devices. This review critically assesses these research activities and highlights the differences in properties and performance of the magnetic thin films produced by each of these techniques.

The practicality of these rare-earth magnetic thin films as an alternative to permanent magnetic thin films has been debated over the years. Therefore, the requirements for each application in terms of magnetic properties, adaptability and environmental stability will be evaluated and rationalized with respect to rare-earth magnetic thin films. The ongoing challenges of preparing Sm-Co and Nd-Fe-B magnetic thin films as well as the inherent limitations are also addressed.

1. Introduction

1.1. Historical Basis

The first permanent magnet, known as loadstone (magnetite, Fe_3O_4) [1], dates back to the 600 B.C and was found to be magnetic in its natural state. The choice of permanent magnets and their use expanded as knowledge on magnets and their properties increased. In 1931, Mishima [2] discovered the magnetic alloy of Al-Ni-Fe and Al-Ni-Co-Fe which became known as "alnico".

Although widely used as a modern permanent magnet, applications of alnico are limited by brittleness and low magnetic characteristics, leading to a lost in popularity and replacement by ceramic magnets and rare-earths. The development of hard hexagonal ferrite in the 1950s quickly gained the attention of many researchers due to their high coercivity compared to alnico, and most importantly the low processing and material costs of this newly discovered ceramic magnet.

Although both barium and strontium ferrite have inferior magnetic properties to rare-earth magnets that were discovered much later, they still hold an important place in the market due to their reasonable cost-to-performance ratio.

On the other hand, rare-earth (RE) - transition metal (TM) compounds (such as samarium cobalt, $SmCo_5$ or Sm_2Co_{17} and neodymium iron boron, $Nd_2Fe_{14}B$) are the most powerful modern magnets known. $SmCo_5$ discovered in the 1960s, proved to have superior properties compared to the earlier magnets. One drawback though, is that it is more costly due to the presence of cobalt in the magnet and also due to the fact that samarium is less abundant compared to other rare-earth elements. The major breakthrough in magnetic industry came in the year 1984 when $Nd_2Fe_{14}B$ was invented [3-5]. It has the best overall magnetic properties among all other permanent magnets and is less costly compared to $SmCo_5$.

Despite cost advantages, $Nd_2Fe_{14}B$ has disadvantages; i.e., relatively lower Curie point and high sensitivity to corrosion. The advantages and disadvantages of two popular RE-TM magnets need to be balanced in determining the appropriate material for a certain application. Table 1. lists the properties of typical industrial hard magnetic materials.

Table 1. Typical industrial hard magnetic materials

Materials	Remanence Br (T)	Coercivity, Hc (kA/m)	Intrinsic coercivity, iHc (kA/m)	Maximum energy product, (BH)max (kJ/m3)	Curie temperature Tc (°C)
Alnico, Al-Ni-Co-Fe	1.12 – 1.25	54.1 – 109.4	-	43.8 – 83.6	880
Hard ferrite, Ba/SrFe12O19	0.38 – 0.40	191.0 – 290.5	202.9 – 318.3	27.1 – 31.8	469
Neodymium iron boron, Nd2Fe14B	1.00 – 1.29	763.9 – 986.8	1830.3 – 3262.7	191.0 – 318.3	310 – 400
Samarium cobalt, SmCo5	0.87 – 1.07	676.4 – 775.9	2069.0 – 2387.3	143.2 – 206.9	720 – 800

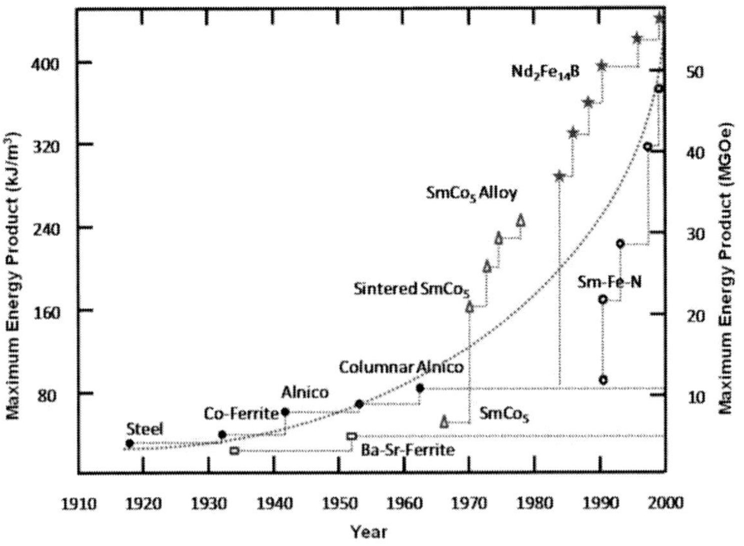

Figure 1. Evolution of maximum energy product of permanent magnets *(Adapted from Ref.* [7], *Copyright (1995), with permission from Elsevier).*

Figure 1. outlines the development of permanent magnets and the improvements in maximum energy product since 1910. The energy product usually specifies the quality of a magnet. Figure 1. indicates that the maximum energy product of hard magnetic materials has almost doubled every 12 years and follows an exponential trend. A more detailed early history of permanent magnets has been given by da Costa Andrade [1] and more recently by Stijntjes and Van Loon [6]. Prior to the discovery of rare-earth magnets, the applications of hard magnetic thin films were limited and the majority of magnets were confined to conventional bulk materials due to their weak magnetic properties. However, since the introduction of samarium cobalt and neodymium iron boron, the research and development of hard magnetic thin films has advanced. Various techniques are available for preparing these hard magnetic thin films and patterning them into desired geometries for micro-electromechanical system (MEMS) applications. Commercial recognition and interest in MEMS technology is growing rapidly because micro-devices can be manufactured in large volumes at a more attractive cost. For certain applications, the successfully prepared hard magnetic thin films will potentially lead to the incorporation of micro-systems or devices, and enable design and development of high energy and low power MEMS actuators with micrometer-sized features. These complex micro-machines can have many functions, including sensing, communication and actuation. Extensive applications for these micro devices exist in both commercial and defence systems. Actuators with micrometer dimensions will be more commercially viable in industrial and domestic markets since the potential consumer units are significantly greater.

1.2. Crystal Structures

There are two important magnetic Sm-Co compounds, namely $SmCo_5$ and Sm_2Co_{17}. These compounds are distinguished from all other compounds, such as Sm_3Co, Sm_9Co_4,

$SmCo_2$, $SmCo_3$ and Sm_2Co_7 due to their extraordinary magnetism behaviour. The crystal structure of these two compounds confers their physical and magnetic properties and their alignment is extremely important when it comes to achieving a high magnetic energy density. It is emphasized that the crystal symmetry could deviate immediately below the Curie temperature; and that this deviation will become more apparent as the temperature approaches the spin-reorientation temperature.

$SmCo_5$ has the same crystal structure as $CaCu_5$, with the space group *P*6/mmm [8, 9]. The hexagonal crystal symmetry of $SmCo_5$ is shown in Figure 2. by Sayama *et al.* [10]. The Sm atoms are located in (0,0,0) while the Co atoms are located in two crystallographic sites as follows: two Co(I) in ± (1/3,2/3,0) and three Co(II) in (1/2,0,1/2), (0,1/2,1/2), (1/2,1/2,1/2). The *a* and *c* lattice constants for $SmCo_5$ are 5.002 and 3.964 Å, respectively [8]. On the other hand, Sm_2Co_{17} is rhombohedral with the crystal structure similar to Th_2Zn_{17} and space group of $R\bar{3}m$. The structures of Sm_2Co_{17} are derived by substituting each third Sm atom in the basal plane from $SmCo_5$ with a dumbbell pair of Co atoms. The Th_2Zn_{17} rhombohedral structure results when the substituted layers stack in an '*abcabc...*' sequence.

A different rare-earth element instead of Sm in the same composition may result in a different stacking sequence which in turn will lead to a different crystal symmetry. The substitution leads to expansion of the *a* and *c* lattice constants, which are 8.395 and 12.216 Å respectively. More details on crystal structure of Sm-Co compounds are available elsewhere [9, 11].

On the contrary, the composition of a tetragonal phase Nd-Fe-B that exhibits remarkable magnetic properties has been debated extensively. Separate studies by Herbst *et al.*[3] and Givord *et al.*[12], who used neutron-diffraction and x-ray measurements, verified that the composition of the compound with tetragonal crystal structure was $Nd_2Fe_{14}B$. This compound consisted of a tetragonal structure with the $P4_2/mnm$ space group that formed an eight layer arrangement perpendicular to the *z*-axis. There are four $Nd_2Fe_{14}B$ units of 68 atoms per unit cell as shown in Figure 3. All the Nd and B atoms together with four Fe (*c*) of the 56 Fe atoms are located at the z = 0 and z = 1/2 planes. The rest of the Fe atoms form three puckered, fully connected hexagonal nets between the planes.

The hexagonal arrangements are distorted slightly by the Fe(k_1), Fe(k_2), Fe(j_1) and Fe(e) sites that form two layers (z = 1/8 and z = 3/8) rotated by approximately 30° to one another. A net of Fe(j^2) sites (z = 1/4) located above or below the centres of the hexagons in neighbouring layers are enclosed by these two layers.

The centre of a trigonal prism (refer to Figure 4) formed by the three closest Fe atoms above and three below the basal (z = 1/2) planes is occupied by each B. The linking of Fe layers contributes to the stability of the structure. Three Nd atoms are bonded to each of the B through the rectangular prism faces. The prisms exist in pair and have a common Fe(*e*)-Fe(*e*) edge, sharing two rare-earth atoms [13, 14]. The *a* and *c* axis lattice parameters appear to be 8.8 and 12.2 Å respectively [14, 15].

The coordination of the atoms and distances between them has been given in several studies [3, 12, 16, 17]. The structure and magnetic properties of $Nd_2Fe_{14}B$ have been compared with those of Nd_2Fe_{17} [18] and contended to have relatively higher T_c (627 K as oppose to 330 K for Nd_2Fe_{17}) due to a stronger exchange interaction between atoms [3].

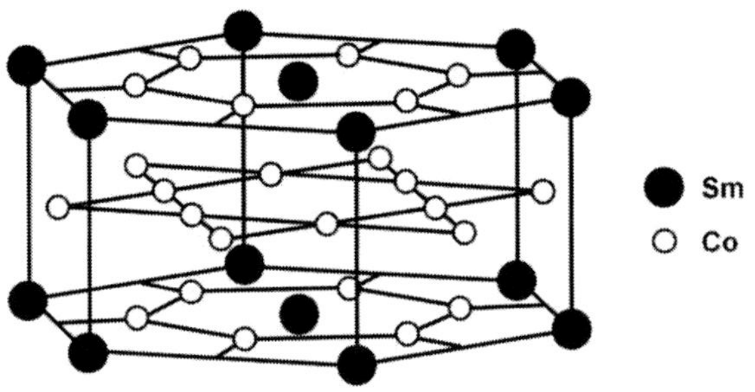

Figure 2. Hexagonal unit cell of $SmCo_5$ crystal structure *(Reprinted with permission from Ref.* [10]. *Copyright (2004) by Institute of Physics Publishing Ltd.).*

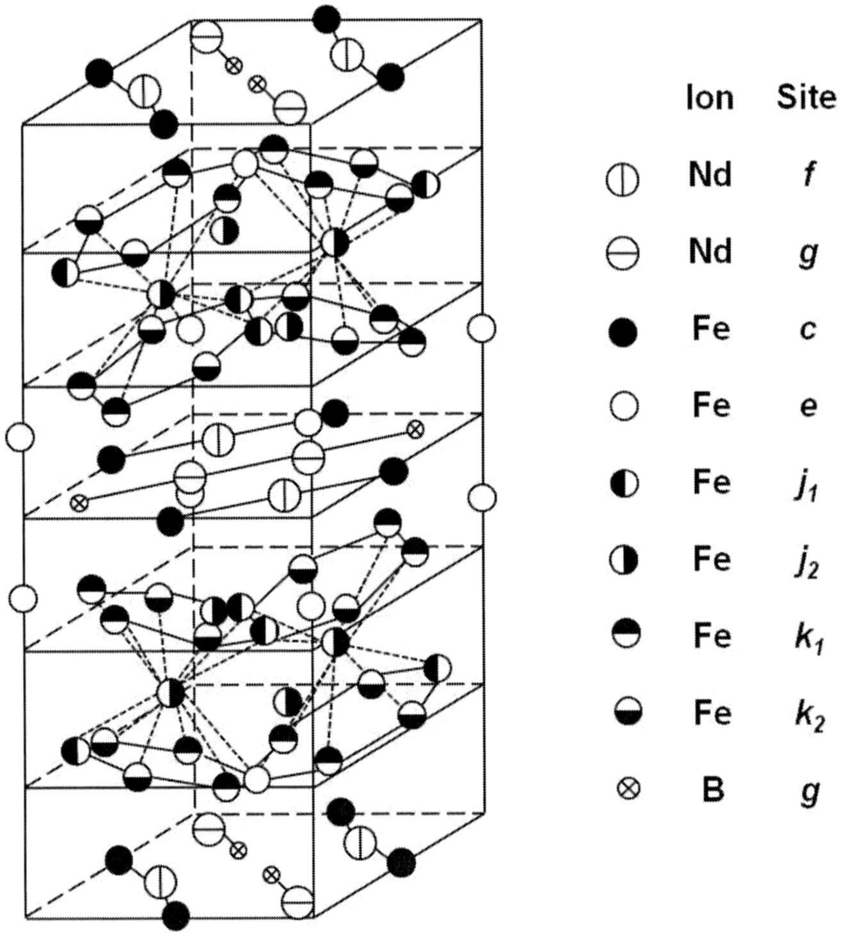

Figure 3. Crystal structure of $Nd_2Fe_{14}B$ showing the different Fe sites *(Reprinted with permission from Ref.* [3]. *Copyright (1984) by the American Physical Society*).*

---

* Readers may view, browse, and/or download material for temporary copying purposes only, provided these uses are for non-commercial personal purposes. Except as provided by law, this material may not be further reproduced, distributed, transmitted, modified, adapted, performed, displayed, published, or old in whole or part, without prior written permission from the American Physical Society

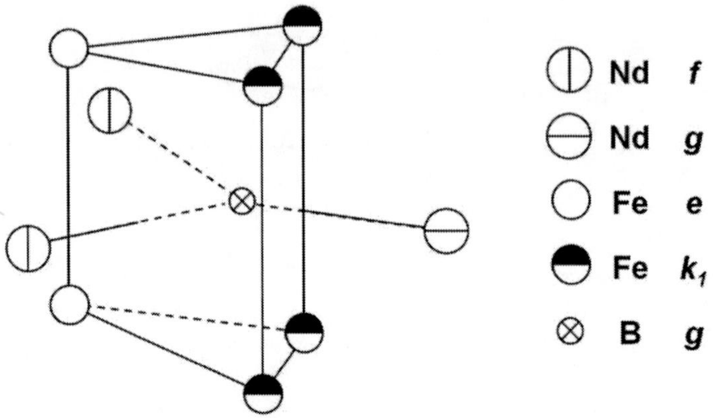

Figure 4. A trigonal prism containing B atom in the $Nd_2Fe_{14}B$ structure *(Reprinted with permission from Ref.* [3]. *Copyright (1984) by the American Physical Society*).*

1.3. Principles of Magnetic Behaviour

Both Sm and Nd are light rare-earth elements. Thus, the magnetism in both Sm-Co and Nd-Fe-B are very much similar, both of which stems from the angular orbital momentum and electron spin alignment. The magnetism of $SmCo_5$, Sm_2Co_{17} and $Nd_2Fe_{14}B$ will be discussed concurrently by referring to them as rare-earth (RE) – transition metal (TM). Although the same principle applies to other RE-TM compounds, it only applies to light rare-earths that include Y, La, Ce, Pr, Nd and Sm. The TM in this context refers to Co and Fe. A detailed discussion on the magnetism of RE-TM has been provided in several reviews. [9, 13, 14, 19]

Generally, the magnetic moment of a system may originate from two sources: (1) motion of electric charges (e.g. electric currents) and/or (2) intrinsic magnetism of elementary particles (e.g. electrons). The latter applies to RE-TM compounds. The total magnetic moment that emanates from intrinsic magnetism is the product of the following: (i) unpaired electron spins (if any), giving a total spin value, (ii) orbital motion of its electrons, giving a total orbital angular momentum value, and (iii) combined magnetic moment of nuclear spins.

Both the 4*f* electron shell of RE and 3*d* electron shell of TM are only partially full. The magnetism of RE-TM arises from these unpaired electrons. The shielding of 4*f* electrons by 5*s* and 5*p* electrons in RE gives rise to the contribution of both spin and orbital to the total magnetic moment; i.e., contributions (i) and (ii) in the preceding paragraph. Total magnetic moment in TM on the other hand is derived mainly from the electron spin of the unpaired 3*d* electrons because its orbital angular momentum is mainly quenched; i.e., only contribution (i) in the preceding paragraph. It is known that the spins of RE and TM atoms are always aligned antiparallel in RE-TM compounds. There are two arguments that explain this spin arrangement.

The first argument, by Wallace [20] and Buschow [21], proposed that the spin arrangement results from the strong positive 3*d* correlations in RE-TM alloy. The 3*d* correlations negatively polarize the *s* conduction electrons. A positive Ruderman-Kittel-Kasuya-Yosida (RKKY) interaction between the conduction electrons and 4*f* electrons of RE leads to the antiparallel alignment.

However, Campbell [22] disagrees since the experimental results are not explained sufficiently. Campbell's coupling scheme stressed on the significance of RE's 5*d* electrons.

It was suggested that the 4*f* spins of RE induce a positive, local 5*d* moment through 4*f*-5*d* ferromagnetic exchange. A direct 5*d*-3*d* (of TM) exchange will in turn create the indirect 3*d*-4*f* interaction. The 5*d*-3*d* interaction is expected to be negative, thus antiferromagnetic 3*d*-4*f* spins. This theoretical background regards REs as early transition metal series with respect of their 5*d* attribute while Fe and Co belong to the second half of such a series. Campbell's argument was supported by Buschow [19, 23].

Due to the antiparallel spins of RE-TM, the total angular momentum given by Hund's rule is $J = L - S$, where L is the orbital angular momentum and S is the spin angular momentum. For light RE, the spin and orbital moments oppose one another, and thus the net moment of RE couples ferromagnetically to that of Fe/Co. The total angular momentum of heavy RE is given by $J = L + S$ (parallel spin), which confers ferrimagnetic behaviour.

1.4. Phase Diagrams

This section discusses the Sm-Co and Nd-Fe-B systems to provide guidance on the phase formation at several compositions and temperatures. The phases of interest in this discussion are $SmCo_5$ and Sm_2Co_{17} for Sm-Co binary systems and $Nd_2Fe_{14}B$ for Nd-Fe-B ternary systems.

The phase diagram of binary Sm-Co systems originally published by Buschow and Van der Goot [8] is as shown in Figure 5. The binary phase diagram is distinguished by a series of intermetallic compounds that show low solubility at low temperatures. The phases that are pertinent for the development of permanent magnets are $SmCo_5$ and Sm_2Co_{17} [9]. $SmCo_5$ is formed by a peritectic reaction where its peritectic melting point is 1320 °C (point A in Figure 5).

Therefore, only very small proportions of other phases co-exist with $SmCo_5$ upon cooling from the melt. On the contrary, Sm_2Co_{17} has been reported to melt congruently in the temperature range of 1335 – 1375 °C (point B in Figure 5). Distinct two phase regions were identified at both high and low temperatures between the homogeneity regions of $SmCo_5$ and Sm_2Co_{17} phases [8]. The phase diagram was later reviewed [24, 25] where the $SmCo_5$ and Sm_2Co_{17} phase regions and boundaries are more clearly defined. The maximum coercivity and energy product of Sm-Co magnets develop at the maximum Sm composition in $SmCo_5$ phase, which is at 82.7 at.% Co [24].

The development of Nd-Fe-B ternary systems has been reviewed by Burzo [14]. $Nd_2Fe_{14}B$ (Φ) phase was first identified in a phase diagram by Chaban *et al.* [26] where the composition of Φ phase was mistakenly identified as $Nd_3Fe_{16}B$, which was later corrected [3, 16]. An isothermal section for Nd-Fe-B systems was given at 600 °C where the liquid phases could be avoided [26]. Phase diagram for Nd-Fe-B system is more complicated than Sm-Co system since it consists of three elements. Figure 6.(a) highlights the major Nd-Fe, Nd-Fe-B and Fe-B phases that are present in the system. A simpler phase diagram for Fe-rich region where Nd:B compositional ratio is 2:1 was constructed by Schneider *et al.* [27] as depicted in Figure 6.(b). The section with Φ phase is dominated by the peritectic reaction of $L + Fe \rightarrow \Phi$ (where L is Nd-rich liquid) at 1180 °C (point C in Figure 6.b). Hence, at composition $Nd_2Fe_{14}B$ (82.4 at.% Fe), Fe crystals comprise the primary phase in the alloy. In addition, a two phase L + Φ (region A in Figure 6.b) and a three phase L + Φ + η, where η is $NdFe_4B_4$ (region B in Figure 6.b) regions exists between the maximum temperature of Φ phase formation at 1180 °C (point C in Figure 6.b) and the ternary eutectic temperature where the residual liquid solidifies at 655 °C (point D in Figure 6.b).

The short range order of Φ clusters in the liquid phase will be disrupted above the decomposition point and Φ crystallization will be restrained from this temperature onwards. Thus, the primary crystallization field of Fe will be extended to lower composition and temperature [28]. The region of primary solidification of Φ phase was found to become very narrow at higher Nd content while the primary crystallization regions of Φ phase were identified to be at 1080, 1050 and 1000 °C in a liquidus projection of Nd-Fe-B system by Knoch *et al.* [29].

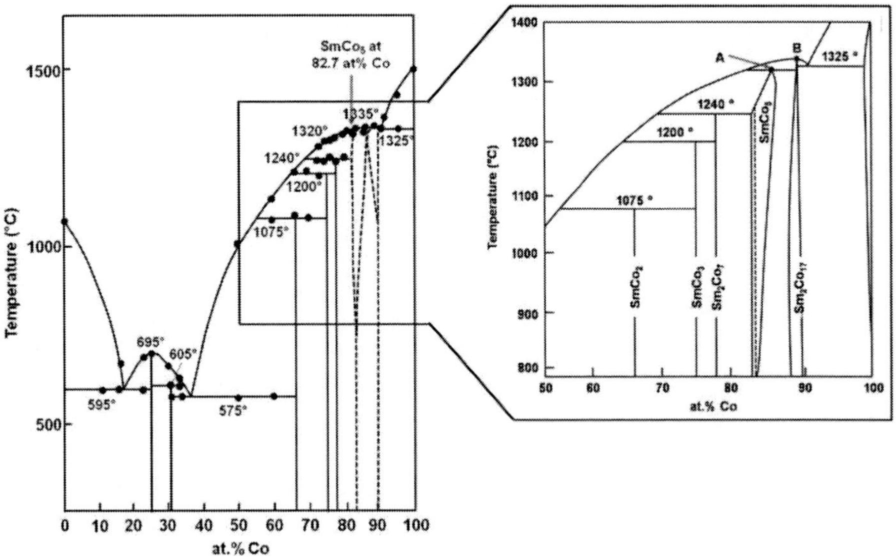

Figure 5. Phase diagram of Sm-Co system *(Adapted from Ref.* [8, 24]*)*.

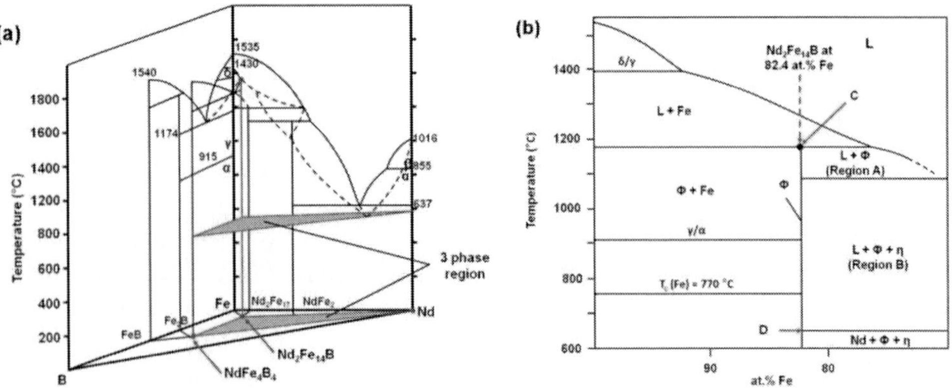

Figure 6. (a) Phase diagram for Nd-Fe-B ternary system and (b) phase diagram of Nd-Fe-B system with Nd:B ratio fixed at 2:1; α, β, γ and δ in the diagrams represent the different phases of Fe while Φ and η are $Nd_2Fe_{14}B$ and $NdFe_4B_4$ respectively *(Adapted from Ref.* [15, 27]*)*.

1.5. Magnetocrystalline Anisotropy

Magnetic anisotropy; that is, the magnetic properties depend on the direction of measurement, influences the shape of a hysteresis loop. In other words, anisotropy is evidenced as preferred crystallographic directions where the moments align themselves in a material. There are several types of magnetic anisotropy: magnetocrystalline anisotropy, shape anisotropy, stress anisotropy, induced anisotropy and exchange anisotropy. This section addresses magnetocrystalline anisotropy since this is intrinsic to a material. The main three components that contribute to magnetocrystalline anisotropy are represented in Figure 7. It is believed that magnetocrystalline anisotropy mainly originates from spin-orbit coupling. The orbit of an electron will tend to reorient when an external field tries to reorient its spin. However, the strong orbit-lattice coupling hinders the spin axis from being rotated. Therefore, the anisotropy energy required to rotate the spin system of a domain away from its easy direction is the energy required to overcome the spin-orbit coupling. Generally, the magnitude of magnetocrystalline anisotropy (also coercivity since they are closely related) decreases with temperature more rapidly than magnetization and disappear at the Curie point [30].

It is known that Sm has uniaxial anisotropy while Nd has basal plane anisotropy [14]. Hence, most Co-ased magnets incorporate Sm to form stable $SmCo_5$ or Sm_2Co_{17}. In addition to the contribution of the Sm sublattice anisotropy, the Co sublattice, which also demonstrates uniaxial anisotropy, further reinforce uniaxial anisotropy of Sm-Co magnets. This is evidenced when the strong uniaxial anisotropy prevails even after one-third of the Sm atoms are substituted by Co. However, the temperature variation of the Co sublattice moment, which is significantly smaller than that of Sm sublattice moment, allows Co sublattice to dominate at high temperature and result in easy plane anisotropy for $SmCo_5$ [9].

On the contrary, Fe atoms, which generally show anisotropy in the basal plane, form the $Nd_2Fe_{14}B$ compound with Nd that which also displays basal plane anisotropy. The anisotropy of $Nd_2Fe_{14}B$ is often described in terms of two contributions. The first would be the 4*f* charge cloud of Nd ions and the second originates from the 3*d* Fe sublattice. Similar to Sm, the contribution of the Nd magnetic sublattice to the anisotropy is dominant at low temperature. The only difference that distinguishes the type of anisotropy in these two magnetic compounds is that the 3*d* sublattice in Fe tends to align its magnetic moments along the *c*-axis while that in Co tends to align perpendicular to the *c*-axis [13]. There is a rich literature on $Nd_2Fe_{14}B$ magnetic anisotropy measurements elsewhere [31, 32].

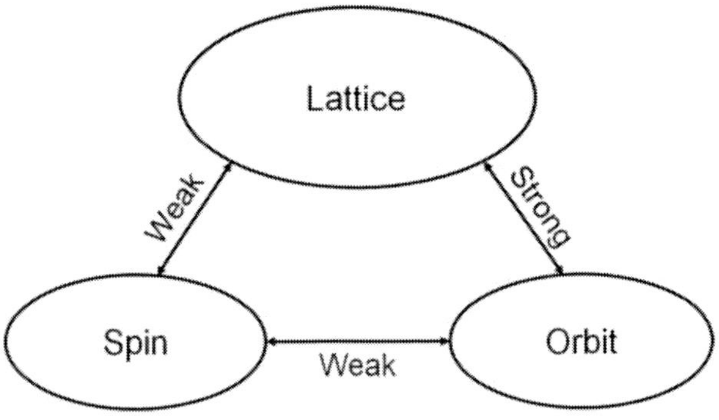

Figure 7. Spin-lattice-orbit interactions *(Reprinted with permission from Ref.* [30]. *Copyright John Wiley and Sons, Inc.).*

2. Film Preparation Techniques

The preparation of thin films can be classified into physical and chemical methods[33-36]. Examples of physical methods include sputtering, evaporation and laser deposition while chemical methods comprise chemical vapour deposition, electroplating and sol-gel techniques. There are numerous references available that discuss thin film deposition techniques in great detail [33-36].

It is not the intention of this review to go in depth into these techniques as the focus of this chapter is to examine the effect of process parameters on the characteristics of magnetic thin films; especially samarium cobalt and neodymium iron boron.

The aim is to provide readers with sufficient knowledge on the processes to understand the operating parameters and at the same time become aware of the research groups associated with each technique.

Preparation of samarium cobalt and neodymium iron boron thin films, summarized in Figure 8, has revolved around physical processes rather than chemical approaches.

2.1. Sputter Deposition

Sputter deposition, referring to A in Figure 8, was originally invented to deposit refractory materials that could not be deposited via evaporation. However, the versatile nature of this method in depositing compound materials has proven valuable with regards to depositing magnetic thin films.

It is the most popular method to deposit thin films because it confers a consistent and regular thickness (about 0.1 – 10 μm); substantial compositional control; good homogeneity and sufficient deposition rate for most applications [37]. Some of the earliest rare-earth magnetic films have been prepared by this method.

A traditional sputter deposition system is comprised of a vacuum chamber, a sputter source, a substrate holder and a pump system as illustrated in Figure 9. A glow discharge (also known as a plasma) is initiated and maintained by means of an applied electric potential between the anode (substrate) and cathode (target) to generate Ar^+ ions. These argon ions are accelerated to the cathode and bombard the target.

Some of the surface atoms of the target will acquire sufficient energy by momentum transfer and can be knocked from the target. The ejected atoms will then be deposited onto the substrate which is placed directly above the target. Manipulation of gas flow into the system as well as the pumping speed is crucial in maintaining the chamber pressure.

Referring to Figure 8, the main sputtering source used can be classified into diode (A1), magnetron (A2) and ion plasma beam (A3). Both diode and magnetron sputtering can be operated with direct current (DC) or radio frequency (RF) potential for plasma forming due to potential difference.

Ion plasma beam sputtering on the other hand generates the incident ions on the ion source instead of in the glow discharge. General configuration of the three sputtering systems is shown in Figure 9.

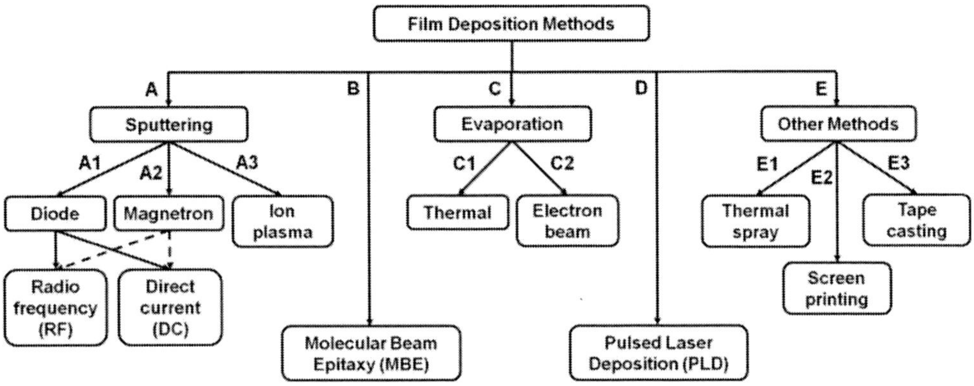

Figure 8. Overview of film preparation techniques for Sm-Co and Nd-Fe-B.

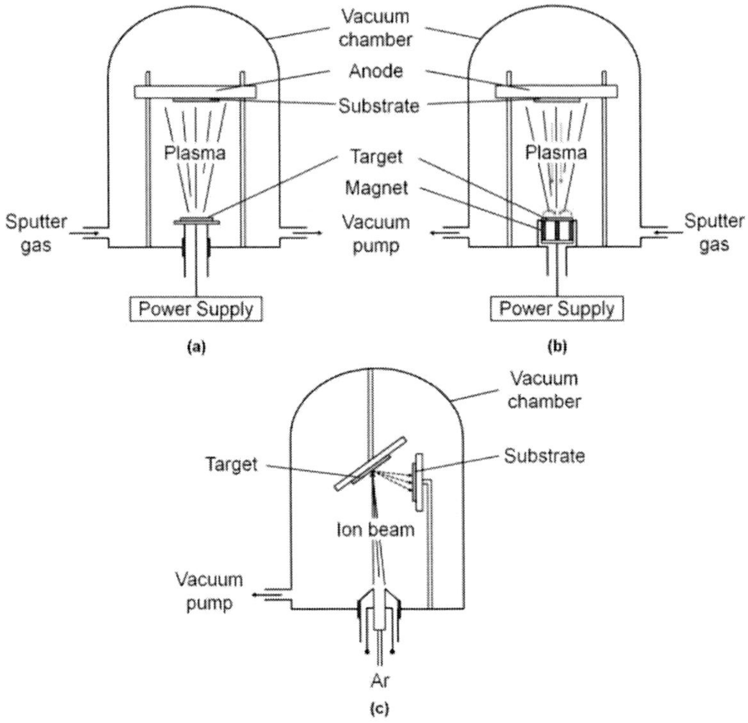

Figure 9. Basic configuration of (a) diode, (b) magnetron and (c) ion plasma beam sputtering systems, which correspond to A1, A2 and A3 respectively in Figure 8.

The DC diode sputtering system is composed of a pair of planar electrodes: a *cathode* which is covered with a target material to be deposited and reverse side that is water-cooled, and an *anode* on which the substrate is placed.

When the vacuum chamber is introduced with the sputtering gas, a glow discharge is initiated by applying a potential of a few kV between the two electrodes in the presence of argon gas, typically at a pressure of 0.1 Torr. The Ar ions generated in the glow discharge are accelerated toward the cathode and sputter the target.

The sputter ejected atoms will deposit onto the substrate and form a thin film. Some studies have used triode instead of diode sputtering. The difference between the two methods is that a separate plasma is generated in front of the target within the triode system. This acts as a source of electrons, typically created by a hot filament or hollow cathode; and magnetic confinement exists along the cathode-anode axis. [35]

The earliest samarium cobalt films have been fabricated by DC (triode) sputtering. Bendson and Judy [38] produced Sm-Co and Gd-Co films of various compositions and their results demonstrated that the magnetic moment densities of these films are in good agreement with those of their bulk counterparts. Walther *et al.* [39, 40] also studied Sm-Co films sputter deposited by a DC triode system.

A later study by Lemke *et al.*, [41] using a DC diode sputtering process, produced a smooth film suitable for micro patterning although the coercivity obtained is relatively lower.

RF sputtering is used to sputter an insulator target, since the sputtering glow discharge cannot be sustained in a DC mode sputtering system due to the immediate build-up of a surface charge of positive ions on the front side of the insulator.

A blocking capacitor that connects a matching network of power supply and vacuum chamber with the target is required in a RF sputtering system. The target area, which is comparatively much smaller than the anode and chamber wall, induces a negative DC bias on the target and causes sputtering [36].

Cadieu *et al.* [42-45] and Aly *et al.* [46] are among the early users of "selectively thermalized" RF sputtering. It is useful because it favours formation of certain textures from an initial random growth pattern during the direct crystallization of films.

Formation of larger grains encourages film growth because these act as seeds for growth of thick films [47]. Subsequent studies on rare-earth magnetic film by RF sputtering have been carried out by Velu *et al.* [48, 49], Shima et al. [50], Tang *et al.* [51] and Serrona *et al.* [52, 53].

By applying an appropriate magnetic field and magnets arrangement in magnetron sputtering (referring to A2 in Figure 8), a high electron flux can be achieved that could sustain a higher density plasma compared to diode sputtering [35]. The two most popular magnetron sources used for thin film deposition are of planar and cylindrical configurations, Figure 10.

In magnetron sputtering, permanent magnets are embedded within the cathode target that results in magnetic fields of several hundred gauss and a circular glow discharge. The magnetic field applied causes the electrons to be deflected and circulate on a closed path near the target surface.

In addition, sputtered particles are able to traverse the discharge space without colliding with each other. This type of sputter deposition technique is widely used for magnetic materials since the permanent magnet used improves the magnetic flux distribution within the system.

The use of this system to deposit other materials tends to shorten the life of cathode target [36]. Magnetic properties of Nd-Fe-B and Sm-Co thin films prepared by magnetron sputtering in several studies (Ref. [54-75]) are summarized in Table 2.

Table 2. Summary on magnetic properties of Nd-Fe-B and Sm-Co films prepared by magnetron sputter deposition

Ref.	Film material	Substrate	Thickness (nm)	Remanence, B_r (T)	Intrinsic coercivity, ${}_iH_c$ (kA/m)	Max energy product, $(BH)_{max}$ (kJ/m^3)	Substrate temperature, T_s (°C)	Annealing temperature, T_a (°C)	Buffer layer and thickness (nm)
[54]	Nd-Fe-B	Si (100)	54 – 540	-	2092.9	-	-	500 – 750	Nb or V (20)
[55]	Nd-Fe-B	Si (100)	400*	-	2070	-	600 – 800	-	Ta (400*)
[56]	Nd-Fe-B	Si (100)	500 – 700	$1.55^{\ddagger}$	$100^{\dagger}$	-	-	400 – 800	-
[57]	Nd-Fe-B	Si (100)	400 – 500	1.22	787.8	278.5	600 – 750	-	Ta (300)
[58]	Nd-Fe-B	Si (100)	550 – 1050	1.10	820	180	450 – 700	-	-
[59]	Nd-Fe-B	Si (100)	$700 – 840^{\ddagger}$	-	-	106	Water cooled	650 – 780	Nb (30)
[60]	Nd-Fe-B	Si (100)	$500 – 10^4$	-	-	235	-	550 – 700	Cr, Ta (20 – 500)
[61]	Nd-Fe-B	Si (100)	800	$0.7^{\dagger}$	1814.4	117	400 – 550	600	Mo, Ti (50)
[62]	Nd-Fe-B	Si (100)	800	$1.2^{\dagger}$	$1352.8^{\dagger}$	143.2	r.t.	450 – 675	Mo (30)
[63]	Nd-Fe-B	Si (100)	800	$1.0^{\dagger}$	1193.7	178.3	25 – 600	-	-
[64]	Nd-Fe-B	Si (100)	600 – 650	$1.1^{\dagger}$	1670	200	490 – 590	-	Nb, Ta (200)
[65]	Nd-Fe-B	Silicon	54 – 180	-	1591.5	82	-	450 – 700	Cr, Mo, Nb, Ta, Ti, and V (20)
[66]	Nd-Fe-B	Silicon	825	0.97	420	136	470 – 800	-	Ta
[67]	Nd-Fe-B	Glass	200	1.30	927	319	550	-	Ta (10)
[68]	Nd-Fe-B	Glass	200	1.44	979	364	500		Ta (10)
[69]	Sm-Co	Si (100)	200	-	636.6	-	350	-	Cr (75 – 300)
[70]	Sm-Co	Si (100)	500	-	795.8	79.6	-	550 – 650	Ta (100)
[71]	Sm-Co	Silicon	230 – 650	0.615	637	-	600	-	Cr
[72]	Sm-Co	Silicon	296 – 310	0.61	676.4	-	600	-	Cr (175 – 180)
[73]	Sm-Co	Silicon	25	-	954.9	-	350	-	Cu, Pt (100)
[74]	Sm-Co	Glass	10 – 20	$0.9^{\dagger}$	875.4	-	350	-	Ru, Cr (4 – 16)
[75]	Sm-Co	Glass	30	-	191.3	-	r.t. – 500		Cr (240)

† Values estimated from figures.

‡ Total thickness of multilayered films.

* Combined thickness of magnetic film and buffer layer.

The values of B_r, H_c and $(BH)_{max}$ given are the maximum values achieved in each studies.

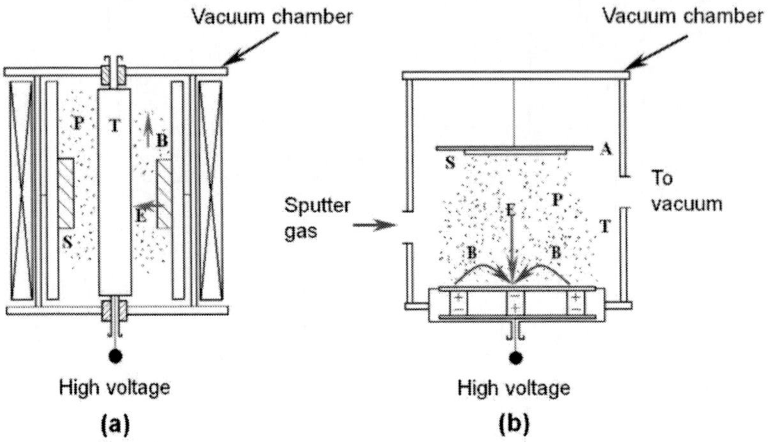

Figure 10. Schematic of (a) cylindrical and (b) planar magnetron sputter deposition systems; (P: plasma, S: substrate, T: target, B: magnetic field, E: electric field and A: anode).

The gas pressure in a glow-discharge system is often high; therefore, the gas molecules tend to irradiate the sputtered films during film growth and cause the gas molecules to become trapped in the sputtered films.

On the other hand, incident ions in the ion beam sputtering system (referring to A3 in Figure 8) are generated at an ion source. The target can be bombarded by the ion beam either in a sputtering chamber separated from the ion source or in a plasma chamber within the ion source with a hot filament cathode [36].

However, ion plasma beam sputtering is not as widely used for permanent magnetic thin film deposition compared to the diode and magnetron sputtering methods.

2.2. Evaporation

Evaporation, referring to C in Figure 8, is considered the most flexible technique for film deposition since it can be used to deposit most materials. This process is simple and relatively easy to use while providing high deposition rates. There are two main sources of evaporation that are used extensively: thermal evaporation (C1) and electron beam evaporation (C2). The evaporation system set up for these two sources are fairly similar, as illustrated in Figure 11. Vacuum evaporation is used most of the time: partly because it helps to control the oxidation of materials being deposited and also partly because the material can be evaporated at temperatures well below its boiling point.

The thicknesses of the films generated by this method are controlled by adjusting the amount or rate of vapour materials generated as well as the distance between the source and the substrate. In thermal evaporation, a crucible, boat or wire coil is typically used to hold the source material.

An electrical current is then passed through the holder that melts the source material. The rate of evaporation can be controlled directly by manipulating the current going through the heating element. However, the maximum temperature that can be achieved by the system is limited and the rate of deposition cannot be adjusted promptly.

Also, there is a risk of contamination if the containment structure is not carefully selected. Gasgnier *et al.* [76] used this method to fabricate $Nd_2Fe_{14}B$ films of ~ 60 nm onto a sodium chloride substrate. The holder used was a tungsten crucible, which exhibits a high melting point of 3422 °C. The film obtained was initially amorphous, but was crystallized by subsequent heat treatment by using an electron beam heat source progressively from room temperature up to about 1227 – 1527 °C. The magnetic properties of this film were not characterized in this study but it is noteworthy that the fabrication method was successful for $Nd_2Fe_{14}B$ films.

On the other hand, electron beam (EB) evaporation uses high energy electron beams that focus on and heat a small localized area of the source material, and evaporate the source material. This technique eliminates the disadvantages exhibited by thermal evaporation.

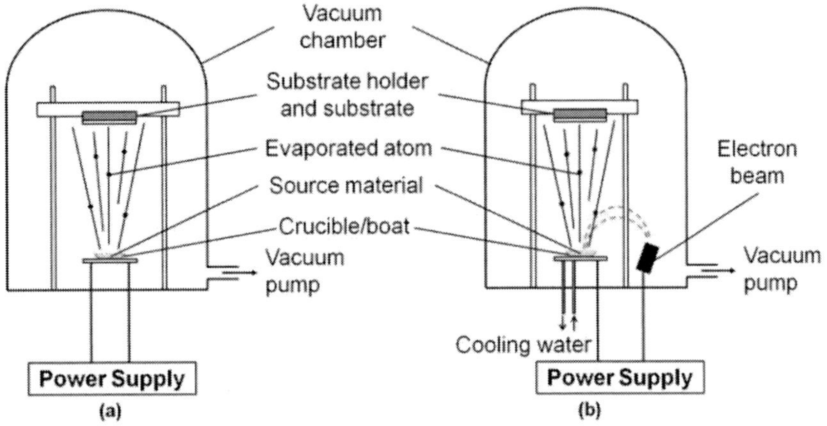

Figure 11. Basic configuration for (a) thermal evaporation and (b) electron beam evaporation.

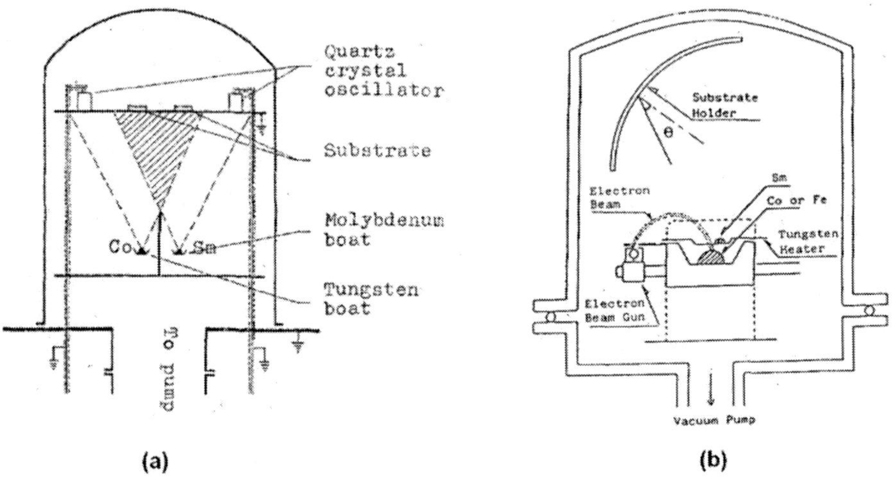

Figure 12. (a) Thermal evaporation used by Lee *et al.* [78] and (b) electron beam evaporation used by Tanaka *et al.* [79] *(Reprinted with permission from Ref.* [78] *and* [79]. *Copyright (1987) and (1983), IEEE).*

Since only a small area of the material is being heated to a high temperature, a wide range of materials can be deposited by this method compared to thermal evaporation. The EB evaporation technique was used by Pereira *et al.* [77] to synthesize multilayer Nd/Fe films. Multilayered films made of alternate Nd and Fe layers of 8 nm each were deposited using two independent EB evaporation sources.

Intermetallic phases of Nd-Fe systems were obtained through interdiffusion between the layers. Introduction of B through ion implantation further enhances the interdiffusion. Vacuum annealing at 600 °C shows a complete redistribution of Fe and Nd atoms in the system with destruction of the multilayer periodic structure.

The use of evaporation process has been limited with regards to depositing compound materials because different elements of the compound have different evaporation rates. This can change the film stoichiometry, and therefore its magnetic properties.

To overcome this problem, Lee *et al.* [78] used two separate heating sources to deposit Sm-Co film on glass substrates. Tanaka *et al.* [79] on the other hand, used a different approach to solve this issue. The source materials of Sm and Co were placed adjacent to each other; with Co, which has higher evaporation point, further away from the electron beam source. The evaporation system used by both Lee and Tanaka is as shown in Figure 12.

2.3. Molecular Beam Epitaxy (MBE)

Epitaxy refers to a film deposition technique characterized by a continuation of crystal structure from substrate to film. Only one single highly ordered crystalline atomic layer can be deposited at one time. Molecular beam epitaxy (MBE), referring to B in Figure 8, is essentially a refined evaporation technique where a source material is heated to produce an evaporated beam of atomic/molecular species.

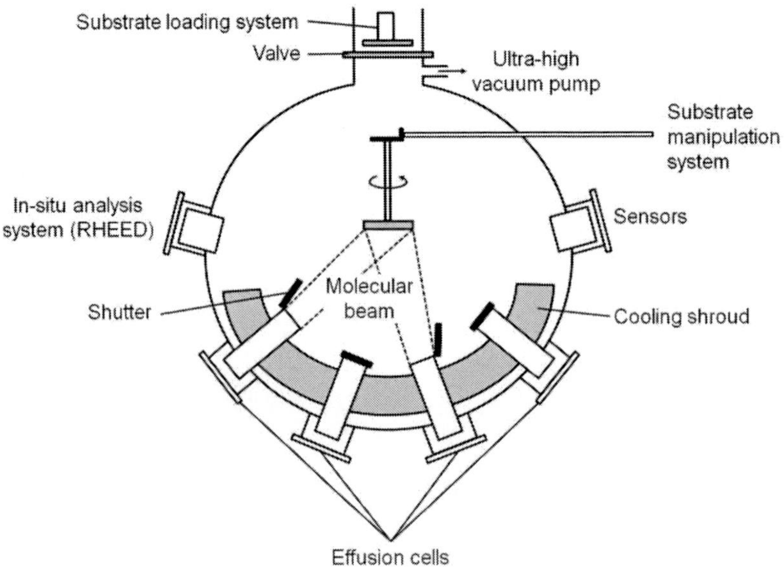

Figure 13. Basic layout of a molecular beam epitaxy (MBE) system.

Condensation of these atomic/molecular species results in the formation of crystalline film. The requirements for film deposition by this method are more demanding than other techniques. MBE needs to be carried out in an ultra-high vacuum environment to achieve high purity films considering the low deposition rate involved.

It is often difficult to establish and maintain an ultra-high vacuum while ensuring a completely clean environment for monolayer deposition. Thus, most MBE systems have separate chambers that execute specific functions, and have an in-situ analysis system that is typically a reflection high-energy electron diffraction (RHEED) unit. A typical MBE system is shown in Figure 13.

This technique is mainly used to prepare III-V semiconductor crystals, the most common being GaAs devices. MBE is often used only if superior epitaxial quality film is essential. A number of magnetic thin films have been grown epitaxially using deposition techniques such as sputtering [80, 81] and pulsed laser deposition [82-84].

However, the many advantages offered by MBE including precise control over the thickness, surface composition and morphology of the films formed have made this an alternative technique to produce magnetic thin films with high magnetocrystalline anisotropy. The in-situ analysis system provides the convenience of monitoring the growth behaviour of the film in real time.

Preparation of $Nd_2Fe_{14}B$ epitaxial film by MBE has been initiated out by Keavney *et al.* [85, 86]. The film thickness range is 30 – 60 nm, which is extremely thin and is not commonly achievable by other deposition methods.

Instead of the conventional co-deposition that is typically employed in sputter deposition and evaporation, Keavney *et al.* employed 'superlattice deposition', followed by solid-state reaction; also known as 'block-by-block deposition'. Each building block consists of three monolayers of Fe, Nd and B. The elemental component layers are deposited so that a compositional path through the ternary Nd-Fe-B phase diagram can be traced. As a result, undesired phase formation can be avoided by choosing appropriate paths.

Unlike Keavney and co-workers who based their works mostly on magnetic materials, Ohtake *et al.* have focused on epitaxial growth of $SmCo_5$ thin films [87, 88]. In comparison to Keavney *et al.* who studied the feasibility of using a block-by-block deposition for magnetic films, Ohtake *et al.* investigated the relationship and interaction between the magnetic films produced by MBE with an applied underlayer.

The use of RHEED analysis provided insights concerning the diffusion of a Cu underlayer into $SmCo_5$ film, forming an alloy compound of $Sm(Co,Cu)_5$. The varying degrees of diffusion with deposition parameters and also the effects of such diffusion will be elaborated in Section 3.4.3.

2.4. Pulsed Laser Deposition (PLD)

Pulsed laser deposition, referring to D in Figure 8, is also known as laser ablation deposition and requires an ultra high vacuum chamber. A small area of target material is vaporized with a high power laser pulse that originates from an external source. A typical configuration of PLD is shown in Figure 14. The laser source is usually an ultraviolet excimer laser (YAG or ArF) that deposits energy in pulses.

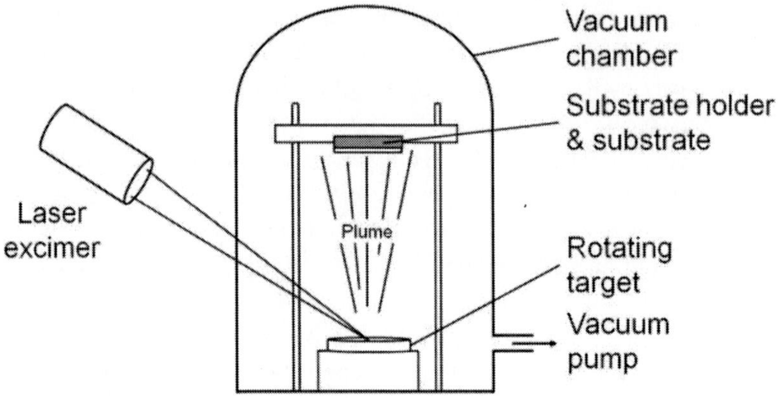

Figure 14. Basic configuration for Pulsed Laser Deposition (PLD) technique.

A cloud of vaporized material will form above the surface of the target where some of the laser energy is adsorbed, followed by ionization and excitation. The ejected material is highly directed, which causes difficulty in manufacturing films of uniform thickness over large areas. This can be overcome by scanning the laser beam across the target while rotating the substrate. It is also a common practise to heat the substrate to promote surface diffusion and film adhesion. The PLD process is desirable in the sense that it can retain the original composition of the target in the vapour as well as in the resulting film. However, the high costs involved in generating the ultra high vacuum environment and the laser source have made this process less popular than the two processes previously discussed. Cadieu *et al.* [89] and Geurtsen *et al.* [90] have studied the preparation of $SmCo_5$ and $Nd_2Fe_{14}B$ films by PLD methods. While Cadieu preferred sputtering to PLD in synthesizing magnetic thin films, Geurtsen obtained $Nd_2Fe_{14}B$ film of ~100 nm thickness that exhibited strong magnetic anisotropy along the film axis. Fähler *et al.* [91] and Hannemann *et al.* [92] employed a high purity elemental target instead of the conventionally used compound target to synthesize Nd-Fe-B films via PLD. By adequate substrate heating, Fähler and his co-workers successfully produced epitaxial films with high coercivity. Constantinescu *et al.* [93] investigated the effect of protective gas on Nd-Fe-B thin film by substituting the traditionally used vacuum in the PLD process with a protective argon environment. This study indicated that there were no differences in film properties fabricated under both conditions. This finding is significant because it shows that argon may be used as an alternative operating environment rather than a costly ultra high vacuum. Nakano *et al.* [94-97] have also investigated Nd-Fe-B films produced by PLD. Relatively thick Nd-Fe-B films have been successfully fabricated and applied to millimetre-sized motors. Nakano controlled the film thickness by varying the deposition rate via the target-substrate distance. Despite Nd loss via high temperature oxidation, high coercivity films were obtained since any Nd lost was compensated by using a target that has slightly higher Nd than the desired final composition.

2.5. Other Deposition Methods

There are other film preparation techniques that are not as popular as those aforementioned; e.g., plasma spray process, screen printing and tape casting. These methods produce considerably thicker films than physical vapour deposition processes such as sputtering, evaporation and PLD.

In fact they produce 'coatings' because the starting materials are often powders with grain sizes within micrometer range that are normally retained in the films formed. Nevertheless, these techniques are interesting because some of the films possess high coercivity. Also, these methods present processes that are comparatively low in cost.

Plasma spray is a branch of thermal spray process, referring to E1 in Figure 8, which involves the use of a high temperature heat source to melt the material of interest. The molten material is then accelerated by gases towards the substrate where a layer is formed as the molten material impacts upon the substrate.

In plasma spray, the heat source is generated by superheating an inert gas, typically argon, by a DC arc. Since considerably high temperature is involved and rare-earth compounds are generally oxidation prone, a vacuum plasma spray (VPS) process is desirable for depositing magnetic films. There have been several studies on Sm-Co films by VPS led by Kumar *et al.* [98-100]. Nd-Fe-B films prepared by the VPS process have shown diverse properties [101-106].

The development of cold spray as a branch of thermal spray has presented an alternative of producing magnetic films since it involves a much lower temperature. A study by King *et al.* [107] indicates that the brittle nature of rare-earth compounds dictates blending with a ductile material, such as aluminium, before the material can be deposited by the cold spray method.

Screen printing and tape casting are two possible options to produce magnetic films that are not constrained by a line of sight process. Screen printing, referring to E2 in Figure 8, involves the use of a woven mesh that is stretched over a frame.

The open areas of the mesh will then transfer the material to be deposited onto a substrate. A roller is moved across the screen to force the material through the mesh onto the substrate. This method has been used by Pawlowski *et al.* [108, 109] and Speliotis *et al.* [110, 111] to prepare Nd-Fe-B films. The ink is made by dispersing magnetic powder in resin that is dissolved by an organic solvent. The films are comparatively thick; i.e., a single print can create a film of 10 μm thickness, and the method has been applied to forming a micro motor [111].

On the other hand, tape casting, referring to E3 in Figure 8, involves spreading a slurry onto an appropriate carrier film that moves constantly under metal knife blades placed at a short distance above the film. A wet, thin sheet of the slurry material is formed as the film and slurry move under the blades. The sheet of material is air dried to remove the solvents and a film is formed. Tape casting has conventionally been used to produce ceramic substrates for thin film applications instead of the thin film itself. In spite of that, Pawlowski *et al.* [108, 112] has proven that it can be used to prepare Nd-Fe-B films.

The films obtained have thicknesses in the range of 100 – 1000 μm. However, the magnetic properties of the films obtained are highly dependent on the volume loading of magnetic powders; that is, the relative amount of magnetic powder and binder.

Table 3 summarizes all the deposition processes that have been discussed.

Table 3. Summary of various deposition processes

	Sputtering			Molecular beam epitaxy (MBE)	Evaporation		Pulsed laser deposition (PLD)	Other Methods		
	Diode	Magnetron	Ion plasma		Thermal	Electron beam (EB)		Thermal spray	Screen printing	Tape casting
Branch[†]	A1	A2	A3	B	C1	C2	D	E1	E2	E3
Deposition source	Electric potential generating glow discharge		Ion beam current	Molecular beam	Electric current	High energy EB	Laser	Combustion, electric arc	Roller and woven mesh	Metal blade
Material form	Disk			Same as evaporation; Vapour	Rod, wire, pellet, piece, foil		Powder, pellet, single crystal	Powder	Dispersion of powder in resin and organic solvent	Dispersion of powder in organic solvent
Environment	Vacuum			Ultra high vacuum	Ultra vacuum		Vacuum/ Ar gas	Vacuum/ atmospheric	Atmospheric	Atmospheric
Film thickness (μm)	0.002 – 1.70	0.01 – 10	0.001 – 2.50	0.03 – 0.06	0.02 – 1.20	0.02 – 0.40	1 - 120	20 – a few mm	8 – 220	5 - 1200
Advantages	Ability to deposit high melting point materials; better film-substrate adhesion	High deposition rates;	Energy and flux of ions can be controlled independently	Precise control of composition, thickness and morphology; in-situ analysis	High deposition rates; less substrate surface damage from impinging atoms; less tendency for unintentional substrate heating		Simple design; ability to retain target composition	High deposition rates; deposit a wide range of materials with low to high melting	Ability to produce large quantities of film; economical; high deposition rate; easy control over film composition	
Issues	Sputter gas molecules are sometimes included in the films		High maintenance	Low deposition rate; difficulty in maintaining ultra-high vacuum and clean environment	Low melting point materials only; different evaporation rates of elements in compound materials	Different evaporation rates of elements in compound materials	Limited area of uniform deposition and ejection of micro-sized globules from target	Risk of oxidation	Labour intensive; equipment cleanup can be time consuming	Stress and shrinkage during drying process
Solutions (if any)	Use alternative sputtering method, i.e. ion plasma			Have separate chambers to execute specific functions	Use alternative evaporation method, i.e. EB; use separate heating source	Place targets at different distances from EB source	Place substrate at off-axis position	Use vacuum or inert chamber		Dry at slower rates and choose appropriate solvents

[†] Refer to Figure 8

3. Properties of Magnetic Thin Films

3.1. Microstructure of Magnetic Thin Films

Rare earth magnetic films are readily prepared in an amorphous or crystalline state by deposition techniques discussed previously. The final microstructure of thin films influences their properties and can vary from a single crystal film to a polycrystalline film, with columnar or equiaxed grains, through to an amorphous film. Modes of film growth play a crucial role in the final film microstructure. In the case of magnetic thin films that are mostly grown by vapour deposition, the film growth initiates from condensation of the vapour formed. The condensate forms adatoms that are bonded to the substrate surface. Diffusion of adatoms over the film surface leads to desorption until an equilibrium position is achieved with these atoms located in the low-energy lattice sites. The four processes involved in the diffusion mechanism are shadowing, surface diffusion, bulk diffusion and desorption.

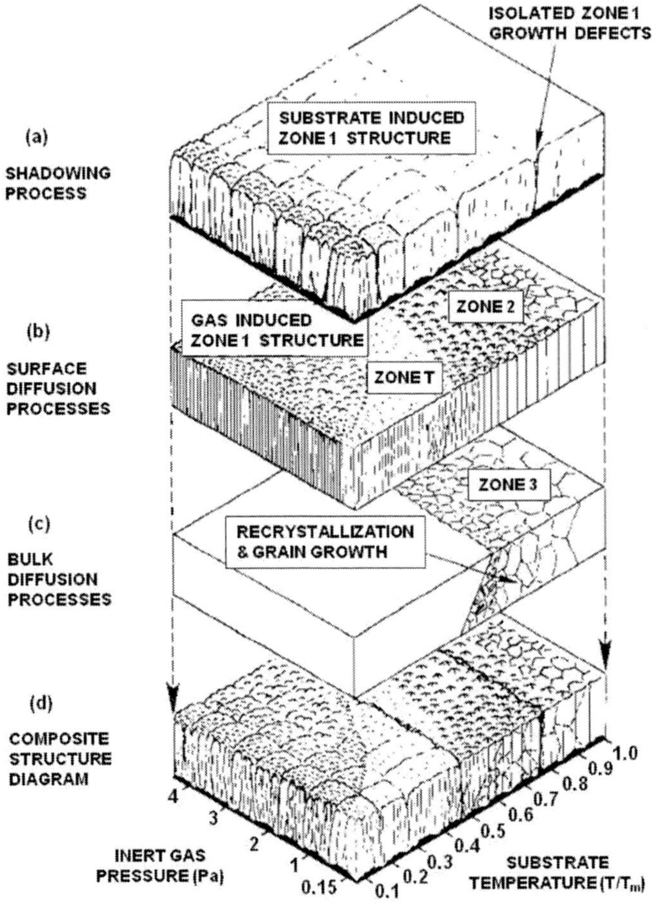

Figure 15. Schematic representation of physical processes involved in film structure evolution as a function of substrate temperature (relative to melting temperature of film material) and inert gas pressure *(Reprinted from Ref.* [113], *Copyright (1977), with permission from Annual Reviews).*

The evolution of polycrystalline film structure as a function of substrate temperature and argon gas pressure was originally depicted by the Thornton diagram, [113] Figure 15. Surface and bulk diffusion as well as desorption are quantified by diffusion and sublimation activation energies, which are directly related to the melting point of the condensate. On the other hand, shadowing arises from geometric constraints posed by film roughness and line-of-sight impingement of arriving atoms. The dominance of these processes is influenced by the substrate temperature, T_s and these are distinguished by observing their different structural morphologies. This forms the basis of structure-zone models used to characterize grain structures of films and coatings [114]. Further elaboration on film growth mechanisms is available elsewhere [113-118]. The two prime factors that control the microstructure of a growing film are the growth flux, also related to the deposition rate, and the substrate temperature. Growth flux represents the flow rate of a film material onto the growth surface and is linked to the degree of vapour supersaturation. Generally, formation of films that exhibit a more refined microstructure is more favoured at higher deposition rates [117]. Deposition at high substrate temperatures leads to more equiaxed microstructures while at low temperature deposition, fine-grained or amorphous structures are more favoured [114]. Figure 16. illustrates thin film formation where grain boundaries are formed through cluster formation, growth and impingement. The degree of film crystallinity depends on deposition parameters as well as post-deposition processing such as heat treatment. Crystallinity is an important attribute for magnetic films since it affects the coercivity and remanence of the film. Crystalline films could be formed by either (i) depositing directly on a heated substrate or (ii) depositing an amorphous film on a substrate at low temperature, and followed by post-deposition annealing. Frequently, Nd-Fe-B thin films are deposited directly onto heated substrates to acquire textured films [45, 86, 90, 92]. However, a deterioration of magnetic properties is often observed due to difficulty in controlling the grain growth and oxidation of RE at prolonged heating. Another technical approach involves post-deposition annealing for films deposited at lower temperatures [120, 121]. Although films prepared at substrate temperatures lower than the crystallization temperature of $Nd_2Fe_{14}B$ are amorphous, annealing will induce formation of a *c*-axis texture of tetragonal $Nd_2Fe_{14}B$ perpendicular to the film plane due to the presence of a weak out-of plane anisotropy in the soft magnetic material [61]. Key characteristics of film microstructure are grain size and morphology [117]. Control of grain size is crucial in magnetic information storage applications because it contributes to the recording capacity and directly influences the noise level. Greater recording densities have been achieved by shrinking the size of recorded bits. In order to preserve a constant signal to noise ratio, the grain sizes of magnetic films need to be reduced [122].

A continuum of microstructures ranging from columnar to equiaxed to amorphous, and also mixtures in between these, can be observed in films [114]. The columnar morphology, which is the typical morphology of Sm-Co and Nd-Fe-B films, can develop in both amorphous and crystalline films. However, this type of morphology often includes internal defects and intercolumnar voids or porosities that will deteriorate film properties. The film morphology is highly dependent on substrate heating as well as the composition of material used. Kim *et al.* [55] found that films with lower Nd content grown on a heated substrate consist of columnar grains perpendicular to the film plane. However, films of higher Nd content do not exhibit such structures, although growth of both films initiate from tiny spherical grains (refer to Figure 17). It was speculated that this dissimilarity is caused by differences in the thermal gradient of the heated substrates for varying Nd content.

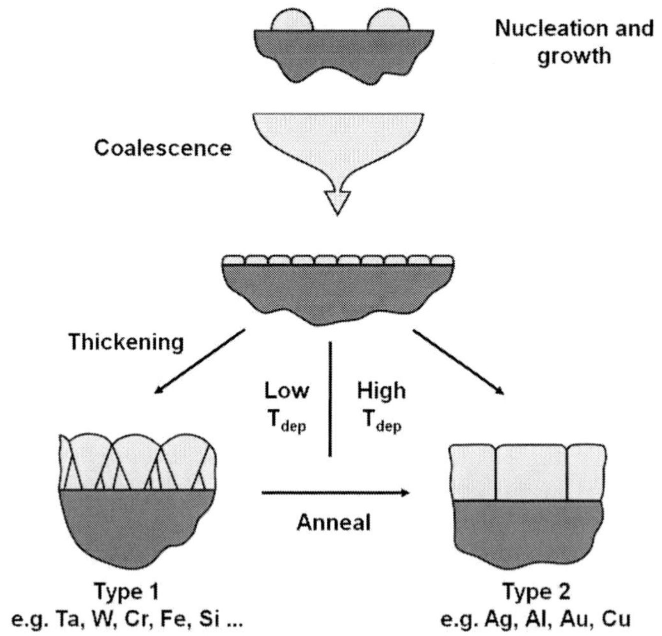

Figure 16. Schematic of grain structure evolution during polycrystalline thin film formation *(Reprinted from Ref.* [119], *Copyright (2000), with permission from Annual Reviews).*

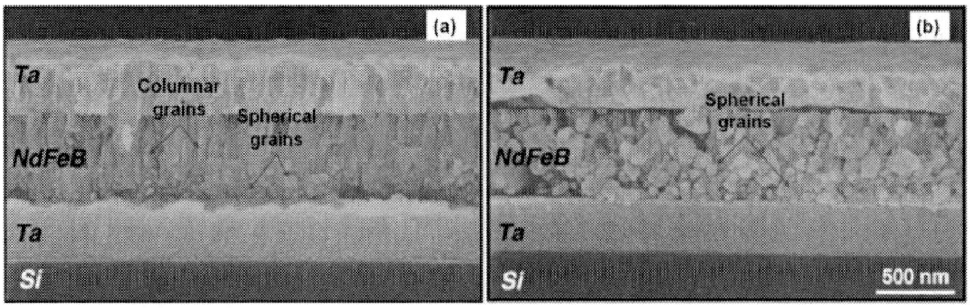

Figure 17. Morphology differences for films with Nd/Fe compositional ratios of (a) 27/73 and (b) 11/69 *(Reprinted from Ref.* [55], *Copyright (2001), with permission from Elsevier).*

A columnar structure is highly desirable since magnetic films with this structure have proved to demonstrate strong perpendicular magnetic anisotropy. Similar columnar structures and perpendicular magnetic anisotropy were observed by Tang *et al.* [123].

On the contrary, Uehara *et al.* [67] observed the formation of equiaxed grains in Nd-Fe-B films, Figure 18, rather than a columnar structure. A relatively rough film surface was observed compared to that of Kim *et al.* [55]. Since both used very similar deposition conditions; that is magnetron sputtering, Ar gas pressure and a buffer layer, the reason behind this anomaly may be attributed to different substrate temperatures. Again, this discrepancy may also be due to the difference in Nd/Fe compositional ratio as discussed previously by Kim *et al.* [55]. Even so, the reasoning comes down to substrate heating since thermal gradient effectiveness is believed to cause distinctive structures for varying compositional ratios. Such a structure is justified by referring to the Thornton diagram in Figure 19. that outlines structure evolution in a film as a function of substrate temperature and inert gas pressure. Thornton diagram in Figure 19. is more useful in comparing the variation of film structures with substrate temperature and/or gas pressure compared to that in Figure 15. as it incorporates the distinctive structures and zones formed by the different physical processes in a single layer rather than separate layers.

Formation of an irregular grain structure roughens the film surface as shown in Figure 18.(a). Attempts to regularize the crystal size by forming multilayered Nd-Fe-B film with a Ta buffer layer in between the layers to interrupt the grain growth of $Nd_2Fe_{14}B$ have been successful [67].

The multilayered film exhibited improved magnetic anisotropy primarily because the equiaxed grain size is smaller than the critical domain size, which affects the magnetization reversal mechanism. Unlike the formation of $SmCo_5$ film, which requires diffusion of a Cu underlayer into the film for phase stabilization, observation by backscattered electron microscopy shows that there is no severe migration between Ta and Nd-Fe-B layer.

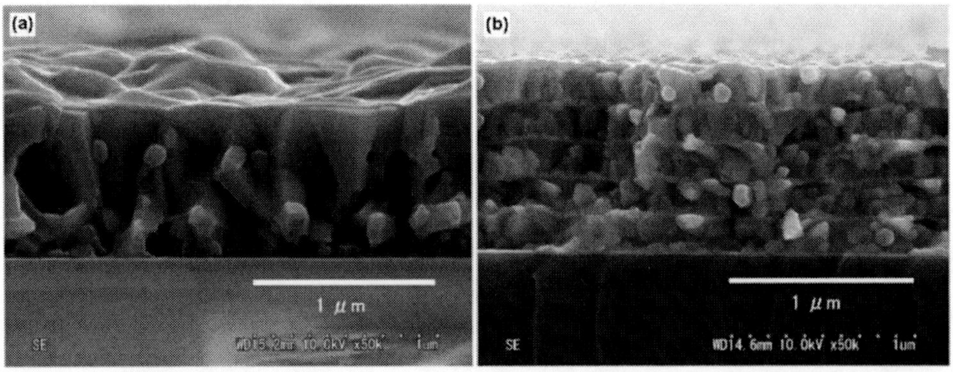

Figure 18. Cross section of (a) Ta (20 nm)/NdFeB (1 mm)/Ta (10 nm) film and (b) Ta (20 nm)/[NdFeB (200 nm)/Ta (10 nm)]$_5$ film, both deposited at 700 °C *(Reprinted from Ref.* [67], *Copyright (2004), with permission from Elsevier).*

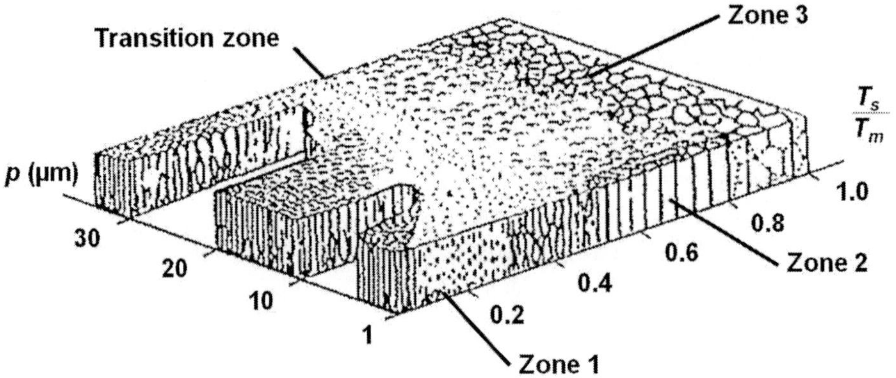

Figure 19. Thornton diagram outlining structure evolution as a function of substrate temperature and inert gas pressure *(Reprinted with permission from Ref.* [124]. *Copyright (1974), American Institute of Physics).*

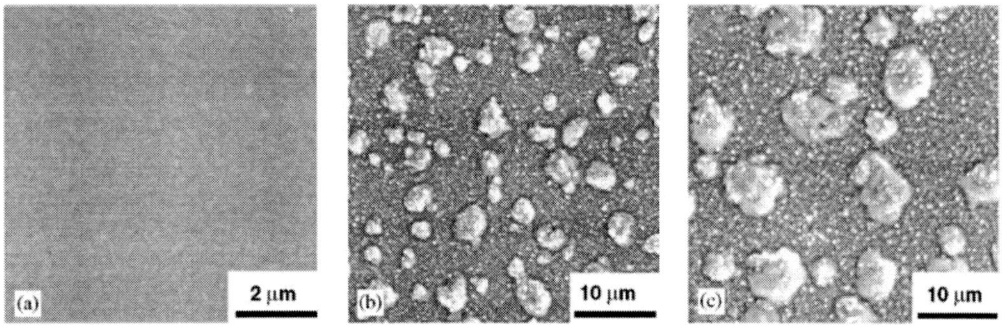

Figure 20. Planar view of Nd–Fe–B thin films sputtered at different deposition rates of (a) 0.5 nm/s, (b) 0.6 nm/s and (c) 0.8 nm/s *(Reprinted from Ref.* [125] *Copyright (2006), with permission from Elsevier).*

Variation of deposition rates shows that unusual big "islands", as phrased by Chen *et al.* [125], tend to form at high deposition rates, Figure 20. The size of the "islands" increases as deposition rate increases and they were identified as a Nd-rich phase. High Nd contents often lead to strong perpendicular anisotropy and high coercivity [51, 55, 126]. Nd-rich areas that form a ridge-like morphology, as shown in Figure 21.(a), have also been observed. Formation of these ridge-like areas results in a "shoulder" in the hysteresis loop because they cannot be completely covered by a subsequent layer, and hence they are exposed and oxidized [61]. Variation of film composition may be attributed to evaporation losses of the heavier Nd atoms at different deposition rates [125].

Prolonged deposition time may lead to further Nd atom losses that lead to a low Nd/Fe ratio and the formation of α-Fe, hence low film coercivity. Conversely, a high Nd/Fe ratio results in formation of a Nd-rich phase and increased coercivity. A decrease of saturation magnetization is also often observed with increasing deposition rate due to the decrease of Fe content in the films.

Another aspect of film morphology concerns the film surface roughness. The absence of a buffer layer or too thick of a buffer layer will result in a more rough surface, Figure 21.(b). The optimum buffer thickness depends on the buffer and film material as well as film thickness; for instance, the optimum thickness for a Ti buffer of 300 nm $Nd_2Fe_{14}B$ film is about 50 nm [127]. Nd-Fe-B films with surface roughness as high as 100 nm has been observed in the absence of a buffer layer. The high surface roughness has been linked to the reaction between Nd-Fe-B and the heated substrate [127]. Similar morphologies have been observed when a Cr buffer is deposited onto a heated Al_2O_3 substrate [128].

The morphology of the buffer layer plays a distinct role in determining the coercivity of magnetic thin films. The same remanence and magnetic texture were observed for films deposited on both smooth and rough buffer layers, but the coercivity increased by 50% by depositing Nd-Fe-B films onto rougher Cr buffer layers [128]. While depositing Sm-Co onto a rougher buffer layer also leads to an increase in coercivity, it has the effect of causing a reduced remanence.

The effect of reduced remanence is probably due to the formation of a soft magnetic phase via the oxygen diffusion from the substrate through the non-wetted buffer area since Co and Cr are immiscible. The increase in coercivity can be explained by a reduction in grain size of the film material when it nucleates in the trenches of the rougher buffer layer [128].

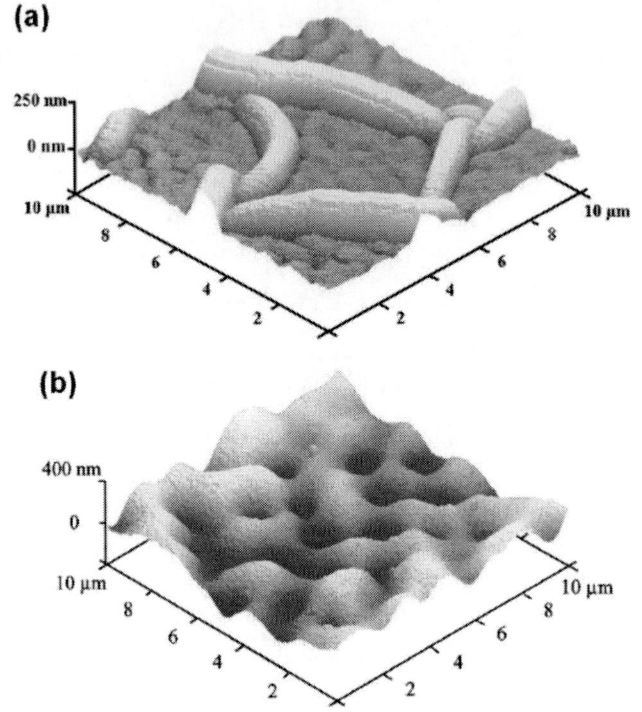

Figure 21. Morphology of (a) [Si/Ti(50 nm)/Nd–Fe–B(800 nm)/Ti(50 nm)] thin film annealed at 600 °C with T_s=470 °C [61] and (b) Nd–Fe–B thin film (300nm) deposited on Si substrate without buffer layer [127] *(Reprinted with permission from Ref.* [61] *and* [127]. *Copyright (2005), American Institute of Physics).*

The choice of buffer layer material is critical since it influences the film morphology and crystallinity. In the case of $SmCo_5$ film, the use of a Cu buffer layer shows higher H_c than a Pt buffer layer due to the presence of an amorphous layer [73]. It can be seen from Figure 22. that $SmCo_5$ film on a Cu buffer layer is rougher than on a Pt buffer. This further reinforces the earlier discussion concerning the influence of roughness on coercivity.

The difference in surface roughness between these two buffers is related to the diffusivity of these elements into $SmCo_5$ film. Diffusivity of Cu is higher than that of Pt due to its lower melting point.

Therefore, the grains of Cu tend to grow faster than Pt grains, resulting in a larger grain size and a columnar structure. Diffusion of Cu into the film lowers the crystallization temperature of $SmCo_5$ and, hence, film deposited on the Cu buffer is completely crystallized as opposed to a partially crystallized film on a Pt buffer.

The crystalline $SmCo_5$ is expected to exchange couple with amorphous $SmCo_5$. Thus, the magnetocrystalline anisotropy at the amorphous/crystalline interface is anticipated to reduce significantly.

The low Pt concentration fluctuation detected in crystalline $SmCo_5$ film indicates that Pt substitution for Co merely decreases the magnetocrystalline anisotropy of $SmCo_5$ without inducing any pinning force for magnetic domain motion, hence resulting in a lower coercivity [73]. The effects of the buffer layer on magnetic properties will be further discussed in Section 3.4.3.

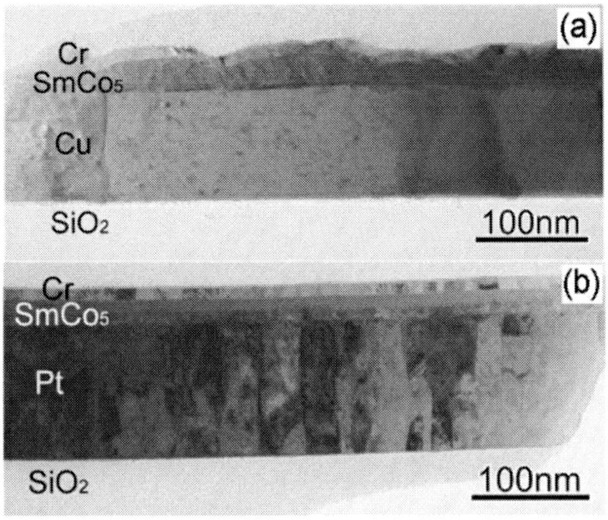

Figure 22. Cross section of $SmCo_5$ thin film with (a) Cu and (b) Pt underlayer *(Reprinted with permission from Ref.* [73]. *Copyright (2006), American Institute of Physics).*

3.2. Structure of Magnetic Thin Films

Film texture is related to the distribution of crystallographic orientations of the film. A film with completely random orientation is said to have no texture, thus no texture develops in an amorphous film. In the case where the orientation is not random but has a certain degree of preferred orientation then the sample is considered to have a weak, moderate or strong texture, depending on the percentage of the crystals that have the preferred orientation. Generally, strongly textured continuous films are expected to show low H_c since pinning sites for magnetic domain wall motion are not anticipated. The magnetization process in strongly textured films occurred mainly through the domain wall motion and hence the whole film layer will be reversed easily once a reversed domain is established. However, Fullerton *et al.* [80] has proven otherwise by showing that epitaxially grown Sm-Co exhibited H_c as high as 4 T. Evaporated films are often more strongly textured and exhibit larger grain sizes than sputter deposited films. However, sputter deposition is the more widely used technique because it offers better control over film stoichiometry and thickness and at the same time has the flexibility to deposit a wide range of crystalline and amorphous materials [117]. Film composition, alongside substrate temperature and buffer layer plays an important role in the formation of columnar structures, which is necessary to obtain highly textured films. At low deposition temperatures, texture is barely observable since there is no adatom mobility or grain boundary migration. In some cases, texture develops randomly in initial deposits, grows into strong orientation of low energy planes and as the film thickens further, evolves into a preferred texture [114]. Strong *c*-axis texture has been observed in Nd-Fe-B films deposited onto a Ta buffer layer at 550 °C and annealed films deposited onto Mo substrates at low temperature [92, 129-131]. Regardless of the substrate temperature used, high intensities of (004), (006) and (008) diffraction peaks and the absence or low intensity of a (410) peak indicate strong [001] texture [64]. An example of an x-ray diffraction pattern of a strongly textured $Nd_2Fe_{14}B$ film is shown in Figure 23. The texture suggests the presence of a preferred *c*-axis orientation perpendicular to the film plane, which is the easy magnetization axis for tetragonal $Nd_2Fe_{14}B$. A deposition temperature that is too high or too low leads to deterioration of film texture. Although the use of a buffer layer has demonstrated improvement in the epitaxy of films, too high a deposition temperature causes oxidation of Nd because the buffer fails to prevent oxygen diffusion from the substrate to the film [129]. It was found that a textured film with a smoother surface and lower Nd content has better long-term stability but its coercivity is reduced [132]. On the contrary, a $SmCo_5$ thin film demonstrates longitudinal anisotropy with its easy axis, i.e. the energetically favourable direction of spontaneous magnetization in a ferromagnetic material, in the film plane direction [133]. Pole-figure analysis, Figure 24, has confirmed this in-plane texture. The sharpness of the (200) pole figure reveals that the films deposited at 350 °C exhibit strong in-plane texture. On the other hand, the sharper (110) pole figure for films deposited at 500 °C indicates that the films are weakly textured [40]. Other studies, however, also designate the (110) texture for $SmCo_5$ and $SmCo_7$ films [134-136]. The predominance of (200) or (110) texture depends on deposition parameters such as deposition rate, deposition temperature and sputtering pressure. Nevertheless, perpendicular anisotropy with (001) texture has still been observed in $SmCo_5$ films prepared by depositing onto a Cu buffer layer thicker than 100 nm or by alternating Sm and Co layers [10, 137]. However, the texture and perpendicular anisotropy were still poor with the use of such seed layers [138, 139]. Films with up to 35 alternating Sm and Co layers demonstrate better perpendicular anisotropy but the complex process has limited application. Considering these two outcomes, Chen *et al.* [140] managed to obtain a highly textured (001) film with high coercivity by incorporating Ta seed layer and co-sputtering Sm and Co targets. Epitaxial $SmCo_5$ film of $[11\bar{2}0]$ texture has been deposited onto MgO [200] substrate with a Cr [200] buffer layer with diffraction pattern shown in Figure 25(a) and (c). The epitaxial relationship between the $SmCo_5$ film, Cr buffer and MgO substrate is given by $SmCo_5$ $(11\bar{2}0)$ <0001> || Cr (200) <011> || MgO (200) <010>. The nanocrystalline structure and strong interaction between grains enhances the in-plane coercivity and remanence of the $SmCo_5$ $(11\bar{2}0)$ film deposited on an amorphous glass substrate [81].

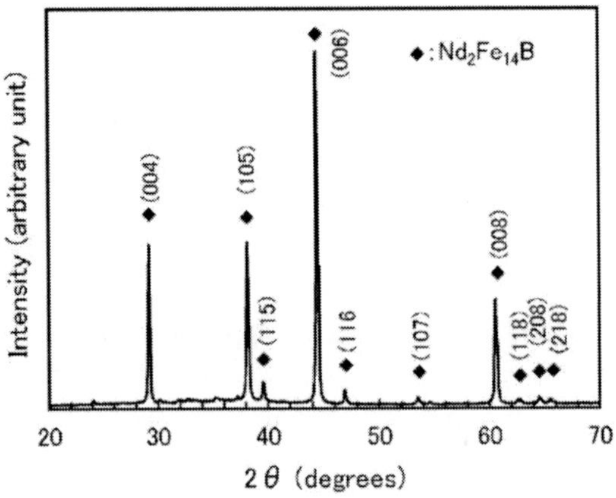

Figure 23. Diffraction pattern of $Nd_2Fe_{14}B$ film formed on glass substrate heated at 550 °C *(Reprinted from Ref.* [67], *Copyright (2004), with permission from Elsevier).*

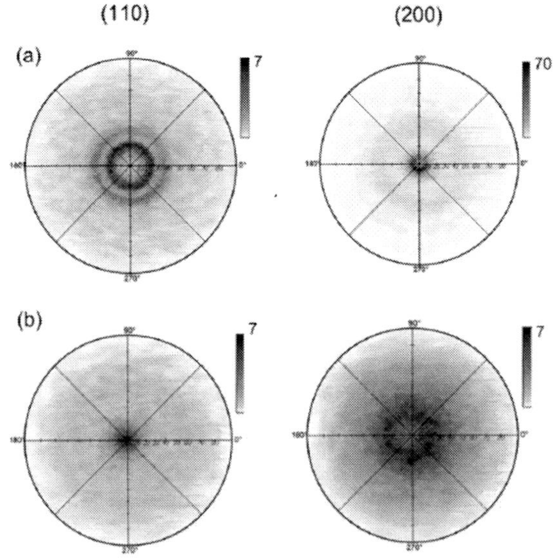

Figure 24. (110) and (200) pole figures of $SmCo_5$ films deposited at (a) 350 °C and (b) 500 °C. The pole intensity of (200) pole figure for sample deposited at 500 °C is ten times greater than the other pole figures, indicating strong in-plane texture *(Reprinted with permission from Ref.* [40]. *Copyright (2008), American Institute of Physics).*

Epitaxial $SmCo_5$ film with $(1\bar{1}00)$ texture has also been observed to grow on the MgO (110) [80] and Cu (111) [73], with the diffraction pattern indicated in Figure 25.(b) and (d).

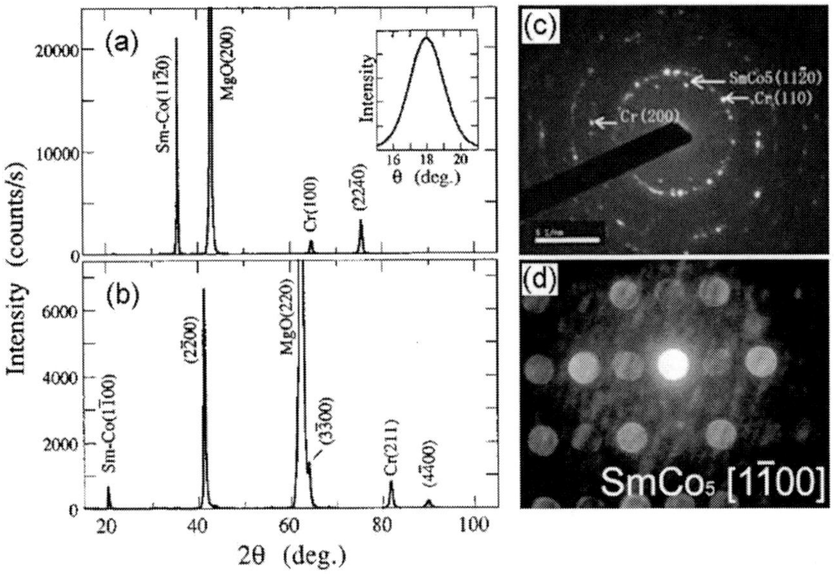

Figure 25. XRD and selected area diffraction pattern for $SmCo_5$ film with $(11\bar{2}0)$ texture for (a) and (c) and $(1\bar{1}00)$ texture for (b) and (d) *[(a) and (b) reprinted with permission from Ref.* [80]. *Copyright (1997), American Institute of Physics; (c) reprinted from Ref.* [81], *Copyright (2009), with permission from Elsevier; (d) reprinted with permission from Ref.* [73]. *Copyright (2006), American Institute of Physics].*

3.3. Magnetic Domain

Features of magnetic domain structure could disclose much information on magnetic properties of thin films. The structure of magnetic domains in thin films could be described by using the structure factor and disorder function calculations. A significant increase in disorder function usually signifies improvement of anisotropic *c*-axis texture and soft to hard magnetic property transitions of the film. Disorderness in magnetic domain is often associated with deviations from the idealized long-range ordered stripe pattern. Such deviations can be seen in the domain walls, point and line defects as well as width and length spatial variations [63]. Generally, a higher level of disorder in magnetic domains leads to harder magnetic properties.

Magnetic thin films deposited at room temperature tend to form magnetic domains with highly regular stripes, Figure 26.(a). Such stripe domains usually indicate soft-magnetic properties of the film. This is supported by the fact that Nd-Fe-B thin films deposited at room temperature without any post-deposition heat treatment are known to comprise mainly amorphous Nd-rich and α-Fe phases, both of which are soft magnets. The magnetization orientation of films that exhibit stripe domains is typically in the film plane [123]. At a slightly higher temperature, irregular bifurcations start to form while the width and length of the stripes are less than those observed at room temperature. While still remaining a soft magnet, this type of domain, together with the change in stripe dimension, indicates the presence of a very weak anisotropy. Further increase in substrate temperature leads to the formation of more curved and forked domains while features such as circles and islands emerge, Figure 26.(c). Magnetic domains with the features indicated in the preceding paragraph, can be described as labyrinthine domains and being maze-like [61, 63]. Maze-like domain structures have been observed in bulk magnets [141, 142] as well as magnetic thin films [61, 63, 127] of Nd-Fe-B and Sm-Co. However, the average domain size in sintered magnets is larger than that in thin films, which is mostly attributed to the difference in microstructure and thickness. As temperature increases, the boundary between the domains becomes less defined and irregular domains develop. At a certain substrate temperature (in the case of Nd-Fe-B film this temperature is 500 °C), highly irregular, cloud-like domains begin to dominate, Figure 26.(f). The boundaries of these cloud-like domains become more visible and the size increases as the substrate temperature rises, Figure 26.(g) and (h). This microstructural behaviour indicates stronger exchange couplings between grains in the film, hence greater magnetic anisotropy. The enhancement of magnetic anisotropy with substrate temperature may also be attributed to the formation of isolated crystallites.

Domain size, wall thickness and coercivity depend largely on surface roughness and the type of domain walls; i.e., either Bloch or Néel walls. The magnetization reversal mechanism shifted from magnetization rotation, which is dominant in smoother films, to domain wall motion that is more prominent in rough films [143]. The surface of the Nd-Fe-B film becomes rougher as deposition temperature increases, Figure 27. The increase in surface roughness causes a change of demagnetizing factors in magnetic thin films. Both Bloch and Néel walls exhibit different trends in the demagnetizing factor for such roughness induced changes. The magnetization of Bloch walls rotates out of the plane when crossing the wall, hence the demagnetizing factor decreases in the direction perpendicular to the film surface. On the other hand, the magnetization of Néel walls rotates within the plane of the wall itself, leading to an increase in demagnetizing factor in the direction parallel to the film surface [144].

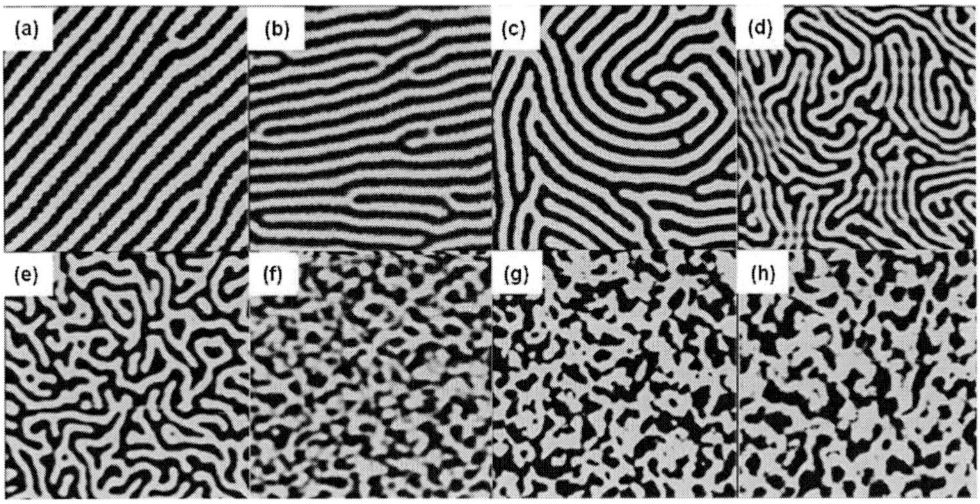

Figure 26. Magnetic domain structure of Nd–Fe–B films prepared at the substrate temperatures (a) 25, (b) 200, (c) 300, (d) 400, (e) 450, (f) 500, (g) 550, and (h) 600 °C *(Reprinted with permission from Ref.* [63]. *Copyright (2008), American Institute of Physics).*

In the case of thin films, Néel walls are the more common domain wall for discussion because the exchange length is much greater in comparison to film thickness. Magnetic properties of the film can be estimated as the surface roughness changes since coercivity, magnetic domain wall thickness and domain sizes are each related to the demagnetizing factor.

For Néel walls, the domain wall thickness tends to decrease as the demagnetizing factor increases; assuming that the anisotropy energy can be neglected. The relationship between local surface roughness and coercivity is given by [144]:

$$H_C = \frac{1}{2M_S}\left(\frac{\delta\sigma_W}{\delta t} + \frac{\sigma_W}{t}\right)\rho_{rms}$$

where H_c is the coercivity, M_s is the saturation magnetization, σ_w is the surface energy of domain wall, t is the thickness and ρ_{rms} is the root mean square (rms) of roughness local slope. If it is now assumed that saturation magnetization depends on film thickness, then the coercivity calculated from domain wall movement is predicted to increase with an increase in ρ_{rms} for a face centred cubic (fcc) film [144]. This interpretation reinforces the idea that magnetic properties of a thin film are enhanced with the increase in deposition temperature since surface roughness increases with an increase in deposition temperature.

The change of magnetic domain structure with annealing is similar to that experienced with an increase in substrate temperature, Figure 28. The maze-like domain is transformed to a highly irregular and disordered pattern after annealing.

The domain pattern of an as-annealed film is similar to that of films deposited at substrate temperatures above the crystallization temperature of the material (refer to Figure 26). Thus, the concept that crystallization of magnetic thin films occurs via two routes is reinforced: (i) direct deposition onto heated substrates, and (ii) post-deposition annealing of films deposited at lower temperatures.

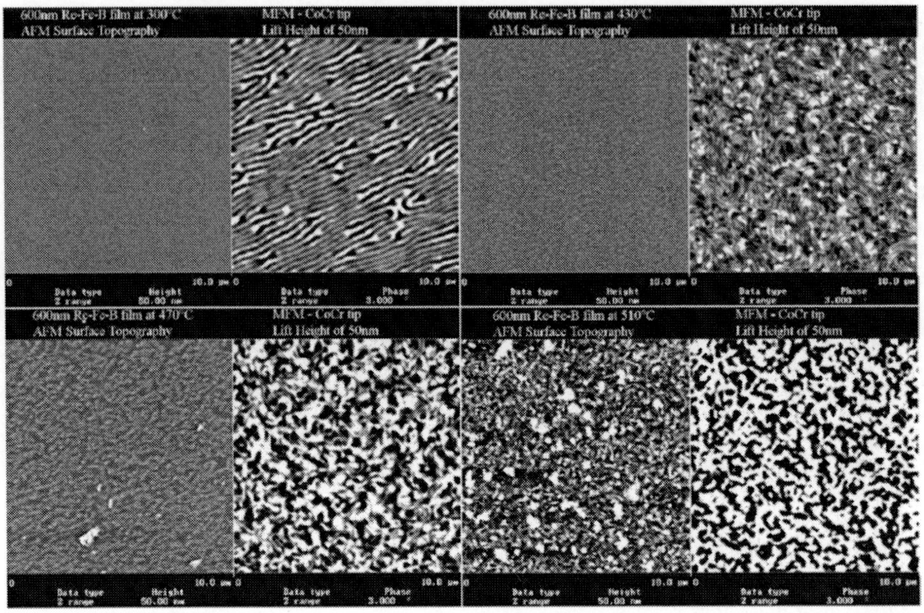

Figure 27. Variation of surface topography and magnetic domain with substrate temperature *(Reprinted with permission from Ref.* [123]. *Copyright (2007), American Institute of Physics).*

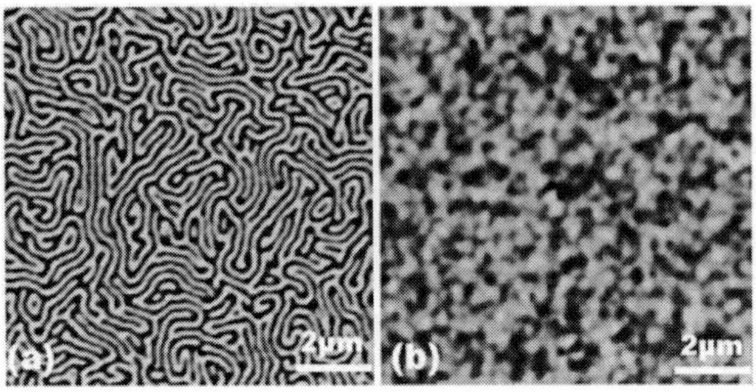

Figure 28. Magnetic domain structure for [Si/Mo(50 nm) / Nd–Fe–B(800 nm)/Mo(50 nm)] thin films (a) deposited at 500 °C and (b) annealed at 600 °C *(Reprinted with permission from Ref.* [61]. *Copyright (2005), American Institute of Physics).*

Variation of the buffer layer material does not induce much difference in the magnetic domain pattern of the films so long as the material used is compatible with the film material [61], as will be discussed later in Section 3.4.3. When copper has been used as a buffer layer for Sm-Co films, the compositional fluctuation within a $Sm(Co,Cu)_5$ layer formed from diffusion of Cu into the film can work as a pinning force for magnetic domain wall motion [73]. The pinning force is maximum when the anisotropy inhomogeneities are at the length scale of the domain wall width [145]. On the other hand, the variation of domain size with buffer thickness depends on the film and buffer material. In the case of Nd-Fe-B films, the domain size increases then decreases with increasing buffer layer thickness as can be seen from Figure 29. [127]. At certain thickness, the domain size is greater than the grain size while at other thickness, the domain size is smaller than grain size. The relationship between domain sizes and grain sizes has not been clearly established.

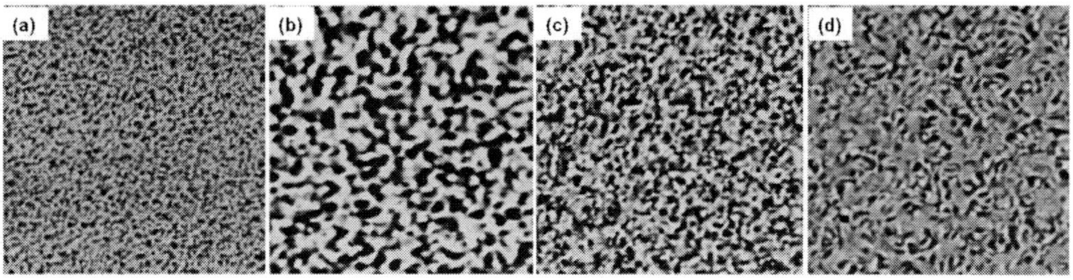

Figure 29. Magnetic force microscopy (MFM) micrographs of [Si/Ti(x nm)/Nd–Fe–B(300 nm)/Ti(20 nm)] thin films with different Ti buffer layer thickness, (a) $x = 20$ nm, (b) $x = 50$ nm, (c) $x = 100$ nm and (d) $x = 200$ nm *(Reprinted with permission from Ref.* [127]. *Copyright (2005), American Institute of Physics).*

3.4. Magnetic Properties

3.4.1. Effects of Sputtering Gas Pressure

As mentioned earlier, sputtering gas pressure is an important technical parameter that influences the microstructural, structural and magnetic properties of a film. The typical sputtering gas used in most systems is argon gas. However, in certain cases xenon is used due to its shorter thermalization distance, and hence reduced reflection of high energy neutral gas particles [146].

As a result, films formed using xenon sputtering gas exhibit an enhanced texture over those formed in argon. In other words, as-sputtered films formed in a xenon environment tend to be crystalline while those formed in an argon environment are prone to be amorphous. A higher pressure of xenon leads to films that are richer in Sm due to the Co-Xe collisions that scatter Co atoms through large scattering angles [47]. Despite the fact that xenon as a sputtering gas produces better quality films, argon is the most widely used sputtering gas due to its lower cost.

A wide range of sputtering gas pressures, ranging from 0.03 to 27 Pa, [37] have been studied in the preparation of RE magnetic thin films. Film composition is highly influenced by the sputtering gas pressure. Generally, films prepared at higher pressures are richer in RE components (refer to Figure 30) [56, 147]. The RE enrichment generally results in significant increases in coercivity and remanence of the films. Similar phenomenon was observed by Zhen *et al.* [75] where coercivity increases with pressure until a point where coercivity decreases while the remanence keeps increasing as shown by hysteresis loops in Figure 31. Strong uniaxial in-plane anisotropy can be observed for Sm-Co films.

Magnetic properties of a film are in part attributed to the microstructure of the film. As highlighted by the Thornton diagram, Figure 19, film microstructure depends on the sputtering gas pressure. At low sputtering gas pressure, the film grain tends to be needle-like and form a smooth and flat surface.

As the pressure increases, the grain is shifted to a columnar shape with increased grain sizes, which is good for enhancing magnetic properties. Figure 32. illustrates the variation of grain structures from needle-like to columnar as a function of sputtering gas pressure. The difference in grain size and shape at different pressures is related to the energy involved while depositing the atoms. At a lower sputtering gas pressure, the atoms have more energy, and hence, form more densely packed and uniform films [75].

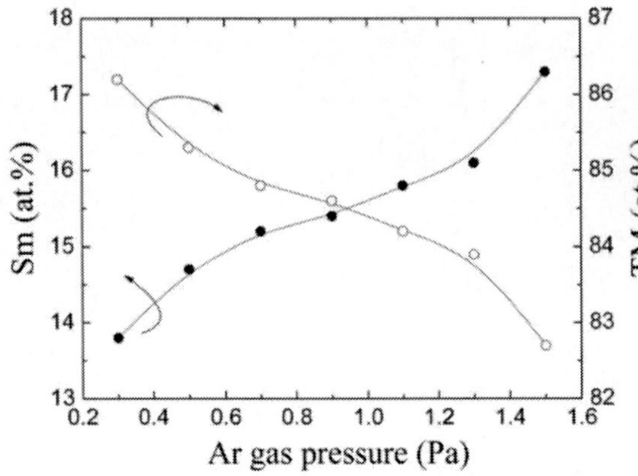

Figure 30. Variation of Sm and transition metal (TM) concentration with Ar gas pressure *(Reprinted with permission from Ref.* [148]. *Copyright (2009), American Institute of Physics).*

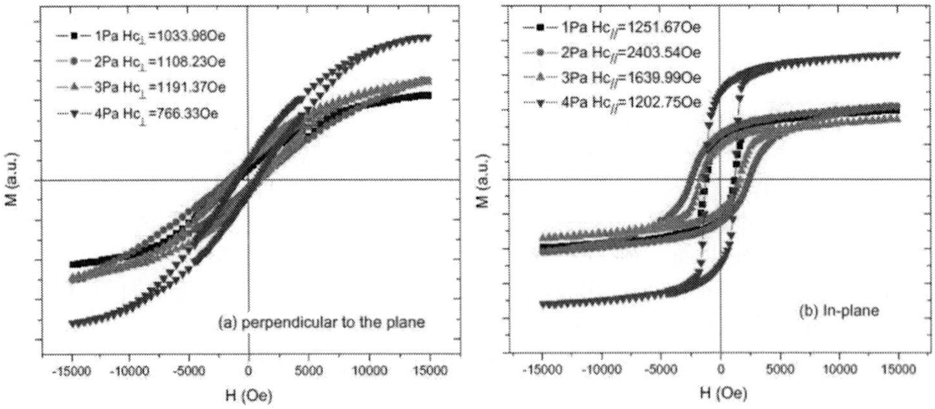

Figure 31. (a) Out of plane and (b) in-plane M-H loops of Sm-Co films prepared at varying Ar pressure *(Reprinted with permission from Ref.* [75]. *Copyright (2008), World Scientific Publishing).*

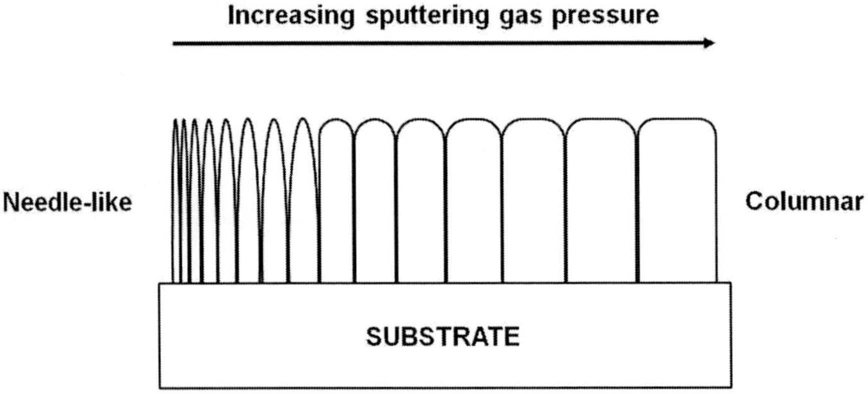

Figure 32. Schematic illustrating the variation of grain structure as a function of sputtering gas pressure.

The increase in coercivity with sputtering pressure can also be related to the increase of surface roughness, which is supported by Hannemann *et al.* [128] and Parhofer *et al.* [147] in separate studies. The increase in coercivity is attributed to the domain wall movement of Néel walls in thin films. Such a variation can be further explained by the change in demagnetizing factor as discussed in Section 3.3. Calculation and experimental data to validate the relationship between coercivity and wall movement is available elsewhere [144]. Two cross-section models have been used to estimate the relationship between the deposition pressure and texture of a film. The *constant* cross-section model estimates a relatively lower total pressure than is used experimentally to acquire complete texturing. On the other hand, the *energy-dependent collision* cross section model predicts that an equivalent thermalization outcome is achieved at pressures of 130 mTorr Ar, 60 mTorr 50% Ar, 50% Xe, and 40 mTorr Xe for a target-substrate distance of 5 cm. At lower deposition pressures, remanence of a film measured perpendicularly has a larger ratio compared to that measured in-plane. Selective thermalization conditions can also be related to magnetic properties of films deposited at higher pressures due to the existence of a lower bound of thermalization conditions as the films start to show enrichment in RE components [47].

3.4.2. Effects of Substrate and Annealing Temperature

There are two alternative routes when it comes to preparing crystallized permanent magnetic films. The first is to deposit an amorphous film which is subsequently heat treated to form a crystallized film. The degree of crystallization depends on the duration and temperature of annealing. The films that are prepared from subsequent crystallization have fine grains and are of single-domain type.

However, the second route is more preferable in order to generate a film with better textured crystal growth which in turn gives higher possible energy density [47]. The second method is to directly crystallize the film by depositing on a heated substrate. Various thermalization methods can be used in accordance to this route to enhance the degree of preferential texturing of the films.

A crystalline film can only form when the material is annealed or deposited on a substrate at a temperature above its crystallization temperature, which in the case of $Nd_2Fe_{14}B$ is about 600 °C. In either case, whether a film is deposited directly or annealed subsequently, a low processing temperature is desirable for film integration with various applications.

Processing temperature that is too high will aggravate interlayer diffusion and at the same time could oxidize RE. The use of suitable underlayer material can lower the crystallization temperature so that the overall processing temperature is also lowered [149]. In the absence of such an underlayer, substrate and annealing temperatures as high as 800 °C and 1500 °C, respectively, have been reported for Nd-Fe-B films [37]. Although there has been some evidence of texturing found in Nd-Fe-B films deposited at 360 °C in an Xe environment [146], the optimum deposition and annealing temperatures revolve around 600 – 650 °C. Films prepared within this temperature range exhibit the best perpendicular anisotropy. Comparison of hysteresis loops that implemented different substrate temperatures in forming Nd-Fe-B film are shown in Figure 33. Deposition at temperatures lower than the optimum temperature produces films with soft magnetic properties while at higher temperatures the coercivity decreases due to an increase in grain size.

Remanence also tends to decrease at higher temperatures due to oxidation of Nd (to form α-Fe_2O_3, Nd_2O_3 and $FeNdO_3$) as well as interdiffusion between the film and substrate to form additional phases (such as $NdSi_2$, $Nd_2Si_2O_7$, $Nd_{23}Fe_{77}$). Deterioration of magnetic properties at processing temperatures beyond the optimum temperature can also be attributed to random film growth [150].

Although $SmCo_5$ exhibits a higher Curie temperature (about 700 °C) than $Nd_2Fe_{14}B$ (about 400 °C), its crystallization temperature is lower than that of $Nd_2Fe_{14}B$ and hence it is typically deposited at 400 to 700 °C [47]. An optimum deposition temperature of 450 °C for preparing $SmCo_5$ films has been reported [151].

However, an optimum temperature as low as 350 °C has also been reported that produces $SmCo_5$ film with a coercivity of 1.3 T and maximum energy product of 112 kJ/m^3. Post-deposition annealing at 750 °C for 10 minutes increases the maximum energy product to 144 kJ/m^3 [40]. Similar trends have been observed for Nd-Fe-B films where both a heated substrate and annealing are implemented. In this instance a coercivity of 1.8 T and a maximum energy density of 238 kJ/m^3 were documented.

Romero *et al.* [152] have introduced the terms "thermal annealing" and "flash annealing" for films prepared by sputter deposition and PLD. Thermal annealing refers to heat treatment in conventional furnaces with an inert atmosphere while flash annealing involves applying current pulses to the substrate under vacuum conditions.

Both of the annealing processes managed to induce magnetic hardening of Sm-Co films although only partial crystallization was achieved. The easy magnetization direction observed was neither in the film plane nor perpendicular to the film, but its coercivity was well over 2 T. However, films that underwent flash annealing were found to have increased surface roughness and reduced adhesion that lead to some films peeling off the substrate.

The optimum deposition and annealing temperature depends strongly on the composition of material being deposited. It has been proven that crystallization temperature decreases as RE content increases [62, 153].

The presence of excessive RE acts as a catalyst to speed up film crystallization while simultaneously encouraging the growth of hard grains [153]. That is, films that exhibit higher RE content demonstrate higher coercivity at the same deposition and annealing temperatures, Figure 34. [62, 151].

However, an increase of coercivity is often achieved at the expense of remanence. The compromise between coercivity and remanence should be determined in each case. It is also believed that a suitable amount of excess boron in Nd-Fe-B films enhances the perpendicular anisotropy of the film [153, 154].

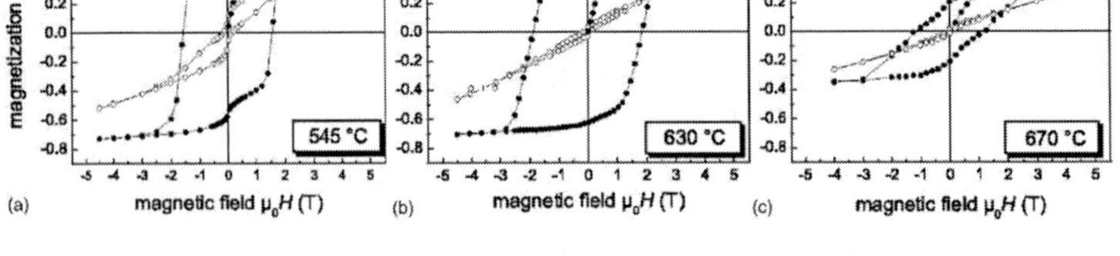

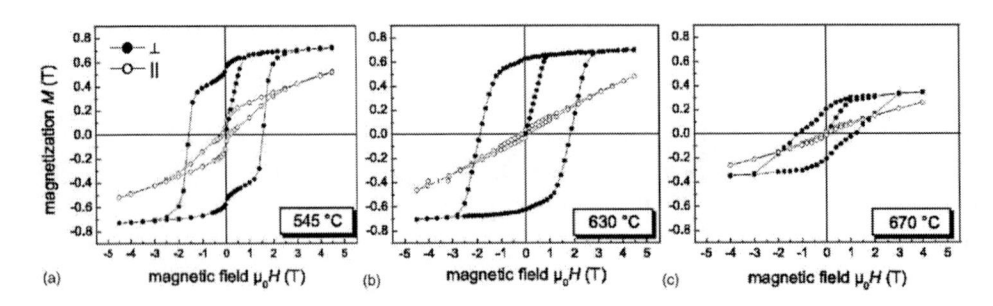

Figure 33. Hysteresis loop measured perpendicular and parallel to the plane of Nd-Fe-B films deposited at (a) 545 °C (b) 630 °C and (c) 670 °C *(Reprinted with permission from Ref.* [129]. *Copyright (2004) by the American Physical Society*).*

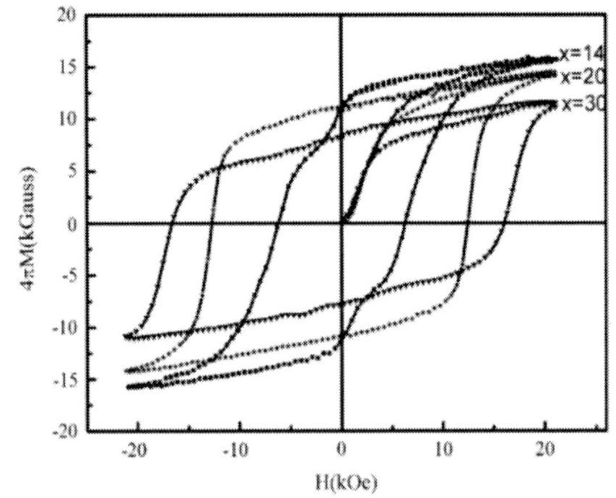

Figure 34. Hysteresis loops for $Nd_xFe_{92-x}B_8$ film at various x after optimum annealing at 600 °C *(Reprinted from Ref.* [62], *Copyright (2006), with permission from Elsevier).*

The role of heating rate during annealing is also crucial. Figure 35. indicates the changes in the magnetic hysteresis loop of Nd-Fe-B films measured parallel and perpendicular to the film as heating rate varies. The improvement of magnetic properties with heating rate is attributed to the change in microstructure. Slow heating rate favours growth of additional phases and may include voids that induce formation of an irregular microstructure. At higher heating rates, more well-defined grains are formed with finer crystallites [39, 52]. It is evident that the columnar structure and homogeneous crystallite distribution leads to the high coercivity and anisotropic behaviour of films. The different heating rate promotes different crystallization behaviour that directly influences coercivity and remanence of the film [52]. The use of high heating rate annealing for a short period of time, termed as rapid annealing, is able to crystallize Nd-Fe-B at low temperatures into single domain nanocrystallites that exhibit a high coercivity and remanence. Such high coercivity and remanence are attributed to the strong exchange coupling among the nanocrystallites and reduced oxidation of the films [155]

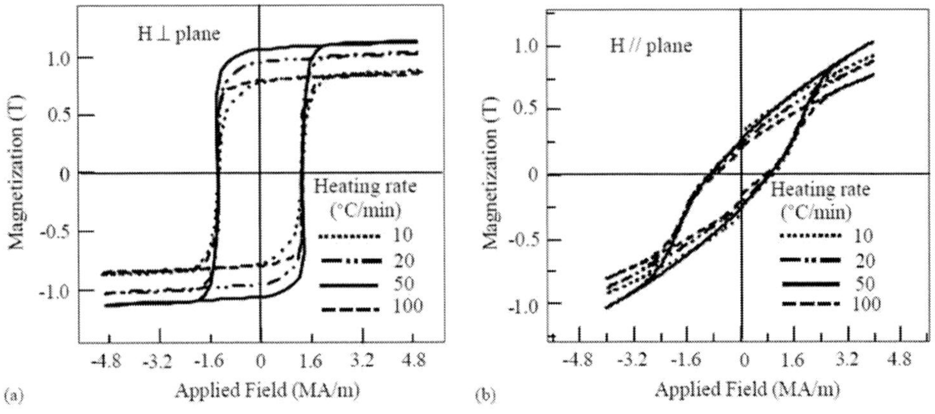

Figure 35. Hysteresis loops of Nd-Fe-B films annealed at various initial heating rate measured (a) perpendicular and (b) parallel to the plane of the film *(Reprinted from Ref.* [52], *Copyright (2003), with permission from Elsevier).*

The dependency of grain size and shape on the heating rate have been pointed out by Walther *et al.* [39]. High heating rate results in equiaxed grains throughout the film while slower rates produce films with a mixed structure where the lower layers are constructed of columnar grains and upper film sections are dominated by equiaxed grains. The equiaxed grains in films heated at lower rates were larger than those rapidly heated. The increment in grain size accounted for the improvement of coercivity since surface roughness increases with an increase in grain size [120, 156, 157]. However, other studies have demonstrated that the grain shape and size have little effect on magnetic properties. The variation of coercivities has instead been assigned to the difference in the Nd-rich intergranular phase and secondary phase distribution [158].

On the other hand, the effects of annealing time are less significant in isotropic films deposited at lower temperatures than anisotropic films deposited on heated substrates. Films deposited at lower temperatures exhibit high coercivity even for a short duration of annealing; while films deposited at higher temperatures exhibit a strong crystallographic texture that is enhanced significantly with annealing time [39].

3.4.3. Effects of Buffer Layer and Seedlayer

The primary reasons for using a buffer layer (also known as an underlayer) are (i) to prevent interdiffusion among substrate and film, (ii) to induce the desired texture as well as (iii) to optimize the microstructure of the film to maximize the magnetic properties. The elements that have been used as buffer layers are gold (Au), chromium (Cr), copper (Cu), iron (Fe), molybdenum (Mo), niobium (Nb), platinum (Pt), tantalum (Ta), titanium (Ti), vanadium (V) and tungsten (W) with thicknesses ranging from 1 – 288 nm [37]. Characteristics such as activity, melting point and high temperature oxidation of an element should be considered when choosing the optimal buffer layer for various materials [159]. Buffer layers are also necessary when growing thicker films [89]. A buffer layer is necessary to avoid oxygen penetration into the film as well as to keep elements of the film from diffusing into the substrate [132].

The use of Cr as a buffer layer for Nd-Fe-B films reduces the film remanence. This deterioration in magnetic properties is induced by interdiffusion between Cr and Fe to form a Cr/Fe rich layer between the Cr buffer and Nd-Fe-B film. The presence of the diffusion layer hinders the epitaxial growth of the film [128, 160]. Nevertheless, the Cr buffer is still effective in improving the film coercivity. The improvement of coercivity can be adjusted by manipulating the roughness of the buffer layer. The use of a more rough Cr buffer is able to increase the coercivity of a magnetic film up to 50% compared to that deposited on a smooth buffer due to reduced Nd-Fe-B grains, which nucleate in the trenches as the buffer becomes rougher [128]. The ability of a Cr buffer to enhance the coercivity of a film more than other buffers, such as Ta, is attributed to the coercivity mechanism of Cr which is based on domain wall pinning as suggested by Melsheimer and Kronmüller [161]. The same trend of coercivity increment with roughness applies to Sm-Co films. However, a soft magnetic phase tends to form in films that incorporate a rough, meander-like buffer surface. Oxygen diffusion from the substrate was suggested to be the reason for this anomaly since both Co and Cr are immiscible [128]. The use of Cr as a buffer layer prevents Sm atoms from reacting with the thermally oxidized Si substrate to form the Sm_2O_3 phase [162] and at the same time reduces the tensile residual stress in the film [72].

Improvement of film crystallinity, which depends directly on buffer crystal orientation, is crucial to obtain high magnetocrystalline anisotropy. A tungsten, W, buffer layer with excellent crystal orientation can be prepared readily and hence is suitable for magnetic thin film deposition.

Although there is about 7.7% lattice mismatch between the tetragonal $Nd_2Fe_{14}B$ plane and W plane, coercivity as high as 12.2 kOe is obtained for Nd-Fe-B films on a W buffer [163]. $Nd_2Fe_{14}B$ film grows heteroepitaxially on a W buffer with the *c*-axis perpendicular to the film orientation; inducing a perpendicular magnetic anisotropy in the film. The W buffer layer also effectively shields the film from oxidation [163]. W has also been used as a buffer for Sm-Co films; however, it is not as effective as Cr and Mo buffers [164]. Tantalum, Ta, is highlighted as a buffer due to its close thermal expansion coefficient to that of Nd-Fe-B; which directly enhances formation of the $Nd_2Fe_{14}B$ phase [159]. However, films with less than 180 nm deposited on a Ta buffer exhibit comparatively lower coercivity than those deposited on Nb and Mo buffers [65].

It has been suggested that there is interdiffusion of the Ta buffer into the film layer and vice versa for film thickness less than 180 nm; and this interdiffusion prevents the formation of the magnetically hard $Nd_2Fe_{14}B$ phase [65, 129]. A similar reduction in magnetic properties, also attributed to interdiffusion, has been observed at film-buffer and buffer-substrate interfaces [165].

The thickness of buffer layers is a crucial factor that must be considered when they are incorporated into film deposition. Generally, remanence and coercivity of a film increase as the thickness of buffer increases up to a point beyond which further increase in thickness diminishes the magnetic properties of the film, Figure 36. At low buffer thicknesses, the buffer could not hinder diffusion between the film and substrate and, hence, the magnetic properties are lowered.

However, at thickness beyond the optimum thickness, the grain size of the buffer grows abnormally and introduces irregularity to the film surface, and therefore the magnetic properties of the film deteriorate [127]. Shima *et al.* [166], Parhofer *et al.* [132] and Chen *et al.* [127] have separately studied the relationship between buffer thickness and Nd-Fe-B film performance and concluded that the optimum buffer thickness is 50 – 80 nm for a film 300 – 350 nm thick. On the contrary, Kato *et al.* [139] and Ohtake *et al.* [88] preferred a 10 nm optimum buffer thickness for a 20 nm of Sm-Co film. Nevertheless, the optimum thickness depends largely on the film material, film thickness as well as the substrate and buffer material used.

Another point concerning the selection of the buffer layer is the lattice match of the buffer material with the film. The lattice mismatch parameter of a film is given by $\delta = |a(u) - a(m)|/a(m)$ where $a(u)$ and $a(m)$ refer to the lattice parameters of the buffer layer and magnetic layer respectively [127].

Depending on the magnetic properties of interest, the buffer material could be tailored to suit various applications. For instance, Mo is a potentially good choice as a buffer to improve the remanence of Nd-Fe-B films due to its perfect lattice match with Nd-Fe-B. A maximum energy density as high as 14.8 MOe was obtained, which is significantly higher than that of the other buffer layer used in the study [127]. The perfect lattice match gives the film a high remanence. Titanium buffers, on the other hand, have less perfect lattice match with Nd-Fe-B and hence provide more pinning sites that hinder magnetic domain movement, resulting in higher coercivity.

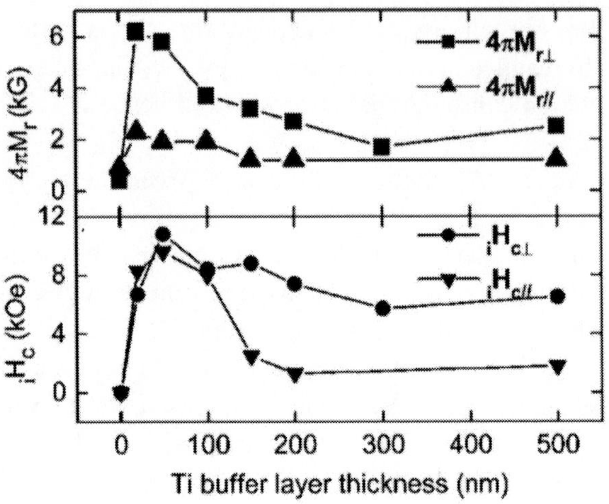

Figure 36. Variation of film remanence and coercivity with Ti buffer thickness *(Reprinted with permission from Ref.* [127]. *Copyright (2005), American Institute of Physics).*

Copper, Cu is a popular choice as buffer layer for Sm-Co films because (i) the lattice misfit between Cu and $SmCo_5$, which is less than 3%, [140] is small and (ii) the $SmCo_5$ phase is stabilized by Cu diffusion into the film. Unlike Nd-Fe-B films, diffusion of Cu into the film is desirable because it forms an alloy compound of composition $SmCo_{5-x}Cu_x$ by substituting Co sites in $SmCo_5$ structure. Formation of this compound stabilizes the $SmCo_5$ phase while destabilizing Sm_2Co_{17} phase [167]. This effect is more intense with higher Cu content. The amount of Cu atoms that diffuse from the buffer layer to the growing $SmCo_5$ film depends largely on the film thickness and substrate temperature. While diffusion decreases with increasing film thickness, a reduction in substrate temperature causes diffusion to decrease as well. Insufficient Cu diffusion destabilizes formation of the ordered phase. Increase in buffer thickness improves the crystallographic quality of the ordered phase and this has been proven by observing an increase in the $SmCo_5$ (0001) peak intensity as shown in Figure 37. It is often difficult to reduce the Cu buffer thickness while retaining good film texture and crystallinity. Especially noteworthy is the use of a dual buffer layer system, where the layer between the substrate and the other buffer layer is sometimes termed as seed layer, in reducing the thickness of the Cu buffer layer, Figure 37. Good texture is maintained with buffer thickness reduction while using Ti or preferably Ta as a seed layer [138, 168]. The improvement in terms of texture, crystallinity and surface roughness of a Cu film on a Ta seed layer was explained by the effects of surface energy and kinetics. A $SmCo_5$ film with 4 nm Ta seed layer exhibited a strong texture with perpendicular coercivity as high as 20 kOe [169]. Although Cu is useful as a phase stabilizer for Sm-Co films, there is constant diffusion of Cu into Si or the SiO_2 substrate when it is used as a buffer layer. Ta, which is a highly stable refractory metal and immiscible with copper, has been used effectively as a diffusion barrier [170, 171]. Hence, Ta is widely used as seed layer whenever Cu is used as a buffer [140, 169]. Ruthenium (Ru) has also been used as a seed layer to reduce the surface roughness and grain size of a Cu buffer used in $SmCo_5$ films. However, Ru serves better as an interlayer rather than a seed layer as evidenced by the increased coercivity and remanence in the 4-layer system of [SmCo (20 nm)/Ru (5 nm)/Cu (10 nm)/Ru (5 nm)] [139]. In this example there is a buffer system of 3 layers consisting of Cu sandwiched between Ru.

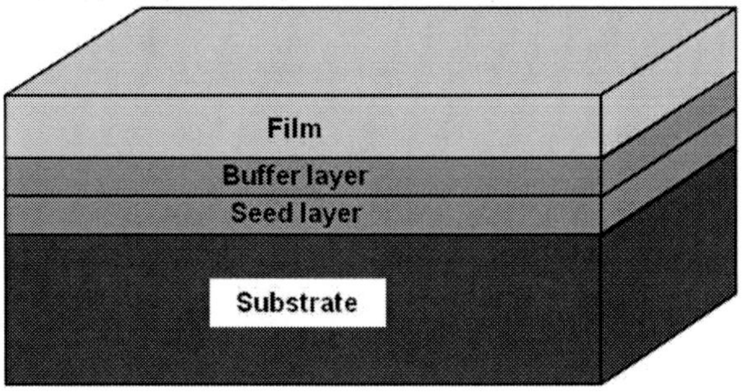

Figure 37. Schematic of film, buffer layer, seed layer and substrate (thickness of each components are not in proportion).

3.4.4. Effects of Film Thickness

A study by Piramanayagam *et al.* [172] shows that Nd-Fe-B films show perpendicular magnetic anisotropy over a wide range of thickness; i.e., 50 – 1500 nm. However, increment in film thickness leads to more random crystalline orientation that causes perpendicular magnetic anisotropy to disappear. Increasing thickness also results in lower coercivity due to weakening of the *c*-axis orientation with development of film thickness. This contradicts the finding of Chandrasekhar *et al.*'s [173] that mentioned a reduction in coercivity and squareness ratio (ratio of remanence to saturation magnetization) for thinner films. The reduction in magnetic properties was attributed to deterioration of crystalline anisotropy as well as degradation of the hard magnetic phase in the film structure.

Generally the magnetic properties improve as the film thickness decreases. Similar to the development of grain size with respect to the buffer layer thickness discussed previously, the grains in a magnetic film also enlarge as the film thickness grows. The average grain size in a film can be controlled by varying the annealing temperature and film thickness. The grain size directly influences the magnetic behaviour and the type of magnetic domains involved. Figure 38. outlines the types of magnetic domain in accordance to grain size and the variation of coercivity with respect to grain size. The maximum coercivity for a particular material is achieved within the single domain range. Coercivity decreases as the grains grow larger due to grains partitioning into domains. As grains become smaller, coercivity decreases again due to the randomizing effects of thermal energy. Furthermore, the magnitudes of exchange, magnetocrystalline and saturation magnetization, which are highly dependent on the deposition temperature and film composition, all play a role in determining the number of domains within a given grain size. This gives rise to the different dependency of domain states on grain size for different materials.

Variation in coercivity with film thickness has also been interpreted in terms of magnetic domain fluctuations. Film thickness fluctuations for two different types of domain walls, zig-zag and straight domain walls have been used to relate coercivity to domain wall motion in thin films [174, 175]. The known "4/3" law that relates the coercivity, H_c to the film thickness, t given by $H_c \propto t^{-4/3}$ was derived by Néel using a similar concept [176]. This relationship assumes constant thickness fluctuation, *dt/dx* (where *x* is the lateral direction along the wall motion). On the contrary, Soohoo [175] took into account the film thickness fluctuation, which increases linearly with film thickness, on the variation of coercivity. Such an increase in *dt/dx* was attributed to changes in surface roughness.

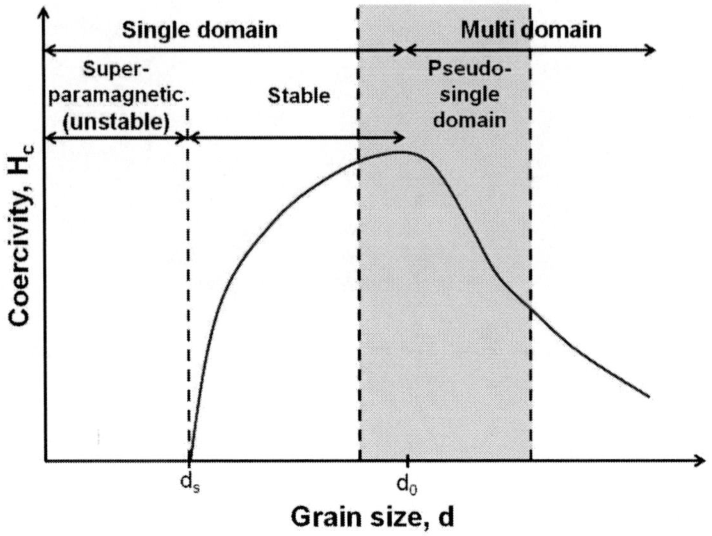

Figure 38. Dependency of coercivity on grain size, which can vary from 5 to 300 nm for Nd-Fe-B and Sm-Co films, depending on the effects of film thickness as well as deposition and annealing temperatures.

In certain applications, a specific film thickness is required. Thus, to control grain growth for a relatively thick film, a multilayered film is recommended. The average grain size of $Nd_2Fe_{14}B$ is reduced by more than six times by preparing multilayered Nd-Fe-B/Nb film instead of a single layer for the same thickness [177]. Improvement of coercivity was also observed in multilayer Sm-Co/Cu film [162]. Irregular grain surface is typically found at the film-substrate or film-capping layer interface in thin films. Such irregularity arises due to interdiffusion and stress between the interfaces during deposition. Multilayered films are able to overcome such distortion and stabilize the grain boundary as well as refine the grain size. Pinning model test by Tsai *et al.* [177] demonstrates that coercivity of multilayered films is not governed by a domain-wall pinning mechanism. A multilayered film of total thickness up to 10 μm with minimal deterioration in magnetic properties can be prepared via magnetron sputtering [68]. A single layered film with similar thickness is often impractical to produce considering not only the magnetic properties but also the stress build-up in the film.

3.4.5. Effects of Deposition Rate

The deposition rate of a film can be controlled by varying the (i) target bias voltage, (ii) sputtering power, (iii) target-substrate distance, or (iv) target surface area. An increase of laser power in the PLD process permits high film deposition rates without deteriorating magnetic properties [178]. Nd-Fe-B films of different thicknesses are manufactured by varying the target-substrate distance so that the local different deposition rates change [95]. The target size can be varied by using triode sputtering to manipulate the deposition rate and at the same time produce a larger film area of homogenous thickness [40]

Studies on the influence of deposition rate on properties of Nd-Fe-B magnetic films have been pioneered by Cadieu *et al.* [179]. Films deposited at deposition rates lower than 0.18 nm/s exhibit perpendicular anisotropy while films deposited at higher rates show in-plane anisotropy. At lower deposition rates, there is more Ar ion bombardment during film growth relative to the sputtered atom flux. This condition favours film growth in a closer packed stacking sequence with *d*-spacings of 0.1973 to 0.2637 nm at the low deposition rate compared to that at high deposition rate which is 0.2715 – 0.3187 nm. The close stacking sequence leads to a *c*-axis preference perpendicular to the film plane. The alignment of the *c*-axis in the film plane direction, for films deposited at a high deposition rate, is consistent with the fact that the lattice exhibits larger *d*-spacings to minimize the stored magnetic energy. This is supported by Homburg *et al.*'s [180] study where an in-plane preferential *c*-axis was observed in the films deposited at high deposition rates.

The majority of permanent magnetic films are grown at lower rates to maintain perpendicular anisotropy. However, Piramanayagam *et al.* [181] found that perpendicular anisotropy still exists at deposition rates as high as 0.5 nm/s; although in-plane anisotropy is more favoured at higher deposition rate. The coercivity of the films is almost independent of the deposition rate. There was no reasoning provided for this disagreement with the result of Cadieu *et al.* [179].

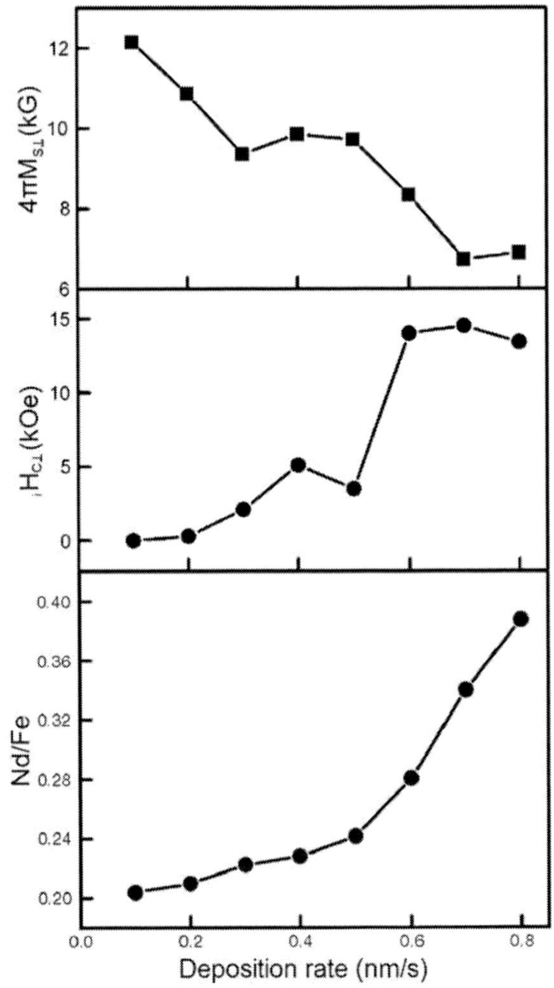

Figure 39. Dependence of saturation magnetization, coercivity and Nd/Fe on deposition rates *(Reprinted from Ref.* [125], *Copyright (2006), with permission from Elsevier).*

On the other hand, Chen *et al.* [125] found that films deposited at 0.2 nm/s or lower do not possess any hard magnetic properties. At deposition rates of 0.6 nm/s and higher, the films begin to exhibit good squareness with coercivity as high as 10 kOe. This contradicts the findings of Cadieu *et al.* [179] and Piramanayagam *et al.* [181] since not only coercivity increases with deposition rates but perpendicular anisotropy actually improves at higher deposition rate. The increase in coercivity can be explained by the increase of the Nd/Fe ratio with deposition rate. At higher deposition rates, less time is required to deposit a film of the same thickness. Hence, less Nd atoms are lost. Higher Nd concentration results in the formation of an Nd-rich phase which leads to strong perpendicular anisotropy and high coercivity. The decrease of Fe concentration with increased deposition rate was confirmed by the reduction of saturation magnetization at higher deposition rates. Figure 39. shows the variation of saturation magnetization, coercivity and film composition on deposition rates. A planar view of film deposited at relatively low deposition rates shows a smooth surface in comparison to those deposited at higher rates. Such data agrees with prior results that indicate a rise in coercivity with respect to surface roughness. There have been few studies on the influence of deposition rates for Sm-Co films. Walther *et al.* [40] used deposition rates of 1 – 5 nm/s; which are higher than those typically used for depositing Sm-Co films (1.39 – 2.78 nm/s). However, this particular study did not focus on the variation of magnetic properties with deposition temperature but more on the topic of substrate temperature. It was reported that at high deposition rates, a higher deposition temperature is required to obtain Sm-Co film for the same magnetic properties as film deposited at lower deposition rates. In other words, films deposited at higher rates have inferior magnetic properties (lower coercivity and squareness) than those deposited at lower rates at the same temperature. More detailed studies are needed in this area to better understand the relationship between deposition rate and performance of the film.

3.4.6. Effects of Target-Substrate Distance

Target-substrate distance (TSD) is an important factor in determining film coercivity [182]. Most studies have this parameter fixed and very few have mentioned its effects on the magnetic properties of films. The least TSD for magnetron sputter deposition of Nd-Fe-B films has been reported to be 37 mm by Parhofer *et al.* [132] while the largest of 115 mm has been reported by Castaldi *et al.* [58] and Tang *et al.* [123]. Thermalization distances of Nd, Fe and B were calculated to be 62 – 75 and 50 – 60 mm at 0.2 Pa for argon and xenon sputtering gases respectively [146]. Hence the average TSDs for sputtering of Nd-Fe-B films have revolved around 50 – 70 mm. On the other hand, TSDs for Sm-Co films deposited by sputtering have been reported to range from 30 mm to 150 mm [182]. On the contrary, films deposited by pulsed laser deposition (PLD) have been prepared at relatively shorter TSDs ranging from 3 – 40 mm. Too small a distance causes ablation of the film due to high plume energy. In addition, partial film crystallization was observed immediately after deposition, which leads to grain coarsening during annealing, and thus, deteriorating the magnetic properties [97]. Degradation of magnetic properties at higher TSDs was attributed to oxidation of Nd. As have been mentioned in the previous section, TSD directly influences the deposition rate. Shorter TSD leads to higher deposition rates which are helpful for the formation of thicker films. The effects of film thickness on film properties have been discussed in Section 3.4.4. Figure 40. depicts the dependence of deposition rate and maximum energy product on TSD. No clear relationship can be concluded between deposition rate and maximum energy product.

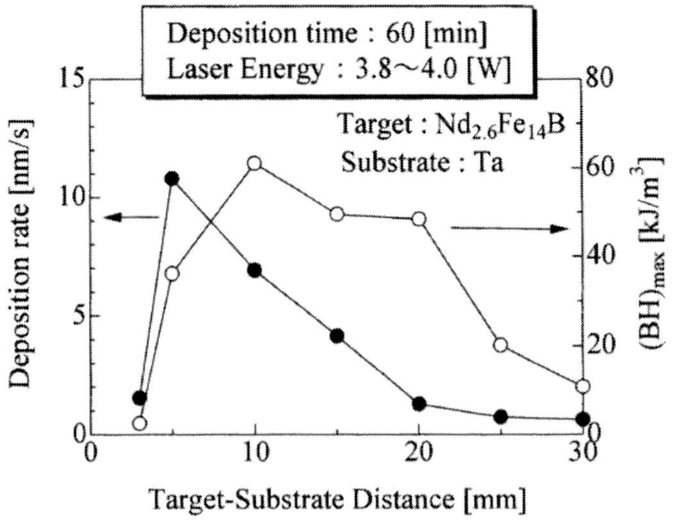

Figure 40. Variation of deposition rate and maximum energy product with target-substrate distance *(Reprinted from Ref.* [97]. *Copyright (2007), with permission from John Wiley and Sons).*

Film composition and structure are readily affected by TSD due to the energy difference of particles near a film surface at different distances. Such a difference arises from the distinctive scattering coefficients of different atom species [147]. In the scattering process, Fe (atomic mass, AM, = 55.847) has a much higher momentum transfer than Nd (AM = 144.24) since it has a more similar mass to the scattering Ar (AM = 39.948) ion. Therefore, Fe atoms can scatter more efficiently and the Fe content in the deposited film can be reduced (that is, an increase of Nd content) by increasing the TSD. With the reduction of Fe content, implying a more Nd-rich phase, the film coercivity increases.

3.4.7. Effects of Substrate

The substrate used for thin film deposition depends on the applications of the film. High temperature resistant substrate materials are usually used since deposition and post deposition treatments need to be carried out at a relatively high temperature to crystallize the film. Such materials include sapphire (Al_2O_3), quartz (SiO_2), mica, Ta, W and Mo. For applications which require epitaxial film growth, single crystal substrate materials such as Si, Al_2O_3 and MgO are used. Single crystal Si substrates are more preferable in micro-electro-mechanical systems (MEMS) devices due to integration issues with the system. Despite the ever increasing studies on optical properties of magnetic recording, the substrates involved remain consistent - glass, Si and oxidized Si. The choice of substrate influences the film properties greatly. For instance, Tsai *et al.* [159] discovered that Nd adatoms have higher reactivity with SiO_2 compared to an Al_2O_3 substrate. Often the high processing temperature calls for the film separation from the substrate after deposition to mitigate interdiffusion and reaction. This was achieved by depositing thin films onto a sodium chloride substrate, which is then dissolved [76]; or depositing thicker films onto water-cooled substrates [183].

However, the practice of substrate removal is not common since (i) it is technically difficult and (ii) diffusion/reaction might have occurred prior to substrate removal. Instead, buffer layers are used to prevent reaction and diffusion between the substrate and film. Nonetheless, inappropriate choice of the buffer will lead to deterioration of magnetic anisotropy [172].

Deposition of Nd-Fe-B films onto metallic substrates such as Ta or Mo could prevent oxidation at the film-substrate interface [185]; the effects of which have been discussed in Section 3.4.3. Stress build-up in the film due to the difference in thermal expansion coefficients between the film and substrate may lead to peeling or cracking of the film. The mechanical stress build-up increases significantly as the substrate temperature increases. The most severe case of film peeling has been reported in the use of mica as substrates [185]; thus mica is not the preferred substrate for RE-TM film deposition. Film adhesion to the substrate has been improved by slow, controlled cooling to room temperature after the deposition process. The topic of mechanical properties of the films will be further discussed in Section 3.5. Formation of rare-earth oxides that deteriorate film magnetic properties is common for films formed on oxygen-containing substrates such as SiO_2. Alternatively, the oxygen could be incorporated within the RE-TM crystal lattice that results in reduction of lattice parameter and relieves the misfit strain across the interface [184].

Different substrate materials could change the texture direction and crystal anisotropy of films deposited under the same conditions [121]. Especially noteworthy is that the role of substrate crystallinity is as important as the material itself. Lemke *et al.* [156] pointed out that coercivity of film deposited on polycrystalline Al_2O_3 is greater than that on single crystal Al_2O_3 due to the extra pinning sites at the surface corrugations. Zhang *et al.* [81] achieved similar results whereby Sm-Co film deposited on polycrystalline glass demonstrates higher in-plane coercivity and remanence enhancement than that on single crystal MgO, Figure 41. The outcome was justified by the stronger exchange coupling between grains in the nanocrystalline $SmCo_5$ film formed on a glass substrate.

In addition, Neu *et al.* [82] established that Sm-Co films deposited on MgO (110) show a preferred orientation perpendicular to the easy axis while films of the same material deposited on MgO (100) appears to be isotropic.

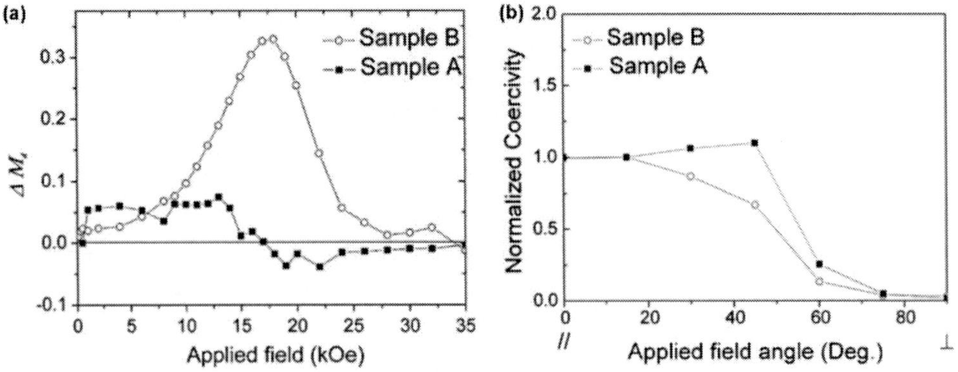

Figure 41. (a) The normalized deviation of demagnetization remanence, $\Delta M_d(H)$ and (b) angular dependence of normalized coercivity of sample A and B (where sample A and B are Sm-Co films deposited on MgO (100) single crystal and Corning glass respectively) *(Reprinted from Ref.* [81], *Copyright (2009), with permission from Elsevier).*

3.4.8. Effects of Magnetizing Field

Amorphous or microcrystalline RE-TM films can be readily deposited at room temperature. Magnetic anisotropy has been observed in such amorphous films with an easy magnetization in the direction perpendicular to the film plane. This anisotropy is associated with the preparation conditions of the film and attributed to deviation from isotropy in the local atomic arrangement in the amorphous films [185]. For example, if an in-plane external magnetic field, H_s, is applied during the film deposition, then this magnetic field is capable of inducing an in-plane magnetic anisotropy in the amorphous or microcrystalline film with easy magnetization in the direction parallel to the direction of H_s applied [186, 187]. Depending on the direction of the applied field H_s, such films can readily saturate in-plane, parallel or perpendicular to the field H_s. The applied field needed to saturate films is within the range of 8 to 10 kOe. In some cases, the H_s needed is only of values less than or equal to the intrinsic coercivity, ${}_iH_c$ of the film material to saturate the film in-plane, parallel to H_s. Such films are said to exhibit switching, which only occurs if ${}_iH_c$ is greater than the anisotropy field, H_A in a specific direction where the measurement is made [47]. On the other hand, inducing such anisotropy in crystallized films deposited directly onto a heated substrate is relatively more difficult. The magnetic alignment energy is hindered from rotating the *c*-axis of the initially grown crystallites by the columnar growth pattern of the directly crystallized films [188].

An externally applied in-plane magnetizing field, H_a, during annealing is able to change the direction of the initial anisotropy formed in the film by H_s. Such reorientation of the easy axis is unique to the field-induced anisotropy in amorphous materials [189]. The original in-plane anisotropy is greatly suppressed after annealing while no other in-plane anisotropy could be introduced through the process. An increase in annealing temperature leads to even greater suppression of the in-plane anisotropy [185].

Several studies have aimed to alter the texture and anisotropy of Nd-Fe-B by applying an external magnetic field ranging from 1.3 kOe to over 4 kOe during deposition, H_s, or annealing, H_a, but with little success. Unlike Sm-Co films, crystallization of Nd-Fe-B films cannot occur below its Curie temperature, T_c, which is the requirement for external H_s and H_a to work [188]. However, Mapps *et al.* [120] has proven otherwise by revealing that nanocrystalline $Nd_2Fe_{14}B$ phase does exist at temperature below T_c. Thus, application of H_s during Nd-Fe-B film deposition may induce magnetic anisotropy in the film. The implementation of an applied field H_s during preparation of Sm-Co films has been proven effective [185, 186].

3.4.9. Effects of Other Added Elements

The magnetic properties of Nd-Fe-B and Sm-Co magnets, such as coercivity, Curie temperature, magnetocrystalline anisotropy and oxidation resistance, can be improved by adding certain amounts of other elements [9, 14]. Based on this, Nd-Fe-B and Sm-Co films with starting materials that have extra elements such as Co, Ti, Cu, Al, Ta, Fe, Zr, Si, Mn, V, Mo and so on, have been produced.

The relatively low Curie temperature has often been pointed out as a drawback of Nd-Fe-B films. It is known that Curie temperature is determined by the Fe-Fe interatomic distance in Nd-Fe-B. Substitution of Fe by Co shows a decrease in the distance between the Fe-Fe pairs, resulting in a decrease in negative exchange interaction [190] and leads to an increase in Curie temperature of $Nd_2Fe_{14}B$ [191]. It was suggested that Co dissolves in $Nd_2Fe_{14}B$ phase and results in enhancement of the *c*-axis texture of Nd-Fe-B films. The improvement of texture in multilayered films is also attributed to the reduction of interface energy between the *c*-plane of Co-doped $Nd_2Fe_{14}B$ phase and Ta layer with alternating Nd-Fe-B films [68]. Deterioration of magnetic properties, typically observed the film thickness increases, is minimized in Co-doped Nd-Fe-B films. Reduction of the optimum substrate temperature with Co-doping demonstrates that Co stimulates crystallization and grain growth of $Nd_2Fe_{14}B$ phase. However, Chandrasekhar *et al.* [173] found a reduction of grain size instead of grain growth in Co- and Ti-doped films. This inconsistency can be explained because Co-doping leads to grain growth but the reduced deposition temperature reduces the average grain size in the film. The same conclusion was deduced for Cu-doped Nd-Fe-B films where the crystallization temperature of $Nd_2Fe_{14}B$ was reduced by almost 100 °C through addition of appropriate amount of Cu [192].

Coercivity of films increases due to accelerated crystallization, but decreases when optimum concentration of dopants is exceeded due to dilution of magnetic elements in films and grain size reduction. Doping with non-magnetic Al reduces the planar anisotropy of tetragonal $Nd_2Fe_{14}B$ while improving its uniaxial anisotropy, which leads to an increase in coercivity [193]. Ta-doped film also shows better grain alignment with its *c*-axis perpendicular to the film. However, magnetization decreases as the concentration of Ta increases [194]. Not many studies have been dedicated to doped-SmCo films but rather more concentrated on the bulk material. $SmCo_7$ phase has been found to be a more suitable candidate for high temperature applications than $SmCo_5$ and Sm_2Co_{17} [40]. Therefore, the use of additives in Sm-Co is directed towards stabilizing the $SmCo_7$ phase, which is sensitive to Sm content. Addition of Fe, Cu and Zr elements that have different sputtering yields presents different compositions and the $SmCo_7$ phase can be obtained when the Sm concentration is around 15.4 at% [148]. Doping with Si stabilized the $SmCo_7$ phase with an exceptional increase in magnetization at low temperatures and a strong uniaxial anisotropy. However, annealing at 1000 °C destabilizes the $SmCo_7$ phase obtained [195]. The doped Si has a distinct preference to occupy the 3*g* site, similar to $CeNi_5Si$ that exhibits the $TbCu_7$-type structure [196]. Similarly, doping with Cu also presents the $SmCo_{7-x}Cu_x$ phase with strong uniaxial anisotropy with an anisotropy field as high as 20 T at x = 0.8. However, further increases in Cu content decrease the saturation magnetization and Curie temperature [197]. On the contrary, vanadium, when used as a doping element, has more preference to occupy the 2*e* site of $SmCo_7$. Occupancy at 2*e* site also enhances the magnetic anisotropy while at the same time reducing the grain size, and thus leads to an increase of coercivity as well as maximum energy product [198]. Further references on doping of Sm-Co with other elements are available elsewhere [199-202].

3.5. Mechanical Properties

There are few publications that mention the mechanical properties of rare-earth transition metal (RE-TM) thin films. Mechanical properties of thin films focus on two prime areas: (i) the effects of stresses in films, and (ii) the hardness and wear resistance of coatings. The first area is of greater concern when it comes to thin film technology and will be the principal discussion in this section. Tensile stress in a typical film is balanced by the compressive stress of the substrate.

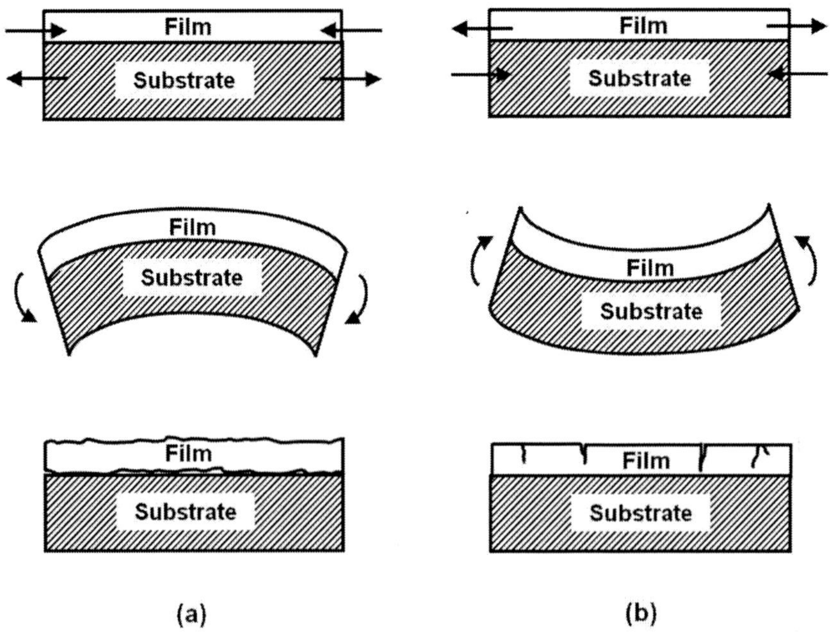

Figure 42. Occurrence of (a) residual compressive stress in film bending the substrate convex outward, causing film wrinkling and de-adhesion and (b) residual tensile stress in film bending the substrate concave upward, causing film fracture.

However, the combination of the stresses is insufficient to place the system in mechanical equilibrium due to uncompensated end moments [114]. Internal compressive stress causes the substrate to curves outward as shown in Figure 42.(a), causing film wrinkling and de-adhesion; while internal tensile stress will bend the substrate inward as shown in Figure 42.(b), causing fracture; Figure 42. Cracking or fracture of film, substrate or film-substrate interface represents the most extreme form of stress relaxation in thin films.

Residual stresses limit the deposition of crack and fracture-free crystalline RE-TM films onto a bare substrate to less than 10 μm thickness. Cracks start to form as the film thickness approaches 10 μm, followed by flaking and peeling of the film from the substrate as the film grows thicker. Such effects were observed for both directly crystallized films and amorphous films that are subsequently crystallized. Cracking does not occur in films deposited in amorphous films until they undergo post-deposition heat treatment for crystallization. It can be concluded that film cracking is caused by stress build-up and difference in thermal expansion coefficient of film and substrate. Deposition of relatively thick films, of at least 50 μm, without the development of cracks can be achieved with the use of a boundary layer that absorbs thermal stresses [203, 204].

The tendency for film peeling with increased thickness is partly due to the increase of tensile stress when deposition flux direction is angled normal to the substrate [205]. As film thickness increases, crack development and propagation throughout the film and substrate increases; hence causing the film to peel from the substrate. A slower cooling rate allows more time for thermal contraction adjustment between the film and substrate. The mechanical properties of as-deposited films depend largely on the substrate temperature. Increase of substrate temperature above 470 °C causes progressive weakening of Nd-Fe-B film adhesion to the substrate, which leads to higher thermal stress in film:

$$\sigma_{th} = E(\alpha_f - \alpha_s)(T_s - T_f)$$

where E is the Young modulus, α_f and α_s are the average thermal expansion coefficients of films and substrate respectively, while $T_s - T_f$ are the substrate temperature and temperature at which the measurements are performed, respectively. Deposition of Nd-Fe-B film at 470 °C for 90 minutes results in the film peeling from the substrate. Reducing the deposition time by 35 minutes, and thus reducing the film thickness, at the same deposition temperature was effective in preventing film delamination [66].

On the contrary, peeling of Sm-Co film begins at a deposition temperature of 500 °C and is initiated by breakage of the Si wafer substrate. Figure 43. shows the influence of substrate temperature on tensile stress of Sm-Co films. Peeling of the film is attributed to stress build-up due to differences in thermal expansion coefficient between the Sm-Co film and the substrate. Failure of the film has been observed to initiate at the Si wafer for all deposition temperatures studied, which explains why thick films are mostly deposited on non-Si substrates. Peeling of the film has also been related to the lateral dimensions of the film where peeling does not occur for Sm-Co films deposited on pre-patterned Si/SiO_2 wafers, which provides the film with topographic relief [40].

Another aspect concerning the mechanical properties of films is the adhesion, which refers to the interaction between surfaces of two closely contiguous, adjacent bodies; that is, the film and substrate. Despite the fact that adhesion is one of the most important attributes of a film, it is the one of the least understood properties due to the lack of quantitative adhesion measurement methods. Adhesion of both Nd-Fe-B and Sm-Co films has not been studied in detail. Mention of good film adhesion is depended on a qualitative judgement where the film does not peel off the substrate; rather than being based on any quantification assessment.

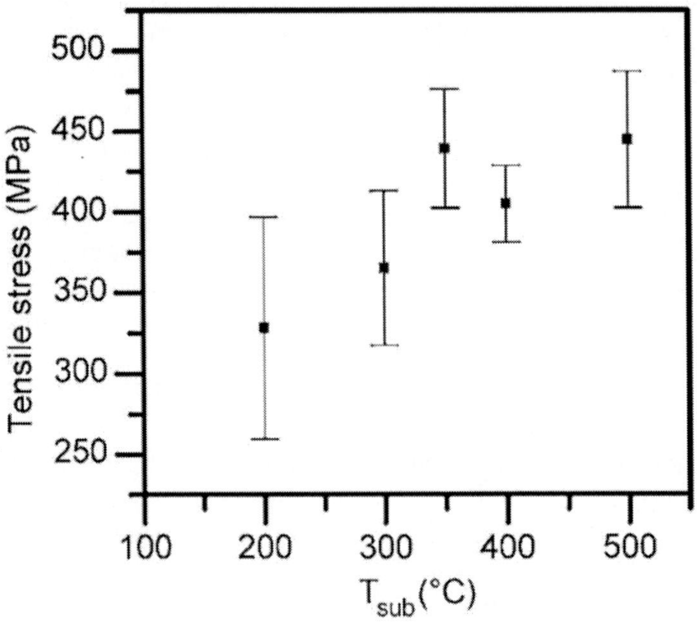

Figure 43. Development of tensile stress in Sm-Co film as a function of deposition temperature *(Reprinted with permission from Ref.* [40]. *Copyright (2008), American Institute of Physics).*

4. Applications

4.1. Magnetic Recording

Magnetic thin films have been the choice for magnetic recording media for more than two decades. The development of magnetic thin films to suit the requirements for recording media has been defined as being divided into three distinct stages: (i) deposition of a film to fit the criteria and enhance existing particulate disk technology, (ii) introduction of magnetoresistive (MR) head technology and (iii) introduction of giant magnetoresistive (GMR) head technology [206]. Figure 44. outlines the growth of areal density for the three stages of thin film media development. There are two parts to any recording media: the recording heads and the film disk. Recording media are different from integrated circuits in the sense that the recording head and film disk are the "actual device" of the media while the substrate is just an inert holder [208]. The recording heads are usually made of soft magnetic materials with high permeability so that magnetization is easy. Thus, the intrinsic property needs are low coercivity (H_c), low remanence (M_r) and high saturation magnetization (M_s). A high saturation magnetization is required since an intense magnetic field is needed to write onto the disk surface. The recording disk, on the other hand, requires high coercivity (H_c), high product of remanence-thickness (M_rT) and low noise. Figure 45. demonstrates the improvement of H_c and M_rT since the mid 1970s. A large coercivity is essential to resist demagnetization and avoid accidental data erasure. However, too large a coercivity leads to a reduction of recording density due to difficulty for the head to write on the medium. Coercivity is largely dependent on the microstructure, presence of impurities and defects; all of which play a part in the magnetization reversal mechanism. Noise reduction can be achieved by isolating grains through a chemical or physical route to reduce intergranular coupling, which is the main cause of noise in recording media.

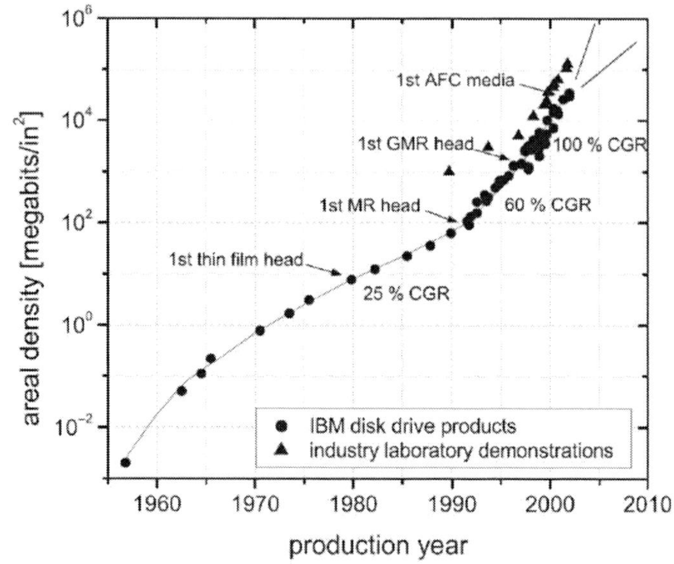

Figure 44. Areal density evolution in magnetic recording since its invention *(Reprinted with permission from Ref.* [207]. *Copyright (2002) by Institute of Physics Publishing Ltd.).*

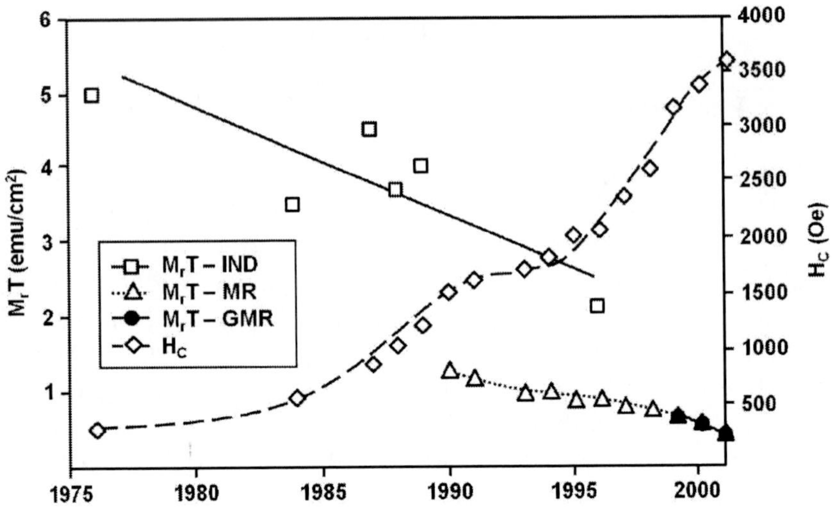

Figure 45. Development of coercivity, H_c and remanence-thickness product, M_rT since mid 1970s *(Reprinted with permission from Ref.* [206]. *Copyright (2000), American Institute of Physics).*

Recording is carried out by the inductive write element in horizontal magnetization reversal patterns while reading is performed by the magnetoresistive (MR) or giant-MR (GMR) sensor element of the same integrated write/read head. The data recorded is characterized by the bit length or transition spacing as well as write and read track width. A film should exhibit high saturation magnetization (M_s), high coercivity (H_c) and a square hysteresis loop to be qualified as a high density longitudinal recording. The product of remanence-thickness (M_rT) is closely related to the maximum packing density of the recording media. Cobalt alloy has been used conventionally for longitudinal magnetic recording. The ferromagnetic nature of cobalt and its high magnetocrystalline anisotropy that arises from its hexagonal close packed (hcp) structure give it the high coercivity values required for recording media. To optimize the longitudinal recording performance of Co alloys that have uniaxial anisotropy, it has become necessary to modify the magnetic thin film to align its c-axis parallel to the disk plane. Formation of ternary compounds by addition of chromium helps enhance the corrosion resistance of the material. At the same time, the use of epitaxial Cr as underlayers helps increase the H_c and M_s of the film by orienting the film in a plane crystallographic direction [48, 49]. Addition of rare-earth elements such as Sm further enhances the coercivity of the film. The invention of MR heads lead to optimization of microstructure and micromagnetic properties of the film to reduce the noise level. It has been proven that noise increases linearly with recording density until a supra-linear region is reached where the noise decreases [209]. The source of noise is associated with the strong magnetic interactions of polycrystalline grains that are too closely packed. Several solutions have been suggested but the most successful is the chemical segregation of non-magnetic Cr at the grain boundaries [206]. The film is initially sputter deposited as a heterogeneous thin film but its structure approaches that of particulate media. There have been few devices that use perpendicular recording due to the firm position held by conventional longitudinal recording in the market. However, the thermal instabilities in longitudinal recording media have allowed development of perpendicular magnetic recording. It has been argued that the larger grain volumes in perpendicular recording will increase its thermal decay and stability ratio [206]. While being the closest alternative to longitudinal recording, this type of recording extends the density limit beyond that of longitudinal recording [210]. The magnetostatic demagnetization forces are reduced in perpendicular recording making its magnetic structure more resistant to demagnetization [206]. Higher bit densities are possible in perpendicular recording due to a narrower transition width, δ, [211] as observed in the two magnetization curves in Figure 46. The schematics in Figure 46. show the differences between longitudinal and perpendicular magnetic recording. Neodymium iron boron and samarium cobalt have proven to show strong magnetocrystalline anisotropy, as discussed earlier. This anisotropy leads to a large coercivity, which permits these materials to be suitable for magnetic recording media application. Sm-Co thin films deposited on Cr and Cu buffer layers have shown oriented microstructure and texture giving them hard magnetic properties suitable for magnetic recording applications [212, 213]. Sm-Co and Nd-Fe-B films are deposited in an amorphous structure, unless deposited on heated substrates or undergo post-deposition annealing. Thus, these materials not as process friendly with the disk substrate materials and multilayered structures due to the high temperatures involved [211]. Nevertheless, the formation of amorphous phase is far more uniform and homogenous compared to polycrystalline films that are limited by intergrain coupling in terms of its recording density [187, 214]. The typical domain structure of films in an amorphous form has been described to form a saw-tooth pattern that is very stable once it is formed [187]. Amorphous Sm-Co has demonstrated higher recording density and better signal-to-noise ratio in comparison to conventional materials used for recording; such as particulate γ-Fe_2O_3 and Co-P plated disks [208].

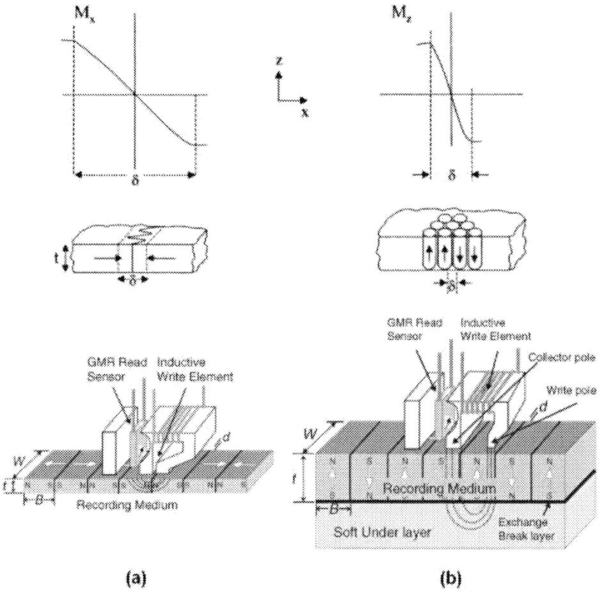

Figure 46. Comparison between (a) longitudinal and (b) perpendicular magnetic recording film where Mx and Mz are magnetization measured parallel and perpendicular to the film respectively, δ is the transition width and t is the thickness of the film; magnetization transition for (a) occur over a shorter distance, δ compared to (b) *(Reprinted with permission from Ref.* [211] *and* [207]. *Copyright (1998) and (2002) by Institute of Physics Publishing Ltd.).*

The research trend on magnetic thin films has shifted from films for longitudinal recording applications towards improving the perpendicular anisotropy of the film for perpendicular recording. Cobalt and titanium substitution in Nd-Fe-B films reduces grain size and coercivity while at the same time improves corrosion resistance of the material [173]. Addition of cobalt has also raised the Curie temperature of rare-earth magnets. Other studies have shown that controlling the thickness of the underlayer alters the magnetic properties of the film can be altered to satisfy magnetic recording requirements [137, 164, 213]. Nevertheless, the use of RE-TM films poses corrosion issues that can be solved by applying a protective coating.

4.2. Micro-Electro-Mechanical Systems (MEMS)

Micro-electro-mechanical systems (MEMS) are miniaturized electromechanical devices such as actuators, sensors and microsystems that are coupled with electric, mechanical, radiant, thermal, magnetic or chemical components. Table 4. shows the commonly used transduction mechanisms in MEMS. Transduction mechanism is the conversion of signals from one form of energy to another form of energy; for instance, a magnetic input can be converted to mechanical energy through a physical effect known as magnetostatics or magnetostriction.

Table 4. Common transduction mechanisms used in MEMS *(Reprinted with permission from Ref.* [217]. *Copyright (2001) by Institute of Physics Publishing Ltd.)*

From \ To	Electrical	Magnetic	Mechanical	Thermal	Chemical	Radiative
Electrical		Ampere's Law	Electrostatic, electro-phoresis	Resistive heating	Electrolysis, ionization	EM transmission
Magnetic	Hall effect, magnetic resistance		Magneto-statics, magneto-striction	Eddy currents, hysteretic loss	Magnetic separation	Magneto-optics
Mechanical	Variable cap, piezo-resistance, piezo-electricity	Magneto-striction		Friction	Phase change	Tribo-luminescence
Thermal	Thermo-electric	Curie point	Thermal expansion		Reaction rate ignition	Thermal radiation
Chemical	Electro-chemical potential	Chemo-magnetic	Phase change	Combustion		Chemo-luminescence
Radiative	Photo-conductor, EM receiving	Magneto-optics	Radiation hardening	Photo-thermal	Photo-chemical	

Magnetic MEMS offers many advantages due to improved magnetic interactions from scale reduction including high energy and force density, stable forces without power supply, remote actuation, levitation, high efficiency electrical generation etc. [215].

MEMS devices also offer advantages over electrostatic actuators due to low voltage and power consumption, large actuation forces over long distances and enhanced reliability in adverse operating conditions [216]. Thus, magnetic MEMS offer new potentials in fields such as bio-medicine, information technology and energy transformation. To begin with, the thickness of permanent magnetic films for MEMS depends on the area on which they are used. Some applications require thickness of a few hundred nm while others demand for thicker films up to a few microns or even to tenths of mm. The selection criteria for permanent magnetic films suitable for MEMS applications can be summarized as follows [218]:

- Magnetic performance: Adequate coercivity; high remanence, maximum energy product and Curie temperature; low temperature coefficients of remanence
- Adaptability to MEMS processing
- Environmental stability: mechanical, chemical and thermal stability

In the recent years, many high performance permanent magnetic films based on RE-TM compounds with promising characteristics for MEMS applications have been developed [40, 158, 219]. The use of films from such compounds would have been ruled out years ago when electroplating process was the only way to form permanent magnetic films. However, with the development of deposition techniques such as sputter deposition and pulsed laser deposition, the use of RE-TM materials has been deemed possible. In addition to their high magnetic performance, Nd-Fe-B and Sm-Co are also promising to be integrated into MEMS devices due to their strong anisotropic behaviour, as discussed earlier in Section 1.5. The strength and direction of anisotropy can be controlled by manipulating the deposition or annealing temperature, application of magnetic fields or substrate crystallinity in epitaxial film growth. Anisotropy is highly desired in such applications due to the preference for *c*-axis textured of films during the application of biasing field. Nevertheless, it is undeniable that a protective coating is required when using RE-TM films to protect the films from corrosion. A new protecting layer needs to be deposited immediately after photolithography or etching. Oxidation of magnetic films is generally more critical than its bulk counterparts due to their higher surface area to thickness ratio. An element of MEMS devices that often implement magnetic thin film is micro actuator. Magnetic micro actuator has been gaining a lot of attention in the past decade as it has the unique ability to create larger forces (hence larger deflections with complementary metal-oxide semiconductor level) and variety means of generating and controlling magnetic forces compared to its electrostatic counterpart. The motions of the micro actuator can be controlled precisely since the exerted magnetic force is accomplished by electromagnetic means. Magnetic micro actuators also offer the possibility of eliminating the need for separate power supplies to control the electronics and MEMS components. Micro actuators which are free of energising coils can be made with the use of permanent magnetic thin film, hence simplifying and compacting the whole geometry of the component. Figure 47. shows an active micro magnetic bearing that utilizes active bearing actuators constructed by electromagnet with laminated magnetic structures.

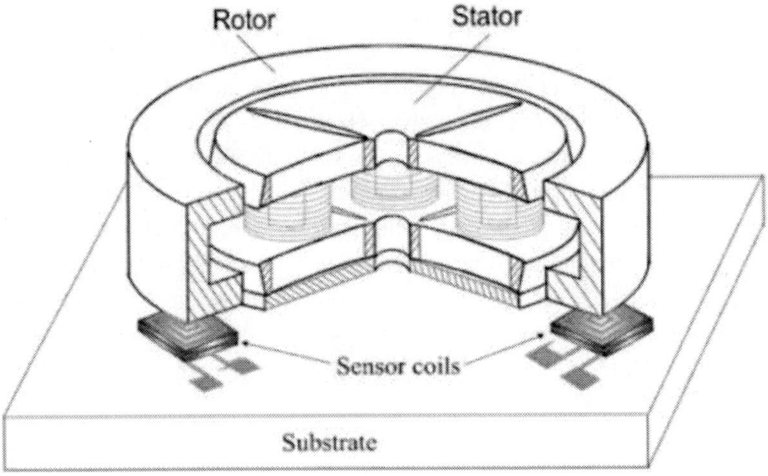

Figure 47. Schematic of a micro magnetic bearing actuator.

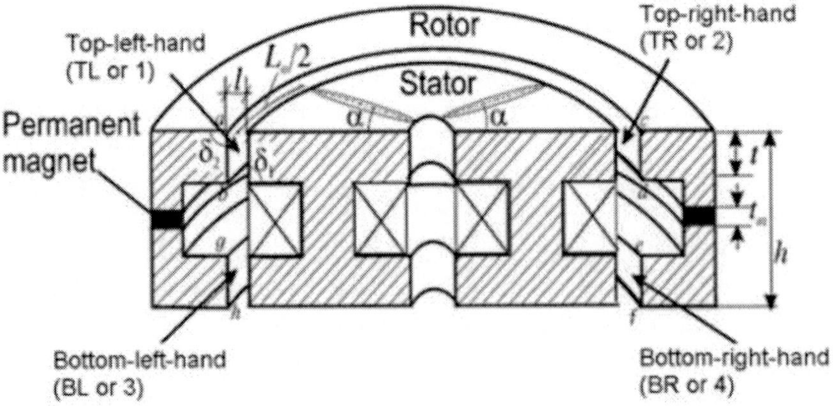

Figure 48. Cross sectional view of a micro magnetic bearing actuator *(Reprinted with permission from Ref.* [220]. *Copyright (2000), IEEE).*

The major components of the actuator include the top and bottom planes of stator, two rotor end planes, three rotor spaces, four yokes as well as bearing coils. This design of actuator not only reduces the motional and transient eddy current effects but also minimizes the friction and increases the speed rotary suspensions. On the other hand, micro actuator made of combination of electromagnet and permanent magnets are generally more complex and require smaller amp-turns per unit force than equivalent electromagnet actuators. However, this type of actuators exhibit higher sensitivity and better linearity as a function of drive coil current. They can be made more compact by using permanent magnetic thin films.

A schematic showing the position of permanent magnetic thin film in a micro magnetic bearing actuator is shown in Figure 48. Permanent magnetic thin films with large coercivity are highly demanded in magnetic bearing actuator in order to simultaneously support the rotor axially. The magnetic thin films are also required to generate a bias magnetic flux in the micro rotor in order to achieve better linear relationship between the force acting on the rotor and the drive coil current. Despite the potential represented by hard magnetic films in MEMS devices, there are still much consideration required with regards to design and fabrication before they can be fully integrated and utilized [221]:

- There is limited knowledge on the unique requirements for hard magnetic films integration
- Different classes of hard magnetic materials represent variation in magnetic performance, temperature and chemical stability and fabrication constraints which requires careful consideration in materials selection
- Sensitivity of magnetic properties to microstructure and chemical composition of hard magnetic films which calls for meticulous control over processing conditions
- Requirement for *in situ* or post-fabrication magnetization steps complicates integration process

4.3. Other Magnetic Devices

Other magnetic devices assembled using permanent magnetic films include micro generator and atom optics. However, these devices utilize relatively thicker films than the two applications discussed previously. The role of magnetic films has become more dominating in small scale power generation due to the limitation in life span of electrochemical batteries. Replacement or recharging of batteries is often inconvenient, especially when it is used in body implantation devices, devices intended for long term usage or systems that are physically remote.

Two such examples are cardiac pacemakers [222] and biomedical sensors [223].Generally, there are three categories of micro magnetic generators, classified based on the concept with which they function: rotational, oscillatory and hybrid type. A detailed review on micro scale magnetic power generation has been nicely done by Arnold [224].

Magnetic mirrors constructed of Sm-Co and Nd-Fe-B permanent magnetic arrays have been adopted for atom optics applications. Sidorov *et al.* [225] utilizes 18 stacked Nd-Fe-B magnets while Meschede *et al.* [226] uses Sm-Co slabs in fabricating the magnetic mirrors. Magnetic arrays typically produce square wave magnetization, which leads to higher harmonics contributions and end effects. These effects can be suppressed by using even number of magnets and at the end of each array, magnets with half the usual width. The resulting mirrors were found to perturb the atoms over longer distance, hence giving them longer storage period.

As a result, the de Broglie wave phase shift reflected from the mirrors is much larger compared to the conventional mirrors. However, the field strength of the magnetic mirror is of higher magnitude, with reciprocating increase in normal velocity that can be reflected [227]. An intricate review article on magnetic atom optics has been published by Hinds and Hughes [227].

Table 5 summarizes the various applications of permanent magnetic thin films, listing the components and material requirements for each application.

Table 5. Summary of permanent magnetic thin film applications

Application	Components	Requirements
Magnetic recording	Recording head	Low coercivity, H_c Low remanence, M_r High saturation magnetization, M_s
	Recording disk	High coercivity (H_c) High product of remanence-thickness (M_rT) Strong magnetic anisotropy Low noise
Magnetic micro-electro-mechanical systems (MEMS)	Actuator, sensor	Adequate coercivity, H_c High remanence, M_r High maximum energy product, $(BH)_{max}$ High Curie temperature, T_C Low temperature coefficients of remanence
Magnetic power generation	Rotational, oscillatory, hybrid	High remanence, M_r High maximum energy product, $(BH)_{max}$ High Curie temperature, T_C Low temperature coefficients of remanence
Magnetic atom optics	Micro magnetic mirror, wave guide, magnetic micro-trap, atom chip	High coercivity, H_c High remanence ratio, M_r/M_s High Curie temperature, T_c and excellent thermal stability Large perpendicular magnetic anisotropy

5. Limitations and Challenges

RE-TM alloys demonstrate exceptional magnetic strength, especially Nd-Fe-B that exhibits the best magnetic performance with a maximum energy product up to 400 kJ/m^3. Sm-Co is more chemically stable and has a higher maximum operating temperature than Nd-Fe-B. However, the high cost of Sm-Co magnets limits their uses to highly sophisticated applications; especially where a high operating temperature (up to 550 °C) is needed. Despite their good magnetic performance, there has been an ongoing debate on whether these materials are alternative choices to other permanent magnetic materials such as hard ferrites in magnetic recording and Co-Ni, Pt-Co or Pt-Fe in MEMS applications. It is undeniable that RE-TM alloys are subjected to corrosion problems and low operating temperature issues. Nevertheless, technical solutions have been proposed to these controversial matters. For instance, a new Sm-Co alloy have been developed that could operate at 550 °C while still maintaining its high maximum energy product and intrinsic coercivity [228].

On the other hand, corrosion problems have raised integration issues with respect to RE-TM magnetic films in MEMS devices. Although film oxidation can be prevented by applying a thin layer of protective coating, such as yttrium oxide or gold, this protective layer is disrupted during patterning or photolithography processes. Often, a new protective layer is applied immediately after etching. A new and more effective protection mitigation strategy is highly sort after in this area, especially since the demand for high performing MEMS devices is increasing.

The limitation in film thickness that can be achieved through conventional means has always been a concern in this field. Stress build-up in film with thickness above a certain value causes the film to crack or peel off the substrate. This becomes a major problem when relatively thicker films are needed in applications that require films with greater magnetic strength. Fabrication techniques such as pulsed laser deposition, tape casting and vacuum plasma spray have been able to produce comparatively thicker films at higher deposition rates. Despite the potentials of these techniques, few studies have explored the possibilities and extent to which they can be used as a means to fabricate magnetic films.

Of great concern are the mechanical properties of the magnetic films. Magnetic films for magnetic recording require high friction and wear resistance due to the high frequency of collision between the recording head and magnetic disk. Also consider MEMS devices that are being constantly challenged to be manufactured with thick dimensions; but this need is restricted by stress build-up in the films that can lead to catastrophic failure. Despite the importance of this feature of the film, it is rarely studied. The majority of studies are qualitative and focus on cracks and peeling of the films without quantification.

Finally, the documentation of data for the magnetic properties of RE-TM films is not consolidated within the open literature; thereby restraining and limiting the expansion of their applications. A reference source that accumulates and compares such data would be highly sought after and advance research in this area.

CONCLUSION

Rare-earth magnets are relatively young in their development and applications compared to other permanent magnets; however, they have gained much attention since their discovery due to their high magnetic performance. Rare-earth magnetic thin films can be prepared by sputter deposition, evaporation, pulsed laser deposition and molecular beam epitaxy. Methods such as screen printing, tape casting and thermal spray have been used to prepare relatively thicker films. Sputter deposition of magnetic thin films has been the most widely studied deposition technique due to its ability to produce a high-quality and even thickness; substantial film composition control; good homogeneity and sufficient deposition rate.

An important aspect that should be considered in magnetic thin films is the microstructure because it relates directly to the magnetic and mechanical properties of the film. Deposition rate and substrate temperature are the two most important factors that govern the film microstructure. Columnar morphology is highly desired in magnetic thin films but it often comes with internal defects and porosity. Irregular grain structure roughens the film surface and increases the film coercivity. Texture of the film depends on the deposition temperature as well as the buffer and substrate materials. On the other hand, highly irregular and disordered magnetic domains with visible boundaries signify improved anisotropy with regards to the *c*-axis texture and the transition from soft to hard magnetic properties. Surface roughness and type of domain wall plays a vital role in determining the domain size, wall thickness and coercivity of the films.

Magnetic properties of films are another significant aspect to consider. Post-deposition treatments are as important as deposition parameters in influencing the magnetic properties. Increased sputtering pressure and deposition rate both contribute to an increase in film coercivity. Deposition rate influences the film composition, hence the anisotropy and coercivity, as well as surface roughness of the film. Optimum substrate temperatures for depositing Nd-Fe-B and Sm-Co films are in the average of 600 and 450 °C respectively while the optimum annealing temperature is about 600 and 750 °C respectively.

Heating rate and duration of post-deposition heat treatment are significant in establishing the film properties. Selection of buffer and substrate material should take into consideration the potential reaction and diffusion between the film and the buffer/substrate materials as well as the thermal expansion coefficient mismatch between the two. Thickness of buffer layer along with film thickness also plays an important role in determining the magnetic properties of the film. Films that are greater than optimal thickness demonstrate deteriorated perpendicular anisotropy and coercivity while too thin of a film leads to a reduction in coercivity due to randomizing effects of thermal energy.

One of the factors that controls the film thickness is the target-substrate distance, which directly affects the composition of the film due to the different energy of particles at the film surface. Applying a magnetic field during deposition induces an in-plane magnetic anisotropy in the film. The direction of this anisotropy changes and is suppressed if another magnetic field is applied during annealing. On the other hand, doping with elements such as Co, Cu, Ti, Ta etc. have the capability of improving coercivity, Curie temperature, magnetic anisotropy and oxidation resistance of the film.

Among the many applications where permanent magnetic thin films can be used, a brief discussion on magnetic recording and MEMS devices has been provided, stressing the criteria of thin films for each application. The requisite of magnetic films for magnetic recording includes high coercivity and remanence-thickness product and low in noise levels, all of which are fulfilled by rare-earth magnets. The role of longitudinal magnetic recording is being replaced by perpendicular magnetic recording, which offers more advantages over its predecessor; such as higher thermal decay stability ratio, extended recording density and lower magnetostatic demagnetization forces. Development in perpendicular recording is accompanied by the constantly improving perpendicular magnetic anisotropy of Nd-Fe-B and Sm-Co films.

Magnetic MEMS offers many advantages over other types of MEMS devices and demonstrated potentials in many fields including biomedical, information technology, and energy transformation. Although rare-earth magnetic thin films are sometimes deemed to be unsuitable for MEMS devices by many critics, the high magnetic performance of rare-earths has attracted commercial attention because higher performing devices are being demanded by the consumers.

References

[1] da Costa Andrade, E. N., *Endeavour* 17 (65), 22-30 (1958).

[2] Mishima, T., Britain Patent No. 378,578 (1931).

[3] Herbst, J. F., Croat, J. J., Pinkerton, F. E. and Yelon, W. B., Phys. Rev. B: Condens. Matter 29 (7), 4176-4178 (1984). http://prb.aps.org/abstract/PRB/v29/i7/p4176_1
[4] Sagawa, M., Fujimura, S., Togawa, N., Yamamoto, H. and Matsuura, Y., *J. Appl. Phys.* 55 (6), 2083-2087 (1984).
[5] Croat, J. J., Herbst, J. F., Lee, R. W. and Pinkerton, F. E., *J. Appl. Phys.* 55 (6), 2078-2082 (1984).
[6] Stijntjes, T. and Van Loon, B., *Proc. IEEE* 96 (5), 900-904 (2008).
[7] Coey, J. M. D., *Endeavour* 19 (4), 146-151 (1995).
[8] Buschow, K. H. J. and Van Der Goot, A. S., J. Less-Common Met. 14 (3), 323-328 (1968).
[9] Kumar, K., *J. Appl. Phys.* 63 (6) (1988).
[10] Sayama, J., Asahi, T., Mizutani, K. and Osaka, T., *J. Phys. D: Appl. Phys.* 37 (1) (2004). doi: 10.1088/0022-3727/37/1/L01
[11] Ervens, W., *Goldschmidt Informiert* 48 (3) (1979).
[12] Givord, D., Li, H. S. and Moreau, J. M., *Solid State Commun.* 50 (6), 497-499 (1984).
[13] Herbst, J. F., *Rev. Mod. Phys.* 63 (4), 819-898 (1991).
[14] http://rmp.aps.org/abstract/RMP/v63/i4/p819_1
[15] Burzo, E., *Rep. Prog. Phys.* 61 (11), 1099-1266 (1998).
[16] Burzo, E. and Kirchmayr, H. R., in *Handbook on the Physics and Chemistry of Rare Earths*, edited by Gschneidner, K. A. and Eyring, L. (Elsevier Science Publishers, 1989), Vol. 12, pp. 71-132.
[17] Shoemaker, C. B., Shoemaker, D. P. and Fruchart, R., *Acta Crystallogr.* C 40, 1665-1668 (1984).
[18] Wolfers, P., Obbade, S., Fruchart, D. and Verhoef, R., *J. Alloys Compd.* 242 (1-2), 74-79 (1996).
[19] Herbst, J. F., Croat, J. *J. and Yelon*, W. B., J. Appl. Phys. 57 (8), 4086-4090 (1985).
[20] Buschow, K. H. J., *Rep. Prog. Phys.* 40 (10), 1179-1256 (1977).
[21] Wallace, W. E., *Prog. Rare Earth Sci. and Tech.* 3, 1 (1968).
[22] Buschow, K. H. J., *Phys. Status Solidi A* 7, 199 (1971).
[23] Campbell, I. A., *J. Phys. F: Met. Phys.* 2 (3) (1972).
[24] Buschow, K. H. J., *Ferromagnetic Materials.* (North-Holland, Amsterdam, 1980).
[25] Williams, K. L., Bartlett, R. W. and Jorgensen, P. J., J. Less-Common Met. 37 (1), 174-176 (1974).
[26] Cataldo, L., Lefèvre, A., Ducret, F., Cohen-Adad, M. T., Allibert, C. and Valignat, N., *J. Alloys Compd.* 241 (1-2), 216-223 (1996).
[27] Chaban, N. F., Kuzma, Y. B., Bilonizhko, N. S., Kachmar, O. O. and Petrov, N. V., Dokl. Akad. Nauk SSSR Ser. A, *Fiz. Mater. Tekh. Nauk* 10, 873-875 (1979).
[28] Schneider, G., Henig, E.-T., Petzow, G. and Stadelmaier, H. H., Z. Metallkd./*Int. J. Mat. Res.* 77 (11), 755-761 (1986).
[29] Grieb, B., Henig, E. T., Schneider, G., Knoch, G., Petzow, G. and de Mooij, D., Powder Metall. 35 (3), 221-227 (1992).
[30] Knoch, K. G., Reinsch, B. and Petzow, G., Proc. 13th Int. Workshop on Rare-Earth Magnets and Their Applications, 503-510 (1992).
[31] Cullity, B. D. and Graham, C. D., edited by IEEE Press Editorial Board (John Wiley and Sons, Inc., New Jersey, 2009).
[32] Otani, Y., Miyajima, H. and Chikazumi, S., J. Appl. Phys. 61 (8), 3436-3438 (1987).

[33] Yehia, S. and Aly, S. H., *J. Magn. Magn. Mater.* 212 (1), 195-200 (2000).
[34] Smith, D. L., *Thin-Film Deposition: Principles and Practice*. (McGraw Hill, New York, 1995).
[35] Barlow, F. D., Elshabini-Riad, A. and Brown, R., in *Thin Film Technology Handbook*, edited by Elshabini-Riad, A. and Barlow, F. D. (McGraw-Hill, New York, 1997).
[36] Mattox, D. M., *Handbook of Physical Vapor Deposition (PVD) Processing*. (Noyes Publication, New Jersey, 1998).
[37] Wasa, K., *Thin Film Materials Technology: Sputtering of Compound Materials*. (William Andrew, Burlington, 2004).
[38] Kruusing, A., *Int. Mater. Rev.* 44 (4), 121-131 (1999).
[39] Bendson, S. A. and Judy, J. H., *IEEE Trans. Magn.* MAG-9 (4), 627-631 (1973).
[40] Walther, A., Khlopkov, K., Gutfleisch, O., Givord, D. and Dempsey, N. M., *J. Magn. Magn. Mater.* 316 (2 SPEC. ISS.), 174-176 (2007).
[41] Walther, A., Givord, D., Dempsey, N. M., Khlopkov, K. and Gutfleisch, O., *J. Appl. Phys.* 103 (4), 043911-043915 (2008).
[42] Lemke, H., Lang, T., Göddenhenrich, T. and Heiden, C., *J. Magn. Magn. Mater.* 148 (3), 426-432 (1995).
[43] Cadieu, F. J., Aly, S. H., Cheung, T. D. and Pirich, R. G., *J. Appl. Phys.* 53 (3), 2401-2403 (1982).
[44] Cadieu, F. J., Cheung, T. D., Aly, S. H., Wickramasekara, L. and Pirich, R. G., *J. Appl. Phys.* 53 (11), 8338-8340 (1982).
[45] Cadieu, F. J., Cheung, T. D., Aly, S. H., Wickramasekara, L. and Pirich, R. G., *IEEE Trans. Magn.* MAG-19 (5), 2038-2040 (1983).
[46] Cadieu, F. J., Cheung, T. D. and Wickramasekara, L., *J. Magn. Magn. Mater.* 54-57 (Part 1), 535-536 (1986).
[47] Aly, S. H., Cheung, T. D., Wickramasekara, L. and Cadieu, F. J., *J. Appl. Phys.* 57 (6), 2147-2154 (1985).
[48] Cadieu, F. J., *Permanent Magnet Thin Films: A Review on Film Synthesis and Properties*. (Academic Press, London, 1992).
[49] Velu, E. M. T. and Lambeth, D. N., *J. Appl. Phys.* 69 (8), 5175-5177 (1991).
[50] Velu, E. M. T. and Lambeth, D. N., *IEEE Trans. Magn.* 28 (5 pt 2), 3249-3254 (1992).
[51] Shima, T., Kamegawa, A. and Fujimori, H., *J. Alloys Compd.* 281 (1), 46-49 (1998).
[52] Tang, W., Jin, Z. Q., Zhang, J. R., Gu, G., Li, J. M. and Du, Y. W., *J. Magn. Magn. Mater.* 185 (2), 241-245 (1998).
[53] Serrona, L. K. E. B., Fujisaki, R., Sugimura, A., Okuda, T., Adachi, N., Ohsato, H., Sakamoto, I., Nakanishi, A., Motokawa, M., Ping, D. H. and Hono, K., *J. Magn. Magn. Mater.* 260 (3), 406-414 (2003).
[54] Serrona, L. K. E. B., Sugimura, A., Fujisaki, R., Okuda, T., Adachi, N., Ohsato, H., Sakamoto, I., Nakanishi, A. and Motokawa, M., *Mater. Sci. Eng.*, B 97 (1), 59-63 (2003).
[55] Jiang, H., Evans, J., O'Shea, M. J. and Du, J., *J. Magn. Magn. Mater.* 224 (3), 233-240 (2001).
[56] Kim, Y. B., Kim, M. J., Yang, J. H., Ryu, K. S., Li, Y. and Kim, T. K., *J. Magn. Magn. Mater.* 234 (3), 489-493 (2001).
[57] Castaldi, L., Davies, H. A. and Gibbs, M. R. J., *J. Magn. Magn. Mater.* 242-245 (PART II), 1284-1286 (2002).

[58] Yang, J. H., Kim, M. J., Cho, S. H., Kim, H. T., Kim, Y. B., Kim, D. H., Kapustin, G. A. and Lee, K. H., *J. Magn. Magn. Mater.* 248 (3), 374-378 (2002).

[59] Castaldi, L., Gibbs, M. R. J. and Davies, H. A., *J. Appl. Phys.* 93 (11), 9165-9169 (2003).

[60] Cui, B. Z. and O'Shea, M. J., *J. Magn. Magn. Mater.* 256 (1-3), 348-354 (2003).

[61] Venkatesan, M., Buschbeck, J., Rhen, F. M. F. and Coey, J. M. D., *J. Magn. Magn. Mater.* 272-276 (SUPPL. 1) (2004).

[62] Chen, S. L., Liu, W., Chen, C. L. and Zhang, Z. D., *J. Appl. Phys.* 98 (11), 1-5 (2005).

[63] Khoa, T. V., Ha, N. D., Hong, S. M., Jin, H. M., Kim, G. W., Hien, T. D., Tai, L. T., Duong, N. P., Lee, K. E., Kim, C. G. and Kim, C. O., *J. Magn. Magn. Mater.* 304 (1), e246-e248 (2006).

[64] Chen, S. L., Liu, W., Zhang, Z. D. and Gunaratne, G. H., *J. Appl. Phys.* 103 (2), 023922 (2008).

[65] Tang, S. L., Gibbs, M. R. J., Liu, Z. W. and Davies, H. A., *J. Alloys Compd.* 455 (1-2), 77-80 (2008).

[66] Jiang, H. and O'Shea, M. J., *J. Magn. Magn. Mater.* 212 (1), 59-68 (2000).

[67] Castaldi, L., Gibbs, M. R. J. and Davies, H. A., *J. Appl. Phys.* 96 (9), 5063-5068 (2004).

[68] Uehara, M., *J. Magn. Magn. Mater.* 284 (1-3), 281-286 (2004).

[69] Uehara, M., Gennai, N., Fujiwara, M. and Tanaka, T., *IEEE Trans. Magn.* 41 (10), 3838-3843 (2005).

[70] Romero, S. A., Cornejo, D. R., Rhen, F. M., Neiva, A. C., Tabacniks, M. H. and Missell, F. P., *J. Appl. Phys.* 87 (9 III), 6965-6967 (2000).

[71] Speliotis, T., Makarona, E., Chouliaras, F., Charitidis, C. A., Tsamis, C. and Niarchos, D., *Phys. Status Solidi C* 5 (12), 3759-3762 (2008).

[72] Wang, J. Y., Sood, D. K., Ghantasala, M. K. and Dytlewski, N., *Vacuum* 75 (1), 17-23 (2004).

[73] Wang, J. Y., Ghantasala, M. K., Sood, D. K. and Evans, P. J., *Thin Solid Films* 489 (1-2), 192-199 (2005).

[74] Takahashi, Y. K., Ohkubo, T. and Hono, K., *J. Appl. Phys.* 100 (5), 053913 (2006).

[75] Liu, X., Zhao, H., Kubota, Y. and Wang, J. P., *J. Phys. D: Appl. Phys.* 41 (23) (2008).

[76] Zhen, C., Zhang, J., Ma, L., Hou, D., Liu, Y. and Li, S., *Surf. Rev. Lett.* 15 (1-2), 105-109 (2008).

[77] Gasgnier, M., Colliex, C. and Manoubi, T., *J. Appl. Phys.* 59 (3), 989-992 (1986).

[78] Pereira, L. G., Teixeira, S. R., Schreiner, W. H., Missell, F. P. and Baumvol, I. J. R., Phys. *Status Solidi A* 125 (2), 625-634 (1991).

[79] Lee, Z. Y., Dai, D. W., Shen, N. F., Hu, Y. S. and Bao, Y. Y., *IEEE Trans. Magn.* MAG-23 (5), 2599-2601 (1987).

[80] Tanaka, T., Miyazaki, T., Kita, E. and Tasaki, A., *IEEE Trans. Magn.* MAG-19 (5), 1650-1652 (1983).

[81] Fullerton, E. E., Jiang, J. S., Rehm, C., Sowers, C. H., Bader, S. D., Patel, J. B. and Wu, X. Z., *Appl. Phys. Lett.* 71 (11), 1579-1581 (1997).

[82] Zhang, L. N., Hu, J. F., Chen, J. S. and Ding, J., *J. Magn. Magn. Mater.* 321 (17), 2643-2647 (2009).

[83] Neu, V., Fähler, S., Singh, A., Kwon, A. R., Patra, A. K., Wolff, U., Häfner, K., Holzapfel, B. and Schultz, L., *J. Iron. Steel Res. Int.* 13 (SUPPL. 1), 102-111 (2006).

[84] Neu, V., Häfner, K., Patra, A. K. and Schultz, L., *J. Phys. D: Appl. Phys.* 39 (24), 5116-5120 (2006).
[85] Seifert, M., Neu, V. and Schultz, L., *Appl. Phys. Lett.* 94 (2), 3 (2009).
[86] Keavney, D. J., Fullerton, E. E., Pearson, J. E. and Bader, S. D., *IEEE Trans. Magn.* 32 (5 PART 2), 4440-4442 (1996).
[87] Keavney, D. J., Fullerton, E. E., Pearson, J. E. and Bader, S. D., *J. Appl. Phys.* 81 (8 PART 2A), 4441-4443 (1997).
[88] Ohtake, M., Nukaga, Y., Kirino, F. and Futamoto, M., *J. Cryst. Growth* 311 (8), 2251-2254 (2009).
[89] Ohtake, M., Nukaga, Y., Kirino, F. and Futamoto, M., *J. Appl. Phys.* 107 (9) (2010).
[90] Cadieu, F. J., Rani, R., Qian, X. R. and Chen, L., *J. Appl. Phys.* 83 (11), 6247-6249 (1998).
[91] Geurtsen, A. J. M., Kools, J. C. S., De Wit, L. and Lodder, J. C., *Appl. Surf. Sci.* 96-98, 887-890 (1996).
[92] Fähler, S., Hannemann, U., Weisheit, M., Neu, V., Melcher, S., Leinert, S., Wimbush, S. C., Singh, A., Kwon, A., Holzapfel, B. and Schultz, L., *Appl. Phys. A: Materials Science and Process*ing 79 (4-6), 1529-1531 (2004).
[93] Hannemann, U., Fähler, S., Neu, V., Holzapfel, B. and Schultz, L., *Appl. Phys. Lett.* 82 (21), 3710-3712 (2003).
[94] Constantinescu, C., Scarisoreanu, N., Moldovan, A., Dinescu, M., Petrescu, L. and Epureanu, G., *Appl. Surf. Sci.* 253 (19), 8192-8196 (2007).
[95] Nakano, M., Tsutsumi, S. and Fukunaga, H., *IEEE Trans. Magn.* 38 (5 I), 2913-2915 (2002).
[96] Nakano, M., Kato, R., Fukunaga, H., Tutsumi, S. and Yamashita, F., *J. Magn. Magn. Mater.* 272-276 (SUPPL. 1) (2004).
[97] Nakano, M., Kato, R., Hoefinger, S., Fidler, J., Yamashita, F. and Fukunaga, H., *J. Alloys Compd.* 408-412, 1422-1425 (2006).
[98] Nakano, M., Kato, R., Fukunaga, H. and Yamashita, F., Electr. Eng. Jpn. (English translation of Denki Gakkai Ronbunshi) 159 (2), 1-6 (2007).
[99] Kumar, K. and Das, D., Thin Solid Films Pap presented at the Int Conf on Metall Coat San Francisco Calif Apr 3-7 1978 54 (3), 263-269 (1978).
[100] Kumar, K., Das, D. and Wettstein, E., *J. Appl. Phys.* 49 (3), 2052-2054 (1978).
[101] Kumar, K. and Das, D., *J. Appl. Phys.* 60 (10), 3779-3781 (1986).
[102] Overfelt, R. A., Anderson, C. D. and Flanagan, W. F., *Appl. Phys. Lett.* 49 (26), 1799-1801 (1986).
[103] Wysłocki, J. J., *J. Mater. Sci.* 27 (14), 3777-3781 (1992).
[104] Asahi, N., Asaka, K., Ueda, K. and Sasaki, M., presented at the First International Conference on Processing Materials for Properties, Honolulu, HI, USA, 1993 (unpublished).
[105] Rieger, G., Wecker, J., Rodewald, W., Sattler, W., Bach, F. W., Duda, T. and Unterberg, W., *J. Appl. Phys.* 87 (9), 5329-5331 (2000).
[106] Willson, M., Bauser, S., Liu, S. and Huang, M., *J. Appl. Phys.* 93 (10), 7987-7989 (2003).
[107] Dzur, B., Zimmer, C., Thomas, G. and Linke, P., presented at the Materials Science and Technology Conference and Exhibition, MS and T'07 - "Exploring Structure,

Processing, and Applications Across Multiple Materials Systems", Detroit, MI, 2007 (unpublished).
[108] King, P. C., Zahiri, S. H. and Jahedi, M. Z., *J. Therm. Spray Technol.* 17 (2), 221-227 (2008).
[109] Pawlowski, B., Beer, H. and Toepfer, J., *Key Eng. Mater.* 132-136 (Pt 2), 1409-1411 (1997).
[110] Pawlowski, B. and Töpfer, J., *J. Mater. Sci.* 39 (4), 1321-1324 (2004).
[111] Speliotis, T., Niarchos, D., Falaras, P., Tsoukleris, D. and Pepin, J., *IEEE Trans. Magn.* 41 (10), 3901-3903 (2005).
[112] Speliotis, T., Niarchos, D., Meneroud, P., Magnac, G., Claeyssen, F., Pepin, J., Fermon, C., Pannetier, M. and Biziere, N., *J. Magn. Magn. Mater.* 316 (2 SPEC. ISS.) (2007).
[113] Pawlowski, B., Schwarzer, S., Rahmig, A. and Töpfer, J., *J. Magn. Magn. Mater.* 265 (3), 337-344 (2003).
[114] Thornton, J. A., *Annu. Rev. Mater. Sci.* 7, 239-260 (1977).
[115] Ohring, M., *Materials Science of Thin Films.* (Elsevier, Burlington, 2001).
[116] Pimpinelli, A. and Villain, J., *Physics of Crystal Growth.* (Cambridge University Press, Cambridge, 1998).
[117] Venables, J. A., *Introduction to Surface and Thin Film Processes.* (Cambridge University Press, Cambridge, 2000).
[118] Freund, L. B. and Suresh, S., *Thin Film Materials: Stress, Defect Formation and Surface Evolution.* . (Cambridge University Press, Cambridge, 2004).
[119] Pelliccione, M. and Lu, T.-M., *Evolution of Thin-film Morphology : Modeling and Simulations.* (Springer-Verlag Berlin Heidelberg, New York, 2007).
[120] Thompson, C. V., *Annu. Rev. Mater. Sci.* 30, 159-190 (2000).
[121] Mapps, D. M. and Chandrasekhar, R., *IEEE Trans. Magn.* 33 (5 PART 1), 3007-3009 (1997).
[122] Lileev, A. S., Parilov, A. A. and Blatov, V. G., *J. Magn. Magn. Mater.* 242-245 (PART II), 1300-1303 (2002).
[123] Vopsaroiu, M., Fernandez, G. V., Thwaites, M. J., Anguita, J., Grundy, P. J. and O'Grady, K., *J. Phys. D: Appl. Phys.* 38 (3), 490-496 (2005).
[124] Tang, S. L., Gibbs, M. R. J., Davies, H. A., Liu, Z. W., Lane, S. C. and Mateen, N. E., *J. Appl. Phys.* 101 (1) (2007).
[125] Thornton, J. A., *J. Vac. Sci. Technol.* 11 (4), 666-670 (1974).
[126] Chen, S. L., Liu, W. and Zhang, Z. D., *J. Magn. Magn. Mater.* 302 (2), 306-309 (2006).
[127] Piramanayagam, S. N., Matsumoto, M., Morisako, A. and Takei, S., *J. Alloys Compd.* 281 (1), 27-31 (1998).
[128] Chen, S. L., Liu, W., Chen, C. L. and Zhang, Z. D., *J. Appl. Phys.* 98 (3), 033907 (2005).
[129] Hannemann, U., Neu, V., Fähler, S., Holzapfel, B. and Schultz, L., *IEEE Trans. Magn.* 38 (5 I), 2949-2951 (2002).
[130] Neu, V., Melcher, S., Hannemann, U., Fähler, S. and Schultz, L., Phys. Rev. B: Condens. Matter 70 (14), 144418 (2004). http://prb.aps.org/abstract/PRB/v70/i14/e144418
[131] Serrona, L. K. E. B., Sugimura, A., Adachi, N., Okuda, T., Ohsato, H., Sakamoto, I., Nakanishi, A., Motokawa, M., Ping, D. H. and Hono, K., *Appl. Phys. Lett.* 82 (11), 1751-1753 (2003).

[132] Okuda, T., Sugimura, A., Eryu, O., Serrona, L. K. E. B., Adachi, N., Sakamoto, I. and Nakanishi, A., Jpn. *J. Appl. Phys.*, Part 1 42 (11), 6859-6864 (2003).
[133] Parhofer, S., Kuhrt, C., Wecker, J., Gieres, G. and Schultz, L., *J. Appl. Phys.* 83 (5), 2735-2741 (1998).
[134] Neu, V. and Shaheen, S. A., *J. Appl. Phys.* 86 (12), 7006-7009 (1999).
[135] Cadieu, F. J., Hegde, H. and Chen, K., *J. Appl. Phys.* 67 (9), 4969-4971 (1990).
[136] Hegde, H., Samarasekara, P., Rani, R., Navarathna, A., Tracy, K. and Cadieu, F. J., *J. Appl. Phys.* 76 (10), 6760-6762 (1994).
[137] Pina, E., García, M. A., Carabias, I., Palomares, F. J., Cebollada, F., de Hoyos, A., Almazán, R., Verdú, M. I., Montojo, M. T., Vergara, G., Hernando, A. and González, J. M., *J. Magn. Magn. Mater.* 272-276 (Supplement 1), E833-E835 (2004).
[138] Takei, S., Morisako, A. and Matsumoto, M., *J. Magn. Magn. Mater.* 272-276 (III), 1703-1705 (2004).
[139] Sayama, J., Mizutani, K., Asahi, T. and Osaka, T., *Appl. Phys. Lett.* 85 (23), 5640-5642 (2004).
[140] Kato, I., Takei, S., Xiaoxi, L. and Morisako, A., *IEEE Trans. Magn.* 42 (10), 2366-2368 (2006).
[141] Chen, J. S., Zhang, L. N., Hu, J. F. and Ding, J., *J. Appl. Phys.* 104 (9) (2008).
[142] Livingston, J. D., *J. Appl. Phys.* 57 (8), 4137-4139 (1985).
[143] Livingston, J. D. and McConnell, M. D., *J. Appl. Phys.* 43 (11), 4756-4762 (1972).
[144] Malyutin, V. I., Osukhovskii, V. E., Vorobiev, Y. D., Shishkov, A. G. and Yudin, V. V., Phys. *Status Solidi A* 65 (1), 45-52 (1981).
[145] Zhao, Y. P., Gamache, R. M., Wang, G. C., Lu, T. M., Palasantzas, G. and De Hosson, J. T. M., *J. Appl. Phys.* 89 (2), 1325-1330 (2001).
[146] Kronmüller, H., Durst, K. D. and Sagawa, M., *J. Magn. Magn. Mater.* 74 (3), 291-302 (1988).
[147] Muralidhar, G. K., Window, B., Sood, D. K. and Zmood, R. B., *J. Mater. Sci.* 33 (5), 1349-1357 (1998).
[148] Parhofer, S., Gieres, G., Wecker, J. and Schultz, L., *J. Magn. Magn. Mater.* 163 (1-2), 32-38 (1996).
[149] Peng, L., Zhang, H., Yang, Q., Li, Y., Song, Y. and Shen, J., *J. Appl. Phys.* 105 (6), 063915 (2009).
[150] Ma, Y. G., Yang, Z., Matsumoto, M., Morisako, A. and Takei, S., *J. Magn. Magn. Mater.* 267 (3), 341-346 (2003).
[151] Zasadzinski, J. F., Segre, C. U. and Rippert, E. D., *J. Appl. Phys.* 61 (8), 4278-4280 (1987).
[152] Speliotis, T. and Niarchos, D., *J. Magn. Magn. Mater.* 290-291 PART 2, 1195-1197 (2005).
[153] Romero, J. J., Palomares, F. J., Pigazo, F., Cuadrado, R., Cebollada, F., Hernando, A. and Gonzalez, J. M., *J. Non-Cryst. Solids* 353 (8-10), 786-789 (2007).
[154] Ma, Y. G., Yang, Z., Matsumoto, M., Morisako, A. and Takei, S., Phys. *Status Solidi A* 199 (3), 491-500 (2003).
[155] Araki, T., Nakanishi, T. and Umemura, T., *J. Appl. Phys.* 85 (8 II A), 4877-4879 (1999).
[156] Yu, M., Liu, Y., Liou, S. H. and Sellmyer, D. J., *J. Appl. Phys.* 83 (11), 6611-6613 (1998).

[157] Lemke, H., Mueller, S., Goeddenhenrich, T. and Heiden, C., *Phys. Status Solidi A* 150 (2), 723-731 (1995).
[158] Lemke, H., Echer, C. and Thomas, G., *IEEE Trans. Magn.* 32 (5 PART 2), 4404-4406 (1996).
[159] Dempsey, N. M., Walther, A., May, F., Givord, D., Khlopkov, K. and Gutfleisch, O., *Appl. Phys. Lett.* 90 (9), 092509-092503 (2007).
[160] Tsai, J. L., Huang, E. Y. and Chin, T. S., *IEEE Trans. Magn.* 33 (5 PART 2), 3646-3648 (1997).
[161] Hannemann, U., Fähler, S., Oswald, S., Holzapfel, B. and Schultz, L., *J. Magn. Magn. Mater.* 242-245 (PART II), 1294-1296 (2002).
[162] Melsheimer, A. and Kronmüller, H., *Physica B* 299 (3-4), 251-259 (2001).
[163] Zhang, J., Takahashi, Y. K., Gopalan, R. and Hono, K., *J. Magn. Magn. Mater.* 310 (1), 1-7 (2007).
[164] Liu, X., Okumoto, T., Matsumoto, M. and Morisako, A., *J. Appl. Phys.* 97 (10), 1-3 (2005).
[165] Takei, S., Morisako, A. and Matsumoto, M., *J. Appl. Phys.* 87 (9 III), 6968-6970 (2000).
[166] Hannemann, U., Fähler, S., Neu, V., Holzapfel, B. and Schultz, L., *IEEE Trans. Magn.* 38 (5 I), 2805-2807 (2002).
[167] Shima, T., Kamegawa, A., Aoyagi, E., Hayasaka, Y. and Fujimori, H., *J. Magn. Magn. Mater.* 177-181 (PART 2), 911-912 (1998).
[168] Hofer, F., *IEEE Trans. Magn.* MAG-6 (2), 221-224 (1970).
[169] Roy, A. G. and Laughlin, D. E., *J. Appl. Phys.* 91 (10 III), 8076 (2002).
[170] Zhang, L. N., Hu, J. F., Chen, J. S. and Ding, J., *J. Appl. Phys.* 105 (7) (2009).
[171] Holloway, K. and Fryer, P. M., *Appl. Phys. Lett.* 57 (17), 1736-1738 (1990).
[172] Chen, G. S., Lee, P. Y. and Chen, S. T., *Thin Solid Films* 353 (1), 264-273 (1999).
[173] Piramanayagam, S. N., Matsumoto, M. and Morisako, A., *J. Magn. Magn. Mater.* 212 (1) (2000).
[174] Chandrasekhar, R., Mapps, D. J., O'Grady, K., Cambridge, J., Petford-Long, A. and Doole, R., *J. Magn. Magn. Mater.* 196, 104-106 (1999).
[175] Middelhoek, S., *Ferromagnetic Domains in Thin Ni-Fe Films.* (G. Van Soest, 1961).
[176] Soohoo, R. F., *J. Appl. Phys.* 52 (3), 2459-2461 (1981).
[177] Néel, L., *Académie des Sciences* 241, 533-584 (1955).
[178] Tsai, J. L., Chin, T. S., Yao, Y. D. and Kronmüller, H., *J. Appl. Phys.* 99 (5) (2006).
[179] Nakano, M., Takeda, H., Yamashita, F. and Fukunaga, H., *IEEJ T. Electr. Electr.* 2 (4), 450-452 (2007).
[180] Cadieu, F. J., Cheung, T. D., Wickramasekara, L. and Kamprath, N., *IEEE Trans. Magn.* MAG-22 (5) (1986).
[181] Homburg, H., Sinnemann, T., Methfessel, S., Rosenberg, M. and Gu, B. X., *J. Magn. Magn. Mater.* 83 (1-3), 231-233 (1990).
[182] Piramanayagam, S. N., Matsumoto, M. and Morisako, A., *J. Appl. Phys.* 85 (8 II B), 5898-5900 (1999).
[183] Xu, X. H., Wu, H. S., Duan, J. F., Wang, F., Jin, F. and Lee, Z. Y., *Physica B* 334 (1-2), 207-211 (2003).
[184] Linetsky, Y. L., Raigorodsky, V. M. and Tsvetkov, V. Y., *J. Alloys Compd.* 184 (1), 35-42 (1992).

[185] Nazareth, A. S., Chopra, H. D., Sood, D. K. and Zmood, R. B., presented at the Mater. *Res. Soc. Symp. Proc.*, Boston, MA, USA, 1995 (unpublished).
[186] Chen, K., Hegde, H., Jen, S. U. and Cadieu, F. J., *J. Appl. Phys.* 73 (10), 5923-5925 (1993).
[187] Cheung, T. D., Wickramasekara, L. and Cadieue, F. J., *J. Appl. Phys.* 57 (8), 3598-3600 (1985).
[188] Kullmann, U., Koester, E. and Dorsch, C., *IEEE Trans. Magn.* MAG-20 (2), 420-424 (1984).
[189] Cadieu, F. J., Cheung, T. D., Wickramasekara, L., Kamprath, N., Hegde, H. and Liu, N. C., *J. Appl. Phys.* 62 (9), 3866-3872 (1987).
[190] Ho, K. Y., *J. Appl. Phys.* 53 (11), 7828-7830 (1982).
[191] Morrish, A. H., Li, Z. W., Zhou, X. Z. and Dai, S., *J. Phys. D: Appl. Phys.* 29 (9), 2290-2296 (1996).
[192] Buschow, K. H. J., *Rep. Prog. Phys.* 54 (9), 1123-1213 (1991).
[193] Ma, Y., Yang, Z., Matsumoto, M., Morisako, A. and Takei, S., *Mater. Sci. Eng.*, B 110 (2), 190-194 (2004).
[194] Ma, Y. G., Yang, Z., Wei, F. L., Matsumoto, M., Morisako, A. and Takei, S., *Phys. Status Solidi A* 201 (9), 2112-2118 (2004).
[195] Aylesworth, K. D., Sellmyer, D. J. and Hadjipanayis, G. C., *J. Magn. Magn. Mater.* 98 (1-2), 65-70 (1991).
[196] Luo, J., Liang, J. K., Guo, Y. Q., Liu, Q. L., Yang, L. T., Liu, F. S. and Rao, G. H., *Appl. Phys. Lett.* 84 (16), 3094-3096 (2004).
[197] Moze, O., Brück, E. and Buschow, K. H. J., J. Alloys Compd. 281 (2), 123-125 (1998).
[198] Luo, J., Liang, J. K., Guo, Y. Q., Liu, Q. L., Yang, L. T., Liu, F. S., Rao, G. H. and Li, W., *J. Phys. Condens. Matter* 15 (32), 5621-5628 (2003).
[199] Hsieh, C. C., Chang, H. W., Chang, C. W., Guo, Z. H., Yang, C. C. and Chang, W. C., *J. Appl. Phys.* 105 (7) (2009).
[200] Gjoka, M., Kalogirou, O., Sarafidis, C., Niarchos, D. and Hadjipanayis, G. C., *J. Magn. Magn. Mater.* 242-245 (PART II), 844-846 (2002).
[201] Yao, Z., Li, P. and Jiang, C., *J. Magn. Magn. Mater.* 321 (3), 203-206 (2009).
[202] Chang, H. W., Huang, S. T., Chang, C. W., Chang, W. C., Sun, A. C. and Yao, Y. D., *Solid State Commun.* 147 (1-2), 69-73 (2008).
[203] Chang, H. W., Huang, S. T., Chang, C. W., Chiu, C. H., Chen, I. W., Chang, W. C., Sun, A. C. and Yao, Y. D., *J. Alloys Compd.* 455 (1-2), 506-509 (2008).
[204] Cadieu, F. J., Hegde, H. and Chen, K., *IEEE Trans. Magn.* 25 (5), 3788-3790 (1989).
[205] Cadieu, F. J., Hegde, H., Schloemann, E. and Van Hook, H. J., *J. Appl. Phys.* 76 (10), 6059-6061 (1994).
[206] Yamamura, Y. and Ishida, M., *J. Vac. Sci. Technol.*, A 13 (1), 101-112 (1995).
[207] Johnson, K. E., *J. Appl. Phys.* 87 (9 II), 5365-5370 (2000).
[208] Moser, A., Takano, K., Margulies, D. T., Albrecht, M., Sonobe, Y., Ikeda, Y., Sun, S. and Fullerton, E. E., *J. Phys. D: Appl. Phys.* 35 (19), R157-R167 (2002). doi: 10.1088/0022-3727/35/19/201.
[209] Howard, J. K., *J. Vac. Sci. Technol.*, A 4 (1), 1-13 (1986).
[210] Baugh, R. A., Murdock, E. S. and Natarajan, B. R., *IEEE Trans. Magn.* MAG-19 (5), 1722-1724 (1983).

[211] Khizroev, S. and Litvinov, D., *Perpendicular Magnetic Recording*. (Kluwer Academic Publishers, Dordrecht, 2004).

[212] Grundy, P. J., *J. Phys. D: Appl. Phys.* 31 (21), 2975-2990 (1998). doi: 10.1088/0022-3727/31/21/001.

[213] Okumura, Y., Fujimori, H., Suzuki, O., Hosoya, N., Yang, X. B. and Morita, H., *IEEE Trans. Magn.* 30 (6 pt 1), 4038-4040 (1994).

[214] Sayama, J., Mizutani, K., Yamashita, Y., Asahi, T. and Osaka, T., *IEEE Trans. Magn.* 41 (10), 3133-3135 (2005).

[215] Chen, T., *IEEE Transactions on Magnetics MAG*-17 (2), 1181-1191 (1981).

[216] Cugat, O., Delamare, J. and Reyne, G., in *Magnetic Nanostructures in Modern Technology: Spintronics, Magnetic MEMS and Recording*, edited by Azzerboni, B., Asti, G., Pareti, L. and Ghidini, M. (Springer, Netherlands, 2007), pp. 105-125.

[217] Gutfleisch, O. and Dempsey, N. M., in *Magnetic Nanostructures in Modern Technology: Spintronics, Magnetic MEMS and Recording*, edited by Azzerboni, B., Asti, G., Pareti, L. and Ghidini, M. (Springer, Netherlands, 2007), pp. 167-194.

[218] Judy, J. W., *Smart Mater. Struct.* 10 (6), 1115-1134 (2001). doi: 10.1088/0964-1726/10/6/301.

[219] Chin, T. S., *J. Magn. Magn. Mater.* 209 (1-3), 75-79 (2000).

[220] Budde, T. and Gatzen, H. H., *J. Magn. Magn. Mater.* 242-245 (PART II), 1146-1148 (2002).

[221] Qin, L. and Zmood, R. B., *IEEE Trans. Magn.* 36 (5 II), 3833-3845 (2000).

[222] Arnold, D. P. and Wang, N., *J. Microelectromech. Syst.* (2009).

[223] Martinez-Quijada, J. and Chowdhury, S., presented at the Proceedings - IEEE International Symposium on Circuits and Systems, Seattle, WA, 2008 (unpublished).

[224] Romero, E., Warrington, R. O. and Neuman, M. R., *Physiol. Meas.* 30 (9) (2009).

[225] Arnold, D. P., *IEEE Trans. Magn.* 43 (11), 3940-3951 (2007).

[226] Sidorov, A. I., McLean, R. J., Rowlands, W. J., Lau, D. C., Murphy, J. E., Walkiewicz, M., Opat, G. I. and Hannaford, P., *Journal of Optics B: Quantum and Semiclassical Optics* 8 (3), 713-725 (1996).

[227] Meschede, D., Bloch, I., Goepfert, A., Haubrich, D., Kreis, M., Lison, F., SchÃ¼tze, R. and Wynands, R., presented at the Proc. *SPIE Int. Soc. Opt. Eng.*, San Jose, CA, 1997 (unpublished).

[228] Hinds, E. A. and Hughes, I. G., *J. Phys. D: Appl. Phys.* 32 (18) (1999). Electron Energy Corporation, (2010).

In: Magnetic Thin Films
Editor: John P. Volkerts

ISBN: 978-1-61209-302-4

Chapter 2

MICROSTRUCTURE DEPENDENT MAGNETORESISTANCE AND ELECTRORESISTANCE PROPERTIES OF $LA_{0.6}PB_{0.4}MNO_3$ THIN FILMS

Ajay Singh, A. K. Debnath, D. K. Aswal, S. K. Gupta and J. V. Yakhmi

Technical Physics Division, Bhabha Atomic Research Center, Mumbai 400 085, India

ABSTRACT

Doped manganites i.e. $RE_{1-x}A_xMnO_3$ (RE is a rare earth ion, A is a bivalent cation e.g. Sr, Ca, Pb etc) are known to exhibit a large change in the resistivity on application of magnetic field. This phenomenon is known as magnetoresistance (MR). These manganites also exhibit a change in resistivity as a function of electrical field, known as electroresistance (ER). Both MR and ER are known to depend on the microstructure of the samples. In this chapter we present our results on the microstructure dependence of MR and ER in $La_{0.6}Pb_{0.4}MnO_3$ (LPMO) thin films with different cyrstallinity namely, single crystalline (SC), nano-crystalline (NC) and polycrystalline (PC) deposited using pulsed laser deposition technique. The inter-granular and intra-granular magnetic and transport properties of these films were investigated using ferromagnetic resonance (FMR) technique, temperature dependent magnetization, and low temperature charge transport measurements. The SC, NC and PC films exhibits the maximum MR values of 60%, 30% and 100% respectively at 1 Tesla magnetic field. Using FMR technique, we demonstrated that large numbers of spin glassy grain boundaries present in the NC films contributes to high field MR (HFMR), however the antiferomagnetic ordering of Mn spins present at grain boundaries in PC films contribute to low field MR (LFMR). To further delineate the role of grain boundaries from the bulk single crystalline regions, we prepared a bi-crystal grain boundary junction of LPMO. The current-voltage characteristics of LPMO grain boundary junction show that transport is dominated by multistep inelastic tunneling along with a small contribution of elastic tunneling, which is responsible for LFMR. It has been observed that only NC and PC films exhibits the ER property, it indicates that ER is an extrinsic properties of CMR materials and arises only due to magnetic disorder present at grain boundaries.

1. INTRODUCTION

The magnetoresistance (MR) of a material is a measure of the change in its electrical resistivity (ρ) on application of an external magnetic field (H) and is expressed by:

$$MR = \frac{\rho(0) - \rho(H)}{\rho(H)} \tag{1}$$

Where ρ (0) and ρ(H) refers to the resistivity in zero field and at a field value of H.

The MR in metals, semiconductors and granular alloys is of the order of few percent, but it has been a field of an active research for several decades because of the commercial use in magnetic recording and read heads [1, 2]. In 1988, Baibich et al, discovered a very high negative MR (i.e. a decrease of resistivity on application of magnetic field) of >50% in artificially grown multilayers comprising of alternate ferromagnetic and nonmagnetic multilayers e.g., Fe/Cr, Co/Cu etc [3]. These multilayer structures are now popularly known as "Giant Magnetoresistance (GMR)" structures. The GMR effect in these multilayer-structures has been understood on the basis of spin dependent scattering of charge carriers. It may be noted that, the thickness of the individual layers is typically tens of angstrom, which is much lower than the mean free path of electrons. In the absence of magnetic field, the alternate ferromagnetic layers get antiferromagnetically coupled through "Ruderman-Kittel-Kasuya-Yosida (RKKY)" interaction via non-magnetic spacers [3]. Thus, both spin up and down electrons suffer scattering owing to antiferromagnetic coupling of alternate layers. This results in high resistivity of the multilayer-structures. On application of a small magnetic field, the ferromagnetic layers get aligned in the direction of the field. In this case, while electrons of one spin, say down, are scattered, the up spin electrons pass without any scattering, and thereby lower the resistivity of the multilayer.

In the early nineties, a very high MR exceeding 70% was discovered in hole-doped perovskite manganites with chemical formula $R_{1-x}A_xMnO_3$, where R is a trivalent rare earth (La, Nd etc) and A is a bivalent cation (Ca, Sr, Pb) [4, 5]. In order to distinguish high MR manganites from GMR multilayers they were termed as "Colossal Magnetoresistance (CMR)" materials. These manganites exhibit paramagnetic-to-ferromagnetic transition concomitant with an insulator-to-metal transition with lowering temperature. These magnetic and transport transitions are found to be extremely sensitive to various inhomgeneties, such as, oxygen defects, grain boundaries and cation vacancies. In the recent years the phenomena of electroresistance (ER) is investigated in these materials; where an electric field or current was found to decrease the resistivity significantly.

In the present chapter we will present and an overview of microstructure dependence of MR and ER properties of the CMR thin films. In the section 5, the results obtained from ferromagnetic resonance techniques, which sheds lights on the magnetic properties of grain boundaries, will be presented. The magnetic nature of grain boundaries gives an insight of the mechanism governing the low field MR and ER. To delineate the role of single grain boundary, preparation and characterization of bi-crystal grain boundary junctions of CMR thin films will be presented in section 6. The main conclusions are presented in the section 7.

2. STRUCTURAL AND MAGNETIC PROPERTIES OF CMR MANGANITES

The compounds of the form $R_{1-x}A_xMnO_3$ fall in the class of distorted ABO_3 type perovskite oxides, as schematically shown in Figure 1. The distortion of the ideal perovskite cell is caused by the mismatch between the sizes of the A and B site cations. In $R_{1-x}A_xMnO_3$, the Mn-O-Mn bonds form a three dimensional network, with Mn ion sitting inside the oxygen octahedron (Figure 1. a).

The cubic perovskite adjusts to $t<1$ by a co-operative rotation of the MnO_6 octahedron that buckles the Mn-O-Mn bond angle from its ideal 180° to (180° -ϕ). With an increasing ϕ, the space group symmetry lowers from cubic via tetragonal to rhombohedral or orthorhombic. In the parent $LaMnO_3$, the symmetry is lowered to orthorhombic by a cooperative rotation of the MnO_6 octahedron around the (110) crystal axis. As shown in Figure 1.(c), the Jahn-Teller (J-T) deformation causes the MnO_6 octahedron to get compressed in the ab-plane and get elongated along c-axis. This results in elongation of Mn-O-Mn bonds in c-plane; while shortening of ab-plane Mn-O-Mn bonds [6]. The J-T distortion, however, does not cause any further reduction in the symmetry of the crystal but increases the c/a ratio of the unit cell. As shown in Figure 2, the five fold degenerate d^5 orbital of the Mn atoms sitting in the MnO_6 octahedron in these compound splits due to the crystal field of the O^{2-} ions into a two fold degenerate e_g and a three-fold degenerate t_{2g} orbitals separated by few electron volts.

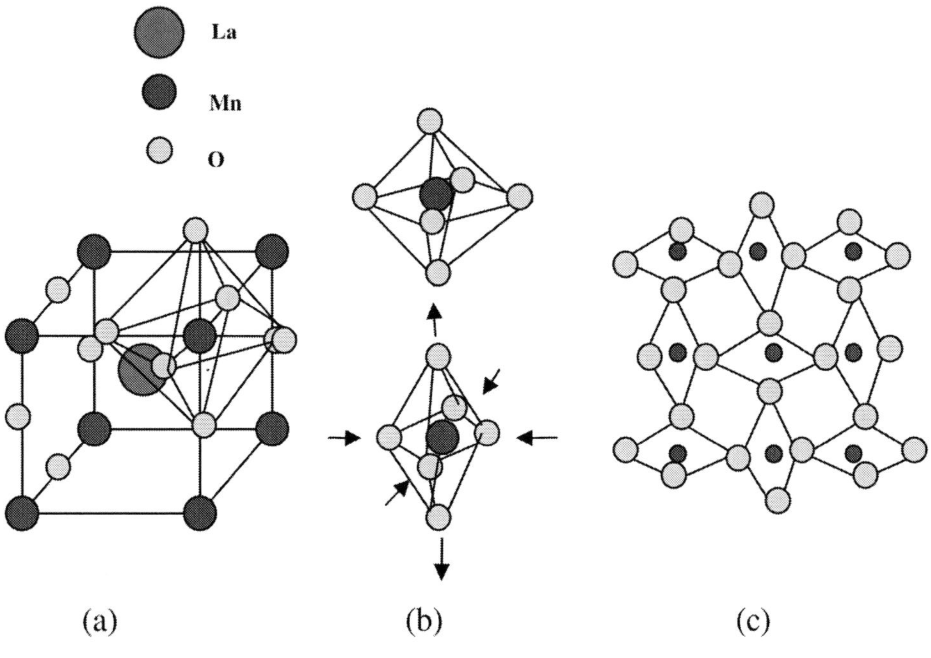

Figure 1.(a) Ideal perovskite unit cell showing the oxygen octahedron, (b) Undistorted oxygen octahedron around a Mn^{4+} ion and Jahn-Teller (J-T) distorted Octahedron around a Mn^{3+} ion. The J-T distortion causes ab-plane contraction (shown by inward arrows) of the Mn-O bond length and along c-axis elongation of Mn-O bond length (shown by outward arrows) (c) The J-T distortion yielding long and short Mn-O bonds in ac-plane.

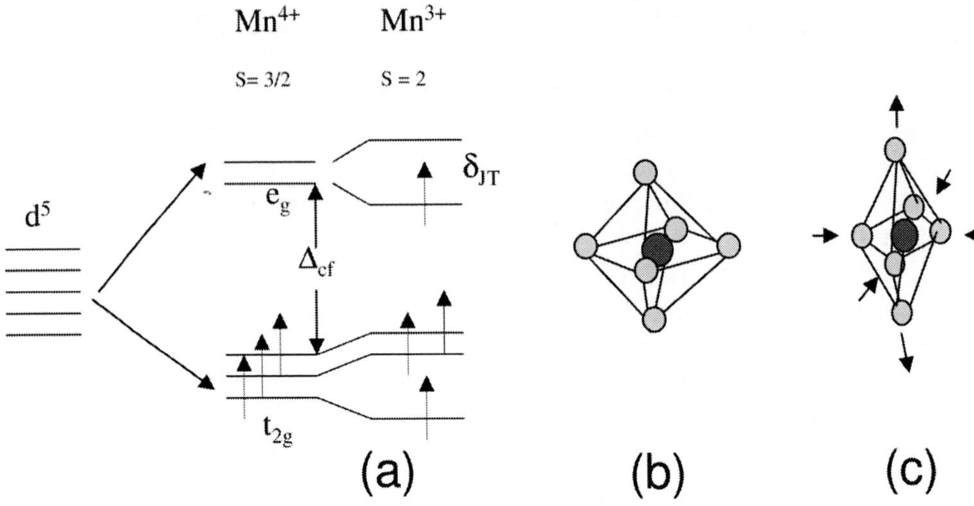

Figure 2. (a) Splitting of d^5 orbital of Mn atoms due to crystal field and J-T distortions of the oxygen octahedron in $R_{1-x}A_xMnO_3$ (b) The undistorted oxygen octahedron around a Mn^{4+} (c) The J-T distortion around a Mn^{3+} causes the octahedron to get compressed (shown by inward arrows) in ab–plane and elongated (shown by outward arrows) in the direction of c-axis.

The J-T distortion further splits the e_g band of Mn^{3+} ion into two sub-bands. The Mn ions in parent $LaMnO_3$ are in the Mn^{3+} state and, therefore have an electronic configuration of t_{2g}^3 e_g^1. This configuration of the electronic state as well as the structure stabilizes the A-type antiferromagnetic structure (with S=2), via anisotropic exchange coupling in the ab-plane and along c-axis [7].

Along the c-axis, the t_{2g} - t_{2g} superexchange interaction stabilizes the antiferromagnetic ordering. In the ab-plane the additional e_g-t_{2g} overlap stabilizes the ferromagnetic coupling of the Mn ions. When doped with bivalent ions like Pb^{2+} or Ca^{2+} at rare earth site, for doping level >20%, $LaMnO_3$ undergoes a change to a metallic ferromagnetic phase at low temperatures.

At low doping levels, initially a canting in the antiferromagnetic lattice appears before the ordering becomes ferromagnetic. Doping of bivalent cation drives the Mn ions into a mixture of Mn^{3+} and Mn^{4+} (for which Jahn-Teller distortion is absent) valence states, making the Mn orbital configuration to t_{2g}^3 $e_g^{1-\delta}$. This is equivalent to doping holes in the parent compound. As a result the electron in the partially filled Mn e_g orbital can hop from site to site via the intermediate oxygen while conserving its spin. This process is responsible for the ferromagnetic order as well as large magnetoresistance in doped rare earth manganites. This mechanism was first proposed by C. Zener [8] and is known as "double-exchange mechanism", which is schematically depicted in Figure 3.

According to double-exchange mechanism, the hopping of electron is governed by the strong on site Hund's rule coupling between manganese e_g and t_{2g} electrons. Hund's rule states that the lowest energy state is when the e_g and t_{2g} spins are oriented parallel to each other. Since this process of hopping conserves the spin of electron, the hopping of an electron from a Mn^{3+} ion to a neighboring Mn^{4+} ion via the Mn^{3+}-O^{2-}-Mn^{4+} bond is favored when the two manganese spins are oriented parallel to each other.

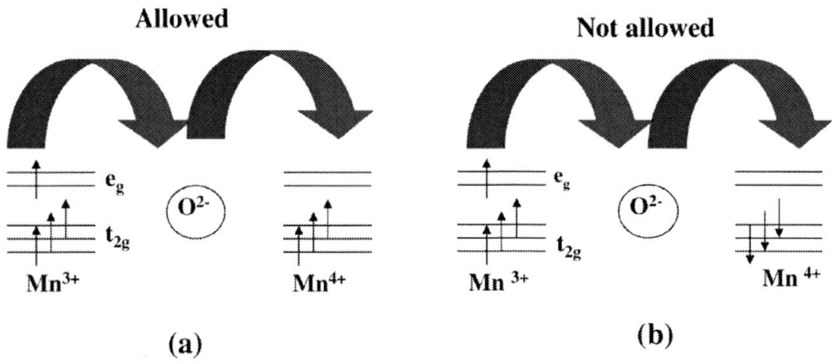

Figure 3. Schematic diagram showing Zener double exchange, where e_g electron from Mn^{3+} hops to unfilled e_g orbital in Mn^{4+}. The process is (a) allowed only when the two Mn spins are parallel to each other (b) not allowed when the two Mn spins are antiparallel to each other.

Detailed calculation by Anderson and Hasegawa shows that the hoping probability of an electron between two neighboring Mn^{3+} / Mn^{4+} is proportional to cos ($\vartheta/2$), where ϑ is the angle between two Mn spins. The double-exchange process is described by the Hamiltonian

$$H = -t\sum_{ij} C_{i\sigma}^{\dagger} C_{j\sigma} - J_H \sum_{i} \sigma_i . S_i \tag{2}$$

Where the t is the nearest neighbor hopping integral, S_i is the e_g electron ½ spin, σ_I is the t_{2g} 3/2 spin and J_H is the ferromagnetic Hund's rule coupling constant. The second term in the Hamiltonian gives the on-site Hund's rule coupling which ensures that the lowest energy state is obtained when the e_g and t_{2g} spins are aligned parallel to each other.

Qualitatively, the double-exchange mechanism explains the insulator–metal transition associated with the paramagnetic–ferromagnetic phase transition. The alignment of the manganese spins below the ferromagnetic phase transition results in an increase in the mobility of electrons, which in turn enhances the electrical conductivity. On the other hand, the same mechanism, which is responsible for the metallicity in the ferromagnetic phase, also ensures that the ground state of the system is ferromagnetic. This mechanism also explains CMR qualitatively, as discussed in the next section.

3. MAGNETORESISTANCE

Qualitatively, the phenomena of colossal magnetoresistance in crystalline samples e.g. single crystals or epitaxial thin films can be understood within the framework of double-exchange mechanism. At temperature close to T_C, the dynamic spin disorder can be most efficiently suppressed by the application of a magnetic field, thereby increasing the conductivity. At temperature well below T_C the spins are already ferromagnetically ordered, hence an application of magnetic field has little effect in further increasing the conductivity. It has, however, been shown that double-exchange alone cannot explain the magnitude of the magnetoresistance observed in these systems [9]. Also the negative coefficient of resistivity as observed in the paramagnetic regime cannot be explained from double-exchange mechanism even with maximal spin disorder. It was found that electron phonon coupling could be crucial to account for the insulating properties above T_C. Intrinsic magnetoresistance (MR) observed in the single crystal samples has nearly linear dependence on magnetic field and is quite small at low temperature. On the other hand polycrystalline samples exhibit higher resistivity and an extrinsic MR (in addition to intrinsic CMR) that yields a sharp drop in the resistance at low magnetic fields [5]. This MR seen in polycrystalline samples does not reduce at low temperatures and on the contrary is often seen to increase. Hwang et al attributed this phenomenon to spin polarized tunneling through insulating grain boundaries. In polycrystalline as well as single crystalline samples phase separation in concomitant with percolation behavior has been supposed to the core of CMR effect. It has been proved by several studies that largest magnetoresistance (MR) is associated with spatial inhomogenities related to the multiphase coexistence, which generally causes the sensitivity of physical properties to the external disturbances, such as application of magnetic field, current bias and pressure etc [5].

The CMR materials are very susceptible to disorders, such as, oxygen and cation vacancies. These disorders may occur during the sample preparation. In thin films, film-substrate lattice mismatch induces strains in the films. These strains influence the easy axis of magnetization, which could be either in-plane or out-of-plane. As a result, a magnetic anisotropy is introduced in the films, which profoundly affects the magnetic and transport properties. However, not much work has been done so far in this direction.

4. Half Metallic Character of CMR Manganites

The half-metallic character is the property of a material because of which current is carried by majority spin channel i.e. the carriers at fermi level are either spin up or spin down. In order to demonstrate the half-metallic character of CMR materials, the energy level diagram of $R_{1-X}A_XMnO_3$ in the vicinity of Fermi level is schematically shown in Figure 4.

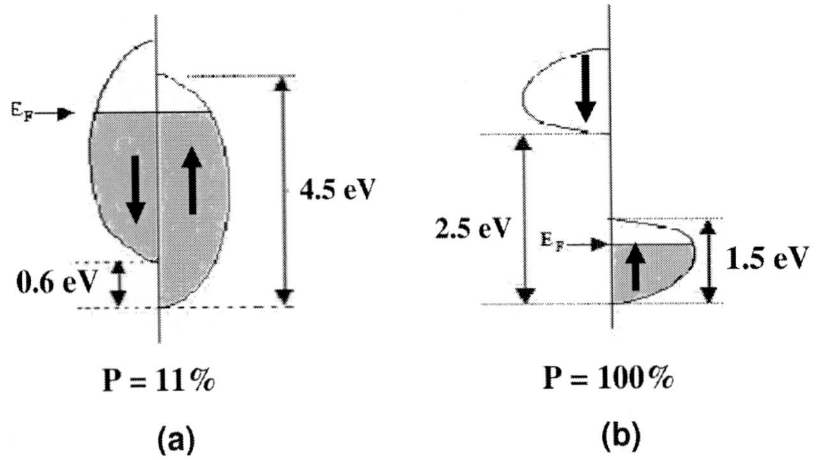

Figure 4. Schematic showing the spin polarization of conduction band in (a) conventional ferromagnet (b) and $R_{1-X}A_XMnO_3$ [10].

For comparison, the energy level diagram of conventional ferromagnet is also drawn. In CMR materials e.g. $La_{2/3}Sr_{1/3}MnO_3$, at $T = 0$ K the Hund's rule coupling energy (given by J_H σ, where σ is 3/2 t_{2g} spin) is of the order of 2.5 eV, whereas the width of the e_g conduction band is around 1.5 eV [10]. This relatively narrow conduction band therefore gets fully spin split due to the double-exchange coupling, giving the conduction electrons a pure spin-up or spin-down character. Therefore, in CMR the carriers are 100% polarized. On the other hand, for conventional ferromagnets e.g. Ni, the width of conduction band is 4.5 eV while Hund's rule coupling energy is 0.6 eV. This leads to only 11% polarization of the conduction band.

The 100% polarization of conduction band in CMR materials makes them attractive for potential application in "Spin-electronics" or "Spintronics". In this rapidly emerging field of electronic devices, one essentially makes use of spin for controlling the devices.

5. Techniques to Prepare Thin Films of CMR Material

Soon after the discovery of manganites, research efforts started on the realization of textured and epitaxial thin films, largely motivated by potential device applications. Several techniques have been employed to grow thin films of CMR, including pulsed laser deposition (PLD), molecular beam epitaxy (MBE), sputtering, and metal-organic chemical vapor deposition technique (MOCVD) [5]. In the following paragraphs, we briefly review prominent deposition techniques employed for the preparation of CMR thin films.

Pulsed laser deposition (PLD): Laser ablation is the most widely used technique for deposition of CMR thin films. Film deposition by PLD is carried out by irradiation of a CMR target by a focused laser beam under a suitable oxygen partial pressure. The laser beam ablates the material from target, which gets deposited onto the hot substrates. In order to get stoichiometric, single-phase and crystalline films by PLD various process parameters need to be optimized e.g. wavelength of laser and its energy density, target to substrate distance and oxygen partial pressure. It is most preferred technique for CMR thin films because it allows reproducible preparation of high-quality epitaxial films with least processing time. For example, the total processing time to deposit 100 nm LPMO thin films would be only 2 hr. However, PLD is not a suitable technique for deposition over area larger than $10mm^2$. This is because the angular distributions of the particles in the plume follow a $cos^n\theta$ behavior; here θ is angle with respect to target normal and n has values between 4 and 16. In a multicomponent system, the value of n is different for each constituent element. Therefore, due to varying angular distribution of elements, the film stoichiometry and uniformity is affected at angles $\theta > 15^0$ [11]. Another problem of PLD technique is the deposition of particulates at the film surface, which however, can be minimized by using a target of high density, properly optimizing the energy of incident laser pulse and target substrate distance and by mechanically chopping a part of plume.

Molecular beam epitaxy: The molecular-beam-epitaxy (MBE) is the most sophisticated technique for growing thin films of semiconductors. In MBE, the constituents are evaporated using effusion cells or electron beam and atoms are deposited on the substrate layer by layer. Owing to base vacuum of 10^{-10} mbar, MBE produces cleanest films. In addition, in-situ characterization tools, such as, reflection high-energy electron diffraction (RHEED), give a continuous signature of the perfection of the structure being grown. However, MBE is not an appropriate technique for growing CMR thin films. This is because CMR need an oxygen partial pressure of $>10^{-3}$ mbar for phase formation, and at such a high-pressure the molecular beam character of other metallic constituents cannot be maintained. To overcome this problem, reactive oxygen sources, such as, atomic oxygen, ozone and NO_2 have been employed for formation of proper phase at lower oxygen pressures. Although very high-quality CMR films have been grown by several groups employing MBE technique, this technique is not usually preferred because of complexity, time consumed and cost.

Sputtering: Besides PLD, sputtering is the preferred technology of the preparation of single-phase films of CMR materials [12]. In this technique, plasma of inert gas mostly Argon (and a small concentration of Oxygen) is created either by DC, RF or AC high voltages. Positive ions strike the target placed at cathode and ejects the atoms of material by momentum transfer, which eventually get deposited on a substrate facing the target. The advantages of sputtering technique are: (i) high reproducibility, (ii) compatible with high oxygen pressure, (iii) easy control of deposition rates, (iv) homogeneous compositions over a larger area, (v) reliable and cheap, and (vi) doesn't suffer from the droplet problem of PLD. However, the main drawback of this technique is the re-sputtering of the growing film, which causes the deviation in the film composition. The re-sputtering of film takes place due to bombardment of films by electrons and negative oxygen ions. The problem of re-sputtering has, however, been minimized by keeping substrates off-axis i.e. perpendicular to the target, where electrons or oxygen ions do not bombard the growing film

Metal organic chemical vapor deposition (MOCVD): In this technique, the carrier gases take the vapor of metal organic precursor to a reaction cell where the substrate is heated to enable the decomposition and reaction of precursor to yield desired film [13]. In comparison with physical vapor deposition methods, MOCVD process does not require ultra high vacuum conditions and therefore allows deposition at much higher partial pressure of Oxygen. Hence films are of high quality and exhibit good MR properties without the need for post-deposition annealing [8]. For reproducibility, good control over gas flow, source and substrate temperature is needed. The main difficulty with this technique is the availability of appropriate precursor and high purity gases.

In summary, sputtering and PLD are the two most widely used techniques for deposition of CMR thin films. However, CMR thin films deposited by either of the two techniques usually requires a post-deposition oxygen annealing treatment. The post deposition oxygen treatments are carried out at higher temperatures as compared to deposition temperature [14]. This leads to strong inter-diffusion between film and substrate, which in turn causes compositional deviations as well as uncontrolled grain growth. Consequently, the film surfaces are usually found to be rough. Therefore, for growth of high quality thin films of CMR, in-situ process is preferred.In the next section we will discuss the growth of $La_{0.6}Pb_{0.4}MnO_3$ (LPMO) thin films using PLD.

6. Growth and Characterisation of $La_{0.6}Pb_{0.4}MnO_3$ (LPMO) Thin Films with Controlled Microstructure

The LPMO thin films were grown on (100) $SrTiO_3$ (STO) single-crystal substrates by pulsed-laser deposition technique.

A brief overview of the PLD system employed for the growth of LPMO films is given here. The top view of the PLD system is shown in Figure 5. It consists of two major parts: (i) Excimer laser and (ii) Vacuum deposition chamber. The Excimer laser (model: Lambda Physics COMPAC 801) is based on KrF, which gives laser light of 248 nm wavelength having a pulse width of 20 ns. The energy and repetition rate of laser can be easily varied from the control panel. The cylindrical stainless steel deposition chamber (height = 400 mm and diameter = 500 mm) consists of six utility ports. The two ports, located across a diameter of the chamber are used for mounting target- and substrate-holders, which allow target and substrate to face each other. The third port is located at an angle of 45^0 with respect to the line joining target and substrate ports. This port consists of a quartz window through which the beam of excimer laser is focused onto the target using a lens. The fourth port is used for providing a shutter, which separates target and substrates when pre-deposition target cleaning is carried out by laser ablation process. Two additional ports, one perpendicular to the line joining target and substrate ports and another located at top lid of the chambers are used for viewing the laser ablation process and aligning the laser induced plume and substrates. The target holder is designed in such a way that four targets can be mounted on it using stainless steel clips. This target holder is mounted on the port using a Wilson seal, which allows rotation of targets under vacuum. To vary the distance between target and substrates, the substrate heater can be moved in forward or backward direction in a stainless stud mounted on the diameter line connecting target and substrate holders. The substrates are mounted on a home made heater (made by winding of super canthal wire of 1mm diameter to a grooved ceramic block and covering it by stainless steel sheets). A chromel-alumel thermocouple is mounted in the vicinity of the substrate for accurate measurement of the deposition temperature of the film. The temperature of the substrates could be controlled to within ± 1°C using Eurotherm temperature controller (model: 902 series).The deposition chamber is evacuated to a base vacuum of ~10^{-6} Torr using a diffusion pump based system. To provide controlled oxygen ambient for thin film deposition, oxygen gas is let in through a needle valve.

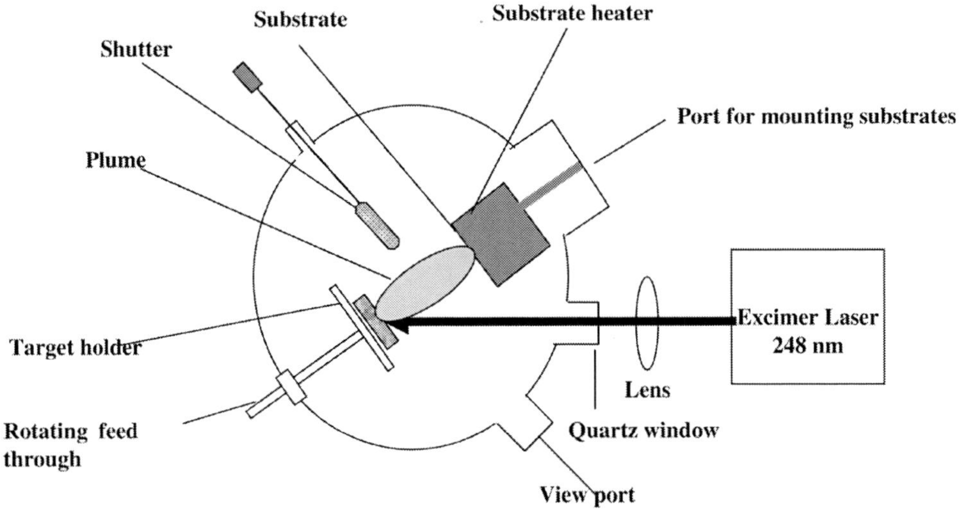

Figure 5. Top view of pulsed laser deposition system employed for synthesis of LPMO thin films.

Optimization of several growth parameters, such as, target-substrate distance, target rotation rate and laser energy indicated that the optimum values of these parameters are essentially necessary [15]. Briefly, a laser beam from a KrF excimer laser of wavelength 248 nm, pulse width 20 ns and repetition rate of 5 Hz was focused onto a rotating target of nominal composition $La_{0.6}Pb_{0.4}MnO_3$. The optimized target substrate distance was found to be 4.5 cm. Foir an optimized laser energy density of 3.5 J/cm^2, the deposition rate was found to be 2 Å/s. However, the substrate temperature (*T*s) and oxygen partial pressure (pO_2) were found to profoundly influence the morphological and structural properties of the LPMO films. The results of the optimization of *T*s is divided into two parts (i) Ts $\geq$ 600ºC and (ii) Ts $<$ 600ºC.

6.1. Growth at Temperatures ≥ 600ºC

In order to investigate the effects of *T*s and pO_2 on the morphological and structural properties of LPMO films, the deposition was carried out under two different sets of growth conditions: (i) Ts was varied between 600 and 850°C, while pO_2 was kept fix at 0.2 torr, and (ii) pO_2 was varied between 0.02 and 4 torr while keeping *T*s at a fix value of 600°C. All other growth parameters, such as, laser energy (300 mJ), target-substrate distance (4.5 cm), target rotation rate (2 rpm) were kept constant in all the experiments. The thickness of films was also kept constant at 100 nm in all experiments.

Effect of Ts: The AFM images of LPMO films grown at different Ts but at fixed pO_2 of 0.2 torr are shown in Figure 6.

Typical XRD patterns for the LPMO films grown at 600°C and 730°C under pO_2 of 0.2 torr are shown in Figure 7. The presence of only (00*l*) reflections of the LPMO films, which are superimposed on (*h*00) peaks of STO substrate having nearly the same lattice parameter, is observed. The overlapping of (002) LPMO and (200) STO peaks is resolved clearly in the expanded plots, as shown in the insets of Figure 7.

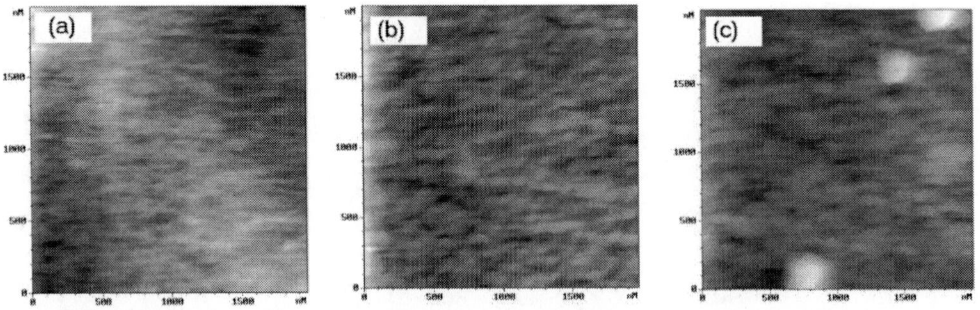

Figure 6. 2000 nm x 2000 nm AFM scan of LPMO films grown on (100) $SrTiO_3$ substrates at different temperatures; (a) 600, (b) 650 and (c) 730°C. The pO_2 was kept fixed at 0.2 Torr in all the cases. Surface roughness of the film increases with increase in growth temperature.The films are found to be very smooth, except for a few particulates on the surface. However, the average surface roughness was found to increase with increasing substrate temperature. The highest average surface roughness of about 1.5 nm was measured for a film grown at 730°C. The AFM images reveal that LPMO films grows over STO substrate via Frank-vander Merve mechanism i.e. 2D layer-by-layer growth mode, and this is expected as both LPMO and STO have perovskite structures with matching lattice parameters [15].

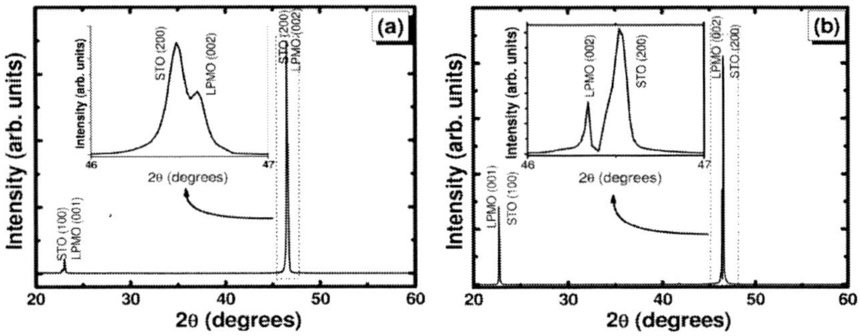

Figure 7. XRD pattern recorded for LPMO films grown on (100) STO substrates at (a) 600 and (b) 730°C under pO_2 of 0.2 Torr. The insets show the expanded plot in the vicinity of (200) peak of STO. Note that there is relative shift in the position of LPMO (002) peaks in inset in Figure (a) and (b); its implications are discussed in the text.

Absence of any additional impurity peak(s) in the XRD pattern indicates growth of single-crystalline LPMO films with *c*-axis perpendicular to the substrate plane. For all films, the full-width at half-maximum of (002) peak is found to be less than 0.12 degree. This, along with smooth surface of the films, indicates that LPMO films grow epitaxially on STO substrates. As shown in the insets of Figure 7, the position of (002) LPMO peak is shifted from the right to the left side of (200) STO peak when the growth temperature is increased from 600°C to 730°C. This suggests that the *c*-lattice parameter of LPMO film changes with growth temperature [15]. The variation in *c*-lattice parameter (computed from the *d*-value of (002) reflection) with growth temperature is plotted in Figure 8. Also shown in Figure 8. is the Pb content of films, determined from EDXA analysis, grown at different temperatures. The sharp reduction in Pb content at higher growth temperatures is expected because of very high vapor pressure of Pb. The data of Figure 8. clearly show that the increase in *c*-parameter is intimately related to decrease in Pb content of the films. This is understood as follows. $La_{1-x}Pb_xMnO_3$ has a pervoskite crystal-structure, in which La is partially substituted by Pb. The loss of Pb atoms from the lattice would create vacancies and therefore a reduction in number of bonds between Pb and O atoms. This results in weakening of chemical bonds in the crystalline film and causes expansion of lattice constant. The loss of Pb content also influences the electrical properties of the films as discussed below [15].

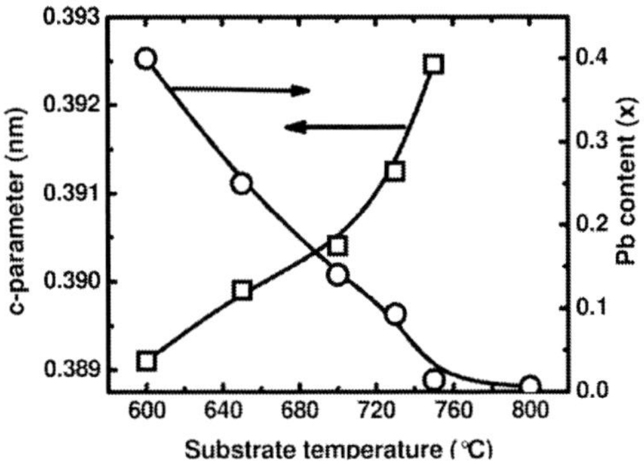

Figure 8. Variation of *c*-lattice parameter and Pb content of LPMO films as a function of growth temperature.

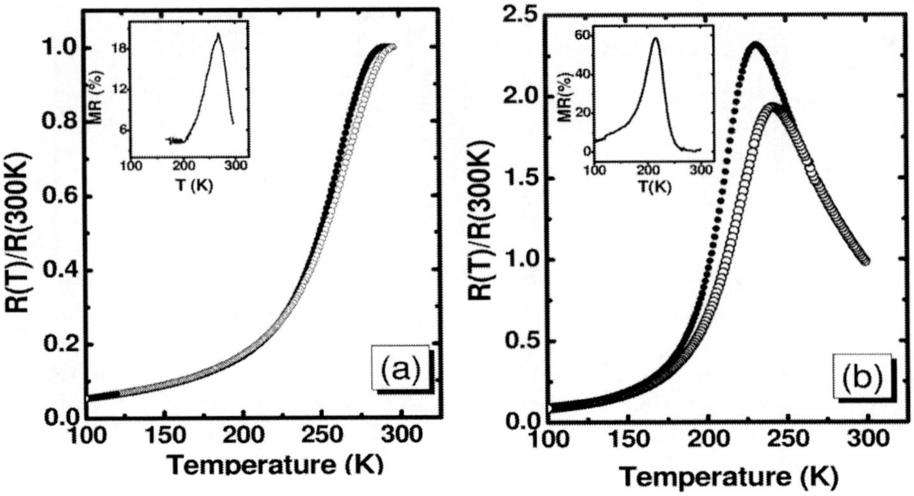

Figure 9. Temperature dependence of the normalized resistance measured in zero-field (solid circle) and in an applied magnetic field of 1 T (empty circle) for LPMO films grown at (a) 600°C (b) 730°C.

In Figure 9, we show the temperature dependence of the normalized resistance measured in zero-field and in an applied magnetic field of 1 T for LPMO films grown at two different temperatures, namely, 600 and 730^{0}C. All the films, grown at substrate temperatures <800°C, exhibit similar qualitative magnetotransport behavior. This includes (i) presence of a peak at T_{IM} corresponding to an insulator-to-metal (IM) transition and (ii) shift in T_{IM} to higher temperatures and reduction in film resistance (in the vicinity of T_{IM}) when a magnetic field is applied. On the other hand, films grown at temperatures >800°C exhibit insulating behavior and show no effect of magnetic field as revealed by data of Figure 10. (a).

It is observed from Figure 11, that the value of T_{IM} and the magnetoresistance ratio (expressed as: MR% = $(R_0\text{-}R_H)/R_H$ X 100) varies with growth temperature. The variation of T_{IM} and highest MR% (determined from the temperature dependence of MR, as shown in the Figure 10.(b)) growth temperature are plotted in Figure 10.(b). It is seen that T_{IM} decreases monotonically with growth temperature, while highest MR% increases. The reduction of T_{IM} is explained in terms of the Pb loss in the films as follows [15].

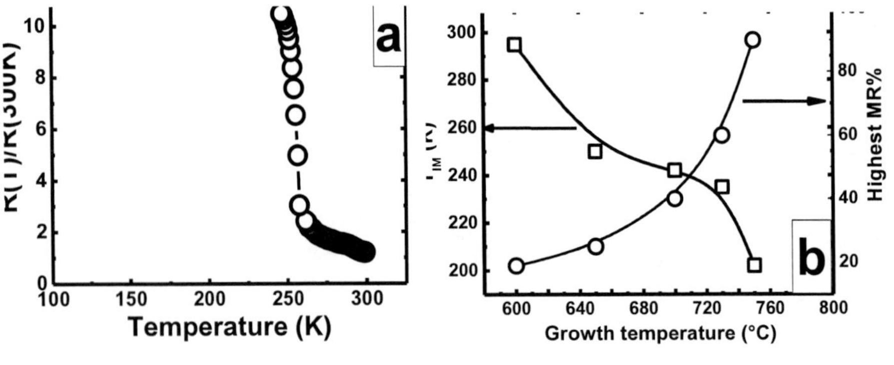

Figure 10. (a)Temperature dependence of the normalized resistance for film grown at 800°C (b) Variation of T_{IM} and highest MR% measured for LPMO films as a function of growth temperature.

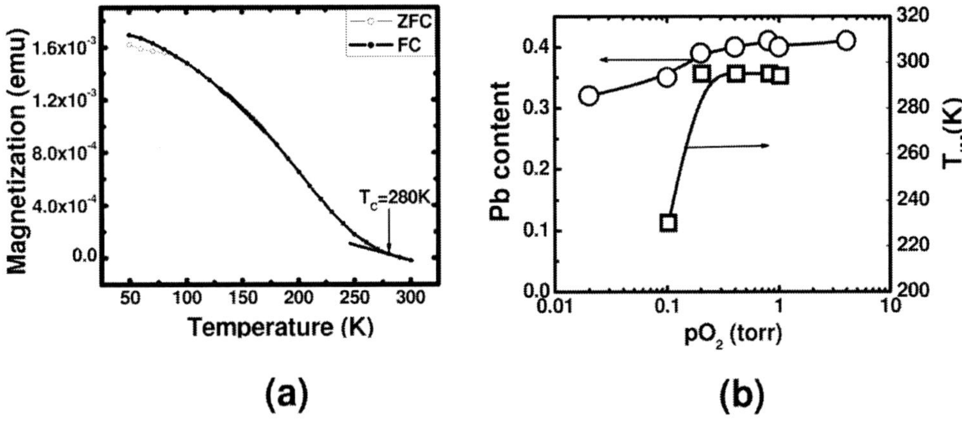

Figure 11. Temperature dependence of magnetization (at 1000 Gauss) recorded for LPMO film grown at 600°C (b) Variation of Pb content and T_{IM} measured for LPMO films as a function of pO_2. The growth temperature was kept constant at 600°C.

Substitution of divalent ion Pb^{2+} for Mn^{3+} in an insulating and antiferromagnetic $LaMnO_3$ leads to hole-doping conductive and ferromagnetic properties. The Pb^{2+} substitution produces Mn^{4+}, which brings itinerant, e_g holes into the system. The hopping of e_g electrons between Mn^{3+} and Mn^{4+} not only induces compound to show metallic conduction below a critical temperature, but also enhances the ferromagnetic interaction. The temperature dependence of magnetization recorded for film grown at 600°C, as shown in Figure 11.(a), confirms this. A paramagnetic-to-ferromagnetic transition at Curie temperature (T_c), coinciding with T_{IM}, suggests double-exchange mechanism for the observed magnetoresistance [5, 15]. As shown in Figure 8, the Pb content in the film decreases with increasing growth temperature leading to neutral vacancies in the lattice. This reduces the number of conduction electrons in the system and therefore explains lowering of T_{IM} with increasing growth temperature.

The monotonic increase in MR%, with growth temperature as shown in Figure 10.(b), is attributed to the reduced Pb content and enhanced grain growth (as revealed by AFM images presented in Figure 6). The Pb loss at the grain boundaries is likely to be higher, which results in weaker grain boundaries. These weak grain boundaries act as scattering centers for the polarized carriers and yield an additional magnetoresistance.

Effect of oxygen pressure: As seen above, the highest T_{IM} and stoichiometric films are obtained at growth temperature of 600°C and pO_2 of 0.2 torr. Thus to investigate the effect of pO_2 on film properties, the substrate temperature was kept constant at 600°C and pO_2 was varied between 0.02 and 4 torr. The observed variation in Pb-content and T_{IM} as a function of pO_2 are presented in Figure 10.(b).

It is seen that the Pb-content is nearly independent of the pO_2, except at very low oxygen pressures. On the other hand, films exhibited insulator-to-metal transition only in a small pO_2 regime of 0.1-1 torr. At low pO_2's, a sharp reduction in T_{IM} is attributed to oxygen deficiency in the films. The low oxygen content would reduce Mn^{4+} ions in to Mn^{3+} leading to decrease in conduction electrons. At high pO_2's, the size of laser-induced plume became much smaller than the target-substrate distance resulting in ultra thin films, which are always found to be insulating [15].

6.2. Growth at Temperatures < 600°C

As seen from the data in the earlier section, films grown at temperatures ≥ 600°C are always single crystalline in nature. Here we present the nature and properties of the films grown at temperatures < 600°C.

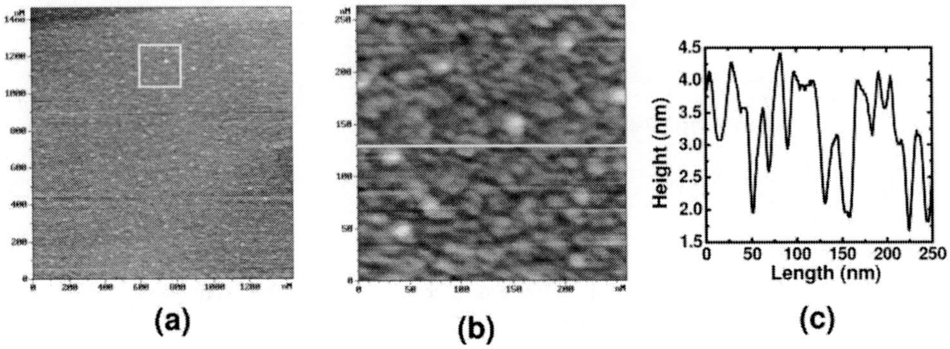

Figure 13. (a) 1400 nm x 1400 nm AFM scan of a NC film, (b) 250nm x 250 nm AFM scan corresponding to the white square drawn in (a), and (c) the height profile across the white line drawn in (b).

Growth at 550°C: A typical 1400 nm x 1400 nm AFM-scan of a LPMO films grown at a substrate temperature of 550°C under an oxygen partial pressure of 0.2 Torr is shown in Figure 13.(a). The film is composed of small-sized grains. In order to measure the size of such grains, a more resolved AFM scan of 250 nm x 250 nm, corresponding to the square drawn in Figure 13.(a), was taken and is presented in Figure 13.(b). The height profile, recorded across a white line drawn in Figure 13(b), is presented in Figure 13.(c). An analysis of the height profile revealed that (i) the average surface roughness of the film is ~ 0.7 nm indicating very smooth surface of the films, and (ii) a uniform distribution of grains with an average size of ~17 nm. These results therefore indicate that the film is nanocrystalline (NC) in nature [16].

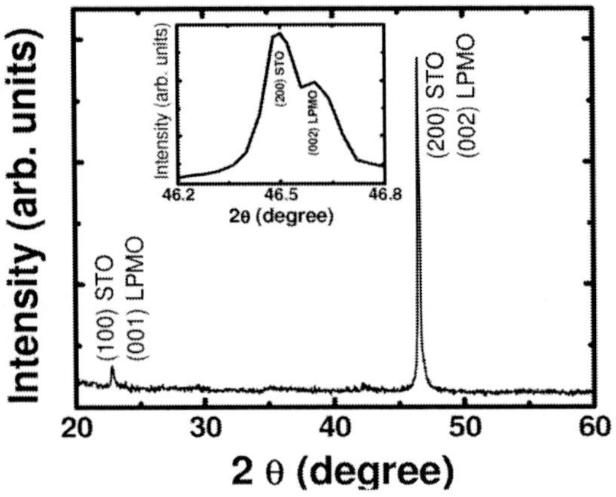

Figure 14. XRD pattern recorded for a NC film. The inset shows the expanded plot in the vicinity of (200) peak of STO.

A typical XRD pattern for the grown NC LPMO film is shown in Figure 14. The presence of only (00*l*) reflections of the LPMO films, which are superimposed on (*h*00) peaks of STO substrate, indicates that all the crystallites or grains have *c*-axis perpendicular to the substrate plane. The overlapping of (002) LPMO and (200) STO peaks is resolved clearly in the expanded plots, as shown in the inset of Figure 14. The *c*-lattice parameter of the NC film (computed from the *d*-value of (002) reflection) was 0.389 nm, which is same as that for SC film grown at 600°C [16]. This suggests that the Pb content in both NC and SC films is close to that in target [15, 16]. The temperature dependence of the normalized resistance (measured in zero-field and in 1 T) for a typical nanocrystalline film is shown in Figure 15.(a). It is seen that the $R(T)$ plot exhibits a sharp insulator-to-metal transition (marked by T_{IM} in the figure) at around 156 K and an application of magnetic field sharply reduces the film resistance. The computed vale of MR%, defined as: $(R_0\text{-}R_H)/R_H$ x100 (R_0 and R_H are the film resistance in zero and 1T magnetic field), is also plotted in Figure 15.(a). It is seen that the magnetoresistance starts increasing from 226 K, exhibits a peak at 132 K and shows a jump at 27 K. The value of magnetorersistance at the peak is ~103%. This behavior is similar to those reported for epitaxial thin films or single crystals, except the fact that the T_{IM} is reduced and the MR% is much enhanced. This result suggests that grain boundaries in NC films are playing an important role [16]. In order to delineate the role of grain boundaries, we have measured the magnetization of nanocrystalline films in zero-field-cooled (ZFC) and field-cooled conditions. For ZFC data, the samples are first cooled to 5 K. Subsequently, a field of 100G is applied and magnetic moment is recorded as a function of slowly increasing temperature. To obtain FC data, the sample is cooled from room temperature under a fixed field of 100G and magnetic moment is measured. The measured ZFC and FC data are plotted in Figure 15.(b).

It is seen that the paramagnetic-to-ferromagnetic transition i.e. Curie temperature (T_c, as marked by arrows in the figure) takes place at ~230 K. In addition, both ZFC and FC magnetization data exhibit an upturn at ~27 K, which has been attributed to the grain boundaries exhibiting a ferromagnetic transition. This suggests that grain boundaries have crystalline structure with suppressed T_c. It may be noted that disordered grain boundaries would have led to a cusp in ZFC data [16].

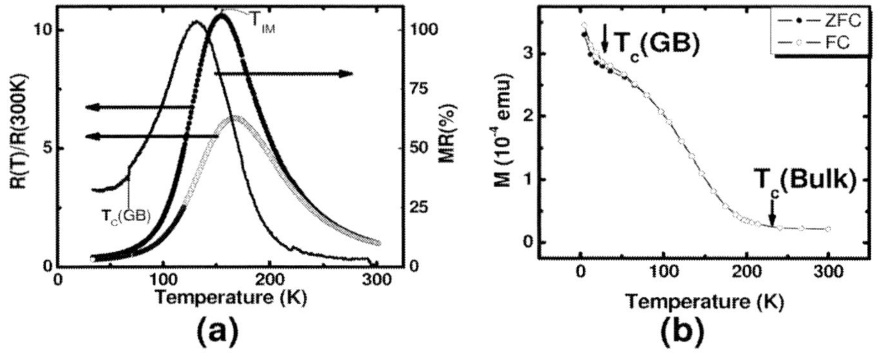

Figure 15. (a) Temperature dependence of normalized resistance (solid circles in zero and open circles in 1T magnetic field) and MR (solid line) recorded for a NC LPMO film. The T_{IM} is the insulator-to-metal transition temperature and T_C (GB) is the temperature at which jump in MR appears (b) Temperature dependence of zero-field-cooled (ZFC) and field-cooled (FC) magnetization recorded for a NC film. T_c and T_c(GB) are the Curie temperature for the grains and grain boundaries, respectively.

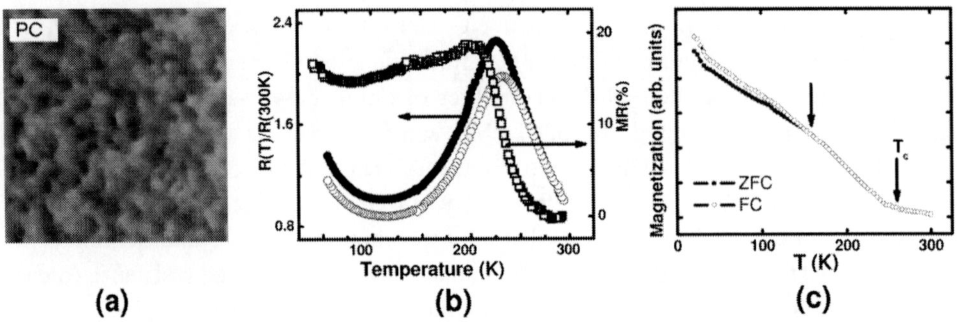

Figure 16. (a) 200nm x 200 nm AFM scan for a LPMO film grown at 500°C. The morphology reveals polycrystalline (PC) nature of the film (b) Temperature dependence of normalized resistance (solid circles in zero and open circles in 1 T magnetic field) and MR (Solid line) recorded for a PC film. (c) Temperature dependence of zero-field-cooled (ZFC) and field-cooled FC magnetization recorded for a PC film.

Some of the features observed here for NC films are quite different from those of epitaxial LPMO films grown at 600°C, despite the fact that both the films have identical chemical composition and *c*-lattice parameters. For NC films, the Curie temperature has reduces to 226 K as compared to 295 K for epitaxial films; while the highest MR% has increased to ~103% from 19%. This indicates that in nanocrystalline films the charge is confined to the individual crystallites or grains and the grain boundaries act as source of spin scattering. Evidence to this comes from the upturns in ZFC and FC magnetization data and jump in MR data at 27K, as shown in Figure 11 and 12, which is indicative of grain boundaries exhibiting paramagnetic to ferromagnetic ordering at ~27 K. This implies that above 27 K the grain boundaries are paramagnetic and therefore acts like source of spin scattering. The observed magnetoresistance of 103% at 1T in the grown NC films is the highest reported value so far as compared to any other CMR films [17].

Growth at 500°C: The growth morphology and properties of LPMO films grown at a substrate temperature of 500^0C under a pO_2 of 0.2 Torr are found to be remarkably different from those of SC and NC films as described earlier. It was found that at 500 ^{0}C the grown films have a polycrystalline (PC) character, which is evident from the AFM image shown in Figure 16.(a). Polycrystallinity of the film was confirmed by XRD measurements.

The temperature dependences of the resistance and magnetoresistance for a PC film is shown in Figure 16.(b). The behavior of MR for PC film is remarkably different than that observed for SC and NC films, whereby the MR shows a peak near metal-to-insulator transition temperature. In the case of PC films, MR increases monotonically with lowering of the temperature, which tends to saturate in the temperature range of 200 K-100 K and again starts increasing when the films enters in the semiconducting state below 120 K [18].

Here it may be noted that DE mechanism alone does not explain the temperature dependence of MR in PC films. The MR of polycrystalline samples at low temperatures and high fields has been attributed to the spin dependent scattering of carriers in the weakly magnetized regions [18]. For a fixed applied field, lowering the temperature, i.e. reducing the thermal fluctuations, would allow more and more spins to align parallel to the field; hence magnetization will increase as shown in Figure 16.(c). This results in a drop in resistance or an increase in MR with decreasing temperature. However well below curie temperature there are two competing phenomena one is the intrinsic MR, which will decrease with the lowering of temperature and which arises from strongly magnetized grains and another is extrinsic MR (arises due to spin dependent scattering, as discussed above), which will increase with the lowering of temperature. These two competing phenomena will gives arises to saturation of MR in the temperature zone of 200K -100K.

Now we discuss an interesting MR aspect of polycrystalline sample at low field. In the polycrystalline samples, MR at low fields and at low temperatures have been attributed to the spin polarized tunneling (SPT) of charge carriers between the magnetized grains through insulating grain boundaries [5, 18]. Within grains, the e_g electrons move easily between pairs of Mn^{+3} and Mn^{+4} ions. However, when these electrons travel across the grains, and when the magnetic moment of the neighboring grains are anti-parallel it leads to a very high resistance at zero fields. When a small field is applied, it aligns the magnetic moment of grains along the field direction, and there is a SPT of charge carrier takes place between the magnetized grains, it leads to a fast drop of the resistance termed as low field magnetoresistance (LFMR) [5]. With further increase in the field, the resistance drops linearly with increasing the field due to enhancement in the overall magnetization (i.e. an intrinsic MR). This fact is supported by the field dependence of MR for SC, NC and PC films (as shown in Figure) at 50K. It can be seen that for SC and NC films the MR increases linearly with the field, while for PC there is a sharp increase of MR at low field followed by the linear increase of MR with field (as seen more clearly in Figure 17.(b)). Therefore we conclude that LFMR is a characteristics feature of polycrystalline CMR samples having insulating grain boundaries [19].We may like to add that although the NC films are granular in nature, the absence of LFMR for NC films indicates in these films the grain boundaries are not insulating. Figure 17.(c) shows the temperature dependence of LFMR. It cab be seen that LFMR increases with the lowering of temperature. In the PC films the insulating nature of grain boundaries is indicated by temperature dependence of resistance in low temperature region (< 120 K) as shown in Figure 16. The resistivity in this region may be understood in terms of coulomb blockade model, where the metallic grains are isolated via insulating grain boundaries and carriers must overcome the charging energy barrier. It terms of this model the resistivity is given by the expression $\rho\,(\mathrm{T}) = \exp\,(\Delta/T)^{1/2}$ [18]. In Figure 18, we have plotted $\ln(\rho)$ versus $T^{1/2}$. This plot exhibits a linear dependence upto 100 K with $\Delta \sim 8.9$ K. Therefore semiconducting state at low temperature ($T < 120$ K) in the film has its origin in a small Coulomb barrier ($\Delta \sim 8.9$ K) of electrostatic origin indicating the insulating nature of grain boundaries.

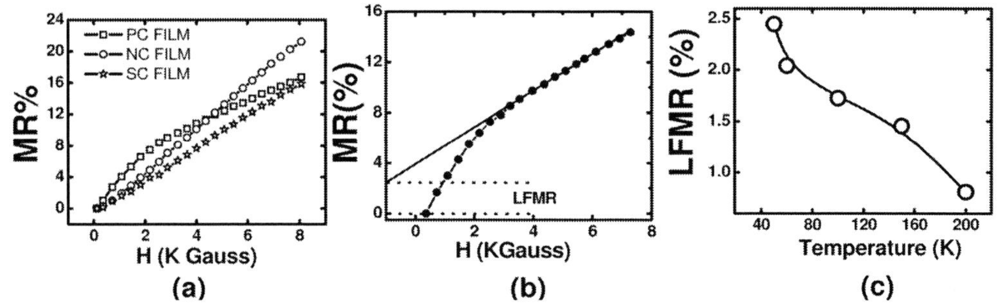

Figures 17. (a) Magnetic field dependence of normalized resistance of MR for the SC, NC and PC films at 50 K. (b) For clarity the separate plot of Low field MR (LFMR) is shown for PC at 50 K. (b) Temperature dependence of LFMR shows that it increases at low temperatures as expected for insulating grain boundaries.

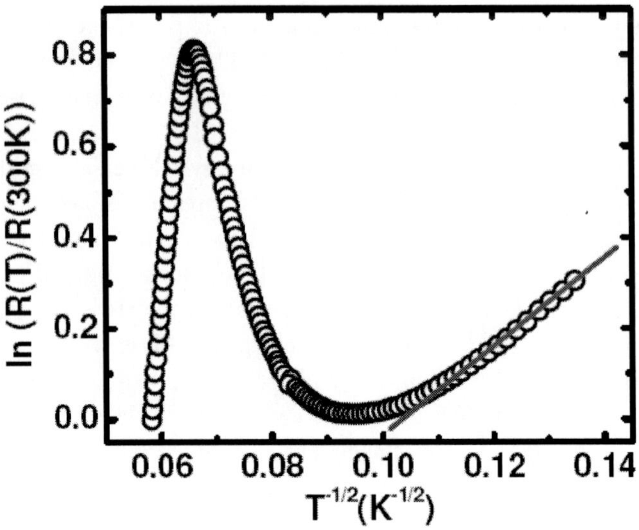

Figures 18. Temperature dependence of normalized resistance vs $T^{-1/2}$ for PC films. Solid line shows linear behavior of the curve at low temperatures.

7. An Insight of the Magnetic Nature of LPMO Grain Boundaries as Revealed by Ferromagnetic Resonance (FMR) Technique

One of the most powerful methods for exploring magnetic homogeneities in ferromagnetic materials is via magnetic field dependent microwave power absorption studies, which results in a ferromagnetic resonance (FMR) phenomenon. In the presence of an external magnetic field (H), the magnetization (M) of a ferromagnetic precesses according to the Landau-Lifschitz equation [20]

$$\frac{dM}{dt} = -\gamma \vec{M} \times \vec{H}_{eff} - \frac{\gamma \alpha}{M_s} \vec{M} \times (\vec{M} \times \vec{H}_{eff}) \tag{3}$$

where $\gamma = (g\mu_B / \hbar)$ the gyromagnetic ratio (g being the Lande factor), H_{eff} the effective local field, M_s the saturation magnetization, and α is damping constant. This causes a precession of the moment about the direction of the field at a frequency ω. An ac magnetic field H_{ac} at this resonant frequency applied perpendicular to H will couple to a uniform precession of M about the direction of H, resulting in absorption of energy from the ac field. FMR is characterized by two parameters: the resonance field (Hr) and the linewidth (Γ). Measurement of these two parameters reveals vital information on the nature of magnetic inhomogeneities, which otherwise cannot be obtained by any other technique [20]. In our study the field dependent microwave absorption measurements were performed on a Bruker ESP-300 spectrometer operating at 9.7 GHz. The FMR spectra were recorded using conventional modulation and lock-in detection techniques over a temperature range of 77-300 K.

7.1. Studies on Nanocrystalline Thin Films:

As discussed in earlier sections the magnetic state of the nanocrystalline (NC) LPMO films is expected to be very complex on following two accounts [20].

(i) Within the bulk of the nanocrystallites magnetic inhomogeneities may arise due to chemical nonuniformity. Therefore, one can have antiferromagnetic, metallic or insulating ferromagnetic, charge and orbital ordered regions within the crystallites owing to the strong coupling of spin, charge and orbitals in manganites [5].
(ii) The magnetic state of the surfaces of the nanocrystallites is expected to be very different from that of the bulk because spins at the surface are subjected to relatively weak magnetic interactions due to lack of nearest neighbors [20].

It is, therefore, natural to investigate the nature of bulk and surface magnetic inhomogeneities in the technologically important nanocrystalline LPMO thin films. We have carried out magnetic field dependent microwave absorption studies on nanocrystalline LPMO films to probe the bulk as well as surface magnetic inhomogeneities of the nanocrystallites. Data for nanocrystalline LPMO films were taken with applied field (H) parallel to the film plane.

Figure 19.(a) shows FMR spectra (field dependence of the microwave power absorption derivative (dP/dH)) for a nanocrystalline LPMO film recorded at different temperatures. One can see that the evolution of FMR peak takes place at 230 K, which happens to be the Curie temperature of the film. For temperatures >170K, only a single FMR line of Lorentzian shape is observed. However, for temperatures ≤170K, the an additional features appear in the spectra i.e. broadening of FMR line, which essentially arises due to appearance of another FMR line at higher resonance field (shown by an arrow in Figure 19.(a)).

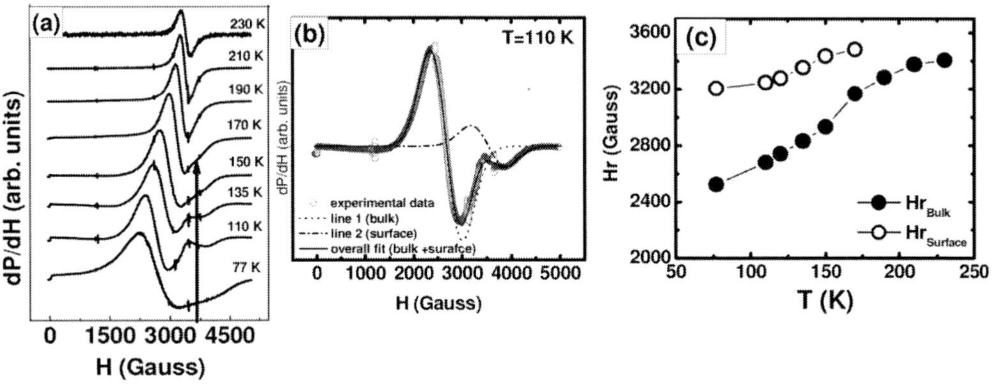

Figure 19. (a) FMR spectra of a nanocrystalline LPMO thin film measured at different temperatures. Note the appearance of second FMR line (indicated by an arrow) at temperatures ≤170K. The superimposed signals are due to the magnetic impurities present in the silver paste, which was applied on the backside of the substrates for sticking to the heater during film deposition. (b) Fitting of data at 110 K using Lorentzian shapes. The full line is a fit with a sum of two Lorentzians represented by dotted (line 1) and dash-dotted (line 2) lines. (c) Variation of bulk (solid circles) and surface (open circles) resonance fields with temperature.

To see these two effects more clearly, the spectrum recorded at 110 K is reproduced in Figure 19.(b). It is seen that the spectrum can be well fitted by a sum of two Lorentzian lines with different resonance fields: one occurring at lower resonance field, and another at slightly higher field. Since the first FMR signal appears right from 230 K, we attribute it to arise from the bulk of the nanocrystallites. The second FMR line appears only at temperatures ≤170 K and is attributed to surface magnetiç ordering. The resonance field (H_r) and linewidth (Γ) of the bulk and surface FMR lines have been deduced by fitting the data with Lorentzian function, and their temperature dependences are plotted in Figs. 19 (c) and 20(a), respectively. From Figure 19.(c) it is evident that with decreasing temperature Hr_{Bulk} decreases sharply in comparison to the $Hr_{Surface}$. On the other hand, the temperature dependence of Γ_{Bulk} and $\Gamma_{Surface}$ are functionally different. It may be noted that the temperature dependence of linewidth provides useful information on the magnetic inhomogeneities in the film, and this is elaborated below. As can be seen from Figure 20.(a), the Γ_{Bulk} initially increases with decreasing temperature and exhibits a local maximum near the Curie temperature. However, at still lower temperatures the Γ_{Bulk} increases again. Similar kind of behavior for the linewidth of crystalline CMR films have been observed in literature, and reported to be a characteristic feature of manganite samples with magnetoresisitve effect [20]. It may be noted that for a homogeneous ferromagnet the Γ remains constant at a low value (typically <100 Gauss) for temperatures <0.8Tc [20]. The various mechanisms responsible for the broadening of the FMR linewidth in CMR films include (i) spin-lattice relaxation process via the lattice orbitals of band electrons [20], (ii) Joule losses caused by eddy currents as the thickness of skin layer is likely to be greater than the film thickness, (ii) increase in film conductivity (σ) at low temperatures as linewidth is proportional to σ[20], and (iv) chemical inhomogeneities leading to a spread in Tc; the linewidth is then the sum of intrinsic part and broadening due to spread in *T*c [20]. The anomalous broadening in Γ_{Bulk} near Tc is reported to be due to presence of antiferromagnetic (AFM) clusters [20]. In our nanocrystalline LPMO films, the temperature dependence of Γ_{Bulk} therefore indicates presence of both AFM clusters and chemical inhomogeneities within the bulk of nanocrystallites.

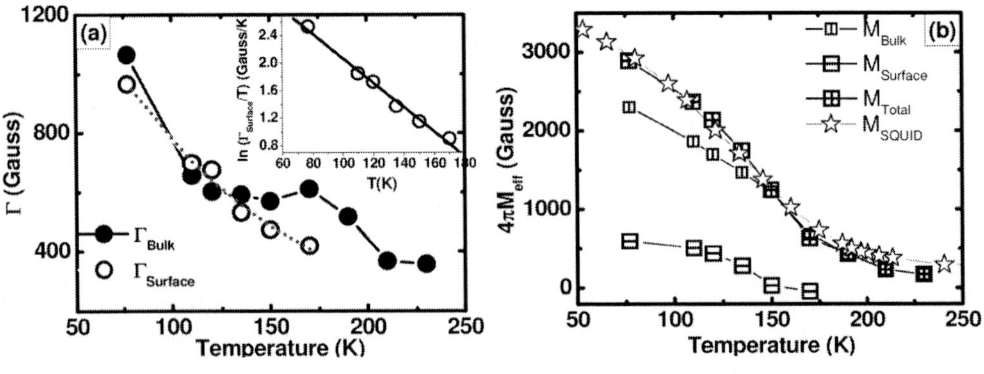

Figure 20. (a) Temperature dependence of bulk (solid circles) and surface (open circles) linewidths. The dotted line is fit of equation 2 to data of surface line width (($\Gamma_{Surface}$). The inset shows a plot of ln($\square_{Surface}$) vs T, and its implication is discussed in the text.(b) The variation of the magnetization of bulk and surface components computed using equation 4. The magnetization measured from SQUID magnetometer is also shown.

On the other hand, the temperature dependence of $\Gamma_{Surface}$ is completely different from that of Γ_{Bulk} (see Figure 20.(a)). As shown by dotted line, the data of $\Gamma_{Surface}$ vs T is satisfactorily fitted by the expression of a spin glass ferromagnet [20], that is,

$$\Gamma_{Surface} = \Gamma_1 T \exp(-T/T_0) \quad (4)$$

where Γ_1 and T_0 are the fitting parameter. These two parameters were determined by least square fitting to the linear equation

$$\ln(\Gamma_{Surface}/T) = \ln\Gamma_1 - T/T_0 \quad (5)$$

The goodness of the fit can be judged from the $\ln(\Gamma_{Surface}/T)$ vs T plot shown in the inset of the Figure 20.(a). The fitted parameters Γ_1 and T_0 were 46±1 Gauss and 56±2 K, respectively. Therefore, the temperature dependence of the $\Gamma_{Surface}$ reveals that surfaces of LPMO nanocrystallites have a magnetic spin glass character.

The solution of equation 3, for the microwave power absorbed in the parallel geometry show that the conditions for resonance in the narrow linewidths is given by the equation [21]

$$\frac{\omega}{\gamma} = \left[H_r(H_r + 4\pi M_{eff})\right]^{1/2} \quad (6)$$

where ω is the microwave frequency, H_r the resonance field in the parallel to the substrate plane geometry, and M_{eff} the effective magnetization. Using equation 6 and data of resonance fields from Figure 19.(c), the values of $4\pi M_{eff}$ are calculated for both bulk and surface components. The resultant temperature dependences of $4\pi M_{eff}$ (Bulk), $4\pi M_{eff}$(Surface) and $4\pi M_{eff}$ (Total = Bulk + Surface) have been plotted in Figure 20.(b). For comparison the data of $4\pi M_{eff}$ measured from SQUID magnetometer is also plotted in the figure. It is clear that the value of $4\pi M_{eff}$ (Total) derived from FMR measurements matched well with that measured from SQUID magnetometer.

7.2. Studies on Polycrystalline Thin Films:

We now present results of microwave study to understand the nature of grain boundaries of PC films. Figure 21.(a) shows the FMR spectra recorded at different temperatures. As seen in Figure 21.(b), these FMR spectra may be resolved into two gaussian peaks at all temperatures. The observation of gaussian peaks is attributed to some inhomogeneities in the sample [18, 20] as is expected for a polycrystalline material. Based on temperature dependence of H_r, the two peaks are attributed to intra-grain and surface (grain boundary) signals. The temperature dependences of resonance field (H_r) and line width (Γ) of both components of microwave absorption peak is plotted in the Figure 22.(a). Component attributed to bulk grains has been identified on the basis of decreasing H_r and increasing Γ at low temperatures as expected for ferromagnetic grains [20]. The second component attributed to surface shows a rapid increase in H_r as temperatures reduce below 160 K. An increase in resonance field as observed here has been earlier attributed to AFM regions in CMR materials [20]. The results indicate that grain boundary region undergoes a transition to AFM state at 160 K. We may note that in terms of double exchange mechanism, AFM state is insulating and therefore this is in agreement with observation of insulating grain boundaries leading to semiconducting nature at low temperatures as discussed earlier. The study of the nature of grain boundaries also explains contrasting nature of MR in nanocrystalline films [18, 20] and polycrystalline films as discussed earlier. In case of nanocrystalline films MR is very large at $T \sim T_{IM}$, but reduces significantly at low temperatures similar to the case for single crystals [15, 16]. For polycrystalline films, MR is not so large at $T \sim T_{IM}$, but it continues to be significant at low temperatures. In case of nano-crystalline films, the grain boundaries are non-magnetic near T_{IM} but have increasing magnetization and therefore conductivity at low temperatures leading to reduced effectiveness of tunnel barriers and thereby reduced MR. For polycrystalline films of this study, the grain boundaries are easily magnetized at temperatures near T_{IM} but are in AFM insulating state at low temperatures leading to high MR even well below *T*c.

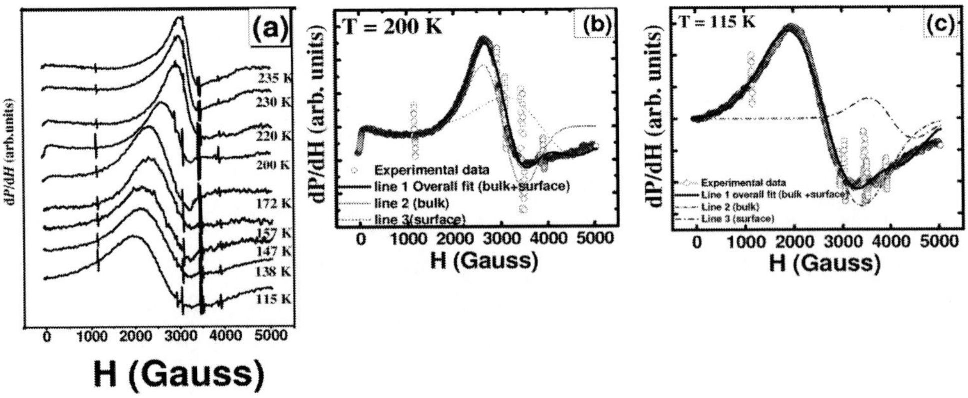

Figure 21. (a) FMR spectra of a polycrystalline film taken at different temperatures and (b) fitting of FMR spectra at $T = 200$ K using Gaussisan peaks (c) fitting of FMR spectra at $T = 115$ K using Gaussisan peaks. The full line is a fit with a sum of two Gaussians represented by dotted (line 1) and dash-dotted (line 2) lines.

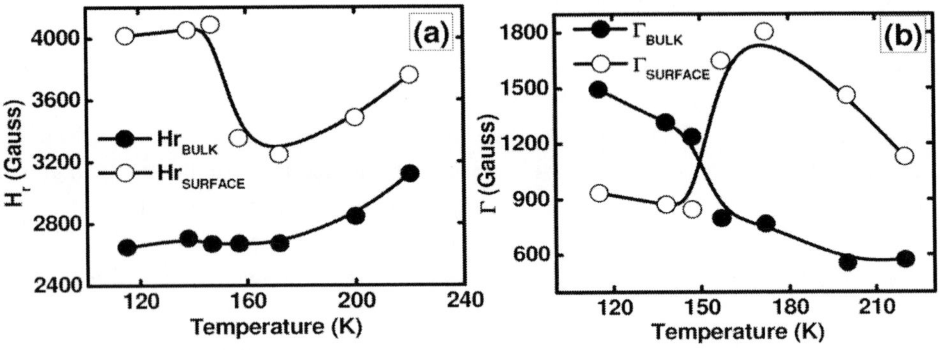

Figure 22. Temperature dependence of (a) resonance field H_r and (b) line width Γ for PC film.

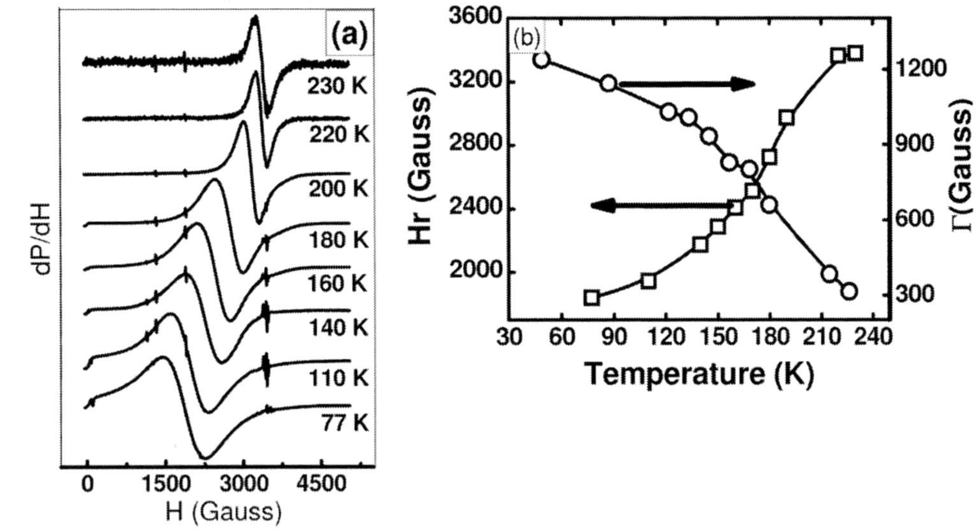

Figure 23. (a) FMR spectra of a single crystalline (SC) film taken at different temperatures (b) Temperature dependence of resonance field H_r and line width Γ.

Our observation of AFM nature of grain boundaries at low temperatures is in agreement with theoretical model of Calderon et al [21]. In this model it has been found that due to absence of cubic symmetry on outermost layer of grains, double exchange mechanism is weakened. This leads to modification in charge state of Mn ions and AFM ordering of spins.

7.3. Studies on Single-Crystalline Thin Films:

The FMR spectra for SC films appear only below 230 K, i.e. below the curie temperature of the films. It can be seen from the Figure 23.(a), the in the entire range FMR spectra is very symmetric and can be fitted with a single Lorenzian function, indicating the presence of signal from bulk only [22]. As seen from Figure 23.(b), the temperature dependence of Γ_{SC} shows a tendency towards the saturation with lowering of temperature. The initial temperature dependent broadening of Γ_{SC} is largely due to the ferromagnetic zone of slightly different *Tc* in the SC film. It may be noted that the resonance field *H*r decreases systematically with the lowering of temperature. It reflects that enhancement of the internal field (ferromagnetic ordering) in the sample; as a result less external field will be required to match the resonance conditions [22].

8. Electroresistance Behaviour of LPMO Thin Films

In the last few years, it has been observed that in CMR materials the resistivity also reduces sharply with an increase in electric current – an effect known as giant electroresistance (ER) [23-29]. In the literature ER has been defined in two ways: absolute ER and relative ER. Absolute ER can be calculated from current (I) -voltage (V) characteristics and defined as $[-\{(dV/dI)(I)-(dV/dI)(0))/(dV/dI)(0)\}\text{x}100\%]$ and the relative ER (%) is defined as 100*(R_{I1} – R_{I2})/ R_{I2}, where R_{I1} and R_{I2} are the resistance of sample at different current values $I1$ and $I2$ respectively, and is highest at the temperature of the resistance peak (T_{IM}) at which metal insulator transition takes place. The strong dependence of resistance on a small current is of great interest for basic physics as well as technological applications of CMR materials. In order to understand the origin of ER, two different explanations have been put forward. (i) Here, the phase separation together with concomitant percolation behavior of CMR materials is taken into consideration. An applied current bias induces a transition from the electrical insulating charge-ordered (CO) state to a ferromagnetic metallic state; even a large magnetic field (~ 40 T) has no effect on the charge ordered state (ii) In this case, ER has been found to correlate with current induced magnetic field. For example, in $La_{0.82}Ca_{0.18}MnO_3$ single crystals an electric current of 0.3 mA is shown equivalent to 1.5 Tesla below Tc [30]. This mechanism results in a stunning correlation between the temperature dependences of ER and MR [31].

To understand whether ER is an intrinsic or extrinsic property of the CMR materials, we recorded the temperature dependence of normalized electrical resistance at different current value for NC, PC and SC films are shown in Figure 24. In the inset of each figure the temperature dependence of relative ER (I1=1μA; I2=100μA) is plotted. From Figure 24, we can summaries the following:

(i) The ER for NC and PC film increases with the lowering of temperature < 230 K, i.e. below the temperature where ferromagnetic correlations start. In addition to this resistance drops with increasing current up to certain value .For example in NC films after 100 μA there is no reduction in the resistance. Similarly for PC films for current values > 50 μA, there is no further decrease in the resistance. It indicates after a particular current value the ER saturates [22, 32].

(ii) The ER for single-crystalline film is clearly absent in the entire measured temperature range. It indicates that films which are magnetically homogenous are not suitable for observing the ER [22].

The above results clearly show that differences in ER behavior of nanocrystalline, polycrystalline and single-crystalline films are possibly due to the grain-boundaries present in the films. The following mechanism is purposed to explain the mechanism of ER for PC and NC films. Below curie temperature (T_c) the grains are ferromagnetic metallic (FMM), while the grain boundaries are either anti-ferromagnetic insulating (AFI) or spin glassy in nature.The FMM phase is half metallic; the electrons injected from FMM into AFI grain boundaries are spin polarized. The injected spin polarized electron will force anti-ferromagnetically ordered / spin glassy phase to become ferromagnetic phase (due to strong Hunds coupling) in the grain boundaries. The spin polarized carriers maintains its spin up to a certain depth that may be called as spin flip scattering length. On increasing the magnitude a longer length of the grain boundaries (GB) regions may be converted to the metallic. The enhancement of the feroomagnetic phase will lead to a decrease in the resistivity. Once the material upto the scattering length (around the ferromagnetic metallic phase) is converted into the metallic phase by injection spin polarized carriers, the rate of increase of ER with a current will reduce and ER will show a saturation behavior [32]. To confirm this proposed mechanism we recorded the current dependence of absolute ER (as defined earlier) for PC films at different temperatures, the results are shown in Figure 25.(a). The absolute ER is obtained by measuring the *I-V* characteristics of the films and then obtained differential resistance *dV/dI* from these data as a function of current. The salient features of Figure 25.(a) are, ER increases with the lowering of temperature, however there is no ER observed at 220K (i.e close to curie temperature). The ER shows a sharp increase at lower current value (<50 μA) followed by a nearly constant ER at higher current. It confirms our earlier data (temperature dependence of resistance at different current values) that higher current is ineffective in further reducing the resistance. We have also measured the field dependence of MR and coulomb barrier for PC films at different current values and data is plotted in Figure 25 (b) and Figure 25.(c) respectively. As seen from Figure 25.(b) and Figure 25.(c), we observed a decrease of low field MR (LFMR) and coulomb barrier with the higher current. As discussed earlier the LFMR and coulomb barrier is result of insulating grain boundaries present in the PC film. By conversion of insulating grain boundaries into metallic one, we expect the decrease of LFMR and coulomb barrier height [32].

The absence of ER for SC films also support the above proposed hypothesis, as there is no scope of ER due to homogenous magnetic nature of the samples. The above study confirms that ER is an extrinsic feature of the CMR materials and arises due to magnetic inhomogenity in the sample.

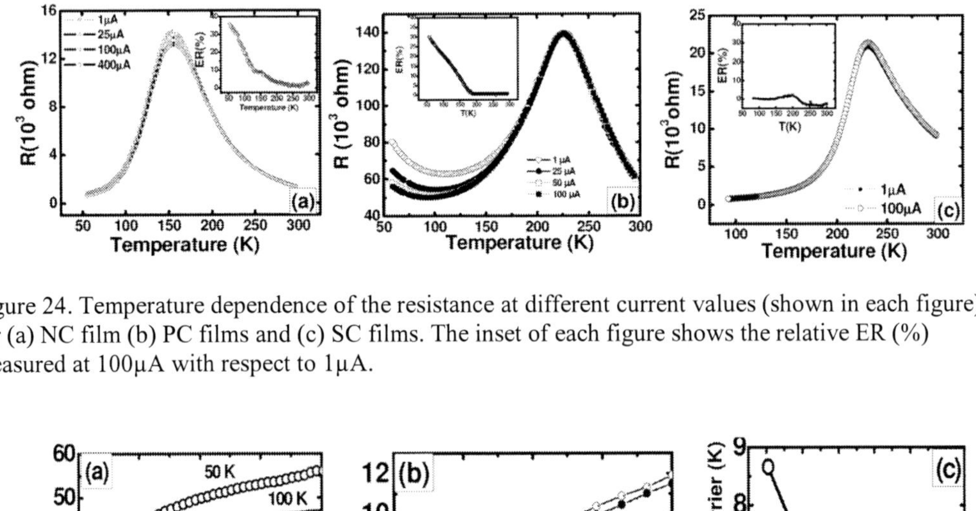

Figure 24. Temperature dependence of the resistance at different current values (shown in each figure) for (a) NC film (b) PC films and (c) SC films. The inset of each figure shows the relative ER (%) measured at 100μA with respect to 1μA.

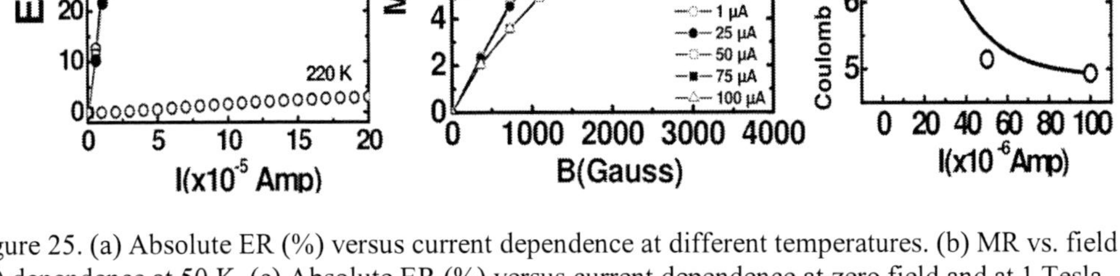

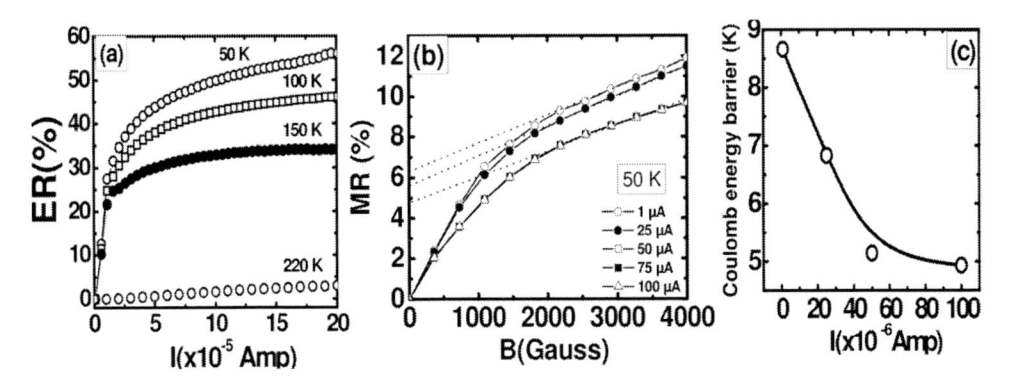

Figure 25. (a) Absolute ER (%) versus current dependence at different temperatures. (b) MR vs. field (B) dependence at 50 K. (c) Absolute ER (%) versus current dependence at zero field and at 1 Tesla field. (c) Coulomb barrier energy as a function of injected current.

8. Charge Transport Charcteristics of LPMO Grain Boundaries

So far all magnetotransport investigations made above on PC and NC samples is the manifestations of average feature of large numbers of the grain boundaries present in the sample. However to understand the transport across single grain boundary in the CMR material, we have prepared the SC films on a (100) oriented $SrTiO_3$ (STO) bi-crystal substrates with 36.8^o grain boundary. As grown films were patterned using conventional photolithography and chemical etching, to yield microbridges, of width ~10 μm and length ~ 500 μm across the grain boundary. A schematic diagram of a typical bridge is shown in the inset of Figure 26. (bottom right corner). After patterning, the films were annealed at a temperature of 700°C in oxygen atmosphere for 1 h for oxygenation. Silver pads for voltage measurement across the GB (bridge-A) and across a similar bridge without GB (bridge-B) were deposited by vacuum evaporation technique as shown in Figure 1. Simultaneous measurement of voltages across bridges A and B was carried out at different values of current to determine *I-V* characteristics. This structure of CMR films separated by a single grain boundaries is also known as magnetic tunnel junctions.Several investigations on transport characteristics of magnetic tunnel junctions (MTJ) comprising of two ferromagnetic metal (FM) electrodes separated by a thin insulating barrier have been reported in the literature. These studies are important due to both the underlying physics and potential device applications. Charge transport through these junctions depends on the relative spin orientation of FM electrodes. A magnetic field applied to such a junction leads to change in relative spin orientation of the electrodes and thereby resistance of the junction. Effect of magnetic field on ideal ferromagnetic-insulator-ferromagneic (FIF) junctions has been described by Julliere model [33]. In terms of this model the tunneling magnetoresistance (TMR), $\Delta R/R = (R_{ap}-R_p)/R_p$ is given by $2PP'/(1-PP')$ where R_p and R_{ap} are resistances with spins of two FM electrodes parallel and antiparallel respectively and P and P' are conduction electron spin polarizations of two electrodes. The higher value of spin polarization in CMR materials is expected to yield better TMR in terms of Julliere model. Experimental investigations on bi-crystal grain boundary junctions have shown that the TMR values can be up to 300 % at 4.2 K [34], whereas, in tri-layer junctions made of $La_{2/3}Sr_{1/3}MnO_3$(LSMO)/ $SrTiO_3/La_{2/3}Sr_{1/3}MnO_3$, a TMR ratio of 1800 % at 4.2 K was obtained corresponding to an electron spin polarization of 95 % [35]. The conduction mechanism through these tunnel junctions has been studied by measurement of *I-V* characteristic [36] that are seen to be non linear. In some of these measurements, it was observed that the inelastic tunneling via localised states is a dominant mechanism that leads to an apparent polarization loss at the interface [37]. In other studies non-linear *I-V* characteristics have been explained using Simmons model [38] for elastic tunneling across an insulating barrier [39]. Paranjape et al. [40] have used multistep tunneling theory proposed by Glazman and Matveev (GM) [41]. In this study, zero bias term is attributed to elastic tunneling and single impurity tunneling. Only elastic tunneling is expected to yield magnetoresistance. Temperature and magnetic field dependence of zero bias term as well as two and three step tunneling processes has been obtained by fitting of *I-V* characteristics to GM model. Magnetic field is reported to enhance the zero bias term as well as affect other terms due to multi step tunneling. They have also reported a small contribution of disordered region across the grain boundary. It is seen that in most of the studies reported in the literature, *I-V* characteristics of GB have been fitted to GM or Simmons theory or combination of both. In GM theory, elastic tunneling term gives linear *I-V* characteristics and is combined with tunneling via single impurity scattering, but in Simmons model elastic tunneling also has non-linear behavior which is difficult to distinguish from inelastic tunneling characteristics by fitting of data particularly in presence of large inelastic tunneling contribution. It may also be noted that GM theory has been developed to describe non-linear characteristics for multistep tunneling in disordered solids and may not well describe elastic tunneling across a barrier. Therefore there is a need for further investigations to understand the transport characteristics of grain boundaries. Before going to the experimental details, here is a summary of the Simmons theory of charge carrier tunneling across a junction.

Simmons Theory of Charge Carrier Tunneling across a Junction:

Simmons [42] proposed a theory for elastic tunneling through a junction formed by two metal electrodes with a thin insulating barrier between them. For a rectangular barrier including the image forces, the current density *J* flowing through a barrier can be expressed in terms of the corresponding voltage drop *V* using the following relations.

(a) For very small voltages ($V \sim 0$)

$$J = \left(3.16\times10^{10}/\Delta S\right)\phi_L^{1/2}\exp(-1.025\Delta S\phi_L^{1/2})V, \tag{8}$$

where

$$\phi_L = \phi_0 - [5.75/(K\Delta S)]\ln\left[\frac{S_2(t-S_1)}{S_1(t-S_2)}\right]$$

$\Delta S = S_2 - S_1$, $S_1 = 6/(K\phi_0)$, and $S_2 = t - 6/(K\phi_0)$

(b) For both intermediate ($0 < V < \phi_0/e$) and high voltages ($V > \phi_0/e$),

$$J = \left[6.2\times10^{10}/(\Delta S)^2\right]\{\phi_L\exp(-1.025\Delta S\phi_L^{1/2}) - (\phi_L+V)\exp[-1.025\Delta S(\phi_L+V)^{1/2}]\} \tag{9}$$

where, $\phi_L = \phi_0 - [V/2t](S_1+S_2) - [5.75/(K\Delta S)]\ln\left[\frac{S_2(t-S_1)}{S_1(t-S_2)}\right]$

$\Delta S = S_2 - S_1$,

$S_1 = 6/(K\phi_0)$, and $S_2 = t[1-46/(3K\phi_0 t + 20 - 2VKt)] + 6/K\phi_0$, for $V < \phi_0/e$

$S_1 = 6/(K\phi_0)$, and $S_2 = (\phi_0 Kt - 28)/KV$, for $V > \phi_0/e$

Here J is current density expressed in units of A/cm^2, K is dielectric constant, φ_0 is barrier height in eV and t is barrier thickness in Å. To describe the non-linear *I-V* characteristics data of grain boundary junctions, some authors [43-45] have used the low voltage expansion, which can be written as $J = J_1 (V + \beta V^3)$, where β and J_1 are constants [42].

Glazman and Mateev (GM) Theory of Charge Carrier Tunneling across a Junction

Glazman and Matveev (GM) theory [41] has been used in some of the studies [8,18] to describe inelastic tunneling contribution to current through the GB. According to this theory, (for $eV >> k_B T$), the *I-V* characteristics are given by

$$J = J_0 V + J_{7/3} V^{7/3} + J_{7/2} V^{7/2} + \dots\dots\dots \quad (10)$$

Voltage dependence of conductance (G) for inelastic tunneling may be written as:

$$G = G_0 + G_{4/3} V^{4/3} + G_{5/2} V^{5/2} + \dots\dots\dots \quad (11)$$

where J_0, $J_{7/3}$ and $J_{7/2}$ in Eqn.(10) and G_0 $G_{4/3}$ $G_{5/2}$ in Eqn.(11) depend on the radius of the localized states, density and thickness of the barrier. The first three terms in these expressions show tunneling via one, two and three impurity states respectively. Elastic tunneling contribution is similar to that via one impurity state and is included in the first term.

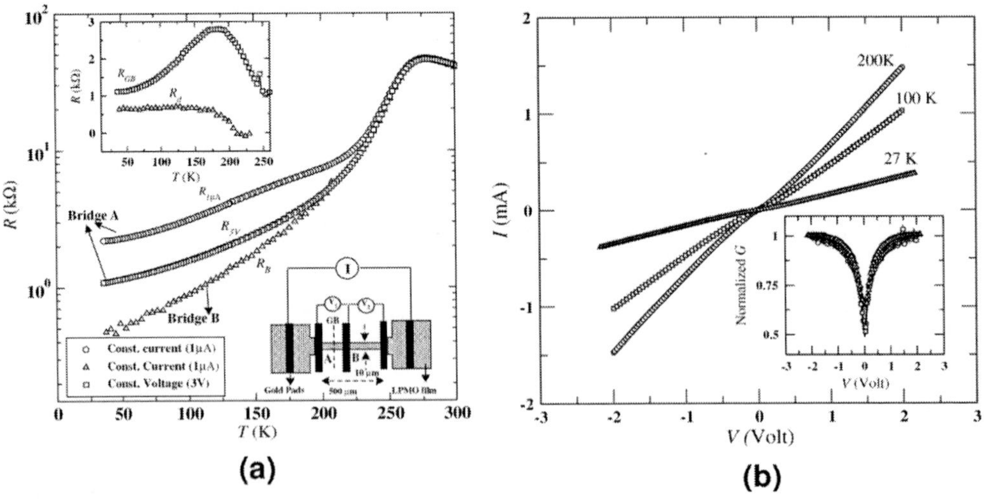

Figure 26. (a) The temperature dependence of resistances, $R_{1\mu A}$ and R_{3V} of bridge A (with grain boundary) measured at applied current of 1 μA and voltage of 3 Volts respectively. Curve R_B gives resistance of bridge B measured at 1 μA. Lower inset shows schematic diagram of the micro-bridge and the contact pads. Upper inset shows temperature dependence of resistances of GB (R_{GB}) and disordered region (R_d) around the GB.(b) *I-V* characteristics of bridge A measured at temperatures of 27, 100 and 200 K. Inset shows the dependence of normalized conductance, $G = dI/dV$ on the biasing voltage for the three temperatures.

Temperature dependence of resistance of bridge A ($R_{1\mu A}$) and bridge B (R_B) measured at fixed current of 1 μA (corresponding to very low bias voltage) is shown in Figure 26(a). Resistance of bridge A at a fixed voltage of 3 volts (R_{3V}) is also shown in this figure. Resistance of bridge A may be written as $R = R_{bulk}+R_{GBR}$ where R_{bulk} and R_{GBR} are the resistances of bulk film and GB region respectively. Resistance of bridge B (R_B) is used to determine the bulk part of bridge A resistance.

Bridge B showed linear *I-V* characteristics (as demonstrated by absence of ER in SC films) and its resistance was found to be independent of current as also reported earlier [22]. The resistance of bridge B shown in Figure 26(a) has been multiplied with a constant k nearly equal to 1 (to take into account a small difference in lengths of bridges A and B) such that resistance of both bridges is equal above metal to insulator transition temperature of 275 K. With the assumption of zero GB resistances at high temperatures, as reported in earlier studies [46], this curve gives temperature dependence of bridge B bulk resistance (R_{bulk}). Using the evidence from *I-V* characteristics (given below) we model the GB region as having two contributions i.e. $R_{GBR} = R_{GB}+R_d$ where R_{GB} is resistance due to tunneling barrier at the GB and R_d is an additional resistance of a small disordered region near the GB. As will be discussed later, resistance of GB region at 3V is solely due to disordered region whereas at low currents both contributions to GB resistance exist (i.e. $R_d = R_{3V}-R_B$). Tunneling barrier resistance at low currents may be determined as $R_{GB} = R_{1\mu A}-R_{3V}$. Temperature dependences of barrier resistance R_{GB} (at low currents) and disordered region resistance R_d is shown in upper inset of Figure 1. It is seen that R_{GB} has a peak at temperature of 175 K (i.e. less than metal to insulator transition temperature of bulk film) as also reported in earlier studies [40] while R_d increases at low temperatures and saturates at 175 K.

I-V characteristics of bridge A measured at temperatures of 27, 100 and 200 K are shown in Figure 26.(b). Voltage dependence of the conductance, G (= dV/dI) at different temperatures, obtained by differentiation of the *I-V* curves, is shown in the inset of Figure 2 and it is clear from this figure that G saturates at some constant value for voltages greater than ~ 1.5 volts. Assuming that the bridge resistance is series combination of tunneling barrier resistance R_{GB} and bulk film resistance R_{bulk}, the conductance of bridge may be obtained from relation ($1/G = 1/G_{GB} + 1/G_{bulk}$).

As the barrier conductance (G_{GB}) continuously increases with voltages, independent of the model (Eqs 8-11), and G_{bulk} is constant, the saturation conductance is expected to be equal to bulk film conductance and should be inverse of resistance of bridge B at same temperature.

However in actual case, saturation conductance was seen to be much less than conductance of bridge B. This was found to be the case for several GB junctions studied by us. This difference in conductance corresponds to a resistance in barrier region that does not reduce with voltage and may be modeled as an additional resistance in series with the barrier.

Physically such a resistance could arise due to distortion of lattice structure or disordered film near the barrier that has higher resistance than bulk of the film. It was also found that the saturation conductance was nearly same as that measured at 3 volts. Therefore the difference of resistances of bridge A (at 3 volts) and bridge B is attributed to disordered region resistance in Figure 26.(a). Possibility of a disordered region around the barrier has been considered in earlier studies [40].

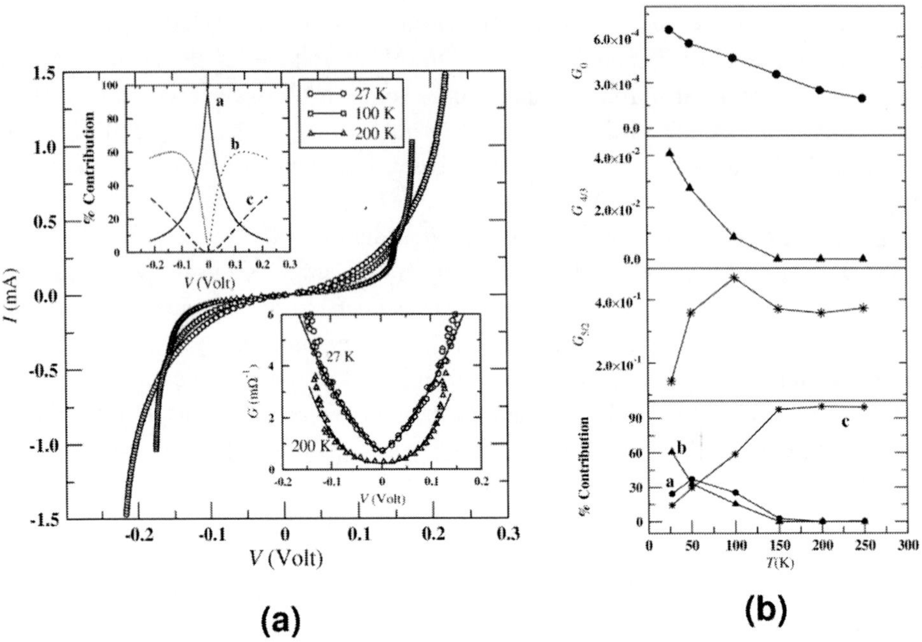

Figure 27. (a) I-V characteristics of the GB in bridge A obtained after removing bulk and disordered region contributions from data of Figure 2. Lower inset shows conductance, G as a function of voltage at two temperatures. The solid lines in this inset are curves obtained by fitting of the data to the GM model. The upper inset shows the voltage dependence of relative contributions of the three terms in the GM model i.e, (a) G_0, (b) $G_{4/3}V^{4/3}$ and (c) $G_{5/2}V^{5/2}$. (b) The temperature dependences of three coefficients, G_0, $G_{4/3}$ and $G_{5/2}$ and relative magnitude of three terms in the GM model i.e., (a) G_0, (b) $G_{4/3}V^{4/3}$ and (c) $G_{5/2}V^{5/2}$ at a voltage of 0.1 V. Temperature dependence of low field magnetoresistance (LFMR) measured at current of 1 μA is also shown in the lower panel.

Based on above we may write resistance of bridge A as $R = R_{GB}+R_d+R_{bulk}$ and the voltage across bridge is given by $V = IR_{GB}+IR_d+IR_{bulk} = V_{GB}+V_d+V_B$. Value of R_d+R_{bulk} is given by the resistance of I-V curves in Figure 26.(b) at high voltages. Voltage across GB barrier (V_{GB}) for each current was determined by subtracting $I(R_d+R_{bulk})$ from measured voltage V to yield actual GB characteristics as shown in Figure 27.(a). The I-Vs seen in Figure 27.(a) are qualitatively quite typical of tunneling barrier with sharp increase in current at some voltage (corresponding to barrier voltage). Voltage dependence of GB conductance for data of Figure 27.(a) is shown in the lower inset of this figure for two different temperatures. We also show fitting of the data to GM model in the inset. A good fit is seen except at higher voltages. As seen later, the deviation arises from elastic tunneling and can be fitted to Simmons model. Relative contributions of elastic (G_0), two step ($G_{4/3}V^{4/3}$) and three step ($G_{5/2}V^{5/2}$) inelastic tunneling terms to conductance at different voltages are shown in upper inset of Figure 27.(a). It is seen that the relative contribution of elastic scattering term reduces at higher voltages. Temperature dependence of parameters G_0, $G_{4/3}$ and $G_{5/2}$ is shown in Figure 27(b) along with the relative contributions of three terms at fixed biasing voltage of 0.1 volt. Temperature dependence of $G_{4/3}$ and $G_{5/2}$ is in agreement with earlier studies [40]. Increase in $G_{5/2}$ term contribution at high temperatures and voltages may arise as this term is due to three step tunneling that being higher energy process has increased contribution with increasing temperature or voltage. Reduction in relative elastic tunneling contribution at higher voltages and higher temperatures is in agreement with reduction in tunneling magnetoresistance at higher temperatures and voltages reported in many studies [40, 47] and seen in our data of Figure 28.(a) (as discussed below).

To understand the deviations at higher voltages, we have measured hysteresis in magnetoresistance (MR = [*R*(*H*)- *R*(0)/ *R*(0)] as function of field at different temperatures and the results are shown in Figure 28.(a). Here, the resistance, *R*(0), is the resistance at *H*=0, obtained by extrapolating the linear portion (at high field) of hysteric MR curve and *R*(*H*) is the resistance at field *H*. Two contributions of MR i.e low field magnetoresistance (LFMR) that shows hysteresis and rapidly falls at high fields and high field magnetoresistance (HFMR) that has linear dependence on field are clearly seen here. LFMR is seen to reduce with temperature. Similar reduction in LFMR was also seen with increasing voltage (data not shown here).

As reported in earlier studies [48], the magnetization on two sides of barrier has maximum anti-ferromagnetic alignment at peak in MR. At large field the two sides of GB are ferromagnetically aligned. The measurement of *I-V*s in these two states, help us to directly determine relative importance of different tunneling current components to magnetoresistance. We have measured *I-V* characteristics at peak in MR as well as at high fields above onset of linear dependence of MR on field. Typical results on a bridge at peak field of 0.05 T (the peak state reached by applying maximum reverse field and then field in forward direction to reach peak in MR) and 0.27 T where LFMR reaches constant minimum value are shown in Figure 28.(b).

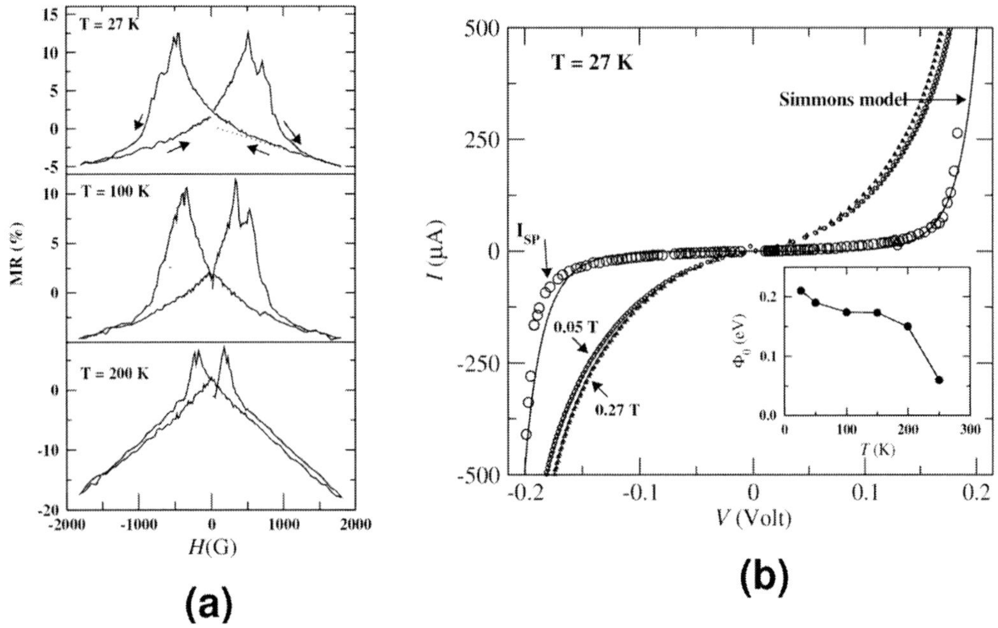

Figure 28. (a) Hysteresis in magnetic field dependence of magnetoresistance at temperatures of 27, 100 and 200 K. (b) The *I-V* curves measured at magnetic fields of 0.05T (peak position in Figure 5) and 0.27 T. The spin polarized current (I_{SP}) at each voltage is obtained as difference of currents at two fields. The solid line shows fitting of the experimental data to the Simmons model. Inset shows temperature dependence of barrier voltage obtained by fitting Simmons model to data at different temperatures.

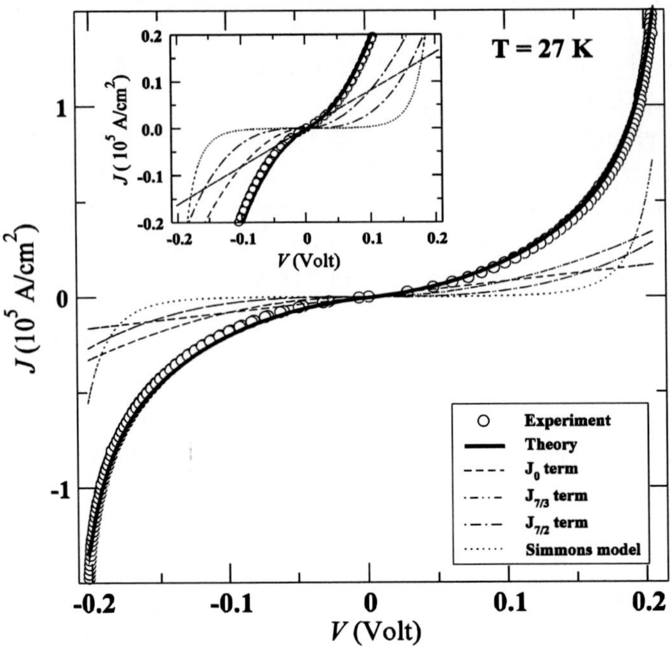

Figure 29. The experimental *I-V* characteristic curve of barrier determined at 27 K (after taking into account disordered region and bulk contributions) is shown by open circles. The lines show contributions of different terms in the GM model and elastic tunneling contribution of the Simmons model and total theoretical current due to all contributions (solid line). Inset shows expanded view in low current region.

We may add that the curves shown here have been arrived at after removing the contributions of bulk film and disordered region near GB as discussed earlier. The difference in these curves (i.e. difference, I_{SP} in two current values at each voltage) arises solely due to change in magnetization on two sides of the barrier. This difference curve along with fitting to Simmons model is also shown in Figure 28.(b). An excellent fit is obtained with fitting parameters of barrier thickness, t = 55 Å and barrier height Φ_0 = 0.20 eV. We may add that this curve could not be fitted to GM model. These values of t and Φ_0 are comparable to those reported in earlier studies [47, 48].

Finally using the direct tunneling contribution as given by Simmons model, data of Figure 3 can be fitted to theoretical curve over full voltage range as shown in Figure 29. It is seen that the direct tunneling has relatively smaller contribution, and this explains the relatively small MR observed for tunnel junctions [47]. Preparation of thinner and defect free junctions where scattering from impurities is minimized may help in improvement in MR. Disordered region near grain boundary as seen here is another source of reduction in MR from predictions of Julliere model that is applicable in ideal case.

CONCLUSIONS

The microstructure and crystallinity of LPMO films is found to be strongly influenced by the substrate temperature. It is observed that by just controlling the substrate temperature, one can grow films with different microstructures i.e. polycrystalline (PC), nanocrystalline (NC) and single crystalline (SC). The grown NC films, having an average crystallite size of 17nm, exhibited a record magnetoresistance (MR) value of 103% at a field of 1 Tesla. The magnetic nature of the grain boundaries of PC and NC was explored using ferromagnetic resonance techniques, which revealed that in PC films the grain boundaries have antiferromagnetic ordering while in NC films grain boundaries are spin glassy at low temperature. It has been shown that only PC and NC samples exhibits the electroresistance property, which can be explained by lowering of resistivity by conversion of disorder magnetic region into the ferromagnetic region by the injection of spin polarized carriers. The temperature dependences of resistivity, magnetoresistance and current-voltage (*I-V*) characteristics of bi-crystal grain boundary junctions of LPMO films have been studied. By taking into account resistance of disordered region near the grain boundary, we obtain grain boundary *I-V* characteristics, with very small current at low voltages followed by sharp increase near barrier voltage, as typically observed for good tunneling barriers. Methodology for direct determination of elastic contribution to transport current has been developed and its voltage dependence is in agreement with Simmons model. It is found that transport through the grain boundary is dominated by multistep inelastic tunneling. However, low field magnetoresistance originates from component of current due elastic tunneling of charge carriers. Small value of elastic tunneling contribution to total current is responsible for small value of low field magnetoresistance.

References

[1] E.H. Putley, *Hall effect and related phenomena*, (ButterWorth (London))1960.

[2] J.S. Hallman and B. Ables, *Phys. Rev. Lett.* 37 (1976) 1429.

[3] M.N. Baibich, J.M. Broto, A.Fert, F.N. Vandau, F. Petroff, P.Etienne, G. Creuget, A. Freiderich and J. Chagelas, *Phys. Rev. Lett.* 61 (1988) 2472.

[4] S. Jin, T.H. Tiefel, M. McCormack, R.A. Fastnacht, R. Ramesh, and L.H. Chen, *Science* 264 (1994) 413.

[5] J. M. D. Coey, M. Viret, S. von Molnar, *Advances in Physics* 48 (1999)167-293.

[6] J.B. Goodenough, *Phys. Rev.* 100 (1955) 564.

[7] E.O. Wollan and W.C. Koehler, *Phys. Rev.* 100 (1955) 545.

[8] C. Zener, *Phys. Rev.* 82 (1951) 403.

[9] A.J. Millis, P.B. Littlewood and B.J. Shraiman, *Phys. Rev. Lett.* 74 (1995) 5144.

[10] H. Y. Hwang, S-W. Cheong, N.P. Ong, and B. Batlogg, *Phys. Rev. Lett.* 77 (1996) 2041.

[11] T. Venkatesan, X.D. Wu, A. Inam, J.B. Wachtman, *Appl. Phys. Lett.* 52 (1988) 1193.

[12] H.Q. Li, Q.F. Fang, Z.G. Zhu, *Mat. Lett.* 52 (2002) 120.

[13] H. Y. Hwang, S-W. Cheong, N.P. Ong, and B. Batlogg, *Phys. Rev. Lett.* 77 (1996) 2041.

[14] H.Q. Li, Q.F. Fang, Z.G. Zhu, *Mat. Lett.* 52 (2002) 120.

[15] A.Singh, D. K. Aswal, S. Sen, C. S. Vishwanadham, G. L. Goswami, L. C. Gupta, S. K. Gupta, J. V. Yakhmi, V. C. Sahni, *Journal of Crystal Growth* 243 (2002)134.

[16] Singh, D. K. Aswal, C. S. Viswanadham, G. L. Goswami, L. C. Gupta, S. K. Gupta, J. V. Yakhmi, *Journal of Crystal Growth* 244 (2002) 313.
[17] T. Kanki, H. Tanaka, T. Kawai, *Solid State Commun.* 114 (2000) 267.
[18] A.Singh, P. Chowdhury, N. Padma, D. K. Aswal, R. M. Kadam, Y. Babu, M. Jayanth Kumar, C. S. Vishwanadham, G. L. Goswami, S. K. Gupta, J. Y. Yakhmi, *Solid State Communications* 137 (2006) 456.
[19] Singh, D. K. Aswal, P. Chowdhury, N. Padma, C. S. Vishwanadham, S. Kumar, S. K. Gupta, J. V. Yakhmi, *Journal of Magnetism and Magnetic Materials* 313 (2007) 115.
[20] D. K. Aswal, A. Singh, R. M. Kadam, M. K. Bhide, A. G. Page, S. Bhattacharya, S. K. Gupta, J. V. Yakhmi, V. C. Sahni, Materials Letters 59 (2005) 728.
[21] M. J. Calderon, L. Brey and F. Guinea, *Phys Rev. B* 60 (1999) 6698.
[22] A.Singh, D. K. Aswal, P. Chowdhury, N. Padma, R. M. Kadam, Y. Babu Y., M.. Jayanth Kumar, C. S. Vishwanadham, S. Kumar, S. K. Gupta, J. V. Yakhmi, *Solid State Communications* 138 (2006) 430.
[23] A.N. R. Rao, A. R. Raju, V. Ponnambalam, S. Parashar, N. Kumar, *Phys. Rev. B* 61 (2000) 594.
[24] K. Debnath, J. G. Lin, *Phys. Rev. B* 67 (2003) 064412.
[25] Y. G. Zhao, Y. H. Wang, G. M. Zhang, B. Zhang, X. P. Zhang, C. X. Yang, P. L. Lang, and M. H. Zhu , P. C. Guan, *Appl. Phys. Lett.* 86 (2005) 122502.
[26] F. X. Hu, J. Gao, *Phys. Rev. B* 69 (2004) 212413.
[27] V. Markovich, E. Rozenberg, Y. Yuzhelevski, G. Jung, G. Gorodetsky D. A. Shulyatev Ya. M. Mukovskii , *Appl. Phys. Lett.* 78 (2001) 3499.
[28] V. Markovich, E. S. Vlakhov, Y. Yuzhelevskii, B. Blagoev, K. A. Nenkov,G. Gorodetsky , *Phys. Rev. B* 72 (2005) 134414 (2005).
[29] F.X. Hu, J. Gao, X. S. Wu, *Phys. Rev. B* 72, 064428 (2005).
[30] V. Markovich, E. Rozenberg, Y. Yuzhelevski, G. Jung, G. Gorodetsky D. A. Shulyatev Ya. M. Mukovskii , *Appl. Phys. Lett.* 78, 3499 (2001).
[31] T. Wu, S. B. Ogale, J. E. Garrison, B. Nagraj, Amlan Biswas, Z. Chen, R. L. Grenne, R. Ramesh, and T. Venkatesan, *Phys. Rev. Lett.* 86, 5998 (2001).
[32] A.Singh, D. K. Aswal , P. Chowdhury, N. Padma, S. K. Gupta, J. V. Yakhmi, *J. Appl. Phys.* 102 (2007) 043907.
[33] M. Julliere, *Phys. Lett. A* 54, 225 (1975).
[34] J.B. Philipp, C.Hofener, S.Thienhaus, J.Klein, L.Alff, and R.Gross, *Phys. Rev. B* 62 (2000) R9248.
[35] M.Bowen, M.Bibes, A.Barthelemy, J.-P. Contour, A.Anane, Y.Lemaitre, and A.Fert, *Appl. Phys. Lett.* 82(2003) 233.
[36] N.Khare, U.P. Moharil, and A.K. Gupta, *Appl. Phys. Lett.* 81(2002) 325.
[37] E. Yu. Tsymbal and D. G. Pettifor, *Phys. Rev. B* 58(1998) 432.
[38] J.G. Simmons, *J. Appl. Phys.* 34 (1963) 1793.
[39] N.K. Todd, N.D. Mathur, S.P. Issac, J.E. Evetts, and M.G. Blamire, *J. Appl. Phys.* 85(1999)7263.
[40] M.Paranjape, J.Mitra, A.K. Raychaudhuri, N.K. Todd, N.Mathur, and M.G. Blamire, *Phys. Rev. B* 68 (2003) 144409.
[41] L.Glazman and K.Mateev, *Sov. Phys. JETP* 67(1988) 1276.
[42] J.G. Simmons, *J. Appl. Phys.* 34(1963) 1793.

[43] N.K. Todd, N.D. Mathur, S.P. Issac, J.E. Evetts, and M.G. Blamire, *J. Appl. Phys.* 85(1999) 7263.

[44] D.Niebieskikwiat, R.D. Sanchez, D.G. Lamas, A.Caneiro, L.E. Hueso, and J.Rivas, *J. Appl. Phys.* 93(2003) 6305.

[45] P.Rottlander, M.Hehn, and A.Schuhl, *Phys. Rev. B* 65(2002) 054422.

[46] J.Klein, C.Hofner, S.Uhlenbruck, L.Alff, B.Buchner, and R.Gross, *Europhys. Lett.* 47 (1999) 371.

[47] P. Chowdhury, S. K. Gupta, N. Padma, C. S. Vishwanadham, S. Kumar, A. Singh, and J. V. Yakhmi, *Phys. Rev. B* 73(2006) 104437.

[48] M. Ziese, *Rep. Prog. Phys.* 65(2002) 143.

In: Magnetic Thin Films
Editor: John P. Volkerts

ISBN: 978-1-61209-302-4

Chapter 3

ROLE OF DEFECTS, SURFACE AND INTERFACE DISORDERS AND CLUSTERS ON THE MAGNETIC PROPERTIES OF TRANSITION METAL DOPED SEMICONDUCTING OXIDE THIN FILMS

*C. Sudakar**
Department of Physics, Indian Institute of Technology Madras,
Chennai 600036, India

ABSTRACT:

Thin films exhibit interesting magnetic properties including unusual anisotropy, and surface ferromagnetism. Recently, dilute concentrations of transition metal cation doped in semiconducting oxide thin films showed interesting ferromagnetic properties which has been proposed as a promising candidate for applications including spintronics and magneto-optoelectronics. Several different transition metal doped oxide thin films have shown range of magnetic properties from paramagnetism to spin clusters to weak or strong ferromagnetism. The broad range of properties in these oxide thin films is related to the defects. The defects include oxygen and metal cation vacancies, coherently intergrown secondary phase segregations within the oxide lattice, and surface and interface defects. In this chapter we will discuss the magnetic properties of transition metal doped wide bandgap semiconducting oxide thin films. After including a brief introduction to the research on the oxide thin films, we will discuss the work from our studies with focus on the role of defects, surface and interface disorder, and clusters on the magnetic properties of nanostructured and crystalline thin films. The role of oxygen vacancies in triggering the ferromagnetic properties is proposed. In the case of Co doped ZnO introduction of oxygen defects induces the ferromagnetism. Further the ferromagnetic to nonmagnetic state and vice versa can be switched by introducing and removing oxygen vacancy defects. To unambiguously establish the Co^{2+}/O^{2-} defects which significantly influence the magnetic property, the possible role played by the secondary impurity phase and the surface and interface disorder are studied. We show by the

* E-mail: csudakar@iitm.ac.in

structural and magnetic property correlations of $Co_{3-y}Zn_yO$ thin films that the introduction of oxygen vacancy defects in these films lead to a crystalline Zn:CoO core with a disordered oxygen deficient amorphous Co:ZnO surface. The disordered surface layers are shown to be ferromagnetic which couples with the antiferromagnetic core therefore exhibiting an exchange bias coupling. The oxygen vacancy defects are also shown to induce weak ferromagnetism in Cr doped and a pure In_2O_3 thin film which is intrinsic to the material with the evidence of spin polarized charge carriers present in these thin films. Such materials exhibit O and In vacancies with associated In-In clustering which is predominantly found in the vicinity of the surfaces. With the nonmagnetic doping such as Cu in ZnO thin films ferromagnetic properties are observed with the film magnetization decreasing with the increasing concentration of Cu. Structurally at high concentrations of Cu (>3 at.%) doping, CuO nanophase inclusions within the basal planes of ZnO are evidenced. However the origin of ferromagnetism with large magnetic moment (1.6 μB/Cu) at low concentrations (<1 at.%) remains unclear with possible contributions from the coherently intergrown nanoclusters or due to the doping of Cu at Zn site with associated oxygen defects. Systematic investigations on the magnetic properties of CuO-ZnO heterostructures reveal the observed magnetization cannot be accounted for solely by spins localized near the CuO–ZnO interface or in the CuO layer. The possible role by the oxygen defects in the transition metal doped semiconducting oxides and even in undoped semiconducting oxides such as In_2O_3 and TiO_2 will be discussed in this chapter.

1. Introduction:

Materials that are multi-functional such as semiconducting, optically transparent and ferromagnetic are of high demand for applications that foresee the control of one property by the manipulation of other property [1-4]. There are great challenges to synthesize materials that have interactive combinatorial properties including semiconductivity and strong magnetism [5, 6]. Novel functional properties can be activated in a material system by including small changes in the structure of the materials. Notable example includes doping a semiconductor with a few atomic percent of transition metal (TM) that leads to the ferromagnetic behavior. Researchers have been successful in developing these semiconducting ferromagnets. Few of the renowned examples include Mn doped GaAs and InAs[7, 8]. However these systems develop magnetic ordering at low temperatures well below 300 K. Therefore there has been a great deal of research interest in other new classes of materials, mostly oxide compounds, that are semiconducting (or have wide bandgap of 2 to 4 eV) and can be made ferromagnetic above room temperature by doping with transition metal cations in dilute concentration. Such materials are commonly called as diluted magnetic semiconducting oxides (DMSO)[9, 10]. In the last decade, DMSO materials have been extensively studied to identify doped semiconducting materials with large magnetic moments, M, and with a high Curie temperature T_C [11, 12] and to find the feasibility of these materials for applications to spintronics, a technology in which spin and charge degrees of freedom are exploited unlike the charge only manipulation in the field of electronics [13]. High values of T_C and M are desired for the operational stability and integrity of spintronics based devices at room temperature. The results suggest that semiconducting oxides exhibit ferromagnetism with a Curie temperature (T_C) well above room temperature when doped with small amounts of transition metal ions [14, 15]. In particular, oxides such as ZnO, TiO_2, In_2O_3, SnO_2 [15-18]

have been identified as promising host semiconductor material, exhibiting ferromagnetism when doped with most of the transition metal elements-- V, Cr, Fe, Co and Ni [14, 15]. These metal oxides have been widely studied in pure form and extensively used in several applications. ZnO has attracted a renewed interest because of its relevance to wide range of applications in optical, piezoelectric devices and chemical sensors [19]. ZnO is a candidate material for short-wavelength magneto-optical applications, because of its large bandgap value (3.4 eV). TiO_2 is used in the solar cells and catalytic applications [20, 21]. In_2O_3 and SnO_2 are excellent transparent conductors and are used in gas sensing applications [22]. Further the material properties of these oxides can be tuned to suit them for applications including optoelectronics [19]. While the reports on the above room temperature ferromagnetism are found in many DMSO compounds the interpretation of ferromagnetic property is not straightforward. The complications arise by questions whether the observed magnetic properties are intrinsic or arise from impurity clusters in the lattice [23-25]. The ferromagnetism in these materials is very sensitive to the synthesis or fabrication and processing conditions. The dependence of magnetic properties in semiconducting oxides is believed to arise from the interplay between the transition metal dopants and charge carriers [18]. In this chapter the role of defects including oxygen and metal cation vacancies, coherently intergrown secondary phase segregations within the oxide lattice, and surface and interface defects are discussed. Several different transition metal doped oxides thin films with insulating, semiconducting and metallic characteristics have shown range of magnetic properties from paramagnetism to spin clusters to weak or strong ferromagnetism. The broad range of properties in these oxide thin films is related to the defects in the films. We will discuss the magnetic properties of transition metal doped wide bandgap semiconducting oxide thin films. After including a brief introduction to the research on the DMS oxide thin films, we will discuss the work from our studies with focus on the role of defects, surface and interface disorder, and clusters on the magnetic properties of nanostructured and crystalline thin films. The role of oxygen vacancies in triggering the ferromagnetic properties is proposed. In the case of Co doped ZnO, introduction of oxygen defects induces the ferromagnetism [26]. Further the ferromagnetic to nonmagnetic state and vice versa can be switched by introducing and removing oxygen vacancy defects. To unambiguously establish the Co^2+/O^2-defects which significantly influence the magnetic property, the possible role played by the secondary impurity phase and the surface and interface disorder are studied. We show by the structural and magnetic property correlations of Co_3-yZn_yO thin films that the introduction of oxygen vacancy defects in these films lead to a crystalline Zn:CoO core with a disordered oxygen deficient amorphous Co:ZnO surface [27, 28]. The disordered surface layers are shown to be ferromagnetic which couples with the antiferromagnetic core therefore exhibiting an exchange bias coupling. The oxygen vacancy defects are also shown to induce weak ferromagnetism in Cr doped and a pure In_2O_3 thin film which is intrinsic to the material and is shown with the evidence of spin polarized charge carriers present in these thin films [29]. Such materials exhibit O and In vacancies with associated In-In clustering which is predominantly found in the vicinity of the surfaces. With the nonmagnetic doping such as Cu in ZnO thin films ferromagnetic properties are observed with the film magnetization decreasing with the increasing concentration of Cu [30]. Structurally at high concentrations of Cu (>3 at.%) doping, CuO nanophase inclusions with the basal planes of ZnO are evidenced. However the origin of ferromagnetism with large magnetic moment (1.6 μB/Cu) at low concentrations (<1 at.%) remains unclear with possible contributions from the coherently intergrown nanoclusters or due to the doping of Cu at Zn site with associated oxygen defects. Systematic investigations on the magnetic properties of CuO-ZnO heterostructures reveal that the observed magnetization cannot be accounted for solely by spins localized near the CuO–ZnO interface or in the CuO layer [31]. The possible role by the oxygen defects in the transition metal doped semiconducting oxides and even in undoped semiconducting oxides such as In_2O_3 and TiO_2 will be discussed in this chapter [29, 32, 33].

2. BRIEF REVIEW ON DMSO:

The transition metal III-V semiconductors such as Mn doped InAs with T_C =7.5 K [7] and Mn doped GaAs with T_C=60 K [8] triggered intensive research on the diluted magnetic semiconducting oxides. A prototypical diluted magnetic semiconductor is (Ga,Mn)As, which is made by including ~2% to 6% Mn into a GaAs lattice [34]. Because Mn is not isovalent with Ga, the Mn ions act both as magnetic moments, and p-type acceptor dopants. The ferromagnetic interaction between the randomly distributed Mn moments is mediated by itinerant holes. Several prototypical devices based on (Ga,Mn)As such as spin dependent resonant tunneling diodes, magnetic tunnel junctions and spin polarized light emitting diodes have been demonstrated. However their application in real spintronic devices is limited by the low Curie temperature. The research efforts to increase the Curie temperature to above 300 K did not yield despite significant improvement in the growth techniques and post annealing processes. A maximum T_C of 173 K has been achieved in the semiconducting Mn δ-doped GaAs/Be-doped p-type AlGaAs heterostructures [35].

The effort on the search for novel materials which can exhibit above room temperature ferromagnetism by Dietl and co-workers led to the theoretical prediction that ZnO and GaN based DMS should exhibit room-temperature ferromagnetism by Mn doping with a carrier concentration in the materials ~ 10^{21} /cm^3 [36]. The first experimental observation of room temperature ferromagnetism in laser ablated films of Co doped anatase TiO_2 further spurred a great deal of activity in the field of DMSO [2]. This led to the beginning of many studies on the development of ferromagnetic order at room temperature in a range of transition metal doped semiconducting oxides, including TiO_2, ZnO, SnO_2 and In_2O_3 [10]. Co doped ZnO and TiO_2 were widely studied DMSO material. However there are large variations in the reported magnetic properties of these materials. Co doped TiO_2 prepared by spray pyrolysis showed no ferromagnetism for T>10 K [37]. While ion implantation techniques to dope rutile TiO_2 single crystals with Co showed T_C of 700 K[38]. The ferromagnetism in this ion implanted TiO_2 has been attributed to the exchange interaction mediated by oxygen vacancies. A large number of reports present a saturation magnetic moment between 0.16 μ_B/Co [39] to 1.7 μ_B/Co [40] in Co doped TiO_2. While some reports show the absence of ferromagnetic order [16] others claim a giant magnetic moment of 6.1 μ_B/Co in these oxides [41]. Similar variations in room temperature saturation magnetic moments are reported in Co doped ZnO which ranges from 0.5 μ_B/Co [42] to 2.6 μ_B/Co [43]. Ueda et al [44] prepared laser ablated films of $Zn_{1-x}M_xO$ (M= Cr, Mn, Co, Ni) and found ferromagnetic behavior with T_C= 280 K for Co doped films when the films were n-type. They concluded that ferromagnetic interactions between localized spins of Co are mediated by electrons. Magnetic circular dichroism of Co doped ZnO thin films show large magneto-optical effects that is indicative of some p-d interaction. The effect is significant at low temperature however disappear at room temperature due to the very small magnetization. Apart from ZnO and TiO_2, Co doped SnO_2 exhibiting giant magnetic moment of 7.5 μB/Co has been reported [45]. Other notable works include the Co doped Cu_2O with a magnetic moment of 0.44 μB/Co [46] and Cr doped In_2O_3 with a magnetic moment of 1.5 μB/Cr [18]. The ferromagnetism in Cr doped In_2O_3 films were shown by Philip et al [18] to be strongly dependent on the oxygen partial pressure during synthesis. The net magnetization was shown to vary systematically with carrier concentration. These results suggest that the development of ferromagnetic order is sensitive to the sample fabrication methods. The quality of the thin film samples is highly dependent on the fabrication technique, growth ambient conditions, post-growth processing. In addition to the substrate and the dopant type, concentration also influence the quality of the thin films. The origin of ferromagnetism in these systems remains unclear. Recent studies also show net magnetic moment appearing in undoped oxide thin films posing a great challenge to the theorists and experimentalists to understand the mechanism for the ferromagnetic ordering in these semiconducting oxides [47, 48]. The ferromagnetism observed in several of the undoped nonmagnetic nanoparticle oxides has been attributed to the unpaired electron spins at the defect sites mainly from the oxygen vacancies at the surface of the nanoparticles [49]. The important role of defects on the ferromagnetism has been shown by the studies on Cr doped TiO_2 thin films [50]. High quality thin films obtained by oxygen-plasma assisted molecular beam epitaxy by depositing at a slow rate were found to be non-magnetic compared to the defective thin films prepared at high deposition rates which turns out to be ferromagnetic. It is also interesting to note that in Mn and Co doped ZnO fabricated with a very wide range of carrier densities, there existed two regimes, metallic and insulating, in which the films exhibit ferromagnetic behavior [51]. However in the intermediate carrier density regime the films are non-magnetic.

Due to large differences in the magnetic properties of these materials, no single model has been shown to universally explain the DMSO system. In most of the models the degree of localization of charge carriers is an important parameter. In the early DMS models the strong spin-dependent sp-d exchange interactions between the delocalized band electrons and the magnetic impurity electrons with localized moments in 3d shell occur [36]. The superexchange interaction mechanism in II–VI compounds between the sp–d bands [52], leads to stabilization of the antiferromagnetic state. In III-V compounds such as InMnAs and GaMnAs ferromagnetic state is observed and the superexchange interaction cannot explain as this mechanism only leads to antiferromagnetism. Models based on the Ruderman-Kittel-Kasuya-Yoshida (RKKY) interaction have been developed for systems with delocalized charge carriers interacting with local moments [53, 54] making use of the fact that the first node of the oscillatory RKKY function occurs at distances larger than the average distance between magnetic ions. A modified RKKY interaction was proposed for stabilizing the ferromagnetism in III–V semiconductors [54] taking into account the role played by the free carriers in stabilizing the observed ferromagnetism. For more insulating systems, a bound magnetic polaron model has been proposed [15, 55-57], in which charge carriers localized on oxygen vacancies interact with the magnetic dopants to produce magnetic polarons. In this model, ferromagnetism arises when the magnetic polarons, which interact ferromagnetically, form a percolative network. However, at high doping levels, the antiferromagnetic interactions between the doped local moments suppress ferromagnetism [15]. Recently, a combination of magnetic polaron percolation and the RKKY interaction was proposed by Calderon and Das Sarma [58] to account for high Curie temperatures.

3. Role of Defects, Surface/Interfaces and Clusters

3.1. Role of Oxygen Defects on the Magnetic Properties of Oxides

(i) Co Doped ZnO Thin Films:

The distribution of oxygen vacancies is believed to be a crucial component in the development of ferromagnetism in transition metal cation doped wide band gap semiconducting systems, most notably in the bound magnetic polaron model [15]. We showed systematic changes in the magnetic properties of Co doped ZnO thin films and the Raman modes of Zn-O-Co vibrations which change as oxygen vacancies are created and destroyed [26]. Such correlation of magnetic properties with the Raman vibrational modes which are characteristics of oxygen related defects suggest the important role played by the oxygen vacancy defects in Co doped ZnO. A high vacuum (HV) annealing process is used to introduce oxygen vacancies (Vo). The oxygen vacancies in turn can be removed when the film is annealed in oxygen or air. $Zn_{1-x}Co_xO$ ($0 \leq x < 0.1$) films of thickness ~0.5 – 1 μm were prepared by spin coating technique using a metalorganic precursor which is decomposed after coating on sapphire substrates. The final annealing of the thin film was done at 700 °C in air for 60 minutes. The high-vacuum (HV) annealed films were prepared at a pressure of 10^{-5} to 10^{-6} Torr, and at a temperature of 550 °C for 1 hr. The crystal structure and the phase formation of ZnO and $Zn_{1-x}Co_xO$ thin films characterized by x-ray diffraction (XRD) reveals a single phase polycrystalline ZnO with wurtzite structure. No additional peaks were observed indicating that there are no structural changes and/or formation of additional phases due to the incorporation of Co in ZnO within the limits of XRD detection. The optical transmission spectra of $Zn_{1-x}Co_xO$ thin films annealed in air and in HV show absorption peaks around 652, 611 and 566 nm in addition to the band edge absorption in all compositions of $Zn_{1-x}Co_xO$ thin films (Figure 1.a). These localized absorption bands are attributed to Co^{2+} interatomic d-d transition associated with the crystal-field splitting in ZnO host due to transition from $^4A_2(F)$ to $^2E(G)$, $^4T_1(P)$, and $^2A_1(G)$ [59]. The observation of these transitions in the transmission spectra confirms the substitution of Co^{2+} cations for the Zn^{2+} ions at the tetrahedral sites of ZnO. A significant difference in the band edge slopes between the spectra of air and HV annealed samples is observed. The band edge shows more diffuse transition for air annealed samples, whereas the absorption edge becomes sharper on HV annealing the films. The band tailing is generally attributed to point defects such as Co_{Zn}, interstitial Co^{3+}, and zinc vacancies [60]. The Raman spectra of $Zn_{1-x}Co_xO$ films show all the expected ZnO phonon modes for $x = 0$ [61]. In Co doped ZnO samples, a broad impurity mode at ~690 cm^{-1} shown in Figure 1.b. develops in the air annealed samples.

The intensity of this impurity band increases with x. For films with $x \geq 0.047$ very strong additional bands are observed at 186, 491, 526, 628 and 718 cm^{-1}[26]. These additional bands may be assigned to complexes such as $Zn_xCo_{3-x}O_4$ as these bands are similar to Co_3O_4 [62-64]. Presence of $Zn_xCo_{3-x}O_4$ clusters are too small to be detected by XRD, but show up clearly in Raman scattering.

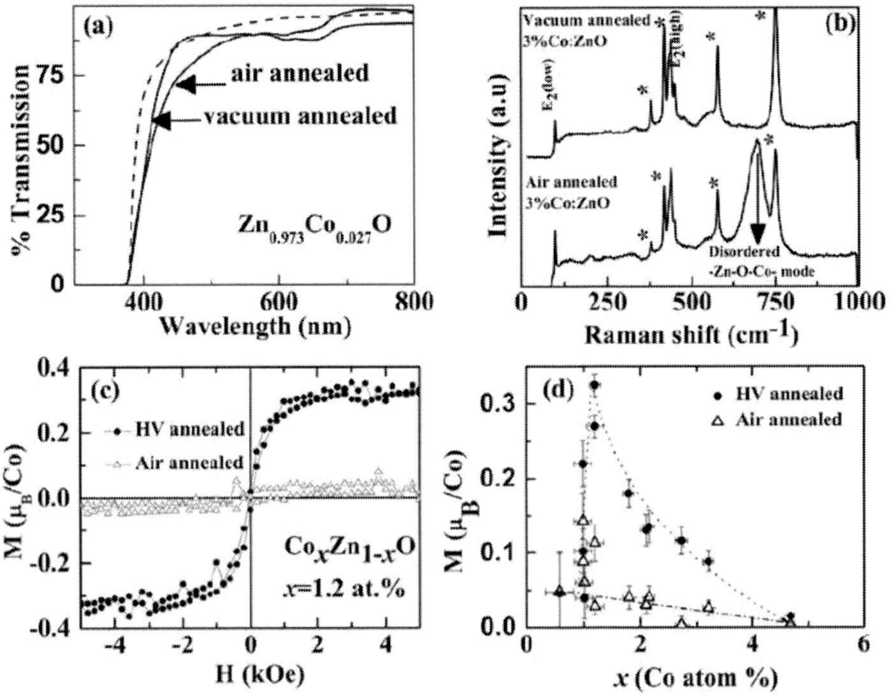

Figure 1. (a) UV-Vis transmission spectra of typical Co doped ZnO system in the air annealed and high vaccum (HV) annealed condition. The dashed line show the spectra of air annealed ZnO thin film. (b) The Raman spectra of 3 % Co doped ZnO recorded with a scattering geometry [z(x +y, x +y)z] using 514.5 nm (2.41 eV) excitation line from an Ar^+−ion laser. The peaks labeled with * originate from the sapphire substrate. (c) M vs H plot at 300 K for a typical Co doped ZnO after HV annealing (solid symbol) and air annealing (open symbols). (d) Magnetic moment per Co^{2+} in $Zn_{1-x}Co_xO$ thin films annealed in air and HV as a function of x. The dashed and dash-dot lines are guide to the eye.

In the low Co concentration ($x < 0.047$) samples the broad impurity band around 690 cm^{-1} can be associated with disordered local vibrational modes (LVM) of Zn-O-Co in ZnO. Based on the observation that this mode is absent in pure ZnO and also the ratio of LVM(690cm^{-1})/E2(high) increases with increasing x, this broad mode can be associated with the vibrations of Co-O-Zn complex. This assignment is consistent with our expectation that the LVM of Co-O-Zn in ZnO should be detected at higher wavenumber than the highest phonon frequency of ZnO because of the smaller mass of Co compared to Zn [65].

On annealing the samples at high vacuum, the impurity band at 690 cm^{-1} in all samples and also the strong peaks vanish, leaving only the phonon modes characteristic of ZnO (Figure 1.b). These bands reappear on annealing in air at 700 ^{0}C for 1hr (albeit with a smaller intensity) and again disappear on second high vacuum annealing, demonstrating a switching behavior between the two states [26]. The switching in $Zn_{1-x}Co_xO$ ($0 \leq x < 0.1$) thin films are highly reproducible. These observations suggest that the 690 cm^{-1} mode is related not only to the presence of the dopant ions, but also to the intrinsic lattice defects surrounding these ions. Oxygen vacancies are created when the films are annealed under high vacuum. A local lattice rearrangement takes place in the sample, leaving a larger concentration of V_O defects next to Co ions in the lattice. The activation energy required to develop and diffuse oxygen vacancies through ZnO is ~0.07 eV [66], so we expect a significant V_O diffusion at 550 ^oC. Postulating that the 690 cm^{-1} Raman active mode is related to disordered Co-O-Zn vibrations, this excitation would be strongly suppressed when Vo defects cluster about Co ions. In the case of Co-doped TiO_2 system, calculations predict that the oxygen vacancies are more likely near the Co sites than Ti sites [67]. A similar mechanism could increase the concentration of V_o close to Co sites as compared to the Zn sites. When the sample is annealed in air, the V_o defects are annealed out as additional oxygen enters the system. These annealing conditions are not sufficiently extreme to promote Co clustering inside the ZnO matrix, as evidenced by the optical absorption spectrum showing that Co remains incorporated in ZnO lattice in the 2+ state.

The investigations on the magnetic properties of these $Co_xZn_{1-x}O$ films (corrected for diamagnetic contribution of substrates) annealed in air show a small magnetic moment of ~ 0.04μB/Co for $x < 0.02$ and even smaller values for $x > 0.03$. The decrease in magnetic moment with increasing x may arise from the formation of an antiferromagnetic $Zn_xCo_{3-x}O_4$ secondary phase, as suggested by Raman spectroscopy. The panel of Figure 1.c. plots the substrate-corrected value of M versus H at T=300 K for a $Co_{0.012}Zn_{0.988}O$ film, under both air annealed and vacuum annealed conditions. Figure 1.d. shows the room temperature M of $Zn_{1-x}Co_xO$ ($0 < x < 0.1$) films annealed in air and in HV. The lines are guide to eye and show the most of the points lie close to the line for the air and HV annealed films.

When annealed in high vacuum, the samples show a large increase of the magnetic moment. The highest magnetic moment of ~0.3 μB/Co is observed for $Zn_{0.988}Co_{0.012}O$ film, which is small in comparison to the full Co^{2+} moment (3μ_B). This dramatic increase in the magnetization of the films occurs simultaneously with the elimination of the 690 cm^{-1} Raman peaks, which allows the possibility of a relationship between the magnetic properties and lattice structural properties of the samples. The possibility of an enhanced magnetic moment arising from the reduction of Co^{2+} to metal clusters has been excluded for the following reasons: If the metal clusters were to form in ZnO, the characteristic optical absorption peaks of Co^{2+} would be absent in the optical spectra. Furthermore, with the formation of metal clusters one would expect an increase in the magnitude of the magnetic moment with increasing Co concentration, contrary to the observations. The origin of the decrease in magnetization for the HV annealed samples at larger values of x is unclear. Raman spectroscopy shows no evidence for $Zn_xCo_{3-x}O_4$ clusters in these samples. At higher concentration of Co, it is possible that the decrease in magnetic moment arises from increasing antiferromagnetic interactions between Co ions in close proximity in the lattice.

The magnetization of the $Zn_{1-x}Co_xO$ samples is very sensitive to the HV annealing treatment. With the first vacuum annealing, the magnetic moment increased from 0.1 to 0.27 μ_B/Co. Subsequently annealing the sample in air is found to lower the magnetization (~0.02 μ_B/Co), while annealing the sample in vacuum increases the magnetization. This is qualitatively what one would expect for bound magnetic polaron induced ferromagnetism, where the presence of oxygen vacancies is crucial for developing ferromagnetic order in semiconducting oxides. The present results are distinct from the previous report Schwartz et. al. [68], where they demonstrated the reversible 300 K ferromagnetic ordering in Co:ZnO by lattice incorporation and removal of the native n-type Zn interstitial defect. Similar observation in Co:TiO_2 has been reported by Griffin et al. [66], however no reversible switching behavior has been demonstrated. A detailed investigation on the possible role played by the secondary impurity phase to the measured magnetic moment is required before the observed ferromagnetic order can be unambiguously attributed to Co^{2+}/oxygen vacancy effects. In fact, the results discussed in section III.2 on $Co_{3-x}Zn_xO_4$ thin films indicate similar enhancement in the magnetization on high vacuum annealing. The role of oxygen deficient amorphous Co:ZnO surface for the ferromagnetic signal is obvious from these studies.

In summary, we have demonstrated the existence of Raman active disordered mode – Zn-O-Co- in Co:ZnO which depends on the concentration of Co ions. The reversible behavior of the Raman band involving this local vibrational mode on air and vacuum annealing strongly suggests that oxygen vacancies play an important role in this excitation. These $Co_xZn_{1-x}O$ films also show a ferromagnetic moment at room temperature, having a maximum value at x=0.012. This moment is greatly enhanced when the samples are vacuum annealed, and is reduced when the samples are again air annealed. While this dependence of magnetic moment on oxygen vacancy concentration, deduced from Raman measurements, is consistent with models of ferromagnetism in dilute magnetic semiconductors, the presence of a secondary Co_3O_4 impurity phase at larger x values precludes us from definitively attributing the observed ferromagnetic order to intrinsic effects in samples with $x < 0.04$.

(ii) Role of Defects in Vacuum Annealed TiO_2 and In_2O_3 thin Films:

As there are several reports on metal oxides that develop new physical properties when doped with transition metal cations in dilute concentrations, room temperature ferromagnetism has also been found in pure (undoped) oxides in thin films [48, 69] and nanostructures [49]. The magnetic properties observed in these intrinsically diamagnetic semiconducting oxides find evidence for the crucial role played by oxygen vacancy defects in the development of ferromagnetic order [3]. While it is difficult to differentiate the effects of defects from the magnetic dopant ion contributions in transition metal doped oxides, the studies from the past several years has shown the apparent role played by the oxygen defects on the development of ferromagnetism in diamagnetic metal oxides. The strong connection between the defects and ferromagnetism in TiO_2 nanostructured thin films [32] and evidence for an intrinsic origin for ferromagnetism by the presence of spin-polarized charge carriers in In_2O_3 thin films exhibiting room-temperature ferromagnetism after high vacuum annealing are discussed below [29, 33]. Thin films of pure TiO_2 have been prepared using both spin coating and sputter deposition techniques on sapphire and quartz substrates. These two vastly different synthetic techniques were used to prepare the samples to see how the magnetic properties are affected which, in general, show significant dependence on the sample preparation method. The structural characteristics of the films have been investigated in detail using Raman spectroscopy and high-resolution transmission electron microscopy. The as-prepared TiO_2 thin film are relatively defect free, as shown by the high resolution transmission electron microscopy image for TiO_2 samples prepared by both spin coating and sputter deposition (Figure 2). Both the air and vacuum annealed films are free of defects in the interior of the nanoparticles. However, the TiO_2 lattice in the air annealed film remains crystalline to the very edge of the sample, whereas the vacuum annealed TiO_2 particles have an amorphous component at the surface (Figure 2.a. and 2.b). This amorphous component has a thickness ranging from approximately 2 nm to 5 nm. Vacuum annealing the TiO_2 films produces significant non-stoichiometry at the surface of the nanoparticles. Similar effects were observed in the sputter deposited thin film which has much larger single crystalline particles ranging from 300 to 500 nm (Figure 2.c. and 2.d). The air annealed sample shows good crystalline order with few defects.

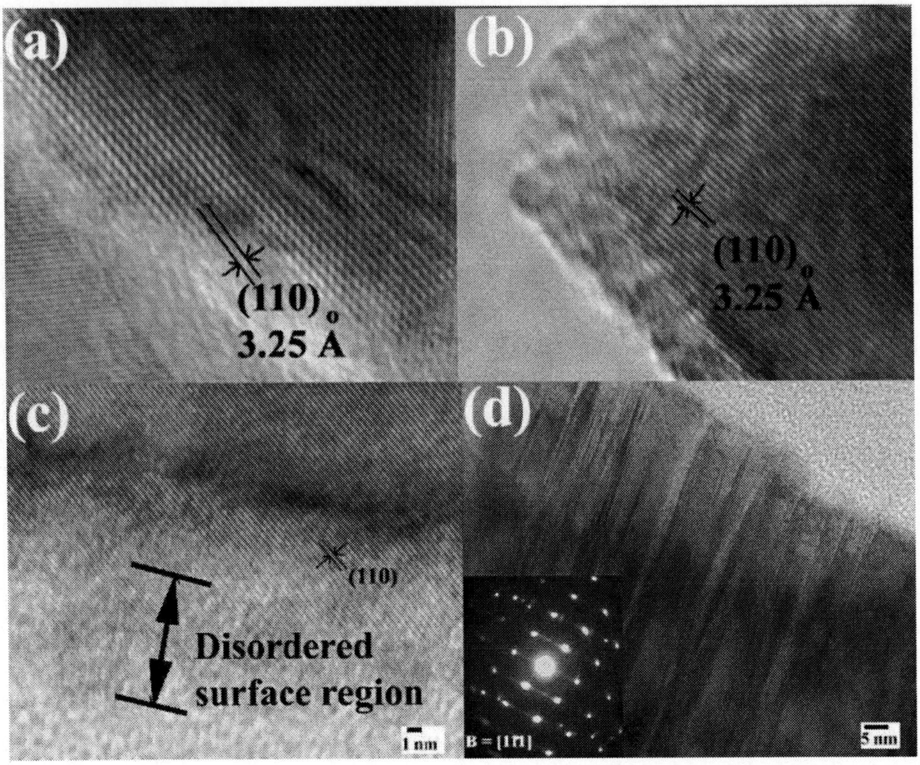

Figure 2. High resolution TEM images of typical surface regions for (a) air, (b) high vacuum annealed TiO2 nanoparticular samples. (c-d) HRTEM images of vacuum- annealed sputter deposited thin films. In figure 2.c. disordered surface regions of the particles are shown. (d) The interior of the crystallites showing large number of parallel twin rung along (011) plane with the inset showing the electron diffraction from the same region.

Conversely, the vacuum annealed films exhibit a highly disordered surface phase a few nanometers thick. Additionally, these particles show numerous crystallographic twin boundaries, with individual grains often containing several parallel twins. Such common microstructural features are observed in non-sotichiometric TiO_2 and are intimately related to the formation of shear structures [70]. The twinning produced in the rutile subcell structure is parallel to (011), which is the common twinning plane for TiO2. The selected area electron diffraction (SAED) also clearly show a diffuse streaks along [011] which arise due to dense twins running along (011) plane. The twinning is not observed in the MOD thin film samples, which suggests that such defects may be more easily removed by annealing samples with smaller particle sizes. Thus, vacuum annealed TiO2 thin films, presumably having much higher concentration of oxygen vacancies, exhibit a much higher concentration of defects. This leads to an amorphous structure at the surface of the particles both in the spin coated and sputter deposited thin films of TiO_2. The effects of this microstructural inhomogeneity are reflected on the magnetic properties of these films. When annealed in vacuum, all films demonstrated room temperature ferromagnetism, while the air annealed samples show much smaller, often negligible, magnetic moments. The fact that vacuum annealed TiO_2 films prepared by both MOD spin-coating and sputter deposition exhibit room temperature ferromagnetism [32], coupled with previous observations of ferromagnetic ordering in TiO2 samples prepared by PLD [71], argue that the development of a net moment is a robust property of this system. The magnitude of the magnetization in the vacuum annealed MOD TiO_2 film of ~ 50 emu/cm^3 (for a film thickness of 1 μm) is slightly larger than previously reported values, but consistent with measurements on PLD TiO_2 films [71]. Thus, introducing oxygen vacancy defects leads to a considerable enhancement in the magnetization. In addition we observe the magnetization of the vacuum annealed sputtered samples depends on film thickness, with the volume magnetization decreasing monotonically with increasing thickness [32].

This observation that the film magnetization appears to vary approximately with the surface area, rather than with film volume, supports the proposal that ferromagnetism in these materials arises from surface or interfacial effects [72, 73]. This point to an intimate connection of microstructure and point defects, with the emergence of ferromagnetism. It is interesting to know if the amorphous layer on TiO_2 films, induced by vacuum annealing, may play a role in the development of ferromagnetism.

The importance of non-crystalline structures on ferromagnetism in Cr:TiO_2 has been discussed previously [74], and the general appearance of ferromagnetic order in oxide nanoparticles has been attributed to unpaired electron spins near defects [50]. Assuming that the ferromagnetic order develops solely in the oxygen-deficient amorphous TiO_2 interface (Figure 2), the measured magnetization corresponds to a magnetic moment of ~0.05 μ_B/Ti, significantly less than the spin-only moment for Ti^{2+} or Ti^{3+}, indicating that only a small fraction of the TM ions are participating in the ferromagnetic order. These results suggest that ferromagnetism in the vacuum annealed TiO_2 films is mediated by surface defects or interfacial effects, but does not arise from stoichiometric crystalline TiO_2. Many investigations on metal oxide nanoparticles such as CeO_2, Al_2O_3, ZnO, In_2O_3, and SnO_2 [49, 75] show evidence for weak ferromagnetism. Nanoparticles are prone to have defect structures, specifically the oxygen vacancy defects, at the surface of the nanoparticles. These defects can be reduced by sintering the oxide nanoparticles at high temperatures in presence of oxygen. It has been shown that weak ferromagnetism is suppressed by the sintering process leading to the conclusion that the ferromagnetism is intimately connected to the defect structures [49]. It is suggested that unpaired electrons trapped on oxygen vacancies may be relevant for the development of ferromagnetism and, furthermore, that such ferromagnetic order may be a general characteristic of all metal oxide nanoparticles [3, 76]. However, a direct probe to identify, if the weak ferromagnetism observed in these metal oxides is an intrinsic property of the material, is essential. The observed moments in these nanostructured metal oxide samples is very small, on the order of only 10^{-4} to 10^{-3} emu/g [3, 49]. But this moment is significantly larger than the completely negligible magnetizations observed in diamagnetic bulk samples. Since such small magnetic moments may arise from the accidental incorporation of ferromagnetic impurities during sample preparation or handling, there is some uncertainty on the origin of ferromagnetism in the undoped metal oxides. Probing the material with techniques which are not sensitive to the trace amounts of ferromagnetic impurity phases will indicate the intrinsic or extrinsic nature of the observed magnetic property. In the following we discuss the evidence for the spin-polarized carriers in In_2O_3 thin films and propose the possible contribution of the anionic and cationic defects to the magnetic property. The undoped In_2O_3 thin films prepared by RF reactive magnetron sputter deposition turns into a conducting and ferromagnetic phase when annealed under high vacuum (10^{-5} to 10^{-6} Torr) at 873 K for 9 h [29]. The electrical conductivity increases by 3 to 4 orders of magnitude on vacuum annealing with the n-type carrier concentration and resistivity of the annealed films reaching ~10^{20} cm^{-3} and 10 mΩcm respectively. The room-temperature magnetization curves of the as-deposited and vacuum annealed In_2O_3 films using a high-sensitivity Quantum Design MPMS magnetometer, corrected for the diamagnetic contribution from the sapphire substrate, are shown in Figure 3. The magnetization of the as-prepared films is negligible, while the magnetization increases significantly on vacuum annealing (~ 0.5 emu/cm^3). It should be noted that sapphire substrate annealed under similar conditions in vacuum at 873 K for 9 hrs shows no evidence for any ferromagnetic moment. Thus, thin films of undoped In_2O_3 exhibit a small but distinct ferromagnetic signature with the inclusion of oxygen vacancy defects, introduced by vacuum annealing [29, 33]. Because these vacuum annealed In_2O_3 films remain conducting to low temperatures, it is possible to probe the spin polarization of the charge carriers using Point Contact Andreev Reflection (PCAR) [173]. A representative result of PCAR measurements made at T=2 K with a Nb tip is shown in the inset of Figure 3.

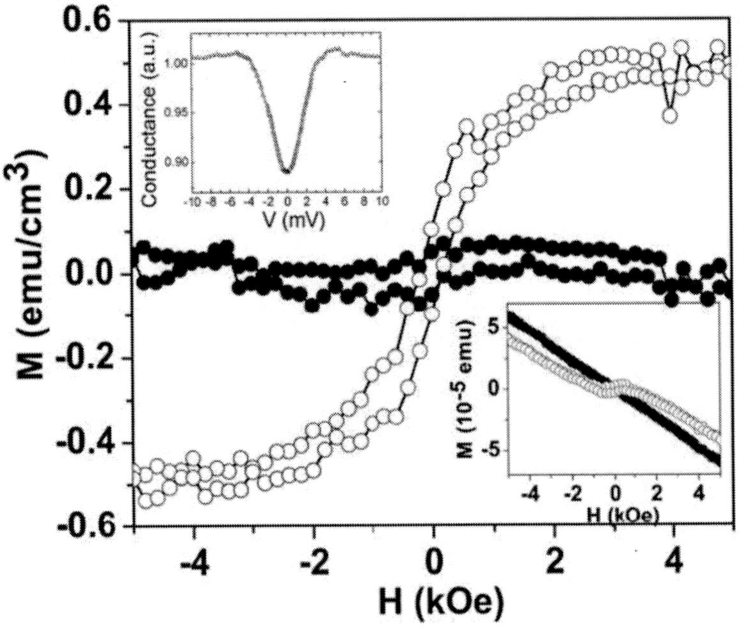

Figure 3. Magnetization curves for as-deposited (filled circle) and HV annealed (open circle) In_2O_3 thin films (corrected for diamagnetic contributions). The bottom inset shows the unprocessed data. The top inset shows a representative point contact Andreev reflection (PCAR) measurement on high vacuum annealed In_2O_3 thin film, measured at 2K using Nb tip. A fit to this curve yields an estimated spin polarization of P=45%.

The zero voltage dip in conductance is characteristic of a finite spin polarization of the In_2O_3 charge carriers. A more careful analysis of the conductance curve yields an estimated spin polarization of approximately 50%, indicative of ferromagnetism in these defect-rich metal oxide nanostructures [33]. While these investigations do not unambiguously prove the existence of intrinsic, carrier mediated ferromagnetic order, they do firmly establish that the measured magnetization is at least strongly coupled to the conduction electrons. Evidence for a finite spin polarization in Co doped ZnO films has also been inferred from low temperature tunneling magnetoresistance measurements on

$Co/Al_2O_3/Co:ZnO$ heterostructures [77]. Recent PCAR measurements show spin polarization of the transport current in pulse laser deposited thin films of ZnO with 1% Al and with and without 2% Mn[78]. Only films with Mn are shown to be ferromagnetic and have a spin polarization of the transport current of up to 55±0.5% at 4.2 K, in sharp contrast to measurements of the nonmagnetic films without Mn where the polarization was zero.

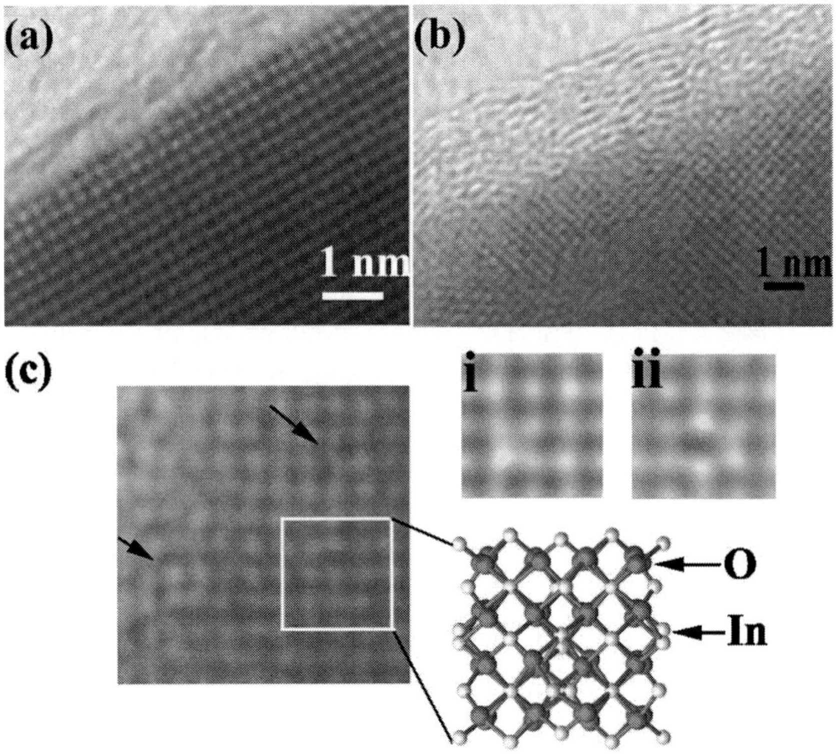

Figure 4. HRTEM of typical (100) surface regions for (a) as-deposited (stoichimetric) and (b) HV annealed (oxygen deficient) In_2O_3 samples. (c) A magnified view of a section of (b) with arrows showing typical of several distortions in the crystal lattice. The square region of HRTEM in (c) corresponds to a unit cell of In_2O_3 shown in the ball-and-stick model projected along (100) plane. The insets (i) and (ii) in (c) are the simulated HRTEM images showing the typical defects with (i) a oxygen vacancy and (ii) a oxygen vacancy with two adjacent In atoms clustering models.

The defect structure of vacuum annealed In_2O_3 (VA-In_2O_3) thin films can be imaged using high resolution transmission electron microscopy (HRTEM) and composition obtained from the analysis of Rutherford backscattering spectra. The RBS spectra shows that VA-In_2O_3 is highly oxygen deficient (~7 at. %) [29]. Further the high-resolution lattice images of air annealed and vacuum annealed In_2O_3 thin films clearly demonstrates the effects of point defects on the nanostructure. Figure 4.a. shows a HRTEM image of an air annealed In_2O_3 nanoparticle, with the beam oriented along the <100> directions of In_2O_3 lattice, showing a well-ordered lattice with no obvious defects. Vacuum annealing this sample introduces oxygen vacancies, as well as other point defects, as shown in Figures 4.b. and 4.c. The agglomeration of point defects on the surface leads to a 2–3 nm thick surface disordered layer, as shown in Figure 4.b. Additional point defects, both V_O and In_I, can be seen in the bulk of the sample. These additional point defects are highlighted by arrows in Figure 4.c.

These crystalline defect states, oxygen vacancies and indium vacancies (produced by In-In clustering), having densities of ~ 10^{13}-10^{14} cm^{-2} (≈3×10^{19}–10^{21} cm^{-3}) are observed predominantly in the vicinity of the surface.

The insets to Figure 4.c. show simulated HRTEM images for an oxygen vacancy defect (i) and an oxygen vacancy with and adjacent cluster of two In (ii). This surface disorder is not observed in as deposited In_2O_3 films, suggesting that these structural defects may be important for understanding many of the novel properties found in non-stoichiometric In_2O_3 thin films, and specifically may be associated with the development of room temperature ferromagnetism. Several studies [79-81] suggest that both oxygen and cation vacancies may be necessary to explain the electrical and magnetic properties found in undoped, but oxygen deficient samples. From these studies we can conclude that vacuum annealing the In_2O_3 films produces significant non-stoichiometry at the surface of the crystallites and this non-stoichiometry persists to a few tens of nanometers below the surface.

Our interest is to explore and understand possible relationship between these microstructural defects and the novel electrical and magnetic properties of these films. Both oxygen and cation vacancies coexist in vacuum annealed In_2O_3 thin films, where the presence of cation vacancy defects is indirectly inferred from the evidence for In-In clustering. These cation vacancies could also provide a mechanism for the charge compensation that has been observed in oxygen deficient In_2O_3 [79]. Furthermore, ferromagnetism has been observed in a number of undoped oxides including HfO_2, TiO_2, and In_2O_3, with many studies highlighting the role of both cation and anion vacancies [49, 69, 80-82]. Many of these studies propose that the simultaneous presence of cation vacancies are necessary for the development of ferromagnetism in these non-stoichiometric oxides [80, 83-86]. Recent theoretical studies also suggest that oxygen depleted In_2O_3 surfaces with In vacancies should become ferromagnetic [86]. These vacuum annealed In_2O_3 thin films having self-doped donors (oxygen vacancies) and self-compensated magnetic acceptors (indium vacancies) could exhibit room temperature carrier-mediated ferromagnetism [33]. The simultaneous presence of these distinct defects in In_2O_3 may provide a mechanism for the partial compensation of the large number of n-type carriers expected from the oxygen vacancy defects as well as providing a pathway for the development of carrier mediated ferromagnetism in In_2O_3 thin films.

3.2. Role of Clusters on the Magnetic Properties of Oxides

In spite of the distinct observations of magnetism in TiO_2 oxides (by Matsumoto et.al. [2]) and in ZnO (by Ueda et.al. [44]) doped with a transition metal cobalt, a major question is whether the observed ferromagnetism is intrinsic or extrinsic nature [87], as some studies reported the phase separation and the formation of ferromagnetic clusters (see Janish, et.al. [88] for a review). Co nanocluster's presence within the oxide lattice was shown to be the origin of ferromagnetism in Co doped TiO_2 and ZnO. While the Mn doped ZnO could evade such contributions from Mn clusters due to its antiferromagnetic nature, still questions were raised about the magnetic contributions from the antiferromagnetic forms of manganese oxides (as uncompensated spins in nanoclusters can give rise to significant magnetic signal) and some of the metastable phases of zinc-manganese oxides [89]. To prove the intrinsic nature of the ferromagnetic ordering anomalous Hall effect (AHE) were shown to exist in DMSO [90]. However, observing AHE is not adequate to conclude the intrinsic nature of

DMSO. AHE response was detected in a Co(5%)-(La,Sr)TiO_3 sample containing scattered Co clusters in the nonmagnetic (La,Sr)TiO_3 matrix [91]. Surprisingly, the AHE were also observed in the nonmagnetic $Z_{n0.985}C_{u0.015}O$ [92] samples and other nonmagnetic materials [90]. Due to high inhomogeneity and the small magnetic moment of the material it is very difficult to prove the intrinsic nature. However, recently magnetic circular dichroism was regarded to be very useful additional check to unravel this ambiguity. A conclusive evidence for intrinsic ferromagnetism in individual ZnO nanoparticle doped with Co was shown by simultaneous magnetic and microstructural characterization using electron magnetic chiral dichroism and channeling-enhanced electron energy loss microanalysis [93].

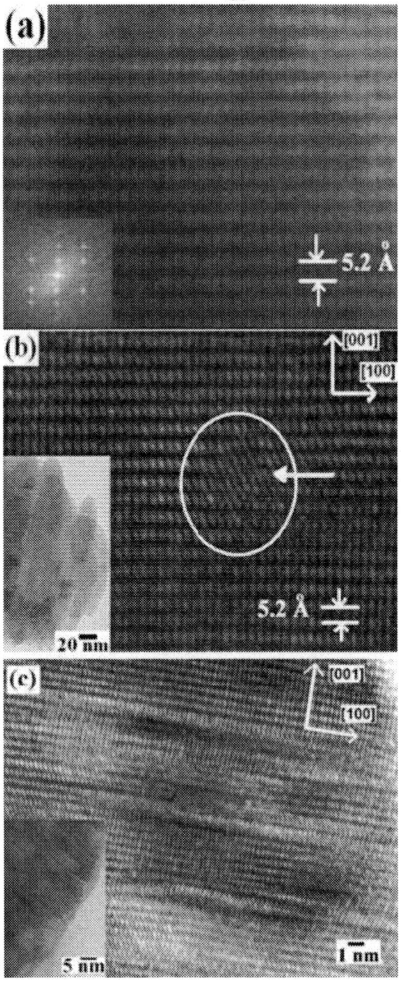

Figure 5. HRTEM of ZnO:xCu films with (a) x=0.06, (b) x=2.7, and (c) x=3.8 at. %. Formation of dislocations are shown with arrows in (b). Planar nanoclusters are clearly visible along the basal planes (001) of ZnO.

The nanoclusters in the metal oxide lattice can exhibit different properties which depend on the type of the matrix, the size of the clusters, the chemical and physical interactions among the clusters and between the clusters and the oxide matrix, etc. Therefore tools to detect such nanoclusters and understand their magnetic contribution is an important issue. Cu doped ZnO is a typical example where the role of Cu-O clusters or the oxygen defect are being discussed as possible reason for the induced ferromagnetism. Sputter deposited ZnO:xCu (x in at% with $0<x<10$) thin films were found to be ferromagnetic above 300 K [30]. Detailed structural analyses indicate the development of CuO clusters due to the poor dissolution of Cu in ZnO.

The HRTEM Figure 5. shows three cases: (i) at low concentration of Cu (x=0.6) grain interior and grain boundary regions show an ordered structure with no discernible segregation, secondary phases, or clustering of copper cations, (ii) occasional dislocations attributable to CuO nanophase segregation within ZnO lattice for x=2.7, and (iii) lattice images exhibiting distortions along (001) planes suggesting the location of copper atoms within the bulk of the zinc oxide crystals along the c-planes for x=3.8 at.%. The CuO nanophase inclusions are present even for $x<3$ where Cu-O structures of a few nm in size are observed. For $x<1$, a large magnetic moment of 1.6 μ_B/Cu was observed, which decreases monotonically with increasing x (Figure 6). The presence of Cu-O planar nanophase may give rise to the ferromagnetism as the small CuO particles are known to acquire ferromagnetic order[94] and the magnetic susceptibility of these CuO nanoparticles increases rapidly when their size decreases. Such strong magnetic behavior in nonmangnetic CuO has been attributed to the uncompensated spins of the surface Cu ions in nanoparticles. Further, the lattice distortion can promote ferromagnetic superexchange interaction in Cu-O-Cu complex lying in the basal plane. When cluster sizes are small, at small Cu doping, the residual spins of the Cu ions at the cluster interface can couple ferromagnetically with planar spins to generate larger values of the moment. The surface field becomes pronounced for small cluster sizes and leads to a dramatic increase in the M values.

As the Cu doping increases, the bulklike Cu-O-Cu antiferromagnetic superexchange interaction begins to dominate [95, 96] which leads to smaller values of M at higher concentration of Cu. While the role of CuO for the magnetic signal remains inconclusive from these studies, our effort to understand the details of the electronic and atomic environments around the spin bearing Cu^{2+} ions (studied for x=2) using x-ray absorption spectroscopy (XAS) show that Cu^{2+} substitutes Zn^{2+}, oxygen vacancies and the presence of local strain fields around Cu^{2+} ions exist [97]. These conditions may be essential in establishment of the RTFM ground state in Cu-doped ZnO films and are consistent with the hypothesis of a bound magnetic polaron percolation mechanism proposed for diluted magnetic oxides. Since the processing conditions can significantly influence the microstructure and chemical state of the materials leading to the formation of spurious unwanted clusters in the system which finally will influence the physical property. Mild hydrogenation can lead to such formation of metal nanoclusters within the oxide martrix [98]. The effect of hydrogenation on the structural and magnetic properties of $Zn_{0.85}Co_{0.05}Li_{0.10}O$ nanoparticles showed weak ferromagnetism at low temperature of hydrogenation processing. However the samples showed robust ferromagnetism at room temperature when hydrided at 400 °C. Interestingly, reheating the sample at 400 °C in air converts it back into the paramagnetic state completely. While the detailed characterization by x-ray and electron diffraction showed that room temperature ferromagnetism observed in the samples hydrogenated at RT could be intrinsic in nature, the magnetization observed in the samples hydrogenated at higher temperature (400°C) is partly due to the cobalt metal clusters [98]. The nm sized metal clusters segregation was identified by HRTEM. These studies point out the need for the careful analyses of microstructure and chemical nature of the material.

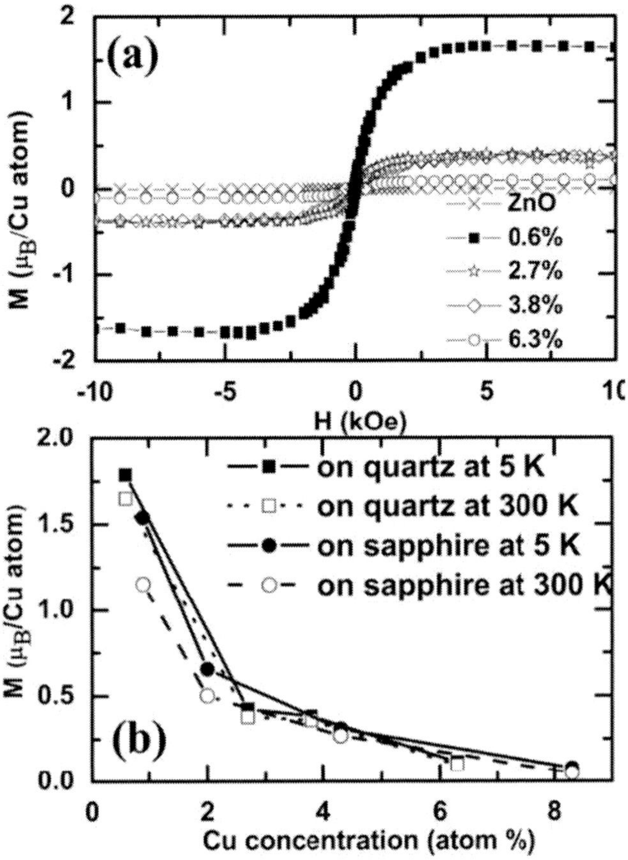

Figure 6. (a) M-H loops at 300K for ZnO:xCu (x in at. %) thin films. (b) Cu concentration dependent magnetization changes measured at 5 and 300 K. The lines are guides to the eyes.

3.3. Role of surface/interfaces on the magnetic properties of oxides

Intriguing magnetic properties in DMSO has been shown to emerge from the surface and interface effects [99, 100]. Since the magnetic properties are observed mostly in nanoparticles and nanostructured thin films, finite size effects of the particles can dominate the magnetic property. The competition between surface magnetic properties and core magnetic properties give rise to this effect [101]. Therefore carefully controlled interfaces in heterogeneous systems can give rise to new properties and functionalities not exhibited by the constituent materials. The development of new types of magnetic devices depends on engineering novel materials exhibiting desirable magnetic effects. Understanding the nature of ferromagnetism at the interface of composite materials may lead to new opportunities for designing magnetic materials and to be important for the development of ferromagnetic semiconductors for spintronics application. Magnetic correlations at the interface are expected to be important in determining the macroscopic properties of such systems. Ferromagnetism developing at the interface has been observed in oxide superlattices composed of antiferromagnetic insulating layers of $CaMnO_3$ and paramagnetic metallic layers of $CaRuO_3$ [102] and has also been seen in superlattices composed of the halfmetallic ferromagnet $La_{2/3}Ca_{1/3}MnO_3$ and the high-temperature superconductor $YBa_2Cu_3O_7$. These latter studies used absorption spectroscopy with circularly polarized X-rays and off-specular neutron reflectometry to probe the interplay between competing electronic order parameters in oxide heterostructures [103]. The observation of ferromagnetism in non-magnetic cation (Cu^{2+}) doped ZnO was surprising [104, 105]. However, the origin of the FM in Cu-doped ZnO is unclear [106], and as we discussed in the previous section the possible role of CuO planar nanoclusters in promoting a net moment is unavoidable. Crystallographically coherent secondary phase nanocrystals present within the host oxide lattice has been suggested for the origin of such puzzling FM in certain DMSO [107]. If the concentration of dopant ions exceeds the solubility limit, spinodal decomposition leads to regions with lower and higher densities of magnetic ions [107, 108]. Such mechanisms are believed to be responsible for Co metal clusters in (Zn,Co)O [109], the ZnMnO metastable phase in (Zn,Mn)O [89], CuO nanoplanar clusters in (Zn,Cu)O [30], and the Cr-rich (Zn,Cr)Te metallic nanocrystals embedded in a Cr deficient (Zn,Cr)Te matrix [108]. These secondary phases have traditionally been ruled out as the origin of the FM moment, as they order antiferromagnetically (AFM). However, as discussed in the latter part of this section uncompensated spins at the surface of Zn rich CoO [28], and Co rich (Zn, Co)O [110], lead to FM, and that AFM nanoparticles exhibit clear FM signatures. It is important to clarify that how finite size and surface/interface effects in AFM can lead to FM signals in order to understand the origin of FM in DMS oxides. The magnetic properties of CuO-ZnO heterostructure multilayers were studied to elucidate the origin of the ferromagnetic signature in Cu doped ZnO. Systematic studies were carried out to analyse the magnetic properties at the interface of antiferromagnetic CuO and diamagnetic ZnO thin layers of three different multilayer samples (Figure 7): (i) CuO (150 nm) on sapphire(0001), (ii) [ZnO(350nm)/CuO(150nm)/ZnO (350nm)] trilayer (ZCZ) on sapphire (0001), (iii) and ten layers of [CuO(~15 nm)/ZnO(~70 nm)] on sapphire (0001) with the top and bottom layers being ZnO $[ZCZ]_{10}$. These three sample geometries were chosen to control the area of the CuO-ZnO interface, while maintaining a fixed amount of CuO and ZnO (for ZCZ and $[ZCZ]_{10}$) in the different multilayers.

This is consistent with few reports of ferromagnetism in ZnO thin films [47]. Similar studies are being carried out in multilayer structures comprising of CoO and ZnO. Our preliminary data shows that unlike the CuO/ZnO, the CoO/ZnO system show significant decrease in magnetization by an order of magnitude with the increase in the surface area of about one order. This clearly shows that CoO/ZnO interface is not contributing to the magnetic moment. A more convincing evidence for the surface ferromagnetism is reported in the nanostructured thin films of $Co_{3-y}Zn_yO_4$ coated on sapphire substrates. Structural compatibility, solubility and variations in the valence state of Co in ZnO which depend strongly on the processing conditions will lead to possible secondary phases existing coherently grown within ZnO lattice.

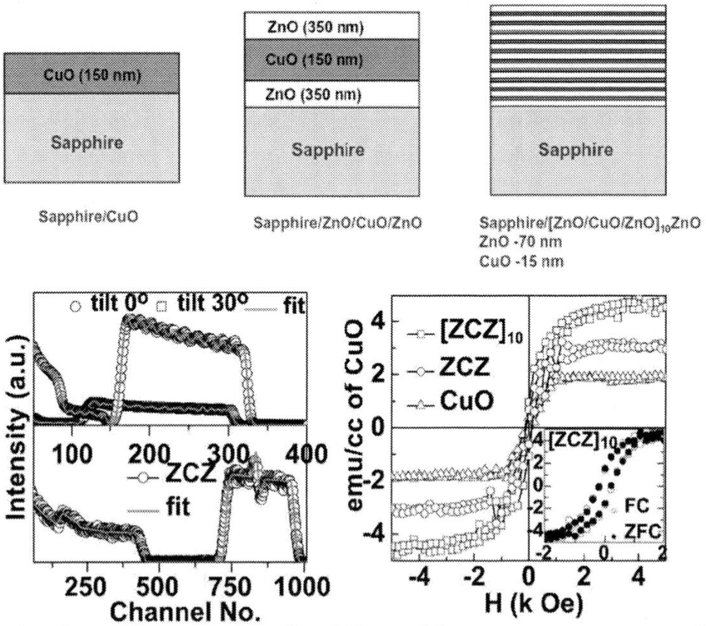

Figure 7. Schematic of sample geometries of multilayered heterostructures to control the area of the CuO-ZnO interface. The amount of CuO and ZnO are fixed in different multilayers, while the interface is increased by an order of magnitude in the [ZCZ]10 compared to ZCZ. (Bottom left) RBS spectra of [ZCZ]10 (top plot) and ZCZ (bottom plot) multilayer samples. (Bottom right). Hysteresis loops of CuO, ZCZ, and [ZCZ]10 samples measured as a function of external applied field at 300 K. Inset shows the hysteresis loop measured at 10 K after ZFC (closed circle) and FC (open circle) in 5 T for [ZCZ]10. The linear field dependence at high field is subtracted from the data.

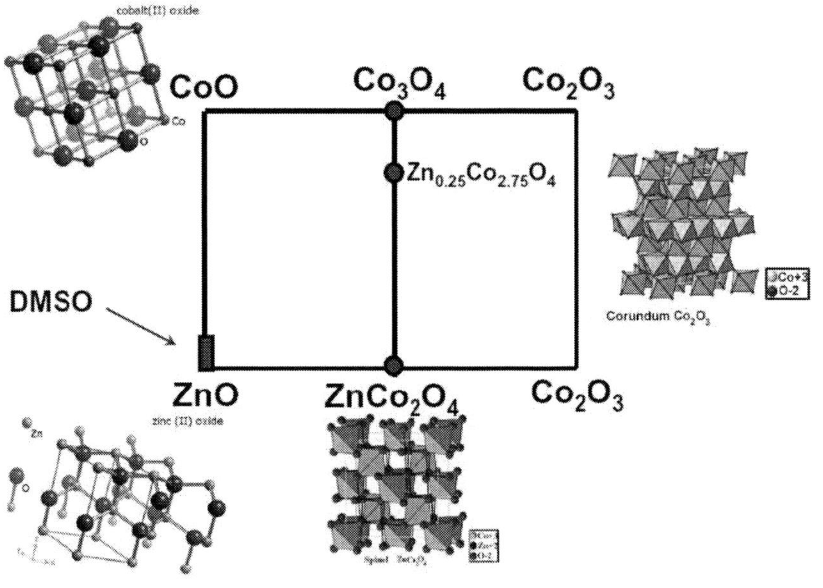

Figure 8. To investigate the role of transition metal oxide secondary phases on the development of ferromagnetism in DMSO the structural and magnetic properties of Zn doped Co_3O_4 are studied. Exploring Zn-Co-O compositions from the Co rich side will complement existing studies on Co doped ZnO by explicitly considering the effect of a Co_3O_4 phase on the magnetic properties. The shaded rectangle region shows the spintronic material composition regions. The shaded dot shows the materials studied to understand the surface and interface role on the magnetic properties.

It will be highly challenging to identify the source of magnetism in such system. Exploring Zn-Co-O compounds from the Co rich side will complement existing studies on Co doped ZnO by explicitly considering the effect of a Co_3O_4 related phase on the magnetic properties (Figure 8). Thin films of $Co_{3-y}Zn_yO_4$ become ferromagnetic with a magnetic moment of ~15 m μ_B/Co on high vacuum annealing from a nonmagnetic state in the as prepared condition (Figure 9.b). This effect is similar to Co doped ZnO thin films as discussed in the section III. In the as-prepared condition $Co_{3-y}Zn_yO_4$ compounds exist in the spinel phase. When annealed in high vacuum (HV), the microstructure of these samples exhibit a crystalline Zn:CoO core with an amorphous Co:ZnO surface component, having a high concentration of oxygen vacancy defects (Figure 9.a). HV annealing reduces the Co in 3+ state to 2+ state and therefore transforming the spinel phase compounds into NaCl type CoO phase [28]. Depending on the Zn concentration in $Co_{3-y}Zn_yO_4$ phase the concentration of Zn in the core and surface may differ (it is highly controlled by the phase stability and miscibility of Zn in CoO). The effect of difference in the Zn in the core can be seen by the change in the Neel temperature of the compounds. Magnetization studies reveal an exchange bias of up to H_E ~ 500 Oe at 5K for the HV annealed Co_2ZnO_4 films, despite the absence of any Co metal clusters Figure 9.c. This exchange anisotropy is due to the coupling between weak ferromagnetism produced by uncompensated surface spins and the antiferromagnetic spins at the surface of the CoO core. H_E is smaller for samples with smaller Zn fractions (for Co_3O_4HV, H_E ~250 Oe). Adding Zn to the system significantly enhances the exchange coupling, but does not significantly modify the magnetic moment per Co. These results clearly demonstrate that if small cluster of secondary phases in the antiferromagnetic state coexist in ZnO it could contribute to the weak ferromagnetism which originate from the interface of the crystalline core with a disordered oxygen deficient amorphous surface layers.

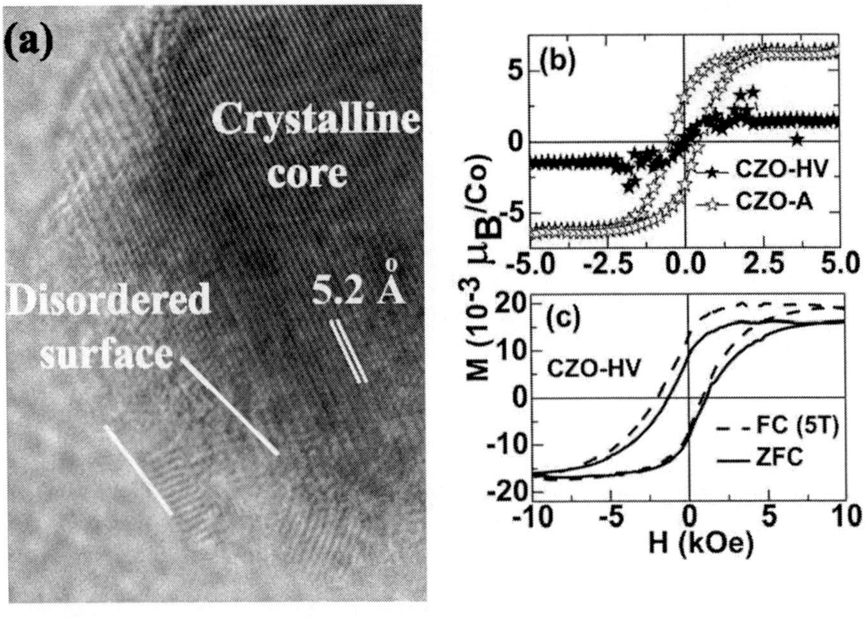

Figure 9. (a) HRTEM image of HV annealed Co_2ZnO_4 nanoparticle surface of disordered nature and crystalline core. (b) Room temperature M-H loop for the air (CZO-A) and HV annealed (CZO-HV) thin films of Co_2ZnO_4 (c). Zero field cooled (ZFC) and field cooled (FC) (with 5T) M-H curves measured at 10K for vacuum annealed Co_2ZnO_4 thin film.

3.4. Magnetism in Undoped Oxides:

Magnetic properties of oxides in the nanostructured form, in general, are very different than their bulk form [49]. Nanostructured materials typically can develop much higher defect concentrations than bulk systems. An interesting example of how point defects can affect the magnetic properties of metal oxides can be found in the observation of ferromagnetism in defect-rich, intrinsically diamagnetic semiconducting oxides [33, 48, 49, 71-73]. Though the studies on the transition metal doped oxide systems showed conflicting reports, subsequent experiments undoubtedly indicated the evidence for the crucial role played by oxygen vacancy defects in the development of ferromagnetic signal, with air annealed films (low oxygen vacancy defect concentration) having negligible magnetizations while vacuum annealed films (high oxygen vacancy defect concentration) exhibiting distinct ferromagnetism [26]. Signatures of ferromagnetic order reported in a range of metal oxides, beginning with undoped HfO_2 [72] and, subsequently in other oxides such as TiO_2 [71, 111], In_2O_3 [33], ZnO [47], and CeO_2 [112, 113], among many others [48, 49], lead to the conclusion that the development of ferromagnetism is solely driven by the presence of point defects. Because most of these systems do not have thermodynamically stable magnetic compositions, it is unlikely that the ferromagnetism arises from the precipitation of ferromagnetic impurity phases. It is found that the magnetic properties of these systems depend strongly on the nature of the point defects present [114], leading to suggestions of defect mediated ferromagnetism in metal oxide nanostructures [76].

It is known that point defects in diamagnetic metal oxides can introduce local moments [115]. The details of this local formation depend sensitively on the compound. For example, oxygen vacancies [116] and zinc interstitials have been predicted to be non-magnetic in wurzite ZnO, although there are reports of Zn interstitials enhancing the magnetic properties in doped ZnO [117], while oxygen interstitials and zinc vacancies are expected to show sizeable moments, ranging from roughly 0.2 μB [116] to 2 μB [118].

However, both oxygen and cerium vacancies are expected to contribute to the magnetic moment in defect-rich CeO_2 [113]. However, in the complete absence of interactions, defect induced moments would be expected to result in paramagnetic, rather than ferromagnetic behaviour, so the emergence of ferromagnetism in these metal oxide materials is rather surprising.

The strong connection between defects and ferromagnetism in metal oxide nanostructures is demonstrated by studies on TiO_2 thin films as discussed in the section III.1 [32]. Investigations on CeO_2, Al_2O_3, ZnO, In_2O_3, and SnO_2 [49, 75] nanoparticles, among others, find evidence for weak ferromagnetism, having small saturation magnetizations but clear hysteresis loops, albeit often with almost negligible coercivities. The moments in these nanostructured samples is very small, on the order of only 10^{-4} to 10^{-3} emu/g [49], but significantly larger than the completely negligible magnetizations observed in diamagnetic bulk samples. Sintering the samples at high temperatures in the presence of oxygen completely suppresses the magnetization [49], leading to the conclusion that the ferromagnetism may be intimately connected with the defect structure, specifically including oxygen vacancy defects, at the surface of the nanoparticles. It is suggested that unpaired electrons trapped on oxygen vacancies may be relevant for the development of ferromagnetism and, furthermore, that such ferromagnetic order may be a general characteristic of all metal oxide nanoparticles [3, 76].

CONCLUSIONS

Several examples of transition metal doped oxide and undoped oxide systems which showed weak ferromagnetism are discussed. The origin of ferromagnetism in these systems is not same and highly depends on the oxide system and dopant type, fabrication methods and processing conditions. Such differences arise due to changes brought in these oxides by the point defects (anion and cation), clusters and the surface disorders as discussed with examples. The oxygen defects seem to play a crucial role in controlling the magnetic properties in common. Controlling the defects in metal oxide nanostructures, understanding the stability and tunability of functional properties in an exquisitely defect controlled system is important to expand the functionality of these materials in developing the next generation of oxide based devices.

ACKNOWLEDGMENTS

Author would like to thank Prof. R. Naik, Prof. V.M. Naik and Prof. G. Lawes and members of their group at Wayne State University (WSU). Special thanks to Dr. P. Kharel, Dr. R.P. Panguluri, Amebsh Dixit, Prof. Suryanarayanan, Prof. J. Thakur, Prof. B. Nadgorny, Dr. B. Kirby, Dr. Sanjiv Kumar for their contributions during these works. The author would like to acknowledge the Jane and Frank Warchol Foundation and the Institute for Manufacturing Research at WSU for support to these works carried out at WSU.

REFERENCES

[1] Bruce, D.W., D. O'Hare, and R.I. Walton, Functional oxides. Inorganic materials series, ed. D.W. Bruce, D. O'Hare, and R.I. Walton. 2010.

[2] Matsumoto, Y., et al., Room-Temperature Ferromagnetism in Transparent Transition Metal-Doped Titanium Dioxide. *Science*, 2001. 291(5505): p. 854-856.

[3] Coey, J.M.D., Dilute magnetic oxides. *Current Opinion in Solid State and Materials Science*, 2006. 10(2): p. 83-92.

[4] Dietl, T., Dilute magnetic semiconductors: Functional ferromagnets. *Nat. Mater*, 2003. 2(10): p. 646-648.

[5] Ogale, S.B., Thin films and heterostructures for oxide electronics. 2005: Springer science.

[6] Ramanathan, S., Thin Film Metal-Oxides;Fundamentals and Applications in Electronics and Energy. 2010: Springer.

[7] Ohno, H., et al., Magnetotransport properties of p-type (In,Mn)As diluted magnetic III-V semiconductors. *Physical Review Letters*, 1992. 68(17): p. 2664.

[8] Ohno, H., et al., (Ga,Mn)As: A new diluted magnetic semiconductor based on GaAs. *Applied Physics Letters*, 1996. 69(3): p. 363-365.

[9] Pearton, S.J., et al., *Dilute magnetic semiconducting oxides Semiconductor Science and Technology*, 2004. 19(10): p. R59.

[10] Ogale, S.B., Dilute Doping, Defects, and Ferromagnetism in Metal Oxide Systems. *Advanced Materials*, 2010. 22(29): p. 3125-3155.
[11] Janisch, R., P. Gopal, and N.A. Spaldin, Transition metal-doped TiO2 and ZnO—present status of the field Journal of Physics: Condensed Matter, 2005. 17(27): p. R657.
[12] Dietl T, O.H., Matsukura F, Cibert J and F. D, Science, 2000. 287(5455): p. 1019.
[13] Wolf, S.A., et al., Spintronics: A Spin-Based Electronics Vision for the Future. *Science*, 2001. 294(5546): p. 1488-1495.
[14] Fukumura T, Y. Yamada, H. Toyosaki, T. Hasegawa, H. Koinuma. and M. Kawasaki, *Appl. Surface. Sci.,* 2004. 223(1-3): p. 62.
[15] Coey, J.M.D., M. Venkatesan, and C.B. Fitzgerald, Donor impurity band exchange in dilute ferromagnetic oxides. *Nat. Mater*, 2005. 4(2): p. 173-179.
[16] Prellier W, F.A. and B. Mercey, *J. Phys.: Condens. Matter*, 2003. 15(37): p. R1583.
[17] Venkatesan, M., et al., Anisotropic Ferromagnetism in Substituted Zinc Oxide. *Physical Review Letters*, 2004. 93(17): p. 177206.
[18] Philip, J., et al., Carrier-controlled ferromagnetism in transparent oxide semiconductors. *Nat. Mater*, 2006. 5(4): p. 298-304.
[19] Ozgur, U., et al., A comprehensive review of ZnO materials and devices. *Journal of Applied Physics,* 2005. 98(4).
[20] Sauvage, F., et al., Hierarchical TiO2 Photoanode for Dye-Sensitized Solar Cells. *Nano Letters, 2010.* 10(7): p. 2562-2567.
[21] Hernandez-Alonso, M.D., et al., Development of alternative photocatalysts to TiO2: Challenges and opportunities. *Energy and Environmental Science*, 2009. 2(12): p. 1231-1257.
[22] Gurlo, A., Nanosensors: towards morphological control of gas sensing activity. SnO2, In2O3, ZnO and WO3 case studies. *Nanoscale*, 2011.
[23] Kane, M.H., et al., Magnetic properties of bulk $Zn_{1-x}Mn_xO$ and $Zn_{1-x}Co_xO$ single crystals. Journal of Applied Physics, 2005. 97(2): p. 023906-6.
[24] Rao, C.N.R. and F.L. Deepak, Absence of ferromagnetism in Mn- and Co-doped ZnO. *Journal of Materials Chemistry*, 2005. 15(5): p. 573-578.
[25] Kolesnik, S. and B. Dabrowski, Absence of room temperature ferromagnetism in bulk Mn-doped ZnO. *Journal of Applied Physics*, 2004. 96(9): p. 5379-5381.
[26] Sudakar, C., et al., Raman spectroscopic studies of oxygen defects in Co-doped ZnO films exhibiting room-temperature ferromagnetism. *Journal of Physics-Condensed Matter*, 2007. 19(2): p. -.
[27] Sudakar, C., et al., Magnetism in $Zn_{1-x}Co_xO$ ($0 <= x < 0.1$) and $Co_{3-y}Zn_yO_4$ ($y = 0$, 0.25, and 1) thin films. *Journal of Applied Physics*, 2007. 101(9): p. 09H118-3.
[28] Sudakar, C., et al., Surface ferromagnetism and exchange bias in vacuum annealed $Co_{3-y}Zn_yO_4$ films. *Applied Physics Letters*, 2008. 92(6): p. 062501-3.
[29] Sudakar, C., et al., Coexistence of anion and cation vacancy defects in vacuum-annealed In2O3 thin films. 2010. 62(2): p. 63-66.
[30] Sudakar, C., et al., Ferromagnetism induced by planar nanoscale CuO inclusions in Cu-doped ZnO thin films. *Physical Review B*, 2007. 75(5): p. 054423.
[31] Sudakar, C., et al., Ferromagnetism in CuO--ZnO multilayers. *Applied Physics Letters*, 2008. 93(4): p. 042502-3.
[32] Sudakar, C., et al., Room temperature ferromagnetism in vacuum-annealed TiO_2 thin films. *Journal of Magnetism and Magnetic Materials*, 2008. 320(5): p. L31-L36.

[33] Panguluri, R.P., et al., Ferromagnetism and spin-polarized charge carriers in In_2O_3 thin films. *Physical Review B*, 2009. 79(Copyright (C) 2010 The American Physical Society): p. 165208.
[34] MacDonald, A.H., P. Schiffer, and N. Samarth, Ferromagnetic semiconductors: moving beyond (Ga,Mn)As. *Nat. Mater,* 2005. 4(3): p. 195-202.
[35] Chiba, D., et al., Effect of low-temperature annealing on (Ga,Mn)As trilayer structures. *Applied Physics Letters*, 2003. 82(18): p. 3020-3022.
[36] Dietl, T., et al., Zener model description of ferromagnetism in zinc-blende magnetic semiconductors. *Science*, 2000. 287: p. 1019-1022.
[37] Manivannan, A., et al., Magnetism of Co-doped titania thin films prepared by spray pyrolysis. *Applied Physics Letters*, 2003. 83(1): p. 111-113.
[38] Khaibullin, R.I. and et al., High Curie-temperature ferromagnetism in cobalt-implanted single-crystalline rutile. Journal of Physics: *Condensed Matter*, 2004. 16(41): p. L443.
[39] Nguyen, H.H., et al., Substrate effects on the room-temperature ferromagnetism in Co-doped TiO[sub 2] thin films grown by pulsed laser deposition. *Journal of Applied Physics*, 2004. 95(11): p. 7378-7380.
[40] Stampe, P.A., et al., Investigation of the cobalt distribution in the room temperature ferromagnet TiO[sub 2]:Co. *Journal of Applied Physics*, 2003. 93(10): p. 7864-7866.
[41] Song, C., et al., Giant magnetic moment in an anomalous ferromagnetic insulator: Co-doped ZnO. *Physical Review B*, 2006. 73(2): p. 024405.
[42] Lee, H.-J., et al., Study of diluted magnetic semiconductor: Co-doped ZnO. *Applied Physics Letters*, 2002. 81(21): p. 4020-4022.
[43] Kim, J.H., et al., Magnetic properties of epitaxially grown semiconducting Zn[sub 1 - x]Co[sub x]O thin films by pulsed laser deposition. *Journal of Applied Physics*, 2002. 92(10): p. 6066-6071.
[44] Ueda, K., H. Tabata, and T. Kawai, Magnetic and electric properties of transition-metal-doped ZnO films. *Applied Physics Letters*, 2001. 79(7): p. 988-990.
[45] Ogale, S.B., et al., High Temperature Ferromagnetism with a Giant Magnetic Moment in Transparent Co-doped $SnO_{2-\delta}$. *Physical Review Letters*, 2003. 91(7): p. 077205.
[46] Kale, S.N., et al., Magnetism in cobalt-doped Cu_2O thin films without and with Al, V, or Zn codopants. *Applied Physics Letters*, 2003. 82(13): p. 2100-2102.
[47] Hong, N.H., J. Sakai, and V. Brize, Observation of ferromagnetism at room temperature in ZnO thin films. *Journal of Physics-Condensed Matter*, 2007. 19(3): p. -.
[48] Hong, N.H., et al., Room-temperature ferromagnetism observed in undoped semiconducting and insulating oxide thin films. Physical Review B, 2006. 73(Copyright (C) 2010 The American Physical Society): p. 132404.
[49] Sundaresan, A., et al., Ferromagnetism as a universal feature of nanoparticles of the otherwise nonmagnetic oxides. *Physical Review B*, 2006. 74(Copyright (C) 2010 The American Physical Society): p. 161306.
[50] Kaspar, T.C., et al., Negligible Magnetism in Excellent Structural Quality Cr_{x}Ti_{1-x}O_{2} Anatase: Contrast with High-T_{C} Ferromagnetism in Structurally Defective $Cr_xTi_{1-x}O_2$. *Physical Review Letters*, 2005. 95(21): p. 217203.
[51] Behan, A.J., et al., Two Magnetic Regimes in Doped ZnO Corresponding to a Dilute Magnetic Semiconductor and a Dilute Magnetic Insulator. *Physical Review Letters*, 2008. 100(4): p. 047206.

[52] Larson, B.E., et al., Theory of exchange interactions and chemical trends in diluted magnetic semiconductors. *Physical Review B*, 1988. 37(8): p. 4137.

[53] Calderon, M.J., G. Gomez-Santos, L. Brey, Impurity-semiconductor band hybridization effects on the critical temperature of diluted magnetic semiconductors. *Phys. Rev. B,* 2002. 66: p. 075218.

[54] Dietl, T., H. Ohno, and F. Matsukura, Hole-mediated ferromagnetism in tetrahedrally coordinated semiconductors. *Physical Review B*, 2001. 63(19): p. 195205.

[55] Dietl, T., A. Haury, Y.M. d'Aubigne, *Phys. Rev. B*, 1997. 55: p. R3347.

[56] Durst, A.C., R.N. Bhatt, P. A. Wolff, Bound magnetic polaron interactions in insulating doped dilute magnetic semiconductors. *Phys. Rev. B*, 2002. 65: p. 235205.

[57] Kaminski, A., Das Sarma, S., Polaron percolation in dilute magnetic semiconductors. *Physical Review Letters*, 2002. 88: p. 247202.

[58] Calderón, M.J. and S. Das Sarma, Theory of carrier mediated ferromagnetism in dilute magnetic oxides. *Annals of Physics,* 2007. 322(11): p. 2618-2634.

[59] Koidl, P., Optical absorption of Co^{2+} in ZnO. *Physical Review B*, 1977. 15(5): p. 2493.

[60] Srikant, V. and D.R. Clarke, Optical absorption edge of ZnO thin films: The effect of substrate. *Journal of Applied Physics*, 1997. 81(9): p. 6357-6364.

[61] Serrano, J., et al., Pressure dependence of the lattice dynamics of ZnO: An ab initio approach. *Physical Review B*, 2004. 69(9): p. 094306.

[62] Zhang, X.N., et al., Phonon scattering and stability of $Na_{0.5}CoO_2$. *Physica B Condensed Matter:*, 2005. 359-361: p. 424-426.

[63] Ohtsuka, H., et al., A study on selective reduction of NOx by propane on Co-Beta. *Catalysis Letters*, 1997. 44(3): p. 265-270.

[64] Shi, Y.G., et al., Raman spectroscopy study of Na_xCoO_2 and superconducting Na_xCoO_2 $.yH_2O$. *Physical Review B*, 2004. 70(5): p. 052502.

[65] R. Merlin, A. Pinczuk, and W.H. Weber, *Raman Scattering in Materials Science*, ed. a.R.M. W.H. Weber. Vol. Vol. 1. 2000, Berlin: Springer.

[66] Griffin, K.A., et al., Cobalt-doped anatase TiO_2: A room temperature dilute magnetic dielectric material. *Journal of Applied Physics*, 2005. 97(10): p. 10D320-3.

[67] Weng, H., et al., Electronic structure and optical properties of the Co-doped anatase TiO_2 studied from first principles. *Physical Review B*, 2004. 69(12): p. 125219.

[68] Schwartz, D.A. and D.R. Gamelin, Reversible 300 K Ferromagnetic Ordering in a Diluted Magnetic Semiconductor. *Advanced Materials*, 2004. 16(23-24): p. 2115-2119.

[69] Venkatesan, M., C.B. Fitzgerald, and J.M.D. Coey, Thin films: Unexpected magnetism in a dielectric oxide. *Nature*, 2004. 430(7000): p. 630-630.

[70] Reece, M. and R. Morrell, Electron microscope study of non-stoichiometric titania. *Journal of Materials Science*, 1991. 26(20): p. 5566-5574.

[71] Yoon, S.D., et al., Oxygen-defect-induced magnetism to 880 K in semiconducting anatase $TiO_{2-\delta}$ films. *Journal of Physics-Condensed Matter*, 2006. 18(27): p. L355-L361.

[72] Coey, J.M.D., et al., Magnetism in hafnium dioxide. *Physical Review B*, 2005. 72(Copyright (C) 2010 The American Physical Society): p. 024450.

[73] Coey, J.M.D., High-temperature ferromagnetism in dilute magnetic oxides. *Journal of Applied Physics,* 2005. 97(10).

[74] Duhalde, S., et al., Appearance of room-temperature ferromagnetism in Cu-doped $TiO_{2-\delta}$ films. *Physical Review B,* 2005. 72(16): p. 161313.
[75] Kharel, P., et al., Room temperature ferromagnetism in Cr-doped In_2O_3 on high vacuum annealing of thin films and bulk samples. *Journal of Applied Physics*, 2007. 101(9): p. 09H117-3.
[76] Coey, J.M.D., d(0) ferromagnetism. *Solid State Sciences*, 2005. 7(6): p. 660-667.
[77] Xu, Q., et al., Spin Manipulation in Co-Doped ZnO. *Physical Review Letters*, 2008. 101(7): p. 076601.
[78] Yates, K.A., et al., Spin-polarized transport current in n-type codoped ZnO thin films measured by Andreev spectroscopy. *Physical Review B*, 2009. 80(24): p. 245207.
[79] Bellingham, J.R., A.P. Mackenzie, and W.A. Phillips, Precise measurements of oxygen content: Oxygen vacancies in transparent conducting indium oxide films. *Applied Physics Letters*, 1991. 58(22): p. 2506-2508.
[80] Elfimov, I.S., S. Yunoki, and G.A. Sawatzky, Possible Path to a New Class of Ferromagnetic and Half-Metallic Ferromagnetic Materials. *Physical Review Letters*, 2002. 89(21): p. 216403.
[81] Rahman, G., et al., Vacancy-induced magnetism in SnO_2 : A density functional study. *Physical Review B*, 2008. 78(18): p. 184404.
[82] Shiming, Y., et al., Effects of carbothermal annealing on structure defects, electrical and magnetic properties in Fe-doped In2O3. 2009. 61(4): p. 387-390.
[83] Osorio, G., et al., Nonstoichiometry as a source of magnetism in otherwise nonmagnetic oxides: Magnetically interacting cation vacancies and their percolation. *Physical Review B*, 2007. 75(18): p. 184421.
[84] Das Pemmaraju, C. and S. Sanvito, Ferromagnetism Driven by Intrinsic Point Defects in HfO_2. *Physical Review Letters*, 2005. 94(21): p. 217205.
[85] Huang, L.M. and et al., Tuning magnetic properties of In_2O_3 by control of intrinsic defects. EPL (Europhysics Letters), 2010. 89(4): p. 47005.
[86] Z.R. Xiao, et al., First-principles study of the magnetization of oxygen-depleted In2O3(001) *surfaces J. Phys.: Condens. Matter*, 2009. 21: p. 272202.
[87] Norton, D.P., Synthesis and properties of epitaxial electronic oxide thin-film materials. *Materials Science and Engineering R-Reports*, 2004. 43(5-6): p. 139-247.
[88] Janisch, R. and et al., Transition metal-doped TiO 2 and ZnO—present status of the field. Journal of Physics: *Condensed Matter*, 2005. 17(27): p. R657.
[89] Kundaliya, D.C., et al., On the origin of high-temperature ferromagnetism in the low-temperature-processed Mn-Zn-O system. *Nat. Mater*, 2004. 3(10): p. 709-714.
[90] Hsu, H.S., et al., The role of anomalous Hall effect in diluted magnetic semiconductors and oxides. *Applied Physics Letters*, 2010. 96(24): p. 242507-3.
[91] Zhang, S.X., et al., Magnetism and anomalous Hall effect in Co-(La,Sr)TiO_3. *Physical Review B*, 2007. 76(8): p. 085323.
[92] Xu, Q., et al., Room temperature ferromagnetism in ZnO films due to defects. *Applied Physics Letters*, 2008. 92(8): p. 082508-3.
[93] Zhang, Z.H., et al., Evidence of intrinsic ferromagnetism in individual dilute magnetic semiconducting nanostructures. *Nat. Nano*, 2009. 4(8): p. 523-527.
[94] Punnoose, A., et al., Bulk to nanoscale magnetism and exchange bias in CuO nanoparticles. *Physical Review B*, 2001. 64(17): p. 174420.

[95] Filippetti, A. and V. Fiorentini, Magnetic Ordering in CuO from First Principles: A Cuprate Antiferromagnet with Fully Three-Dimensional Exchange Interactions. *Physical Review Letters*, 2005. 95(8): p. 086405.
[96] Feng, X., Electronic structures and ferromagnetism of Cu- and Mn-doped ZnO. Journal of Physics: *Condensed Matter*, 2004. 16(24): p. 4251.
[97] Q. Ma, et al., Comparative Study of the Local Structures of Diluted Magnetic Semiconducting Cu:ZnO Thin Films. in communication.
[98] Jayakumar, O.D., et al., Magnetic properties of hydrogenated Li and Co doped ZnO nanoparticles. *Applied Physics Letters*, 2006. 89(20): p. 202507-3.
[99] Hernando, A., et al., Origin of Orbital Ferromagnetism and Giant Magnetic Anisotropy at the Nanoscale. *Physical Review Letters*, 2006. 96(5): p. 057206.
[100] Brinkman, A., et al., Magnetic effects at the interface between non-magnetic oxides. *Nat. Mater*, 2007. 6(7): p. 493-496.
[101] R.H. Kodama and A.E. Berkowitz, surface driven effects on the magnetic behavior of oxide nanoparticles in Surface effects in magnetic nanoparticls, ed. D. Fiorani. 2005: Springer.
[102] Takahashi, K.S., M. Kawasaki, and Y. Tokura, Interface ferromagnetism in oxide superlattices of $CaMnO_3/CaRuO_3$. *Applied Physics Letters*, 2001. 79(9): p. 1324-1326.
[103] Chakhalian, J., et al., Magnetism at the interface between ferromagnetic and superconducting oxides. *Nat. Phys*, 2006. 2(4): p. 244-248.
[104] Buchholz, D.B., et al., Room-temperature ferromagnetism in Cu-doped ZnO thin films. *Applied Physics Letters*, 2005. 87(8): p. 082504-3.
[105] Wang, X., et al., Aggregation-based growth and magnetic properties of inhomogeneous Cu-doped ZnO nanocrystals. *Applied Physics Letters*, 2007. 90(21): p. 212502-3.
[106] Keavney, D.J., et al., Where does the spin reside in ferromagnetic Cu-doped ZnO? Applied Physics Letters, 2007. 91(1): p. 012501-3.
[107] Dietl, T., *Nat. Mater*, 2006. 5: p. 673.
[108] Kuroda, S., et al., Origin and control of high-temperature ferromagnetism in semiconductors. *Nat. Mater*, 2007. 6(6): p. 440-446.
[109] Deka, S., R. Pasricha, and P.A. Joy, Experimental comparison of the structural, magnetic, electronic, and optical properties of ferromagnetic and paramagnetic polycrystalline $Zn_{1-x}Co_xO$ (x=0,0.05,0.1). *Physical Review B*, 2006. 74(3): p. 033201.
[110] Dietl, T., et al., Origin of ferromagnetism in $Zn_{1-x}Co_xO$ from magnetization and spin-dependent magnetoresistance measurements. *Physical Review B*, 2007. 76(15): p. 155312.
[111] Yoon, S.D., et al., Magnetic semiconducting anatase $TiO_{2-\delta}$ grown on (1 0 0) $LaAlO_3$ having magnetic order up to 880 K. *Journal of Magnetism and Magnetic Materials*, 2007. 309(2): p. 171-175.
[112] Liu, Y. and et al., Size dependent ferromagnetism in cerium oxide (CeO_2) nanostructures independent of oxygen vacancies. *Journal of Physics: Condensed Matter*, 2008. 20(16): p. 165201.
[113] Han, X., J. Lee, and H.-I. Yoo, Oxygen-vacancy-induced ferromagnetism in CeO_2 from first principles. *Physical Review B*, 2009. 79(10): p. 100403.

[114] Song, Y.-Q. and et al., Direct evidence of oxygen vacancy mediated ferromagnetism of Co doped CeO_2 thin films on Al_2O_3 (0001) substrates. Journal of Physics: Condensed Matter, 2008. 20(25): p. 255210. Stoneham, M., The strange magnetism of oxides and carbons. *Journal of Physics: Condensed Matter*, 2010. 22(7): p. 074211.
[115] Kim, D. and et al., Ferromagnetism induced by Zn vacancy defect and lattice distortion in ZnO. *Journal of Applied Physics*, 2009. 106(1): p. 013908.
[116] Herng, T.S., et al., Zn-interstitial-enhanced ferromagnetism in Cu-doped ZnO films. *Journal of Magnetism and Magnetic Materials*, 2007. 315(2): p. 107-110.
[117] Zuo, X., et al., Ferromagnetism in pure wurtzite zinc oxide. *Journal of Applied Physics*, 2009. 105(7): p. 07C508-3.

In: Magnetic Thin Films
Editor: John P. Volkerts
ISBN: 978-1-61209-302-4

Chapter 4

DEVELOPMENTS IN GIANT MAGNETORESISTANCE AND TUNNELING MAGNETORESISTANCE BASED SPINTRONIC DEVICES WITH PERPENDICULAR ANISOTROPY

Seongtae Bae and Naganivetha Thiyagarajah
Biomagnetics Laboratory, Department of Electrical and Computer Engineering,
National University of Singapore, 117576, Singapore

INTRODUCTION

In recent years, there has been a dramatic increase in the interest towards application of giant magnetoresistance (GMR) spin-valves and magnetic tunneling junctions (MTJs) with perpendicular anisotropy in spintronics, such as a spin transfer switching (STS) magnetic random access memory (MRAM), ultra high density magnetic information devices, and low field detection spin oscillators. This interest is driven by the fact that spin-valves and MTJs with perpendicular anisotropy are expected to provide technical promises such as high thermal and magnetic stabilities that will allow the realization of extremely low dimensional and high reliability devices in more advanced spintronic applications [1-2]. In this chapter, the recent developments in GMR and tunneling magnetoresistance (TMR) spin valves and devices with perpendicular anisotropy will be reviewed and presented with distinct seven sections to understand their technical roles in advanced spintronics applications. The first section will deal with the physical origins of perpendicular anisotropy of the magnetic materials used for GMR spin-valves studied up to now. The fabrication of the magnetic thin films with perpendicular anisotropy including the optimization of film deposition conditions and the fabrication process of nano-meter sized devices will be included in this section. The GMR performance in various spin-valve structures with perpendicular anisotropy and their magnetic and thermal stabilities will also be disused in this section. The second section will focus on the physical nature of GMR and its correlation with interlayer coupling in different kinds of spin valves with perpendicular anisotropy. A newly proposed physical model of the GMR behavior interpreted in terms of the physical correlation between perpendicular anisotropy and magentostatic energy and its extension to the understanding of underlying physics of perpendicular interlayer coupling including RKKY oscillation and Néel coupling types of indirect exchange coupling will be dealt with in this section. The third section will discuss on a physical model of exchange bias and the effects of nanopatterning on the exchange bias characteristics in perpendicularly magnetized ferrimagentic/anti-ferromagentic thin films with perpendicular anisotropy for optimizing exchange biased GMR spin valves with perpendicular anisotropy. The fourth section looks at anomalous peak behavior observed in Hall effect measurements of exchange biased spin valves with perpendicular anisotropy. The fifth section will focus on magnetic tunnel junctions (MTJs) with perpendicular anisotropy. The basic theories of tunneling magnetoresistance (TMR), the initial and the recent research achievements of MTJs with perpendicular anisotropy in current spintronics will be reviewed and presented in this section. The sixth section will look at the current and potential applications of GMR and TMR devices with perpendicular anisotropy including spin-transfer switched MRAM and spin oscillator devices. The physical mechanisms and the research into the optimization of perpendicular anisotropy materials for these applications will be discussed. Finally, this chapter will be concluded with the survey on the advantages of GMR and TMR devices with perpendicular anisotropy and the future challenges targeting for the further developments in advanced spintronics applications.

1. Magnetic Materials with Perpendicular Anisotropy for Spin-Valves

This section will review perpendicular anisotropy materials considering for spin valves and the physical origin of the perpendicular anisotropy studied up to now. The fabrication of magnetic thin films with perpendicular anisotropy including the optimized deposition conditions, the fabrication process of nano-meter size devices with current-in-plane (CIP) and current-perpendicular-to-the plane (CPP) configurations as well as its structure for CIP measurement will be discussed in this section. In addition, the GMR behavior in spin-valves with perpendicular anisotropy and the magnetic and thermal stability of nano-patterned GMR spin-valves will be described in this section.

1.1. Physical Origin of Perpendicular Magnetic Anisotropy

Perpendicular magnetic anisotropy has been observed in several magnetic materials including multi-layers such as Co/Pt, Co/Pd, Co/Ni, CoFe/Pt, and CoFe/Pd, Co/Cr/Pt, alloys such as CoPt, FePt, and CoCr, and rare-earth transition metal (RE-TM) alloys such as GdFeCo, and TbFeCo [2-9]. According to the previous reports, the effective perpendicular anisotropy of these materials has been generally presented by a combination of crystalline and stress induced anisotropy expressed as Eq. (1):

$$K_{FM,\,eff} = K_{FM,\,crystalline} + K_{FM,\,stress} = (k_{u,crystalline} + 2k_{u,stress}/t_{FM}) - 2\pi M_s^2 \qquad (1)$$

In the case of multi-layered thin films, both crystalline and stress-induced anisotropy contribute to generating the perpendicular anisotropy. Several studies and reviews on the origin of perpendicular anisotropy in various kinds of multi-layers combined of Co, Fe and Ni with Pd, Pt, Au, Cu and Cr have been made for the last few decades [10-15]. As the magnetic layers in these multi-layered strucutres become thinner, the contribution of surfaces and interfaces become dominant in generating the perpendicular anisotropy compared to the crystal structures (bulk properties). It has been demonstrated that smoother interface gives rise to a higher Neel surface anisotropy (interfacial perpendicular anisotropy). The interface anisotropy can be several orders of magnitude larger than magntocrystalline anisotropy and leads to aligning the net magnetization in the perpendicular direction [10]. For most of Co/X multi-layers, the perpendicular anisotropy is higher when the X is a noble metal with a larger lattice constant than Co. For example, Co/Pd (or Pt) multi-layers, the lattice mismatch was found to be more than 10 % resulting in exhibiting a high perpendicular anisotorpy. This indicates that the strain caused by lattice mismatch directly relevant to the stress-induced anisotropy contributes to the perpendicular anisotropy in these systems [11]. In addition, strong crystalline texture associated with crystalline anisotropy as well as Co-Pd or Pt mixture coherently formed at the interfaces of the multi-layered structures contributes to the perpendicular anisotropy [13].

On the other hand, ordered CoPt and FePt single layered thin films with a tetragonal $L1_0$ structure exhibit very high perpendicular magnetic anisotropy. Unlike the multi-layered [Co/Pd (Pt)] thin films, the perpendicular anisotropy in these materials is mainly due to the crystalline anisotropy. The strong magnetocrystalline anisotropy has been found to be attributed to the strong hybridization between the Pt 5-*d* band and Co or Fe 3-*d* band electronic states. Some of the key challenges for CoPt and FePt alloys were the reduction of the ordering temperature, and the control of (001) texture. A highly oriented (001) crystal structure has been achieved using MgO underlayers and thermal annealing at a temperature of 350 – 400 °C [16,17].

However, FePt has been favored for the use in ultrahigh density recording media rather than GMR spin valve materials due to its large perpendicular anisotropy of 7×10^7 erg/cm^3. While, the lower magnetization value of CoPt is favored for spin-transfer driven magnetization switching devices [18-21]. The mechanism of perpendicular anisotropy in RE-TM alloys has not yet been fully understood. RE-TM alloys exist in a mixed crystalline and amorphous state. Several mechanisms including pair ordering, columnar microstructures, single ion anisotropy, exchange anisotropy, bond-orientation anisotropy and anti-parallel dipole energies have been considered for the main physical origin of perpendicular anisotropy in these materials [22-28]. Perpendicular anisotropy, which is material and crystal strucutre dependent, has been found to be sensitive to thin film deposition conditions such as interfacial roughness, microstructures, formation of interfacial alloys and mechanical stress. Therefore, selection of suitable film growth method for special technical purpose such as MBE (Molecular Beam Epitaxial) or other evaporation techniques, sputter deposition, laser ablation deposition and electro-deposition is one of the most important factors to control the perpendicular anisotropy of the materials. Sputter deposition is more commonly used in both research and industrial applications. Higher deposition energies lead to flat multi-layers with fewer defects (dense films) however; there are more instances of inter-diffusion and stresses at the interfaces. In order to reduce these undesirable process-induced magnetic and structural degradation, sputter working parameters such as working gas pressure, input sputtering power, and inert gas are mainly controlled to adjust the stress, the roughness, the kinetic energy of sputtered atoms, and the grain size of the multi-layers. A lower sputtering kinetic energy, a higher pressure or a heavier inert gas is typically preferred to make less energetic atoms at the surface of the films. All the magnetic and non-magnetic thin films considering for spin valve multi-layers with perpendicular anisotropy can be easily controlled their magnetoelectronic and structural properties by manipulating the sputter process conditions.

1.2. Nanometer Scale Device Fabrication

Nano-meter scale patterning and device fabrication techniques are essential in determining the magnetic and thermal stability as well as in actual device realization. Electron beam lithography techniques are required to fabricate devices with nano-meter dimension. Giant magnetoresistance (GMR) with perpendicular anisotropy can be measured simply using current-in-plane (CIP) method with unpatterned films.

For simple device characterization with nanometer dimension, the device area is needed to be defined by electron beam lithography and ion-beam etching with subsequent electrode alignment and deposition on either side of the patterned device as shown in Figure 1. In CIP measurements, there is current shunting through low resistance layers thus reducing the actual resistance signal from the GMR layers. For current-perpendicular to plane (CPP) geometry, the electrodes are above and below the spin-valve or MTJ structure such that the current direction is perpendicular to the film planes. The CPP devices require well defined nano-device fabrication.

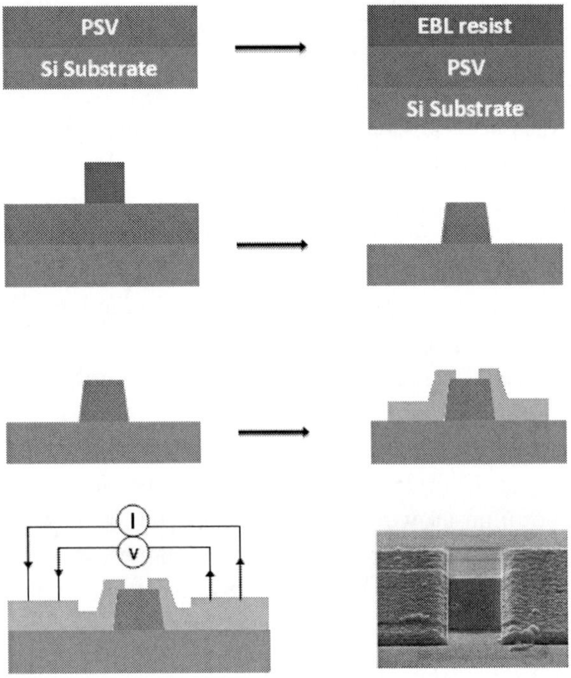

Figure 1. Fabrication process of nano-size controlled spin valves devices with perpendicular anisotropy for CIP measurement.

Generally, the CPP fabrication is based on either subtractive or additive processes. In the subtractive process, the multi-layer structure is first deposited onto the substrate. The pillar shaped structure is subsequently fabricated by masked etching steps to remove parts of the film. After depositing electrical isolation with an oxide film and etch-back or planarization process, the top electrode is deposited. A subtractive process of CPP fabrication is illustrated in Figure 2. An additive process uses a predefined structure on the substrate as a mask to define the CPP structure as illustrated in Figure 3. This process is simpler than the subtractive process in that magnetic materials have problems associated with etching. A method of defining the CPP structure without etching is potentially preferred.

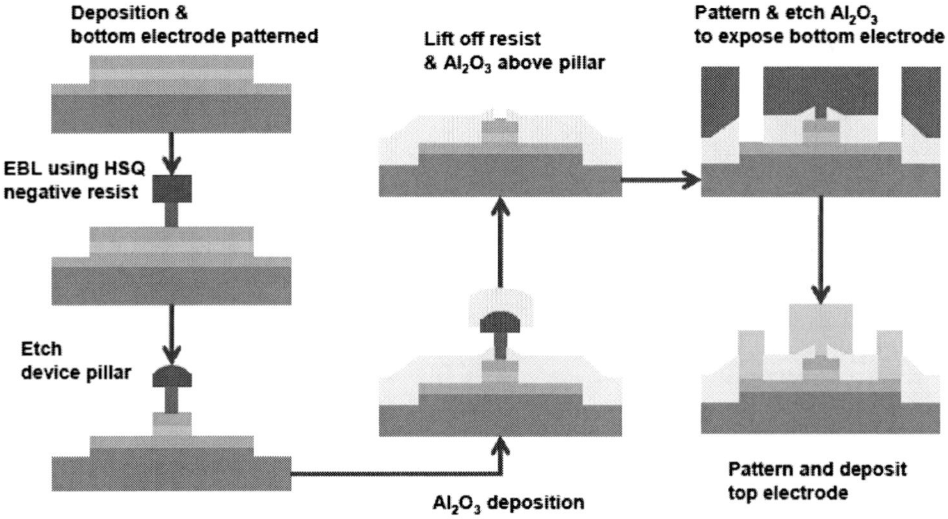

Figure 2.Current-perpendicular to plane (CPP) device fabrication (subtractive).

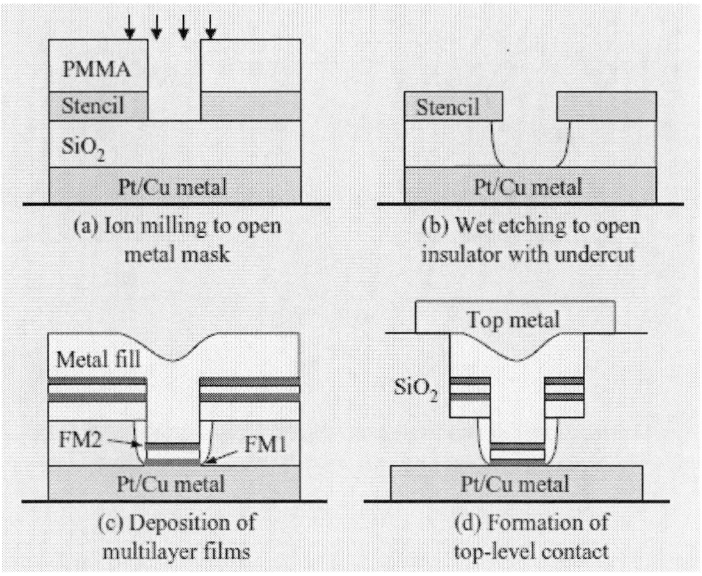

Figure 3. Current-perpendicular to plane (CPP) device fabrication (additive) [29].

1.3. GMR Behaviour in Spin-Valves with Perpendicular Anisotropy

The GMR effect originates from the spin-dependent scattering of majority and minority electrons passing through the magnetic layers. If an electron spin is parallel to the magnetization of the magnetic layers, it experiences weak scattering and hence a low resistance channel, while the electron with the opposing spin forms a high resistance channel. If the magnetic layers are anti-parallel with opposing magnetization directions, each spin direction experiences strong scattering in the magnetic layer whose magnetic moments are opposite to it. This results in a high resistance state. Based on Motts two-current model [30], the GMR phenomenon in a spin-valve system may be described as illustrated in Figure 4.

Based on the two-current model in Figure 4, the parallel and anti-parallel resistance states are given by Eq. (2).

$$R_P = \frac{R_\uparrow R_\downarrow}{R_\uparrow + R_\downarrow} \qquad R_{AP} = \frac{R_\uparrow + R_\downarrow}{2} \tag{2}$$

Accordingly, the GMR can be defined by Eq. (3),

$$GMR = \frac{R_{AP} - R_P}{R_P} = \frac{(R_\uparrow - R_\downarrow)^2}{4R_\uparrow R_\downarrow} \tag{3}$$

The GMR effects may be observed in several spin valve configurations with perpendicular anisotropy. The GMR curves of several [Co/Pd], [Co/Pt] and [Co/Ni] based pseudo spin-valves and exchange biased spin-valves presented in recent works are shown in Figures 5 ~ 8.

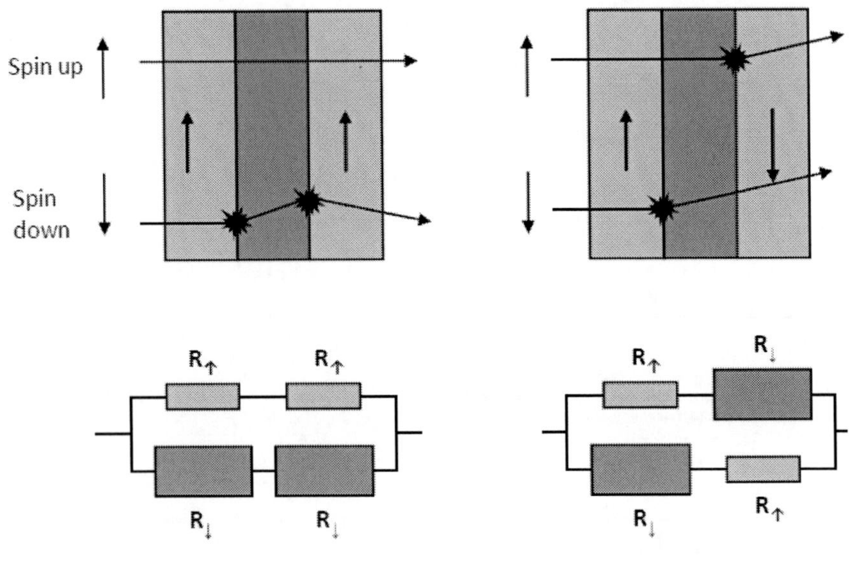

Figure 4. Two current model and equivalent resistor network showing GMR effect.

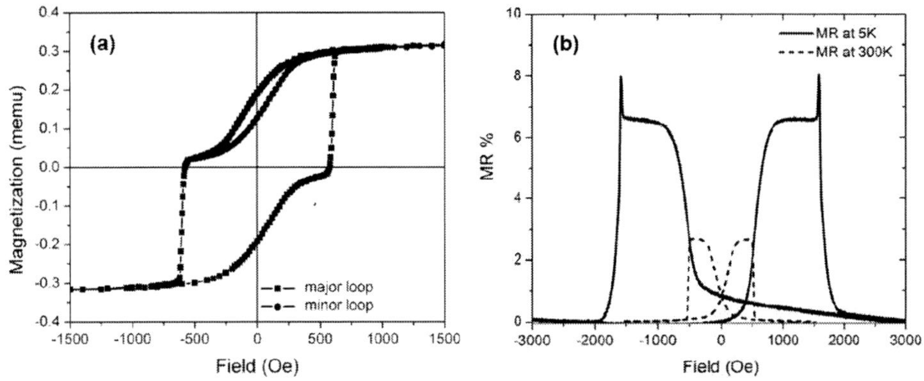

Figure 5.M-H (a) and GMR (b) curves of $[Co/Pd]_2/Cu/[Co/Pd]_4$ pseudo spin valves with perpendicular anisotropy [31].

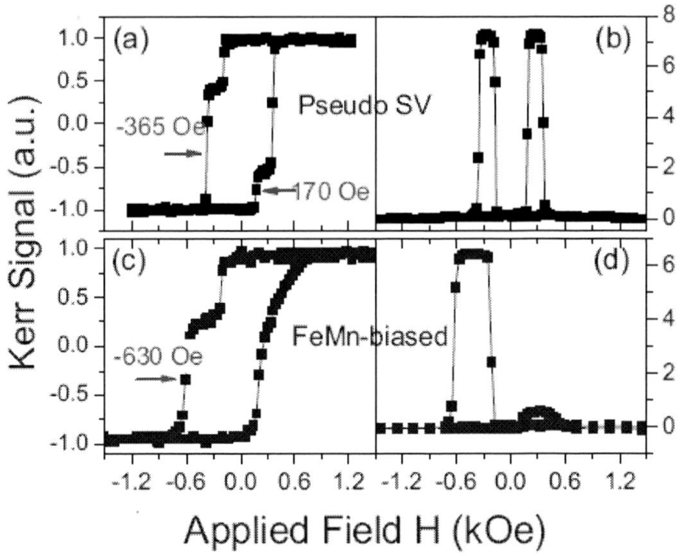

Figure 6. M-H and GMR loops of $[Co/Ni]_3/Cu/[Co/Ni]_5$ pseudo (top) and exchange biased (bottom) spin valves [32].

In a pseudo spin-valve consisting of two magnetic layers separated by a non-magnetic spacer, one of the magnetic layers is defined as a "hard layer" with a higher coercivity and the other is defined as a "soft layer" with a lower coercivity (see Figure 5 and 6). Due to the difference in coercivity, the magnetic moments of the hard and soft layers are magnetically reversed at different values of the applied magnetic field, providing a field range in which they are anti-parallel spin state (high resistance state). However, as shown in Figure 7. and 8, the exchange biased spin valve has an anti-ferromagnetic layer pinning the magnetization of one of the magnetic layers (pinned layer) by direct exchange coupling while, the other is free to rotate responded to the applied magnetic field. Due to the pinning caused by the exchange bias (shown as a hysteresis shift in hysteresis loop, (Figure 7 and 8), the free and pinned perpendicularly magnetized ferromagnetic multi-layers or single layer exhibit anti-parallel spin state leading to a high resistance state. As can be seen in Figure 7.(b), the [Co/Pd] based spin-valves exchange biased by FeMn anti-ferromagnetic layer had a GMR ratio in the range between 4 and 10 % depending on Co layer thickness. The main mechanism of GMR effect in this structure is understood in terms of spin dependent scattering at the Cu/[Co/Pd] interfaces. The strong dependence of Co layer thickness on the GMR ratio indicates that bulk scattering is another physical contribution to the magnetoresistance similar to the spin-valves with in-plane anisotropy.

Several other structures have also been introduced to improve the GMR by improving the spin-dependent scattering. One of these is a dual spin valve in which the free layer is placed between the two pinned layers to increase the number of spin-dependent scattering interfaces (or centers) for enhancing GMR performance. Figure 9. shows the GMR curve of a dual spin valve consisting of [CoFe/Pd] based free layer sandwiched between the two [Co/Pd] based hard layers with different spin-polarization ratio. This structure can be used not only for improving GMR ratio but also for the implementation of a multistate storage device because it can allow for four distinct resistance states under the externally applied switching fields.

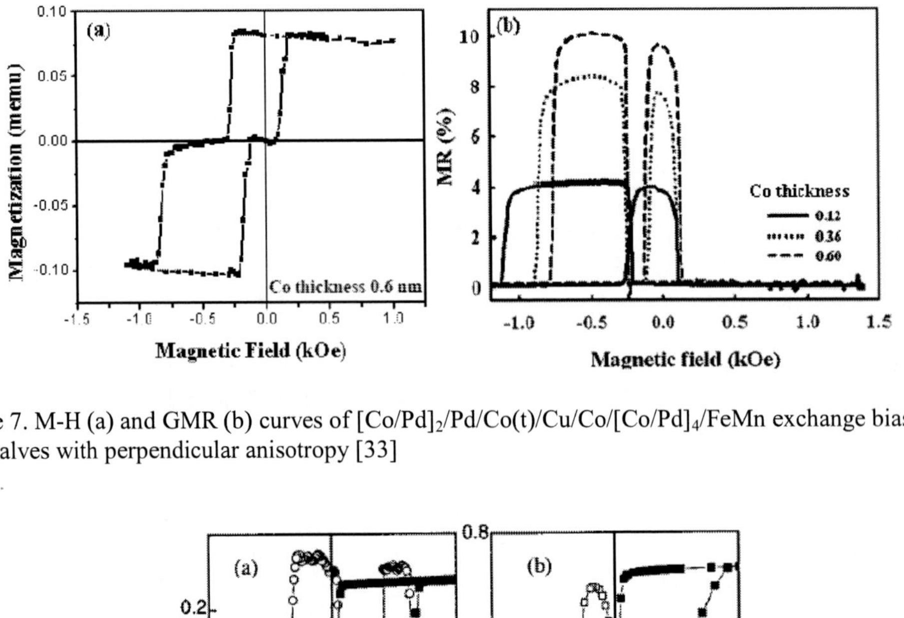

Figure 7. M-H (a) and GMR (b) curves of $[Co/Pd]_2/Pd/Co(t)/Cu/Co/[Co/Pd]_4/FeMn$ exchange biased spin valves with perpendicular anisotropy [33]

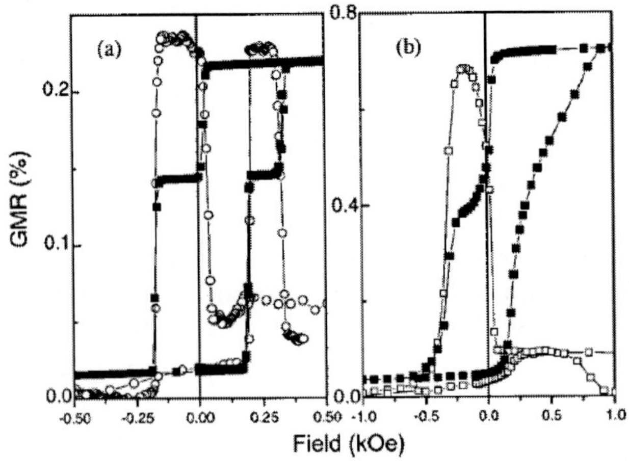

Figure 8. Extraordinary Hall effect (solid) and GMR (open) loops for $[Co/Pt]_5/Co/Cu/Co/[Pt/Co]_5/$ FeMn exchange biased spin valves [34].

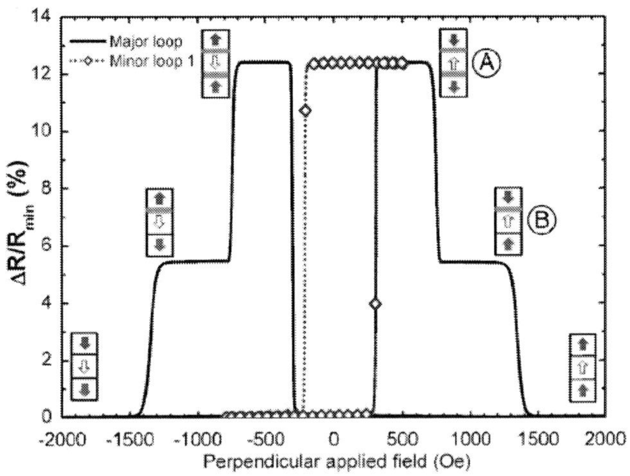

Figure 9. GMR curve of $[Co/Pd]_4/Cu/CoFe/[Pd/CoFe]_3/Cu/Co/[Pd/Co]_4$ dual pseudo spin-valve structure (minor loop in dotted line) [35].

1.4. Magnetic and Thermal Stability

One of the most important properties of perpendicular anisotropy materials is that it can be scaled down to sub-micron or even nano-meter dimensions with high magnetic and thermal stabilities. Nishimura *et al.* [1] demonstrated uniform perpendicular magnetization in 300 x 300 nm patterned GdFe/FeCo layers as shown in Figure 10. Other works [36, 37] have also demonstrated single domain magnetization in Co/Pd multilayer dot arrays.

In [38], the researchers reported on the demonstration of high magnetic and thermal stabilities in Co/Pd based spin-valves with perpendicular anisotropy. Figure 11. (a) and (b) show the GMR behavior of the nano-patterned Co/Pd based spin-valves with perpendicular anisotropy (PPSV), and NiFe/Co based spin-valves with in-plane anisotropy (IPSV), respectively.

Figure 10. Magnetic-force microscopy images for the in-plane magnetization 40-nm-thick 0.5 mm x 0.5 mm square NiFe element (top) and perpendicular magnetization 100-nm-thick square GdFe/ FeCo elements; 0.5 mmx0.5 mm, and 0.3 mmx0.3 mm (bottom) at zero field [1].

As can be seen in Figure 11. (a), the GMR of the IPSV with 500 × 500 nm^2 device size was reduced by 80 % and the soft layer coercivity was increased from 12 Oe to 300 Oe (or by 2400 %). The severe increase in coercivity is due to the increased demagnetization field that results in trapped vortex magnetizations, which require larger switching fields and lead to anomalous switching behavior. They also exhibited a broad switching field characteristic, which indicates incoherent switching in the nano-patterned devices. In contrast, the PPSV with 500 × 500 nm^2 device size showed a 34 % reduction in GMR and a 16 % increase in coercivity. Further reduction down to 90 × 90 nm^2 resulted in an 82 % reduction in GMR and an 85 % increase in coercivity. As confirmed in Figure 12, the increase of coercivity is thought to be not due to the development of vortex magnetizations, but due to the possible physical damage occurring during the nano-patterning process. The reduction in GMR can be explained by device degradation resulting from the nano-patterning process as well as the geometrically induced high current density in the nano-patterned devices during measurement under constant current mode. The results shown in Figures 11. and 12 clearly demonstrate that the nano-patterned PPSV has promising magnetic and GMR switching behavior as well as electrical stability suitable for high density MRAM applications.

In order to study the thermal stability of nano-patterned IPSV and PPSV, their domain configurations were explored by MFM (Magnetic Force Microcsopy). Figure 12. shows the MFM images of the nano-patterned IPSV and PPSV with the device sizes ranging from 500 to 90 nm with an aspect ratio of 1:1. Both nano-patterned PSVs were initially saturated with a +2 kOe of magnetic field along their easy directions and then their remnant states were captured.

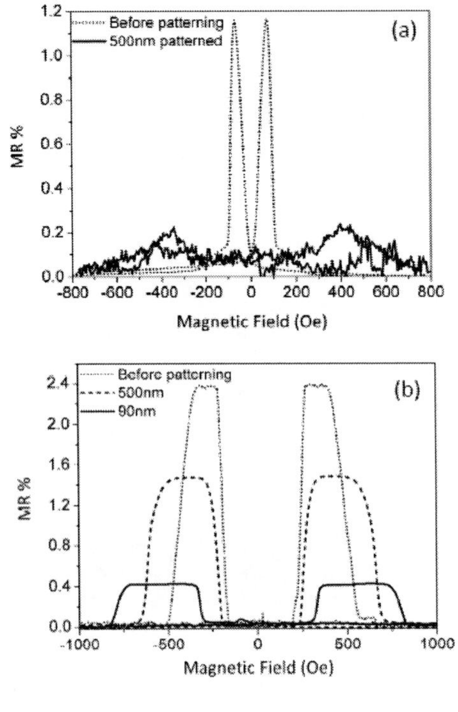

Figure 11.GMR behavior of nano-patterned (a) IPSV, (b) PPSV, and (c) PPSV measured at the different applied current densities [38].

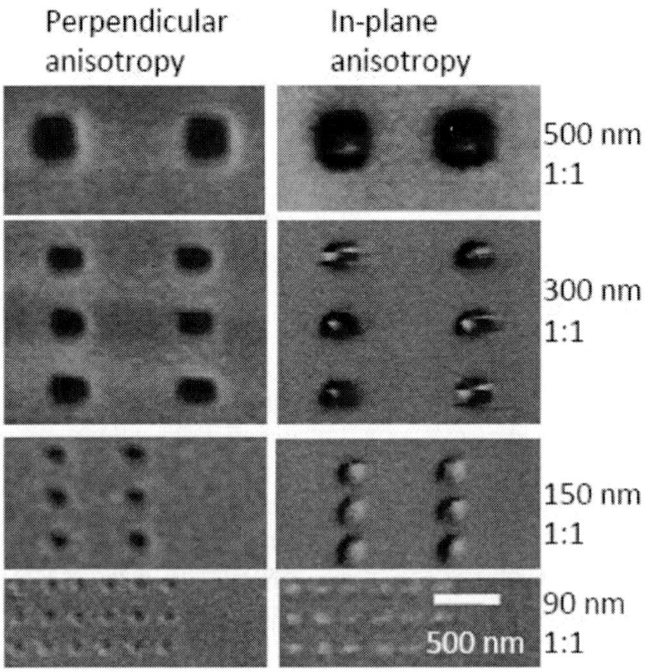

Figure 12. MFM images of nano-patterned IPSV (left) and PPSV (right) for the sizes ranging from 500 × 500 to 90 × 90 nm^2 [38].

As can be seen in Figure 12, the nano-patterned PPSV apparently show single domain structure for all the sizes ranging from 500 × 500 nm^2 down to 90 × 90 nm^2. However, the IPSVs show vortex or "flower" [39] domain structures in their remnant state. When the IPSV is patterned down to sub-micron dimensions, the devices experience curling of magnetization at the edges due to the demagnetization field, and low magnetic anisotropy resulting in the development of vortex magnetization. Thus, to maintain a single domain configuration and to remove the trapped vortices, the patterned IPSV should have a high aspect ratio above 1:5 [40] leading to a reduction in the achievable memory density particularly for MRAM applications.

2. Physical Nature of GMR and Its Correlation with Interlayer Coupling in Spin-Valves with Perpendicular Anisotropy

In order to optimize the GMR performance of the GMR spin-valves with perpendicular anisotropy for real applications in spintronics, the physical nature of GMR behavior and its correlation with interlayer coupling should be studied to precisely understand the underlying physics [41,42]. New physical models describing GMR behavior and interlayer coupling in perpendicular anisotropy systems proposed so far are reviewed and presented in this section.

2.1. A Physical Model of GMR and Interlayer Coupling Characteristics in [Co/Pd] Based Pseudo Spin-Valves with Perpendicular Anisotropy

In order to optimize GMR performance in spin-valves with perpendicular anisotropy, understanding of the physical contribution of the perpendicular interlayer coupling field to the magnetic and magnetoresistance characteristics is essentially required [43,44]. In [31,45], the physical characteristics of interlayer coupling field observed in the perpendicularly magnetized [Pd/Co]/Cu/[Co/Pd] GMR pseudo spin-valves (PSVs) were analyzed in terms of RKKY oscillation and topologically induced interlayer coupling by fitting the experimental results to the calculated values.

Figure 13. shows the dependence of Cu spacer thickness on the perpendicular interlayer coupling field and GMR behavior in Pd (3)/[Pd (1.2)/Co (0.6)]$_2$/Cu (x)/[Co (0.3)/Pd (0.6)]$_4$/Pd (3 nm) PSVs with perpendicular anisotropy. In the region of Cu thickness 1.3 and below, the two layers are strongly ferromagnetic coupled together through pinholes (or defects) and experience simultaneous switching and a correspondingly low MR. Above 1.6 nm, the soft and hard [Co/Pd] layers are ferromagnetically coupled and oscillate with a period of approximately 4.1 nm through the Cu spacer. In addition, the magnetoresistive behavior shows a strong dependence on the interlayer coupling field formed perpendicularly through the Cu spacer. The decrease of interlayer coupling field in the region of moderate interfacial roughness between 1.6 and 4.9 nm is due to the degradation of the perpendicular anisotropy in the soft layer.

Although there is a decreasing trend in the GMR probably due the shunting of current through the thicker Cu, the oscillations follow the periodic oscillation corresponding to that of the interlayer coupling field. This indicates that the GMR behavior in the PSVs with perpendicular anisotropy is dominated by the perpendicular coupling field rather than the topologically-induced magnetic coupling.

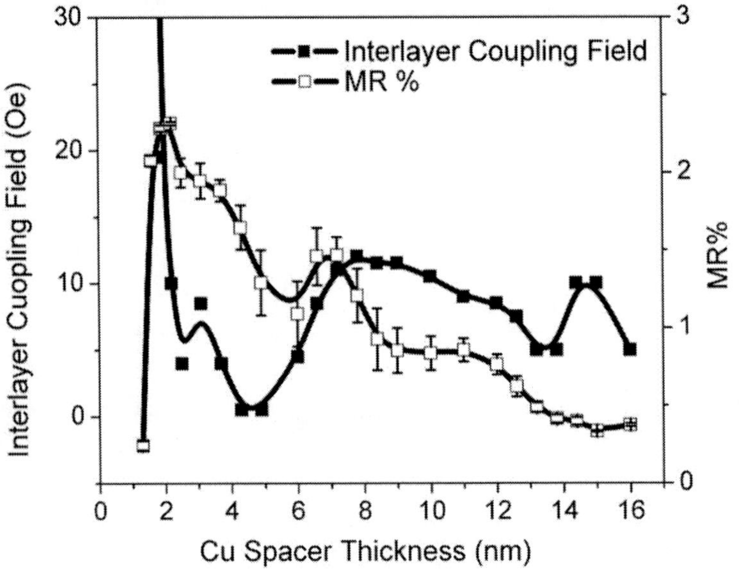

Figure 13. Dependence of interlayer coupling field and GMR ratio on the Cu spacer thickness in the Pd (3)/[Pd (1.2)/Co (0.6)]$_2$/Cu (x)/[Co (0.3)/Pd (0.6)]$_4$/Pd (3 nm) PSVs [31].

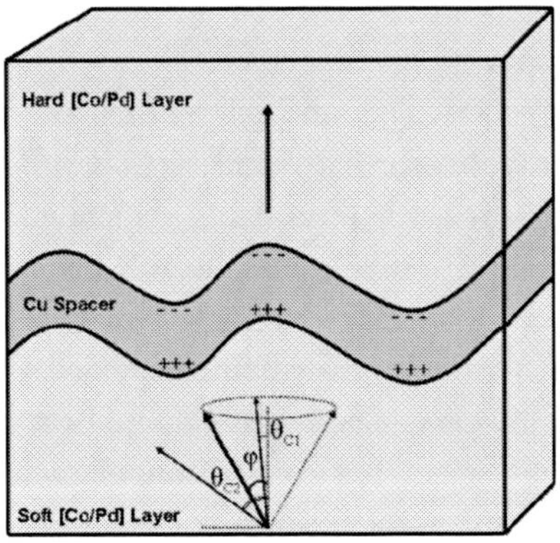

Figure 14. A schematic of Pd/[Pd/Co]$_2$/Cu/[Co/Pd]$_4$/Pd PSV structure illustrating the configurations of magnetization in the perpendicularly magnetized soft and hard [Co/Pd] multi-layers [31].

Based on these results, a new physical model for the GMR behavior in PSVs with perpendicular anisotropy has been proposed that the GMR ratio is proportional to the sine of the angle formed between the soft and hard layer magnetizations along the perpendicular direction during magnetic reversal of the soft layer by the applied magnetic field (Figure 14). Unlike the spin-valve with in-plane anisotropy, perpendicular magnetostatic field induced between the soft and hard [Co/Pd] multi-layers through the Cu spacer caused by the perpendicular anisotropy is directly relevant to the perpendicular interlayer coupling field. However, this model is only applicable once the soft [Co/Pd] multilayer magnetization is slightly tilted against the perpendicular direction. If both layers are perfectly perpendicular and strongly coupled together, they would then switch together upon the application of an external magnetic field.

2.2. Contribution of Topological and Oscillatory RKKY Coupling to the Perpendicular Interlayer Coupling

The topological coupling energy in GMR spin valves with in-plane anisotropy is expressed as Eq. (4),

$$E_{topology} = \frac{\pi}{\sqrt{2}}\frac{h^2}{\lambda} M_1 M_2 \exp\left(\frac{-2\pi\sqrt{2}t_{Cu}}{\lambda}\right) \quad (4)$$

where, h is the waviness amplitude and λ is the wavelength of the surface variations of the spin valve multi-layers, which are determined from AFM (Atomic Force Microscopy) and XTEM (cross sectional Transmission Electron Microscopy) measurements. In addition, the RKKY oscillatory coupling energy of GMR spin valves with in-plane anisotropy is expressed as Eq. (5),

$$\Delta E_{RKKY} = C\frac{m}{\hbar^2\left(2\pi^2\right)}\frac{\sin\left(2k_F t_{Cu}\right)}{t_{Cu}^2}|M_1||M_2|\cos\varphi \quad (5)$$

where, $\hbar$ is the Planks constant, m is the electron mass, and k_F is wave vector at the Fermi surface and energy state **k** [46-48].
However, in the [Co/Pd] PSVs with perpendicular anisotropy, the perpendicular magnetostatic field formed between hard and soft [Co/Pd] multi-layers and spin wave generated from the [Co/Pd] soft layer physically associated with the RKKY oscillation are directly relevant to the magnetization angle of the soft layer deviated from the perpendicular direction.

Hence the general RKKY coupling formula is needed to be modified by adding a $\cos\varphi$ function term where the angle φ is defined as the angle between soft and hard [Co/Pd] multi-layer magnetizations. The results of the calculations shown in Figure 15. indicate that the topological coupling induced by rougher surface roughness is not dominant in determining the variations in the interlayer coupling field. While, the calculated RKKY coupling shows a good fit with the experimental results only if the effect of the soft layer magnetization angle is included. This implies that the deviation of the soft layer magnetization against the perpendicular direction, which is relevant to the perpendicular anisotropy (or perpendicular magnetostatic field formed between soft and hard [Co/Pd] layers), plays the most important role in determining the physical characteristics of interlayer coupling field in the PSV with perpendicular anisotropy.

It is also seen that the PSVs exhibiting only ferromagnetic coupling over the whole range of Cu thickness are corresponded well with previous reports with [Co/Pt] exchange biased spin-valves with perpendicular anisotropy.

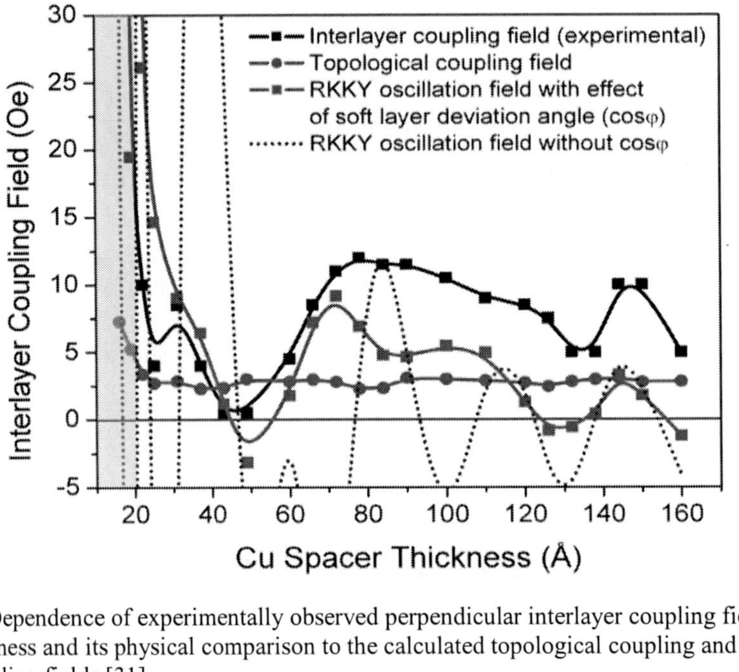

Figure 15. Dependence of experimentally observed perpendicular interlayer coupling field on the Cu spacer thickness and its physical comparison to the calculated topological coupling and oscillatory RKKY coupling fields [31].

2.3. Orange-Peel Coupling in [Co/Pt] Based Exchange Biased Spin-Valves with Perpendicular Anisotropy

Neels theory of magnetostatic coupling in magnetic multi-layers has been extended for multi-layers with perpendicular anisotropy and used to interpret the coupling in [Co/Pt] exchange biased spin-valves with Pt spacer [49].

For two ferromagnetic layers with thickness of t separated by a spacer with thickness b, the roughness of the interfaces can be described by Eq. (6),

$$z = h\cos\left(\frac{2\pi x}{T}\right) \tag{6}$$

where, $2h$ and T represent the peak-to-peak amplitude of the roughness and its wavelength. Furthermore, the local magnetizations can be described by $\psi(x) = \psi_0 \cos px$, where $p = \frac{2\pi}{T}$ and ψ_0 represents the amplitude of the magnetization fluctuations calculated by minimizing the total energy of the system according to situations of parallel and antiparallel magnetization alignments. In addition, θ is the angle difference between the normal to the interface and the z-direction (out-of-plane direction) as defined by $\theta(x) = hp\cos px$. The energy terms consist of exchange energy within each magnetic layer, anisotropy energy at each interface and magnetostatic energy. The exchange energy is given by Eq. (7),

$$E_{ex} = \frac{2t}{T}\int_0^T A\left(\frac{\partial \psi}{\partial x}\right)^2 dx = Atp^2\psi_0^2 \tag{7}$$

where, A is the exchange constant. Assuming that the anisotropy axis is always locally normal to the interface, the anisotropy energy is given by Eq. (8),

$$E_{ani} = -\frac{2t}{T}\int_0^T \cos^2\left(\theta(x) - \psi(x)\right)dx = Kt(hp - \psi_0)^2 \tag{8}$$

The magnetostatic energy eventually expressed as Eq. (9),

$$E_{mag} = \frac{\mu_0}{2}\left[-\varepsilon\frac{\sigma_1^2}{8p}\exp[-2pb]\left[1-\exp[-2pt]^2 + \varepsilon\frac{\sigma_2^2}{4p}\exp[-p(b+t)] + \sigma_0^2 t - \frac{\sigma_1^2}{4p}\exp[-2pt]\right]\right] \tag{9}$$

where, σ_0, σ_1, and σ_2 are related to the interface density of charges σ_s, and bulk density of charges σ_v. The interface density of charges is given by Eq. (10),

$$\sigma_s = M_s \cos[(\theta_0 - \psi_0)\cos px] \approx \sigma_0 - \sigma_1 \cos 2px \tag{10}$$

with $\sigma_0 = M_s\left[1 - \frac{(\theta_0 - \psi_0)^2}{4}\right]$ and $\sigma_1 = \frac{M_s}{4}(\theta_0 - \psi_0)^2$.

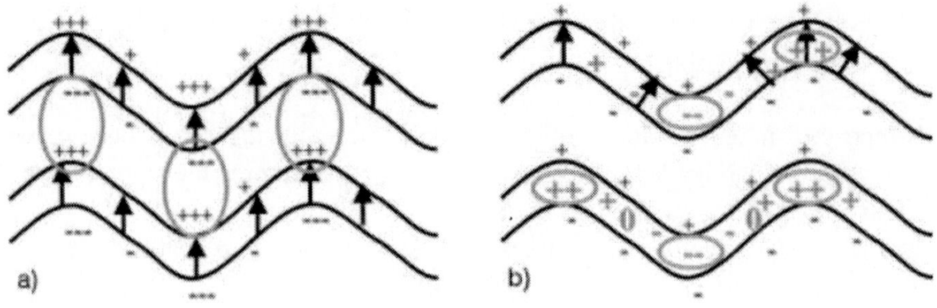

Figure 16.Schematic representation of magnetization in the case of low anisotropy (a) and high anisotropy (b) [49].

The bulk density of charges is given by $\sigma_v = \sigma_2 \sin px$ with $\sigma_2 = pM_s\psi_0 t$. The total energy is the sum of the exchange, anisotropy and the magnetostatic energies, which is then minimized with respect to ψ_0 in the parallel and anti-parallel magnetic configurations. The interlayer coupling is thus given by the difference in the total energies between the parallel and anti-parallel configurations.

The orange-peel interlayer coupling can favor either parallel or anti-parallel coupling depending on the anisotropy. In the case of a low anisotropy, the magnetization is parallel to the perpendicular direction in order to minimize surface charges and due to exchange stiffness and there are no volume charges. This leads to a dominant magnetostatic interaction between the opposite interface charge densities facing each other as shown in Figure 16.a). On the other hand, for a large anisotropy, the magnetization is aligned along the normal to the interface. For this case, the interfaces do not generate any coupling as they are uniformly charged, rather the oscillatory distribution of volume charges are out of phase for anti-parallel alignment, which is magnetostatically favorable as in Figure 16.b).

3. Physical Nature of Exchange Bias in Perpendicularly Magnetized Multi-Layers/Anti-Ferromagnetic Thin Films for GMR Spin-Valves

3.1. A Physical Model of Perpendicular Exchange Bias

Developing a physical model for a PEB system, which can clearly elucidate the underlying physics and predict what physical parameters would more effectively influence on the adjustment of the PEB characteristics has been considered to be the most urgent issue to rapidly extend the application of exchange biased GMR spin-valves with perpendicular aniosotrpy to a wider range of spintronics devices.

In [50], a physical model of PEB established based on the total energy equation per unit area of an exchange bias system by assuming coherent rotation of the magnetization is presented. This model focuses on studying the physical phenomenon of a PEB system in view of the energy competition between the anisotropy energy of AFM layer, $K_{AFM}\times t_{AFM}$, FM multi-layers, $K_{FM,eff}\times t_{FM}$ and the interfacial exchange coupling energy, J_{ex}. Unlike that of an exchange bias system with in-plane anisotropy, this model emphasizes the importance of

$K_{FM,eff}\times t_{FM}$ and the physical contribution of J_{ex} to the PEB system. The energy per unit area of an exchange bias system with in-plane anisotropy is expressed in terms of the anisotropy energy of AFM layer, $K_{AFM}\times t_{AFM}$, FM layer, $K_{FM,eff}\times t_{FM}$ and the interfacial exchange coupling energy, J_{ex}, as given by Eq. (10)[51-53],

$$E = -H_{FM} M_{FM} t_{FM} \cos(\theta - \beta) + K_{FM} t_{FM} \sin^2 \beta + K_{AFM} t_{AFM} \sin^2 \alpha - J_{INT} \cos(\beta - \alpha) \quad (11)$$

where, H is the applied field, M_{FM} the saturation magnetization, t_{FM} the thickness of FM layer, t_{AFM} the thickness of AFM layer, K_{FM} the anisotropy of FM layer, K_{AFM} the anisotropy of AFM layer and J_{INT} the interface coupling constant. By considering Eq. (11) and assuming the spin structure of AFM layer, the angles and the energy terms in a PEB system with perpendicularly magnetized FM multi-layers and AFM layer can be illustrated as shown in Figure 17. The AFM and FM anisotropy axes are assumed to be collinear and aligned in the perpendicular-to-the film direction (out of plane). As indicated in Figure 17, β, α, and θ represent the angles between the anisotropy axis and the FM magnetization, the AFM sub-lattice magnetization, and the applied field, respectively. In addition, from these viewpoints, it can be understood that the first term in Eq. (11) in a PEB system indicates the effect of the applied field on the FM multi-layers with perpendicular anisotropy, the second term is the effect of the FM anisotropy, the third term is the effect of the AFM anisotropy, and the last term accounts for the interfacial exchange coupling.

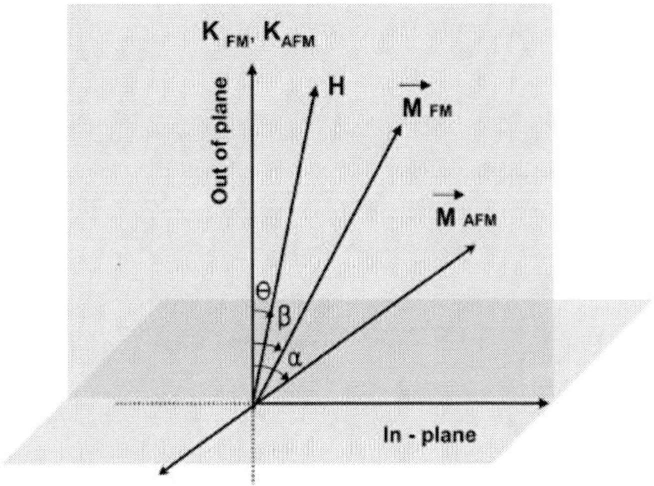

Figure 17. Schematic diagram of angles and magnetizations involved in a PEB system. The AFM and FM anisotropy axes are assumed collinear [50].

The stable state of the system can be obtained by minimizing the energy equation with respect to α and β. The energy minimization with respect to α results in Eq. (12),

$$K_{AFM} t_{AFM} \sin 2\alpha = J_{INT} \sin(\beta - \alpha) \quad (12)$$

Theoretically analyzing the result appearing in Eq. (12) clearly demonstrates that the PEB system should satisfy the critical condition of $K_{AFM} \times t_{AFM} \geq J_{INT}$ to create the exchange bias as with an exchange bias system with in-plane anisotropy. In a general energy equation for an exchange bias system (Eq. (11)), the interfacial exchange coupling energy, J_{ex}, is usually expressed as $J_{ex} = J_{INT}(\vec{S}_{FM} \bullet \vec{S}_{AFM}) = J_{INT}\cos(\beta - \alpha)$ where S_{FM} and S_{AFM} represent the spin vectors of the FM and the AFM layer. For an exchange bias system with in-plane anisotropy, J_{ex} is commonly accepted to be proportional to the cosine of angle difference between the AFM and the FM magnetization at the FM/AFM interface, because all of the spins are aligned in the in-plane direction. However, the physical nature of J_{ex} in a PEB system is different compared to an in-plane exchange bias system in that the FM and the AFM layer have two magnetization components: (1) perpendicular (out of plane) and (2) in-plane components. For a PEB system, although the angle difference between the FM and the AFM magnetizations is very small, if the spins of the FM and the AFM are not aligned in the perpendicular direction, the exchange bias field in the perpendicular direction would be kept small. Referring to Figure 17, which illustrates the magnetization of FM and AFM at the FM/AFM interface, J_{ex} of the system is understood to be proportional to the dot product of FM and AFM spin vectors.

According to the derivation as described in Eq. (13), where S_{FM} and S_{AFM} represent the spin vectors of the FM and the AFM layer, and $S_{\perp}$ and $S_{//}$ represent the scalar magnetization value in the perpendicular and in-plane directions, J_{ex} can be separated into two components: 1) $S_{FM\perp} \times S_{AFM\perp}$ representing the net magnetization of the FM and AFM in the perpendicular direction, which directly relates to the PEB characteristics, and 2) $S_{FM//} \times S_{AFM//}$ representing the net magnetization in the in-plane direction. For a PEB system, $S_{FM//} \times S_{AFM//}$ term is relatively negligible if the perpendicular anisotropy is strong. Therefore, in the exchange biased magnetic thin films with perpendicular anisotropy, i.e. [Pd/Co]n/FeMn PEB thin films, the interfacial exchange coupling is dominantly determined by the net magnetization of [Co/Pd]n and FeMn in the perpendicular direction; that is the cosine of the canting angle of [Co/Pd]n and FeMn magnetizations from the perpendicular direction, $J_{INT} \times (S_{[Pd/Co]_n\perp} \times S_{FeMn\perp}) = J_{INT} \times \cos\beta_{[Pd/Co]_n\perp} \times \cos\alpha_{FeMn\perp}$, as given by Eq. (14).

$$\begin{aligned} J_{ex} &= J_{INT} \times (\vec{S}_{FM} \bullet \vec{S}_{AFM}) = J_{INT} \times \cos(\beta - \alpha) \\ &= J_{INT} \times (S_{FM\perp} + S_{FM//}) \bullet (S_{AFM\perp} + S_{AFM//}) \\ &= J_{INT} \times (S_{FM\perp} \times S_{AFM\perp} + S_{FM//} \times S_{AFM//}) \end{aligned} \quad (13)$$

$$J_{ex} \approx J_{INT} \times (S_{FM\perp} \times S_{AFM\perp}) = J_{INT} \times \cos\beta_{FM} \times \cos\alpha_{AFM} \quad (14)$$

From Eq. (14), it can be seen that J_{ex} is dominated by the cosine of the canting angle of the AFM and the FM magnetizations from the perpendicular direction. In addition, Eq. (14) indicates that the improvement of the exchange bias coupling in a PEB system can be approached by enhancing the net magnetization of either AFM or perpendicularly magnetized FM layer in the perpendicular direction.

By substituting Eq. (14) into Eq. (11), the total energy equation for a PEB system can be more accurately expressed by Eq. (15), where the interfacial exchange coupling term is modified by considering the net magnetization of FM and AFM in the perpendicular direction.

$$E = -H_{FM} M_{FM} t_{FM} \cos(\theta - \beta) + K_{FM} t_{FM} \sin^2 \beta + K_{AFM} t_{AFM} \sin^2 \alpha - J_{INT} \cos\beta \cos\alpha \quad (15)$$

In order to find the exchange bias field (H_{ex}) and to explore what physical parameters significantly influence the physical characteristics of H_{ex} in a PEB system, the energy minimization of Eq. (15) with respect to β was carried out. Eq. (16) expresses the H_{ex} obtained from the energy minimization, which indicates that the net magnetization of the AFM and the FM in the perpendicular direction, relevant to cosα and cosβ, is crucial in determining the exchange bias in a PEB system. In particular, completely different from an in-plane exchange bias system, the ferromagnetic anisotropy (or anisotropy energy), K_{FM} (or $K_{AFM} \times t_{AFM}$) in a PEB system is revealed to be significant in determining the exchange bias characteristics as it directly contributes to the shift of the hysteresis loop as described in Eq. (16).

$$H_{ex} = \frac{J_{INT} \cos\alpha + 2K_{FM} t_{FM} \cos\beta}{M_{FM} t_{FM}} \quad (16)$$

A series of the experimental works using exchange biased $[Pd/Co]_5/FeMn$ thin films with perpendicular anisotropy have been done to verify the physical validity of this model. In order to explore the physical contribution of $K_{AFM} \times t_{AFM}$ to the nature of PEB, two different exchange biased thin film structures with perpendicular anisotropy, $Si/Ta/[Pd/Co]_5/FeMn/Ta$ and $Si/Ta/FeMn/[Pd/Co]_5/Ta$, were compared in terms of the crystalline structure (crystalline magnetic anisotropy) of FeMn AFM layer and the interfacial spin structures based on the 3Q structure model of FeMn.[54]

In addition, in order to study the physical contribution of $K_{FM,eff} \times t_{FM}$ and J_{ex} to the nature of PEB, magnetic annealing was performed at the different magnetic fields applied along the perpendicular or in-plane to the film direction [50]. All the experimental results confirmed that the proposed PEB model is valid to understand the underlying physics of PEB phenomenon in perpendicularly magnetized FM/AFM thin films.

3.2. Effect of Nano-Patterning on Exchange Bias in Perpendicularly Magnetized Multi-Layers/Anti-Ferromagnetic Structures

Studies have demonstrated that the exchange bias can be obtained in nano-patterned exchange biased perpendicularly magnetized FM/AFM thin films for ultra high density spintronics devices as shown in Figure 18. [55-57]. However, it has been undesirably found that although the patterned structures do retain their perpendicular anisotropy, the magnitude of the exchange bias field is progressively degraded and the coercivity is increased by scaling down the pattering size. For the nano-patterned exchange biased thin films, the constraints in the AFM domain size of the patterned dots imposed by the shrunken dimension dots favor and enhance the exchange bias field compared to the sheet films.

However, the reduced coordination of spins at the edges of the patterns makes them more prone to thermal activation, which supports a reduction in the exchange bias field.

Furthermore, it has also been found that the blocking temperature of the system is dramatically reduced after the patterning process.

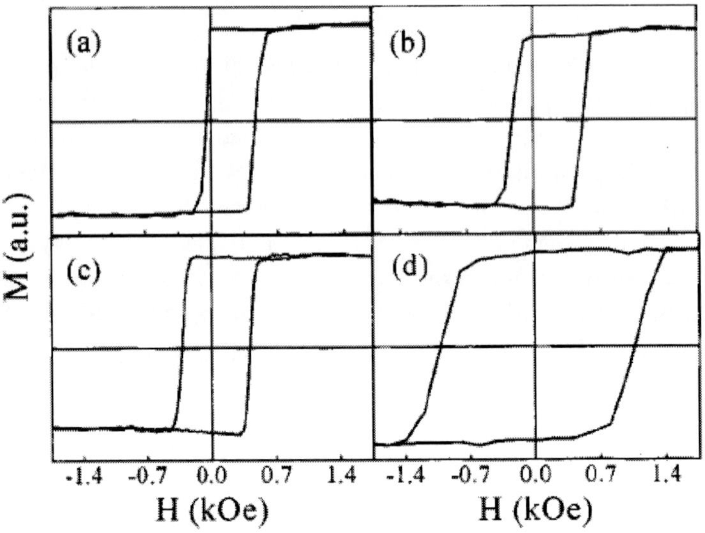

Figure 18. Hysteresis loops for [Co/Pt]/FeMn] continuous film (a), 200nm wide nanowires (b) 200 x 1000 nm stripes (c) and 200 x 200 nm dots fabricated by electron beam lithography [55].

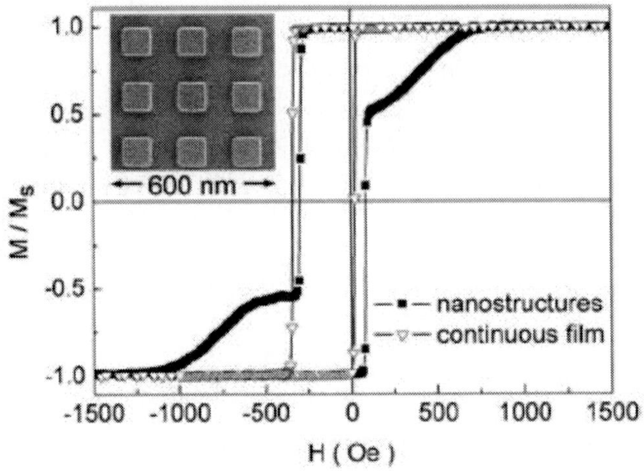

Figure 19. M-H loops of continuous and nanostructures (template of dots and trenches) of $[Pt/Co]_3/IrMn$ films with perpendicular anisotropy [56].

On the other hand, it has been significant to note that in cases where the thin films were deposited on pre-patterned template showed relatively large exchange bias field compared to the post patterning with nanolithography as shown in Figure 19. The pre-patterning method avoids degradation of the process dependent material properties associated with nano-patterning processes [56]. However, as can be seen in Figure 19.a. combination of signals from the magnetic films deposited in the dots and trenches of the pre-patterned substrate has been observed in this strucure. The sharp transitions at the lower fields correspond to the magnetic reversal of the trenches, while the broader transitions at higher fields correspond to that of the dots. The broadening of the transition for the dots is attributed to the inhomogeneities among the dots leading to a larger switching field distribution.

4. Physical Nature of Anomalous Peak Observed in Extraordinary Hall Effect Loops of Exchange Biased Spin-Valves with Perpendicular Anisotropy and Its Applications

The extraordinary Hall effect (EHE) has been widely used to measure the magnetic properties of spin-valves with perpendicular anisotropy [58-61] because the Hall resistivity is proportional to the perpendicular component of the magnetization ($M_\perp$) as shown in Eq. (17).

$$\rho_H = R_0 H + R_S M_\perp \tag{17}$$

where, H is the applied field, R_0 is the ordinary Hall coefficient and R_S is the extraordinary Hall coefficient [62]. It has been experimentally confirmed that the magnitude of the ordinary term, R_0H, is substantially small; the $R_SM_\perp$ term dominates the measured Hall signal. In recent [63], the researchers has reported on the observation of anomalous peaks in EHE loops at the switching field of the free and pinned layers, when the exchange biased [Co/Pd]/Cu/[Co/Pd]/FeMn spin-valves are magnetically reversed by an externally applied field.

Figure 19. shows the measured (a) EHE, (b) M-H, and (c) GMR loops for Ta(20)/[Pd(0.6)/Co(0.4)]$_2$/ Cu(2.2)/Co(0.7)/[Pd(0.6)/Co(0.4)]$_2$/FeMn(10.8)/Ta(20 nm). As can be seen in Figure 20, the EHE loop shows the same magnetic characteristic as the M-H and GMR measurements, with $\rho_H(H)$ being directly related to $M(H)$ as described by Eq. (17). However, the EHE loop exhibits anomalous peaks at the switching field of the free and pinned layers where the magnetization of the spin-valve is reversed by an externally applied field.

Based on the experimental results and numerical calculations, a physical model to understand anomalous peaks and their dependence on the magnetic properties of the spin-valves was proposed. For an exchange biased spin-valve, the magnetostatic energy per unit area may be defined as:

$$\begin{aligned} E = &-H_P M_S t_P \cos(\beta - \theta_P) + K_{UP} t_P \sin^2 \theta_P \\ &- M_S t_P H \cos(\alpha - \theta_P) + K_{Uf} t_f \sin^2 \theta_f \\ &- M_S t_f H \cos(\alpha - \theta_f) - J \cos(\theta_f - \theta_P) \end{aligned} \tag{18}$$

where, t_f and t_P are the free and pinned layer thickness and θ_f and θ_p are the angles between the easy axis and the free and pinned layer magnetizations respectively. M_s is the saturation magnetization of the ferromagnetic layers and K_{uf} and K_{up} are respectively the effective anisotropy constants of the free and pinned layers. H_p is the exchange biasing field, J is the interlayer coupling energy, and β is the angle between the exchange biasing field and the easy axis. H and α are the externally applied field and the angle of the applied field from the easy axis, respectively [64].

For the [Co/Pd] based exchange biased spin-valves, it can be seen that due to the large biasing field and perpendicular anisotropy constant, the total magnetostatic energy of the system is large.

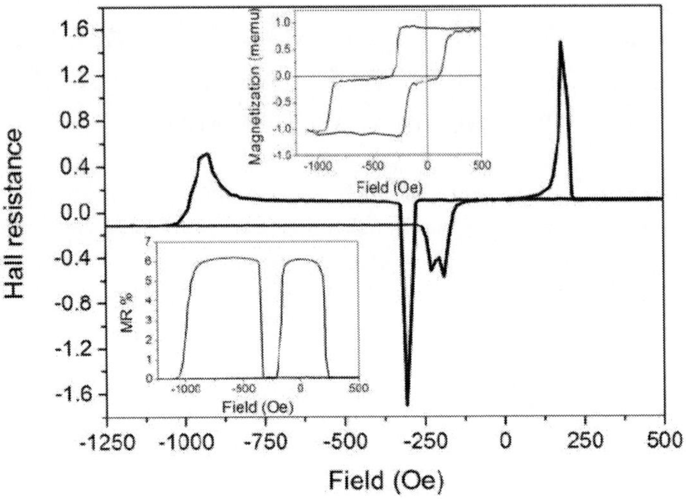

Figure 20. Observed anomalous peaks in EHE measurement and corresponding M-H and GMR measurements (inset) [63].

As an external magnetic field is applied to reverse the free or pinned layer magnetization, there is an abrupt change in the magnetostatic energy. This abrupt change is expected to lead to the appearance of the anomalous peaks in the EHE loops. The highest peak intensity across all the samples studied, is found to be the "negative free peak" which occurs as the external magnetic field is swept from a positive saturation field to negative saturation field and the free layer magnetization switches such that the spin-valve goes from a parallel state to an anti-parallel state. During this magnetization reversal of the free layer, the pinned layer magnetization would oppose the external switching field causing the magnitude of the EHE signal to increase suddenly.

In addition to the magnetostatic energy contribution to the anomalous peak, the effect of GMR on the hall resistivity is also considered. For ordinary ferromagnetic films the field dependence of ρ_H is similar to that of the magnetization M as given in Eq. (17), assuming that R_S is field independent since the field dependence of ρ is not significant. For GMR systems, since ρ is highly field-dependant, the field dependence of R_S must be taken into account. It has been widely accepted [65,66] that extraordinary hall component is due to skew-scattering and side-jump mechanisms, where the skew-scattering contribution is proportional to ρ and both skew-scattering and side-jump mechanism contribute to the ρ^2 term [62,67] as given by Eq. (19),

$$R_S = a\rho + b\rho^2 \tag{19}$$

Taking into account the field dependence of ρ in Eq. (19), the competition between the increasing *M(H)* and decreasing $R_S(H)$ with *H* also leads to the peak in the EHE measurement. The validity of the proposed model was verified by considering the physical relationship between the interlayer coupling, perpendicular anisotropy and the magnetostatic energy, as well as between the giant magnetoresistance (GMR) behavior and the Hall resistance. It was theoretically and experimentally demonstrated that the anomalous peaks provide a way of indirectly determining the magnetostatic energies, interlayer coupling behavior and GMR performance in exchange biased spin-valves with perpendicular anisotropy using EHE measurement [63].

5. Magnetic Tunneling Junctions (MTJ) With Perpendicular Anisotropy

The previous section has been primarily focused on all metal based GMR spin-valves with perpendicular anisotropy. In this section, magnetic tunneling junctions (MTJ) or tunneling magnetoressitance (TMR) with perpendicular anisotropy that has been considered and currently is being developed in spintronics research area will be discussed. This section will begin with the discussion on brief summary of tunneling magnetoresistance effects including general theory of TMR effects, physical model for spin dependent tunneling, and Simmon's theory for tunneling effects. Subsequently, the initial works on the MTJs with perpendicular anisotropy and the recent achievements of TMR spin-valves with perpendicular anisotropy in spintronic devices will be reviewed and presented.

5.1. General Theory of Tunneling Magnetoresistance Effects

Magnetic tunnel junction (MTJ) is consisted of a thin insulating layer (tunnel barrier) separated by two ferromagnetic electrodes. It exhibits tunneling magnetoresistance (TMR) due to spin-dependent electron tunneling through the barrier. The tunneling resistance when the magnetizations of the two ferromagnetic electrodes are parallel is smaller than when they are anti-parallel; this resistance change results in the TMR effect [68]. The TMR resistance change is defined as $TMR = \frac{R_{AP} - R_P}{R_P}$, where the P_1 and the P_2 represent spin-polarization of the ferromagnetic electrodes. The spin polarization is calculated based on the effective density of state, *D*, at the Fermi level expressed as Eq. (30),

$$P = \frac{D_\uparrow - D_\downarrow}{D_\uparrow + D_\downarrow} \qquad (20)$$

and the TMR is defined in terms of the spin polarization given by Eq. (2).

$$TMR = \frac{2P_1P_2}{1 - P_1P_2} \tag{21}$$

Initial research on MTJs was based on Al-O barriers, in which various block states with different states tunnel incoherently through the amorphous barrier. However, this tunneling mechanism was found to lead to the reduction in spin-polarization and accordingly resulted in decreasing TMR ratio. A 70 % of TMR ratio has been achieved in this structure even after optimizing all the fabrication process and materials [69].

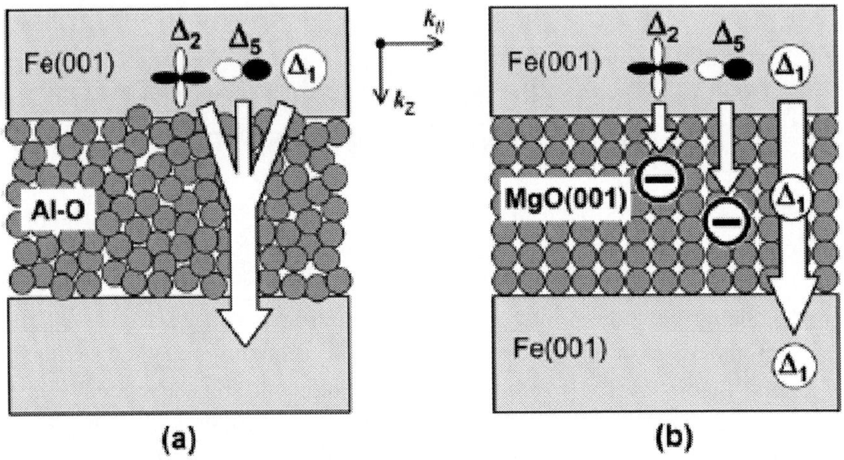

Figure 21. Schematic illustrations of electron tunneling through (a) an amorphous Al–O barrier and (b) a crystalline MgO(001) barrier [69].

More recently, an MgO (001) tunnel barrier instead of AlO_x has been attempted as a tunnel barrier and found to produce a TMR ratio over 600 % at room temperature [70]. With defect-free crystalline MgO barrier, the Block states with $\Delta 1$ symmetry dominantly tunnel through the barrier, because it acts as symmetry filter as illustrated in Figure 21.

Furthermore, MTJs with MgO tunnel barrier and ferromagnetic electrodes composed of Fe or Co and its alloys such as Fe, FeCo, CoFe, CoFeB, which are fully spin-polarized in the [001] direction at the Fermi level, have been recently revealed to exhibit a high TMR value that is preferred to advanced spintronics applications.

5.2. Model for Spin Dependent Tunneling

Spin dependent tunneling between two ferromagnetic layers depending on their relative magnetization has first reported by "Juliere" in 1975 [71]. In this model, he postulated the spin status of electrons is maintained during tunneling and the electrical conductance of spin dependent electrons is proportional to the density of states (DOS) of spin up and spin down electrons in each ferromagnetic electrode.

Based on the postulation on spin dependent tunneling, electrical conductance in metallic tunnel junction is assumed to increase when the magnetizations of two electrodes have parallel state compared to anti-parallel state. Accordingly, electrical conductance under parallel, G_p, and anti-parallel, G_{AP}, states of magnetization can be expressed by Eq. (22),

$$G_p \propto a_1 a_2 + (1 - a_1)(1 - a_2)$$
$$G_{AP} \propto a_1(1 - a_2) + a_2(1 - a_1) \tag{22}$$

where, a_1 and a_2 are the relative ratio of the number of majority spin electrons in two ferromagnetic layers (or electrodes). These two parameters are expressed as $a_{1,2} = n_\uparrow/(n_\uparrow + n_\downarrow)$. From above equation, the ratio of electrical conductance depending on the relative magnetization of two ferromagnetic electrodes is defined by Eq. (23),

$$\frac{\Delta G}{G} = \frac{G_P - G_{AP}}{G_{AP}} = \frac{2P_1P_2}{1 + P_1P_2} \tag{23}$$

The model of spin dependent tunneling suggested by "Julliere" is based on the "Stearns" theoretical model explaining the relationship between spin dependent tunneling current and density of state of spin electrons in metallic ferromagnet.

However, according to the experimental data reported up to now, it is revealed that this theory does not correlate well with experimental data. Different from "Julliere" model, another spin dependent tunneling model was introduced by "Slonczewski" in 1989 [72].

This model is much closer to real model in taking into account for the tunneling phenomenon in tunnel barrier separated by two ferromagnetic layers. "Slonczewski" considered wave vector, k, of electrons, which is dependent on spin states, in metallic ferromagnet and calculated wave function at the interface between metallic ferromagnetic layer and dielectric layer in terms of relative spin directions.

According to "Slonczewski" model, the relative electrical conductance depends on the damping constant of wave factor in dielectric tunnel barrier. Eq. (24) below, shows the relationship between change of conductance and damping constant of wave vector $k_\uparrow$ and $k_\downarrow$.

$$\frac{\Delta G}{G} = 2\left(\frac{k_{1\uparrow} - k_{1\downarrow}}{k_{1\uparrow} + k_{1\downarrow}}\right) \bullet \left(\frac{k^2 - k_{1\uparrow}k_{1\downarrow}}{k^2 + k_{1\uparrow}k_{1\downarrow}}\right) \bullet \left(\frac{k_{2\uparrow} - k_{2\downarrow}}{k_{2\uparrow} + k_{2\downarrow}}\right) \bullet \left(\frac{k^2 - k_{2\uparrow}k_{2\downarrow}}{k^2 + k_{2\uparrow}k_{2\downarrow}}\right) \tag{24}$$

As can be seen in the equation, the change of electrical conductance, $\Delta G/G$ is strongly dependent on the relative amplitude of damping constant of wave vectors.

5.3. Simmon's Theory for Electric Tunnel Effects

In general, there are two theories using a relationship between current and voltage characteristics in confirming electric tunnel effects in tunnel junctions separated by dielectric tunnel barriers. One is called by "Fowler-Nordheim" theory, which has been studied intensively in Metal-Oxide-Semiconductor (MOS) structures to verify tunneling mechanism.

The other is called by "Simmon's Theory", which deals with tunneling theory using a relationship between tunneling current density, J and applied junction voltage, V. The "Simmon's theory" was established by John G. Simmon at the beginning of 1960's in tunneling junction structure, which is composed of two metal electrodes separated by insulating thin films [73].

Figure 22.(a) shows a schematic diagram of general barrier in insulating film between two metal electrodes. The tunneling current density, J is calculated by integrating tunneling electrons through tunnel barrier. The number of tunneling electrons is obtained by using the tunneling probability of electrons at the Fermi-level under different barrier potential heights, which are controlled by the applied junction voltage.

From the relationship between tunneling current density, J and the applied voltage, V, J can be expressed in terms of tunnel barrier height, ϕ, and tunnel barrier thickness S by Eq. (25),

$$J = J_0\{\bar{\phi}\exp(-A\bar{\phi}^{-\frac{1}{2}}) - (\bar{\phi}+eV)\exp[-A(\bar{\phi}+eV)^{\frac{1}{2}}]\} \qquad (25)$$

where, $J_0 = \frac{e}{2\pi h}(\beta\Delta S)^{-2}$, h is plank constant, $A = (\frac{4\pi\beta\Delta S}{h})\sqrt{2m}$, and β is correction factor, which is independent of applied voltage, V and is expressed by Eq. (26),

$$\beta = 1 - [\frac{(eV/S)^2}{8\Delta S}] \int_0^{\Delta S = S\varphi_0/eV} [\frac{(\frac{\Delta S}{2} - x)^2 dx}{(\frac{\varphi_0}{2})^2}] = 1 - 1/24 = 23/24 \approx 0.96 \qquad (26)$$

In the first above equation, $J_0\bar{\phi}\exp(-A\bar{\phi}^{\frac{-1}{2}})$ indicates the current component from electrode 1 to electrode 2 and $J_0(\bar{\phi}+eV)\exp[-A(\bar{\phi}+eV)^{\frac{1}{2}})]$ indicates the current component from electrode 2 to electrode 1. Figure 22.(b) shows pictorial illustration of current flow between two electrodes.

As shown in above eqation, the "Simmon's theory" is considered as a very useful and practical model in that it can provide crucial parameters of tunneling effects. If the mean value of barrier height,ϕ, and barrier thickness, S, are available, the tunneling current density can be easily obtained.

On the contrary, if there is a measured current-voltage characteristic plot, tunnel barrier thickness and height are numerically calculated by fitting measured data to the theoretical formula.

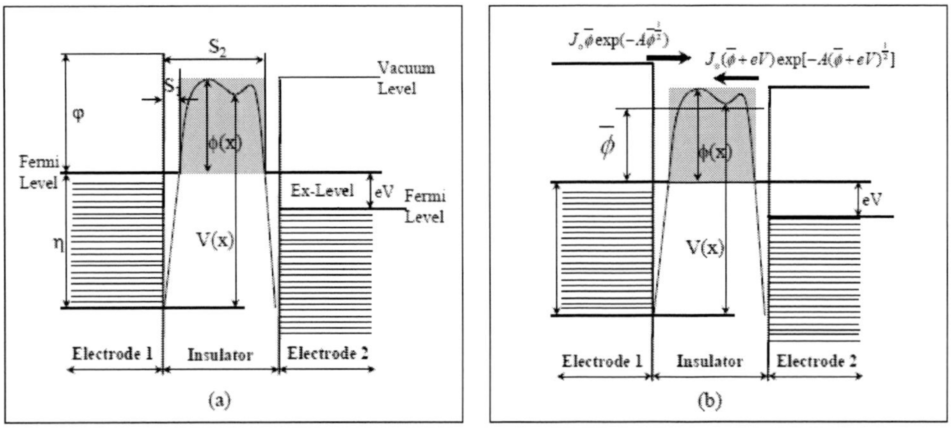

Figure 22. Schematic diagrams of (a) general barrier in insulating film between two metal electrodes, and (b) pictorial illustration of current flow between two electrodes.

5.4. Initial and Recent Works on MTJs with Perpendicular Anisotropy

With the potential advantages promised by perpendicular anisotropy materials in spintronics devices applications at an extremely low dimension, there has been an interest in the studies of MTJs with ferromagnetic electrodes with perpendicular anisotropy.

According to the first report on the TMR in a MTJ with perpendicular anisotropy in 2002, a MTJ device with RE-TM based ferromagnetic electrodes with perpendicular anisotropy and an AlO_x tunneling barrier showed more than 50 % of TMR ratio (Figure 23). In particular, this MTJ structure showed very stable TMR behavior independent of the barrier thickness as shown in Figure 23 [1].

By considering the instability of TMR performance depending on the tunnel barrier observed from the MTJs with in-plane anisotropy, the high magnetic stability as well as good thermal stability of the MTJs with perpendicular anisotropy has triggered a significant attraction to the spintronics research area. Hence, a great deal of research efforts has been intensively made for the past few years to develop various kinds of new functional MTJ systems with perpendicular anisotropy.

A variety of technical approaches in terms of materials science and physics such as using high spin polarization materials, i.e. Fe, CoFe and CoFeB with different compositions as insertion layers between the electrodes with perpendicular anisotropy and MgO tunnel barrier to improve the crystalline texture for coherent tunneling and to reduce lattice mismatch in the MgO barrier and optimizing deposition conditions of MgO tunnel barrier to make perfect (001) texture and to obtain the bulk stoichiometry of MgO have been intensively attempted as main efforts in these research scopes (Figure 24).

As a result, a 200 % of TMR ratio has been recently demonstrated in a MTJ structure with Fe based single layered ferromagnetic electrodes with perpendicular anisotropy and MgO tunneling barrier [74].

Currently, more efforts on the improvement of TMR performance and the development of high density and high speed MRAM devices using MTJs with perpendicular anisotropy are being actively made in both industry and academia for commercialization. Some of the distinct works directly relevant MTJs and MTJ based devices with perpendicular anisotropy are summarized in Table 1.

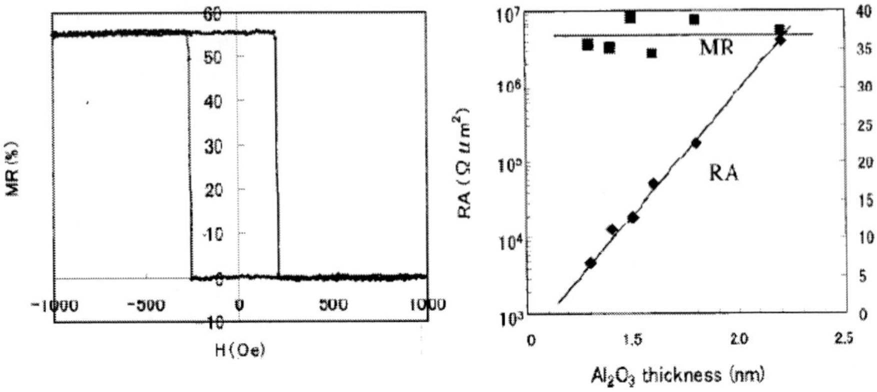

Figure 23. TMR curve of GdFe/CoFe/Al_2O_3/CoFe/TbFeCo MTJ device (left) and the dependence of junction resistance and MR as a function of barrier thickness [1].

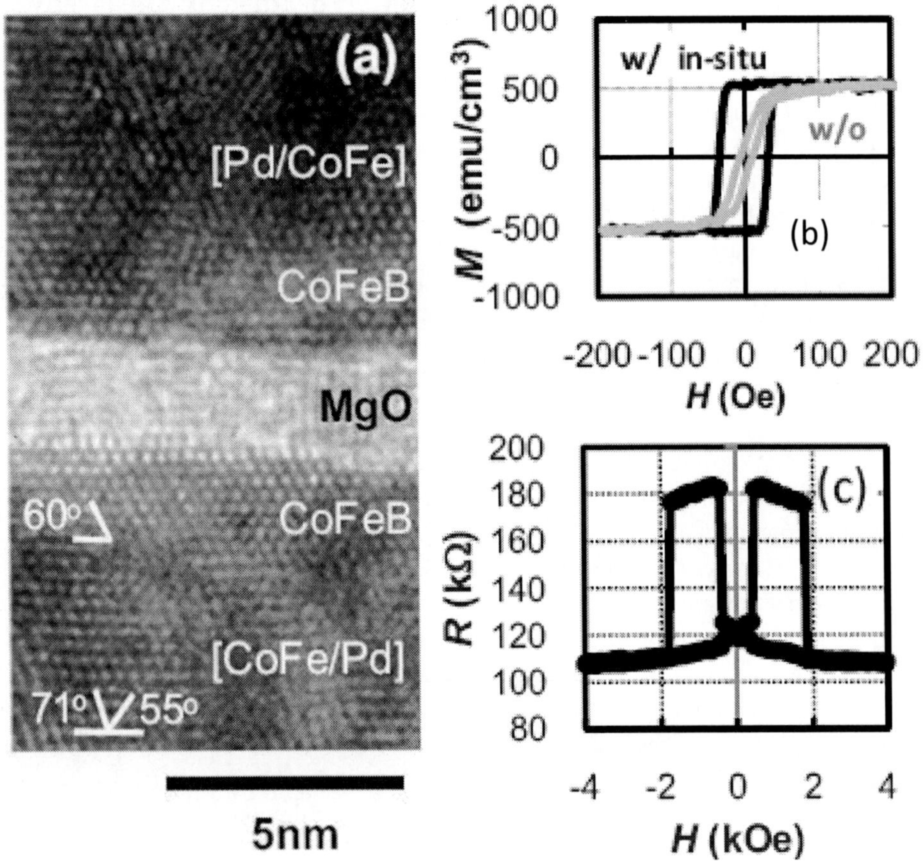

Figure 24. Cross-section TEM (a), M-H (b) and TMR (c) loops for CoFe/Pd based MTJ with MgO barrier and CoFeB insertion layers [79].

Table 1. Summary of recent MTJ works using ferromagnetic electrodes with perpendicular anisotropy

Structure	TMR ratio (%)	Remarks	Reference
[Co/Pd or Ni]$_n$/CoFeB/Mgo /CoFeB/[Co/Pd or Ni]$_n$	10%	TMR not improved by annealing for CoFeB/MgO crystalization	[75]
[Co/Pt]/AlO$_x$/[Co/Pt]	14.7%	Annealing of patterned junctions increases the TMR	[76]
TbFeCo/(Mg/MgO/Mg)/GdFe Co	Not reported	Polycrystalline MgO	[77]
$Co_{90}Fe_{10}$/Pd electrodes with CoFeB insertion, MgO barrier	1.7%	fcc(111) electrodes but imperfect MgO	[78]
with $Co_{50}Fe_{50}$	3%		
CoFe/Pd electrodes with $Co_{20}Fe_{60}B_{20}$/MgO/Fe insertion	78%	Fe above MgO improves interface crystallinity	[79]
$Co_{60}Cr_{20}Pt_{20}$/CoFe/Ru/CoFe/ MgO/CoFe/Ru/Co/CoCrPt	6%		[80]
Structure	TMR ratio (%)	Remarks	Reference
$Co_{60}Cr_{20}Pt_{20}$/CoFe/Ru/CoFe/ MgO/CoFe/Ru/Co/CoCrPt	6%		[80]
GdFeCo/Fe/MgO/Fe/TbFeCo	64% (CIPT)	Perpendicular anisotropy by exchange coupling with RE-TM	[81]
$L1_0$-FePt/MgO/$L1_0$-FePt with Fe and FeCo insertion	105-120% (CIPT)	Lattice mismatch relayed by Fe insertion	[82]
[Co/Pd]/MgO/[Co/Pd]	10-12%		[83]
IrMn/Co/Pd exchange biased	8%		
$L1_0$-$Co_{50}Pt_{50}$/MgO/$L1_0$-$Co_{50}Pt_{50}$	6% , 13% at 10K		[84]

6. Applications of GMR and TMR Devices with Perpendicular Anisotropy

Perpendicular anisotropy materials have been widely considered for the applications in magneto-optical recording, hard-disk media and heat assisted magnetic recording media for the past few years. In more recent years, GMR spin-valve and TMR MTJ devices with perpendicular anisotropy have been intensively studied for their applications in spintronics such as a spin transfer switching MRAM due to their high magnetic stability and a lower operating current density, a spin transfer oscillator, and a spin polarized current-induced domain wall switching memory etc. The physical mechanisms of the devices and the research into the optimization of the devices are discussed in this section.

6.1. Spin Transfer Torque Magnetic Random Access Memory

As illustrated in Figure 25, when conduction electrons pass through a magnetic layer, their spins preferentially align in the direction of the magnetization of that layer (spin polarization).

As these spin polarized electrons encounter a free nano-magnetic material sandwiched between nonmagnetic spacers, the direction of their spins is repolarized to match that of the nano-magnet. This repolarization exerts a torque on the nano-magnet and as a result, its magnetic moment begins to make precession. If the current (or rate of electrons) is below a critical value, the damping torque is larger than the spin torque and the precession is quickly damped with the magnet settling into static equilibrium. If the current is well above this critical current, the spin torque is much larger than the damping torque and the precession increases in amplitude until the magnetization direction is completely reversed. This spin transfer torque exerted on a ferromagnetic layer by a sufficiently large spin polarized current, allows the manipulation of magnetization in a spin valve or magnetic tunneling junction (MTJ) into parallel (P) or anti-parallel (AP) states without the application of an external magnetic field. The AP to P transition takes place due to the spin-torque from the majority electrons polarized by the hard ferromagnetic layer while the P to AP transition takes place due to the spin-torque from the minority electrons scattered by the fixed layer.

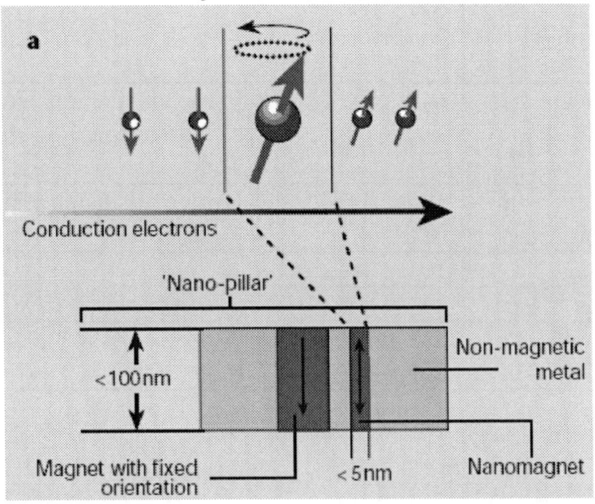

Figure 25.Spin Transfer switching mechanism [29]

Early works on spin transfer torque MRAM (See Figure 26) have been done on spin-valves and MTJs with in-plane anisotropy. However it has been theoretically anticipated that a significant enhancement and a higher thermal stability may be achieved for perpendicular anisotropy elements.

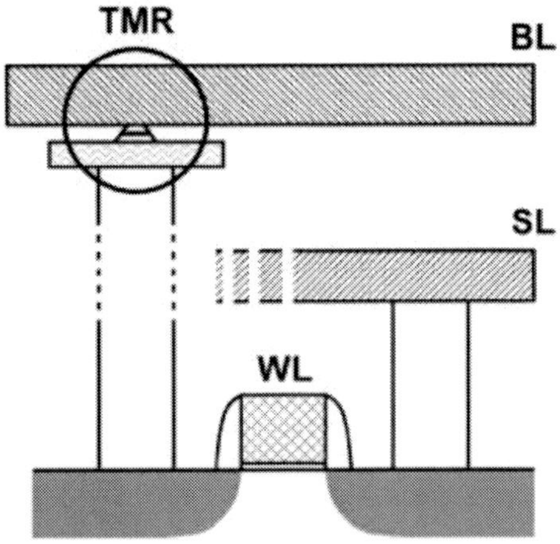

Figure 26. Illustration of spin transfer torque MRAM cell (BL: bit line, SL: source line, WL: word line) [85].

The critical reversal currents can be estimated based on Landau-Lifshitz-Gilbert (LLG) equations including spin-transfer torque term. For an in-plane anisotropy element, the critical current required for switching is given by Eq. (27).

$$I_C^{P-AP(AP-P)} \approx \frac{A\alpha M_S V}{g(0(\pi))p}(H + H_{dip} \pm H_{K\|} \pm 2\pi M_S) \tag{27}$$

where, M_S, V and α are the saturation magnetization, volume and Gilbert damping constant for the free layer, respectively, and p is the spin polarization of the current. H, H_{dip} and $H_{K\|}$ are the in-plane applied field, dipole field from the reference layer acting on the free layer and the in-plane anisotropy field, respectively.

For the spin transfer switching devices with perpendicular anisotropy, the critical currents to induce spin transfer magnetization reversal is given by Eq. (28).

$$I_C^{P-AP(AP-P)} \approx \frac{A\alpha M_S V}{g(0(\pi))p}(-H - H_{dip} \pm H_{K\perp} \mp 4\pi M_S) \tag{28}$$

It can be clearly seen that the energy barrier against thermal fluctuation is $M_S V H_K/2$ for in-plane elements and $M_S V(H_K-4\pi M_S)/2$ for perpendicular elements. This indicates that the critical current of perpendicular elements for switching is directly proportional to the anisotropy and hence the stability of the element. Since the first demonstration of spin transfer switching in full metal spin-valves using Co/Pt and Co/Ni multi-layers with perpendicular anisotropy was done in 2006 (Figure 27) [2], extensive research efforts in the development of spin-valves and MTJs with perpendicular anisotropy for the spin-transfer torque MRAM have been made [86-88] to achieve a higher density and a highly stable MRAM devices in commercialization. Table 2 shows the summary of research progress and achievements of various kinds of spin-valves and MTJ structures as well as their device performance in MRAM applications made so far.

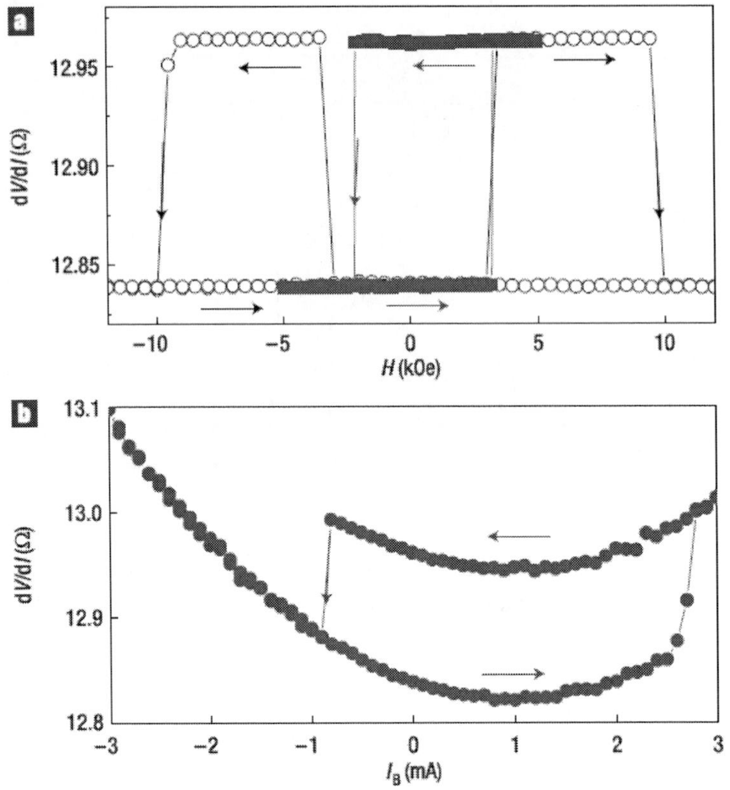

Figure 27. MR curves (solid circles represent minor loop-free layer reversal) (a) and current switching curve (b) for Co/Ni based spin-valve device with perpendicular anisotropy [2].

Table 2. Summary of spin-transfer switching device performance with perpendicular anisotropy

Structure	MR (%)	Free layer Coercivity (kOe)	Switching Current (A/cm^2) / pulse time	Reference
$[Co/Pt]_4/[Co/Ni]_2/Cu/[Co/Ni]_4$	CPP GMR 1%	2.65	P-AP: $7.5x10^7$ AP-P: $2.6x10^7$ 1000ms	[89]
$[CoFe/Pt]_5/Co/Cu/[CoFe/Pt]_7$	CPP GMR 0.47%	0.17	P-AP: 1.3 x10^8 AP-P: 1 x10^8 DC sweep	[90]
[$L1_0$FePt/Au/ $L1_0$FePt]	CPP GMR 0.067%	5.4	AP-P: 1 x10^8 (6.7KOe ext. field, 77K) 100ms	[91]
TbCoFe/CoFeB/ MgO/CoFeB/TbCoFe	TMR 15%	1.2	P-AP: 4.9 x10^6 AP-P: 4.7 x10^6 100ns	[92]
$[Co/Pt]_5/Co/[Ni/Co]_2/Co/Cu/Co/[Ni/Co]_5$	CPP GMR ~1%	0.42	P-AP,AP-P:~7 x10^6	[93]

Structure	MR (%)	Free layer Coercivity (kOe)	Switching Current (A/cm^2) / pulse time	Reference
$[CoFe/Pd]_3/CoFe/Cu/Co/[Pd/Co]_5$	CPP GMR 1.05	0.83	AP-P: 3.6- 3.8 x10^8 10ns	[94]
$CoFe/Cu/[CoFe/Pd]_3/CoFe/Cu/Co/ [Pd/Co]_5$	CPP GMR 0.98	0.13	AP-P: 2.9- 3.2 x10^8 10ns	
[Co/Pt]/Cu/[Co/Pt]	CPP GMR 0.33%	0.5	P-AP: 9.2 x10^7 AP-P: 6.4 x10^7	[95]
$[Co/Pt]_4/Co/[Ni/Co]_2/Cu/[Co/Ni]_2/Co$	CPP CMR 0.3%	0.245	300ps switching	[96]

6.2. Domain Wall Nucleation and Manipulation by Spin Polarized Current in GMR Devices with Perpendicular Anisotropy for Multi-State Storage

As a peculiar feature for GMR devices with perpendicular anisotropy, there have been a few of reports [97,98] of domain wall states, which have been nucleated under a spin polarized current, stabilized and manipulated in nanopillars (Figure 28).

Although this may be a drawback in current MRAM applications where care is taken to avoid domains within the magnetic layers, careful control of the domain wall creation may be of interest in development of multi-bits storage systems.

It was found that although domain wall states could be nucleated and manipulated by a spin-polarized current within a small distribution, the mechanism strongly depended on the presence of structural and magnetic imhomogenities in the device [98].

In order to utilize this phenomenon in spintronics devices, precise control of the magnetic properties of the films and the fabrication of pinning sites are essentially required.

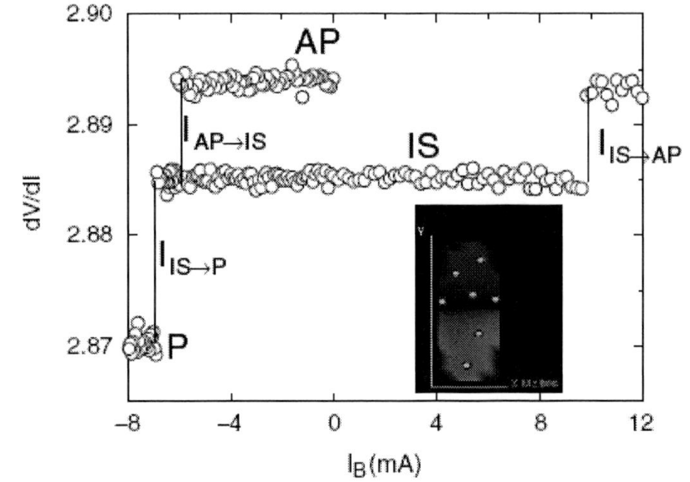

Figure 28. Differential resistance as a function of current for a 100×200 nm^2 pillar. The initial state is the AP state. The current sweeps from the AP to the IS state and then back to the AP state. The transition from the IS to P state is also shown. The inset corresponds to micromagnetic simulations (perpendicular component of the magnetization). The dots are a schematic illustration of the pinning sites. [97]

6.3. Spin Torque Oscillator

Spin transfer torque leads not only to reverse magnetization direction but also to generate high frequency precession of the magnetization. This steady-state precession corresponds to a state where the spin-transfer torque opposes and cancels the damping torque. Precession of the magnetization in spin-valves or MTJs paves the way for applications such as wide-band tunable radio-frequency oscillators [99,100]. Early spin-transfer precession was observed in systems where both the spin polarizer and free layer have an in-plane anisotropy [101]. However a more efficient system would be to have the polarizer with perpendicular anisotropy as seen in Figure 29. Polarizing the spins perpendicular to the free layer has several advantages including higher precession frequencies and lower precession currents [102].

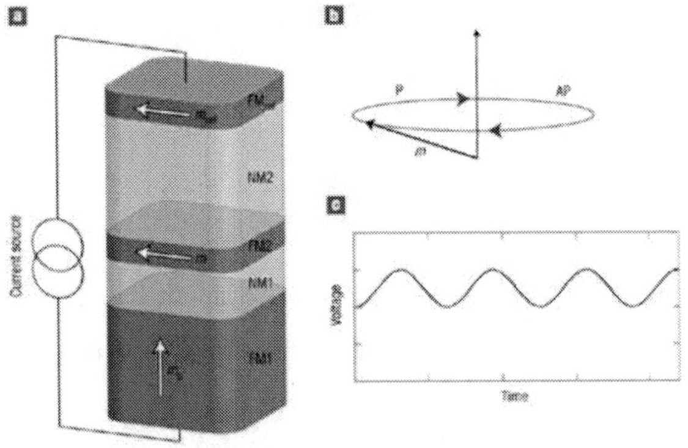

Figure 29. spin-transfer oscillator device structure with perpendicular anisotropy polarizer layer [99].

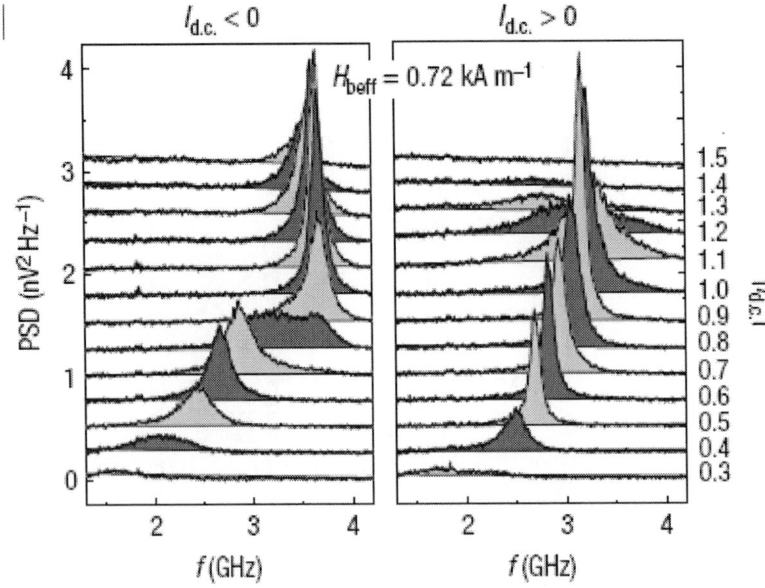

Figure 30. Power spectral density of the output voltage obtained for the spin oscillator device with perpendicular polarizer in the oscillation region for positive and negative injection currents [100].

With the perpendicular polarizer configurations, high frequency oscillations have been observed for both negative and positive injection currents as seen in Figure 30.

The spin torque oscillator provides continuous frequency tenability from zero to several gigahertz depending on the injected current. At low current densities, the oscillating frequency linearly increases with the current density. The relationship between the frequency and the injected current density is derived from the modified LLG equation as given by Eq (24)

$$f = \frac{\gamma}{2\pi\alpha}\left(\frac{P_0 \hbar J}{e M_s \delta}\right) \tag{24}$$

where, f is the oscillating frequency, J is the injected current density, M_s and δ are the saturation magnetization and thickness, respectively, P_0 denotes the polarizing faction of the spin current and α is the Gilbert damping constant [103]. As the current reaches a critical value, the frequency becomes a maximum and then the magnetization of the layers switch irreversible and the oscillations can no longer occur.

CONCLUSION AND FUTURE CHALLENGES

In this chapter, the recent developments in GMR and TMR based spintronic devices with perpendicular anisotropy have been introduced and discussed in terms materials and material process conditions, physical nature of GMR/TMR spin-valves, and the potential applications in advanced electronics. According to the literature reviews and some technical research reports published up to now, the GMR and TMR spin valves with perpendicular anisotropy have technical promises such as high magnetic and thermal stabilities for easy down sizing to a few tens of nano-meter device size, high electrical reliability due to a lower device operating current density, and easy tailoring of magnetic and structural properties by controlling process parameters. Therefore, a variety of device applications in spintronics are expected for advanced electronics such as a low current density spin transfer switching MRAM for ultra high memory density, a low current-induced domain wall switching memory, and a spin oscillator with a high spin polarizer of perpendicular component. However, despite their great technical advantages, the GMR and TMR spin valve devices with perpendicular anisotropy were found to have several technical challenges for further advanced developments in spintronics such as unbalancing of perpendicular anisotropy with shape anisotropy resulting in abrupt increase of operating current density and a larger free layer coercivity than that of GMR and TMR spin valves with in-plane anisotropy [104]. The research efforts to adjust the free layer coercivity using insertion materials such as Co, CoFe, and NiFe or use of angularly magnetized (or canted) free layers and to reduce the critical current density by increasing the spin accumulation within the GMR or TMR system such as using dual spin valves with oppositely magnetized hard layers and using spin scattering layers such as Ru, Ir and W are currently being made in this research area to overcome the technical challenges for further development.

REFERENCES

[1] N. Nishimura, T. Hirai, A Koganei, T. Ikeda, K. Okano, Y. Seguchi, and Y. Osada, *J. Appl. Phys.* 91, 5246 (2002).

[2] S. Mangin, D. Ravelosona, J. A. Katine, M.J. Carey, and E. E. Fullerton, *Nat. Mater.*, 5, 210 (2006).

[3] M. Albrecht, *et. al.*, *J. Appl. Phys.*,97, 103910 (2005).

[4] H. Meng *et al.*, *Appl. Phys. Lett.* 88, 172506 (2006).

[5] T. Seki *et al., Appl. Phys. Lett.* 88, 172504 (2006).

[6] N. Thiyagarajah, *et al.*, *Appl. Phys. Lett.* 92, 062504 (2008).

[7] P.F. Carcia, A.D. Meinhaldt, and A. Suna, *Appl. Phys. Lett.* 47, 178, (1985).

[8] T. Katayama, M. Miyazaki, Y. Nishihara and T. Shibata. *J. Magn. Magn. Mat.* 35 (1983).

[9] G. Kim, Y. Sakuraba, M. Oogane, Y. Ando, and T. Miyazaki, *Appl. Phys. Lett.* 92, 172502 (2008).

[10] M. T. Johnson, P. J. H. Bloemen, F. J. A den Broeder and J. J. de Vries, *Rep. Prog. Phys.* 59, 1409 (1996).

[11] F. J. A den Broeder, W. Hoving and P.J. H. Bloemen, *J. Magn. Magn. Mater.*, 93, 562 (1991).

[12] B. N. Engel, C. D. England, R. A. Van Leeuwen, M. H. Wiedmann, and C. M. Falco, *Phys. Rev. Lett.* 67, 1910 (1991).

[13] S. Hashimoto, Y. Ochai, and K. Aso, *J. App. Phys.* 66, 4909 (1989).

[14] R. H. Victora, and J. M. MacLaren, *J. Appl. Phys.* 73, 6415 (1993).

[15] R. H. Victora, and J. M. MacLaren, *Phys. Rev.* B, 47, 11583 (1993).

[16] S. Jjeong, M. E. McHenry, and D. E. Launghlin, *IEEE Trans. Magn.*, 37, 1309, (2001).

[17] R. Mukai, T. Uzumaki, and A. Tanaka, *IEEE Trans. Magn.*, 39, 1925, (2003).

[18] Y.-N. Hsu, S. Jeong, D. Laughlin, and D. N. Lambeth, *J. Appl. Phys.* 89, 7068.

[19] Y. F. Xu, J. S. Chen, and J. P. Wang, *Appl. Phys. Lett.* 80, 3325.

[20] K. Kang, Z. G. Zhang, C. Papusoi, and T. Suzuki, *Appl. Phys. Lett.* 84, 404.

[21] J. S. Chen, B. C. Lim, and T. J. Zhou, *J. Vac. Sci. Technol. A* 23, 184 (2005).

[22] R. J. Gambino, and J.J. Cuomo, *J. Vac. Sci. Tech.,* 15, 296 (1978).

[23] T. Mixoguchi and G. S. Cargill, *J. Appl. Phys.* 50, 3570 (1979).

[24] R. Sato, N. Saito, and Y. Togami, *Jpn. J. Appl. Phys.* 24, L266 (1985).

[25] Y. Suzuki, S. Takayama, F. Kirino, and N. Ohta, *IEEE Trans. Magn.* 23, 2275 (1987).

[26] W. H. Meiklejohn, *Proc IEEE* 74, 1570 (1986).

[27] Y. Suzuki, J. Haimovich, and T. Egami, *Phys. Rev. B* 35, 2162 (1987).

[28] H. Fu, M. Mansuripur, and P. Meystre, *Phys. Rev. B*. 66, 1086 (1991).

[29] J. Z. Sun, *IBM Res. and Dev.*, 50, 81 (2006).

[30] N. F. Mott, *Proc. R. Soc. A* 153, 699, (1936).

[31] N. Thiyagarajah, and S. Bae, , *J. Appl. Phys.* 104, 113906 (2008).

[32] Z. Li, Z. Zhang, H Zhao, B. Ma, and Q. Y. Jin, *J. Appl. Phys.* 106, 013907 (2009).

[33] H. W. Joo, J. H. An, M. S. Lee, S. D. Choi, K. A. Lee, S. W. Kim, S. S. Lee, and D. G. Hwang, *J. Appl. Phys.* 99, 08R504 (2006).

[34] F. Garcia, F. Fettar, S. Auffret, B. Rodmaq and B. Dieny, *J. Appl. Phys.* 93, 8397 (2003).

[35] R. Law, R. Sbiaa, T. Liew and T.C. Chong, *J. Appl. Phys*. 105, 103911 (2009).
[36] M. Albrecht, G. Hu, A Moser, O. Hellwig, and B. D. Terris, *J. Appl. Phys*., 97, 103910 (2005).
[37] G. Hu, T. Thomson, C.T. Rettner, S. Raoux, and B. D. Terris, *J. Appl. Phys*., 97, 10J702 (2005).
[38] N. Thiyagarajah, H. W. Joo, and S. Bae, *Appl. Phys. Lett*. 95, 232513 (2009).
[39] M. E. Schabes and H. N. Bertram, *J. Appl. Phys*., 64, 1347 (1988).
[40] E. Girgis, J. Schelten, J. Shi, S. Tehrani, and H. Goronkin, *Appl. Phys. Lett*., 76, 3780 (2000).
[41] Z. Y. Liu, G. H. Yu, G. Han, and Z. C. Wang, *J. Magn. Magn. Mater*., 302, 29 (2006).
[42] Z. Y. Liu, L. Yue, D. J. Keavney, and S. Adenwalla, *Phys. Rev. B*, 70, 224423 (2004).
[43] Y. Liu, G. H. Yu, G. Han, and Z. C. Wang, *J. Magn. Magn. Mater*., 302, 29 (2006).
[44] Y. Liu, L. Yue, D. J. Keavney, and S. Adenwalla, *Phys. Rev. B,* 70, 224423 (2004).
[45] N. Thiyagarajah, S. Bae, H. W. Joo, Y. C. Han, and J. Kim, *Appl. Phys. Lett*. 92, 062504 (2008).
[46] Bruno, and C. Chappert, *Phys. Rev. Lett*., 76, 1602 (1991).
[47] Bruno, C. and Chappert, *Phys. Rev. B*, 46, 261 (1992).
[48] Coehoorn, *Phys. Rev. B*, 44, 9331 (1991).
[49] J. Moritz, F. Gacia, J. C. Toussaint, B. Dieny and J. P. Nozieres, *Europhys. Lett*. 65, 123 (2004).
[50] L. Lin, N. Thiyagarajah, H. W. Joo, J. H., K. A. Lee, and S. Bae, ., *J. Appl. Phys*., (2010).
[51] J. Nogués, and I. K. Schuller, *J. Magn. Magn. Mater*. 192, 203 (1999).
[52] H. Umebayashi, Y. Ishikawa, *J. Phys. Soc. Jpn*., 21, 1281 (1966).
[53] W. H. Meiklejohn, *J. Appl. Phys*.. 33, 1328 (1962).
[54] A. E. Berkowitz and K. Takano, *J. Magn. Magn. Mater*. 200, 552 (1999).
[55] J. Sort, B. Dieny, M. Fraune, C. Koenig, Lunnebach and G. Guntherodt, *Appl. Phys. Lett*. 84, 3696 (2004).
[56] A. Bollero, V. Baltz, B. Rodmacq, B. Dieny, and J. Sort, *Appl. Phys. Lett*. 89, 152502 (2006).
[57] J. L. Menedez, D. Ravelosona, C. Chappert, *J. Appl. Phys*. 95, 6726 (2004).
[58] C. L. Canedy, X. W. Li, and G. Xiao, *Phys. Rev. B*, 62, 508 (2000).
[59] D. Rosenblatt, M. Karpovski, and A. Gerber, *Appl. Phys. Lett*., 96, 022512 (2010).
[60] G. Xiang, A. W. Holleitner, B. L. Sheu, F. M. Mendoza, O. Maksimov, M. B. Stone, P. Schiffer, D. D. Awschalom, and N. Samarth, *Phys. Rev. B*, 71, 241307(R) (2005).
[61] D. Chiba, M. Yamaguchi, F. Matsukura, and H. Ohno, *Science*, 301, 943 (2003).
[62] C. M. Hurd, *The Hall Effect in Metals and Alloys* (Plenum, New York, 1973).N. Thiyagarajah, L. Lin, H. W. Joo, and S. Bae to be published (2010).
[63] Z. Q. Lu, G. Pan, J. Li, and W. Y. Lai, *J. Appl. Phys*., 89, 7215 (2001).
[64] A. Gerber, A. Milner, A. Finkler, M. Karpovski and L. Goldsmith, *Phys. Rev. B*, 69, 224403 (2004).
[65] Y. Yao, L. Kleinman, A. H. MacDonald, J. Sinova, T. Jungwirth, D. Wang, E. Wang, and Q. Niu, Phys. *Rev. Lett*., 92, 037204 (2004).
[66] L. Berger and G. Bergmann, in *The Hall Effect and Its Applications*, edited by C. L. Chien and C. R. Westgate (Plenum, New York, 1979).
[67] S. Yuasa and D. D. Djayaprawira, *J. Phys. D: Appl. Phys*. 40, R 337 (2007).

[68] W. D, Nordman C, Daughton J, Qian Z and Fink J *IEEE Trans. Magn.* 40 2269 (2004).
[69] Yuasa S, Fukushima A, Kubota H, Suzuki Y and Ando K , *Appl. Phys. Lett.* 89 042505 (2006).
[70] M. Julliere, *Phys. Rev. B* A84, 225 (1978).
[71] J. C. Slonczewski, *Phys. Rev. B*, 39, 1995 (1989).
[72] J. G. Simmon, *J. Appl. Phys.*, 38, 2655 (1964).
[73] H. Yoda et al. 11th Joint INTERMAG MMM conference, AA-02, (2010).
[74] Z.R. Tadisina, A. Natarajarathinam, B. D. Clark, A. L. Highsmith, T. Mewes, S. Gupta, E. Chen, and S. Wang, *J. Appl. Phys.* 107, 09C703 (2010).
[75] Y. Wang, W. X. Wang, H. X. Wei, B. S. Zhang, W. S. Zhan, X. F. Han, *J. Appl. Phys.* 107, 09C711 (2010).
[76] L. X. Ye, C. M. Lee, J. H, Lai, A. Canizo-Cabrera, W. J. Chen, T. Wu, *J. Magn. Magn. Mater.* 322, L9 (2010).
[77] J.H. Park, S. Ikeda, H. Yamamoto, H. Gan, K. Mizunuma, K. Miura, H. Hasegawa, J. Hayakawa, K. Ito, F. Marsukura and H. Ohno, *IEEE Trans. Magn.* 45, 3476 (2009).
[78] K. Mizunuma, S. Ikeda, J. H. Park, H. Yamamoto, H. Gan, K. Miura, H. Hasegawa, J. Hayakawa, F. Matsukuram and H. Ohno, *Appl. Phys. Lett.*, 95 232516 (2009).
[79] D. Watanabe, S. Mizukami, M. Oogane, H. Naganuma, Y. Ando, T. Miyazaki, *J. Appl. Phys.* 105, 07C911 (2009).
[80] H. Ohmori, T. Hatori, and S. Nakagawa, *J. Appl. Phys.* 103, 07A911 (2008).
[81] M. Yoshikawa, E. Kitagawa, T. Nagase, T. Daibou, M. Nagamine, K. Nishiyama, T. Kishi, H. Yoda. D. Lim, K. Kim, S. Kim, W. Y. Jeun, and S. R. Lee, *IEEE Trans. Magn.* 45, 2407 (2009).
[82] G. Kim, Y. Sakuraba, M. Oogane, Y. Ando, and T. Miyasaki, *Appl. Phys. Lett* 92, 172502 (2008).
[83] T. Kawahara, et al, *IEEE Sol. Stat. Circ.* 43, 109 (2008).
[84] H. Meng, and J. P. Wang, *Appl. Phys. Lett.* 88, 172506 (2005).
[85] T. Seki, S. Mitani, K. Yakushiji and K. Takanashi, Appl. Phys. Lett. 88, 172504 (2006).
[86] M. Yoshikawa et al, AC-01, *Intermag* (2008)
[87] S. Mangin, D. Ravelosona, J. A. Katine, M. J. Carey, B. D. Terris, and E. E. Fullerton, *Nat. Mater.* 5, 210 (2006).
[88] H. Meng, and J. P. Wang, *Appl. Phys. Lett.* 88, 172506 (2006).
[89] T. Seki, S. Mitani, K. Yakushiji and K. Takanashi, *Appl. Phys. Lett.* 88, 172504 (2006)
[90] M. Nakayama, T. Kai, N. Shimomura, M. Amano, E. Kitagawa, T. Nagase, M. Yoshikawa, T. Kishi, S. Ikegawa and H. Yoda, *J. Appl. Phys.* 103, 07A710
[91] S. Mangin, Y. Henry, D. Ravelosona, J. A. Katine, and E. E. Fullerton, *Appl. Phys. Lett.* 94, 012502.
[92] R. Law, E. Tan, R. Sbiaa, T. Liew, and T. C. Chong, *Appl. Phys. Lett.* 94, 062516 (2009).
[93] J. Park, M.T. Moneck, C. Park, J. Zhu, *J. Appl. Phys. Lett.* 105, 07D129 (2009)
[94] D. Bedau, H. Liu. J. J. Bouzaglou, A. D. Kent, J. Z. Sun, J. A. Katine, E. E. Fullerton, and S. Mangin, *Appl. Phys. Lett.* 96 022514 (2010)
[95] D. Ravelosona, S. Mangin, Y. Lemaho, J. Katine, B. Terris and E. E. Fullerton, *Phys. Rev. Lett.* 96, 186604 (2006)
[96] D. Ravelosona, S. Mangin, Y. Lemaho, J. Katine, B. Terris and E. E. Fullerton, *J. Phys. D. Appl. Phys.* 40, 1253 (2007)

[97] A. D. Kent, *Nat. Mater.*, 6 399 (2007)
[98] D. Houssameddine et al., *Nat. Mater.*, 6, 447 (2007)
[99] Kiselev, S. I. et al. *Nature* 425, 380–383 (2004) [102] (2004)
[100] U. Ebels, D. Houssameddine, I. Firastrau, D. Guskova, C. Thirion, B. Dieny, and L. D. Buda-Prejbeanu, *Phys. Rev. B* 78, 024436 (2008)
[101] J. A. Katine, and E.E. Fullerton, *J. Magn. Mag. Mater.* 320, 1217 (2008)

In: Magnetic Thin Films
Editor: John P. Volkerts

ISBN: 978-1-61209-302-4

Chapter 5

CORRELATION OF DOMAIN STRUCTURE WITH MAGNETO-IMPEDANCE EFFECT

T. Nakai, [1] ***K.I. Arai,*** [2] ***J. Yamasaki,*** [3] ***M. Yamaguchi,*** [4] ***K. Ishiyama,*** [5] ***S. Yabukami,*** [6] ***K. Takada*** [1] ***and H. Abe*** [1]

[1]Industrial Technology Institute, Miyagi Prefectural Government, *Sendai 981-3206, Japan*
[2]Research Institute for Electric and Magnetic Materials, *Sendai 982-0807, Japan*
[3]Kyushu Institute of Technology, *Kitakyushu 804-8550, Japan*
[4]Graduate School of Engineering, Tohoku University, *Sendai 980-8579, Japan*
[5]Research Institute of Electrical Communication, Tohoku University, *Sendai 980-8577, Japan*
[6]Tohoku-Gakuin University, *Tagajo 985-8537, Japan*

ABSTRACT

In this chapter, a variation of magnetic domain and correlated magneto-impedance of a micro-fabricated rectangular-shaped magnetic thin film is discussed. In a certain case of the magnetic element, which is made from an amorphous film with in-plane uniaxial anisotropy and low magnetostriction, the Landau-Lifshitz-like magnetic domain is observed. A change in structure of magnetic domain occurs as a function of external magnetic field. The Landau-Lifshitz magnetic domain consists of contiguous domains with anti-parallel magnetic-momentum with existence of closure domains in the edge area of the striped element. The dimensions of the rectangular strip discussed in this chapter, especially the width, must be almost in the same order of the width of a domain area. The high-frequency impedance of the element is determined by the high-frequency permeability, and the permeability is determined by the domain structure, the direction of magnetic moment in each domain and fixed force of the magnetic moment. The variation of the domain structure and correlated magneto-impedance of the element is characterized by a direction of magnetic easy axis. In this chapter, a magnetic phenomenon and an analytical explanation is shown for the rectangular magnetic element with the width of some tens of microns, therefore these element has the

Landau-Lifshitz-like magnetic domain. The phenomenon is based on the variation of magnetic domain and correlated magneto-impedance induced by external magnetic field as a parameter of the in-plane direction of uniaxial magnetic easy axis.

1. CoNbZr Magnetic Thin Film (Fabrication Method and Magnetic Property)

A CoNbZr magnetic thin film fabricated on a glass substrate is useful for a study of the variation of magnetic domain and also the correlated magnetic property of a controlled shaped micro element. The reason is that this film is amorphous, makeable of non-magnetostriction, and possibility of induction of uniaxial magnetic anisotropy in a controlled direction. The amorphous means that there is no grain structure in the film. The grain boundary and related structure of distortion of lattice makes a movement of magnetic domain wall stick on it. The amorphous CoNbZr film has possibility to control the magnetostriction constant, ranging from minus to plus, by controlling the composition of Co, Nb and Zr. The amorphous CoNbZr film has useful property for induction of magnetic anisotropy in a controlled direction. A magnetic annealing makes an uniaxial anisotropy in the direction of applied magnetic field. In this chapter, the composition Co: 85, Nb: 12, and Zr: 3, which is known as almost zero magnetostriction, was used. The saturation magnetization of this composition is 0.93 Tesla.

The single layer amorphous $Co_{85}Nb_{12}Zr_3$ films were RF-sputter-deposited onto a soda glass substrate and then micro-fabricated into rectangular element by using the lift-off process. The thickness of the film was ranging from 0.5 μm to 4 μm. The element was annealed in magnetic field in order to induce uniaxial magnetic anisotropy.

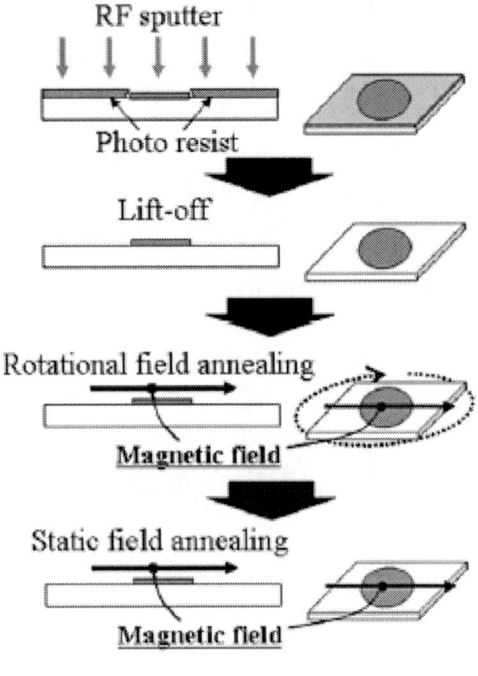

Figure 1.1. Schematic of fabrication of the film.

The direction of the magnetic anisotropy was controlled by the direction of magnetic field. In this chapter, the magnetic field during annealing, 240 kA/m, 673 K, was oriented in a controlled direction, and kept for 1 hour. This annealing process carried out in vacuum atmosphere lower than 6.5×10^{-4} Pa. Figure 1.1 shows a schematic and Figure 1.2. shows a process condition of fabrication of the film. The soda glass substrate was selected as it has almost the same thermal expansion constant as the $Co_{85}Nb_{12}Zr_3$ films, in order not to peel off during annealing process. The condition of RF-sputter is important, such as Ar pressure, distance between target and substrate and also cooling of substrate during sputtering. A surface finish of the substrate and the condition of RF-sputter is an important know-how for fabrication.

Figure 1.3 shows a measured result of hysteresis loop for a disk shaped sample with 5 mm diameter, and 0.5 μm thickness. The sample was processed just after the rotational annealing ended. So it is magnetically isotropy within in-plane direction. This measurement shows a variation of hysteresis as a function of temperature, ranging from room-temperature (R.T.) to 673K. It is confirmed that the $Co_{85}Nb_{12}Zr_3$ film has magnetization even in 673K, which is the temperature of magnetic annealing. The film has low coercivity and also has the "soft" magnetic property until 673K.

1. $Co_{85}Nb_{12}Zr_3$ magnetic thin film

RF sputter deposition (Ar atmosphere)
Patterning was processed by lift-off method

⬇

2. Magnetic field annealing

Rotational field annealing
673 K, 240 A/m, 2 hour (<10^{-3} Pa), 60 rpm
Static field annealing
673 K, 240 A/m, 1 hour (<10^{-3} Pa)

Figure 1.2. process condition of fabrication.

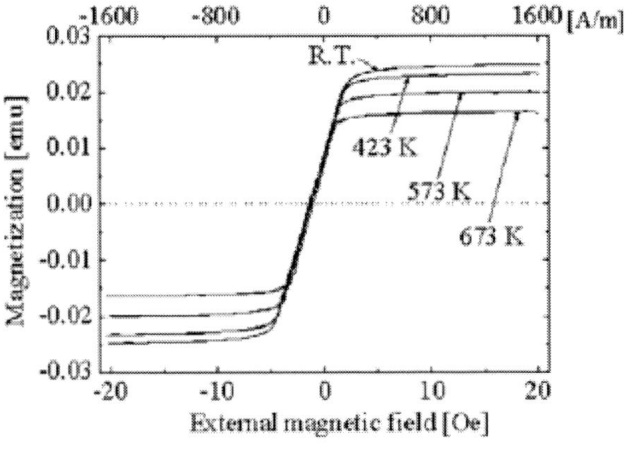

Figure 1.3. Measured result of hysteresis loop.

Figure 1.4 shows a variation of hysteresis as a function of direction of easy axis [1]. In this case, the sample was fabricated in full process shown in Figure 1.1. In Figure 1.4, $\theta=0$ degrees means a hard direction of uniaxial film, therefore easy direction correspond to $\theta=90$ degrees. It is confirmed that this film has in-plane uniaxial anisotropy, because the hysteresis loop changes gradually from hard to easy direction. In this case, the uniaxial anisotropy field is 560 A/m. The anisotropy field changes as a function of annealing temperature, the lower temperature makes the lower anisotropy field. Figure 1.5. shows a measured high-frequency permeability of the film as a function of external magnetic field in various directions of easy axis. This measurement carried out in the frequency of 500 MHz by using the shielded-loop coil method. The high-frequency permeability was measured by applying high-frequency magnetic field in the direction perpendicular to the external field.

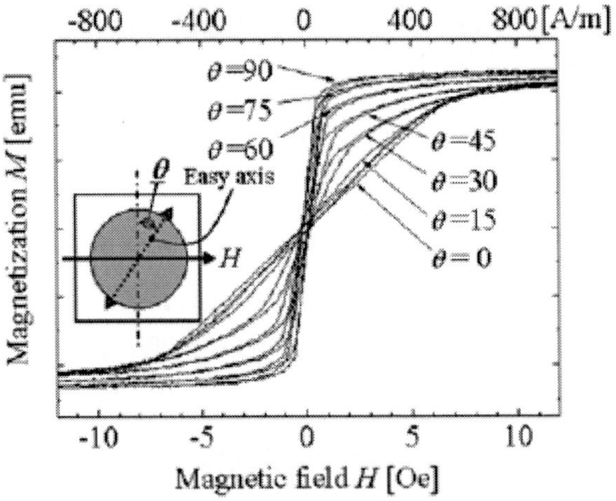

Figure 1.4. variation of hysteresis as a function of direction of easy axis.

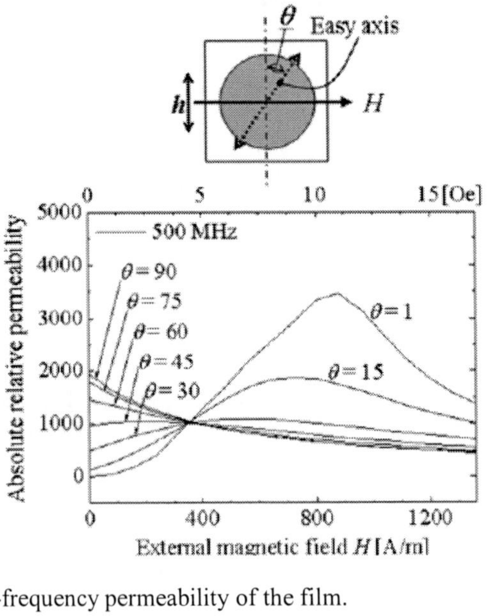

Figure 1.5. Measured high-frequency permeability of the film.

The measurement results was as follows; In case of the measured direction of the high-frequency permeability is easy direction and the applied external field is in the hard direction, the permeability has a maximum value around 900 A/m of external field. Whereas the measured direction of the high-frequency permeability is hard direction and the applied external field is in the easy direction, the permeability has a maximum value around 0 A/m of external field and decreases as increment of the external field. The variation of high-frequency permeability is analytically explained by the bias-susceptibility model of a magnetic thin film. In the next section, the theoretical expression is discussed.

2. HIGH-FREQUENCY PERMEABILITY FOR AN UNIAXIAL MAGNETIC THIN FILM

A theoretical expression of high-frequency permeability is shown by using the bias susceptibility model of a single-domain magnetic thin-film with uniaxial anisotropy. Based on a schematic layout of Figure 2.1. total energy of a single-domain magnetic thin film with uniaxial anisotropy is calculated by eq. (1). Where the easy-axis is along the X axis, both the magnetic thin film and the applied external magnetic field H_{dc} are in XY plane, h the applied high frequency magnetization, K_u an anisotropy coefficient [2].

$$\begin{aligned} E &= K_u \sin^2\phi - M_S\left(H_{dc}\cos\delta + h\cos\beta\right)\cos\varphi\cos\phi \\ &\quad - M_S\left(H_{dc}\sin\delta + h\sin\beta\right)\cos\varphi\sin\phi + \frac{M_S{}^2}{2\mu_0}\sin^2\varphi \end{aligned} \tag{1}$$

$$\frac{d^2\phi}{dt^2} + 4\pi\lambda\frac{d\phi}{dt} + \frac{\gamma^2}{\mu_0}\frac{\partial E}{\partial\phi} = 0 \text{ for } \lambda = \frac{\alpha\gamma M_S}{4\pi\mu_0} \tag{2}$$

Based on the eq. (2), we can obtain a complex high-frequency permeability ($\mu_{ri} = \mu_{ri}' - j\,\mu_{ri}''$) as eq. (3) and eq. (4) [3]. These equations are the permeability resulted from magnetization rotation with slight magnitude around the energy minimum state.

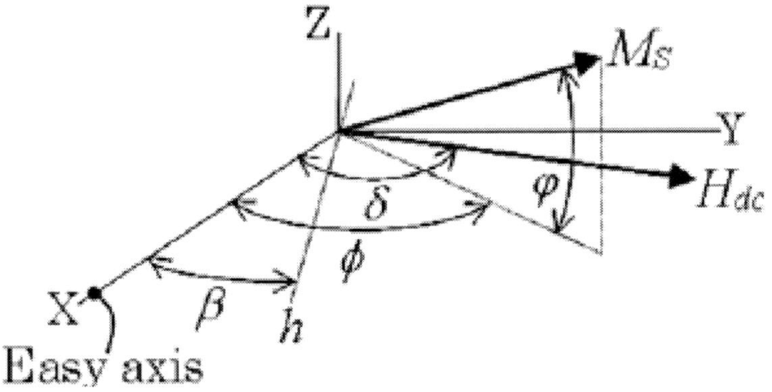

Figure 2.1. Schematic layout of energy analysis.

$$\mu_{ri}' = \frac{\gamma^2}{\mu_0^2} \cdot \frac{M_S^2 \sin^2(\phi_0 - \beta)}{\left(\omega_0^2 - \omega^2\right)^2 + \left(4\pi\lambda\omega\right)^2}\left(\omega_0^2 - \omega^2\right) \tag{3}$$

$$\mu_{ri}'' = \frac{\gamma^2}{\mu_0^2} \cdot \frac{4\pi\lambda\omega M_S^2 \sin^2(\phi_0 - \beta)}{\left(\omega_0^2 - \omega^2\right)^2 + \left(4\pi\lambda\omega\right)^2} \tag{4}$$

Where Ms is a saturation magnetization, μ_0 the permeability of vacuum, γ a gyromagnetic ratio, α Gilbert dumping parameter, and ω angular velocity of high frequency magnetic field. The ω_0 is the magnetic resonant frequency, shown in eq. (5). The ϕ_0 is a stable direction of magnetization without high frequency magnetic field h.

$$\omega_0^2 = \frac{\gamma^2}{\mu_0}\left(\frac{\partial^2 E_0}{\partial \phi^2}\right)_{\substack{\phi=\phi_0 \\ \varphi=0}} \tag{5}$$

Figure 2.2. shows a variation of magnetic energy of uniaxial magnetic thin film calculated from eq. (1). The direction of easy axis is 60 degrees relative to the direction of external magnetic field. The external static field is applied perpendicular to the high-frequency field. There is both a steady state and a quasi-steady state. In Figure 2.2, the solid circle shows a steady state and the open circle shows a quasi-steady state. The quasi-steady state disappears over 320 A/m of external field. This condition of disappearance of quasi-steady state is considered as a condition where the magnetization reversal occurs.

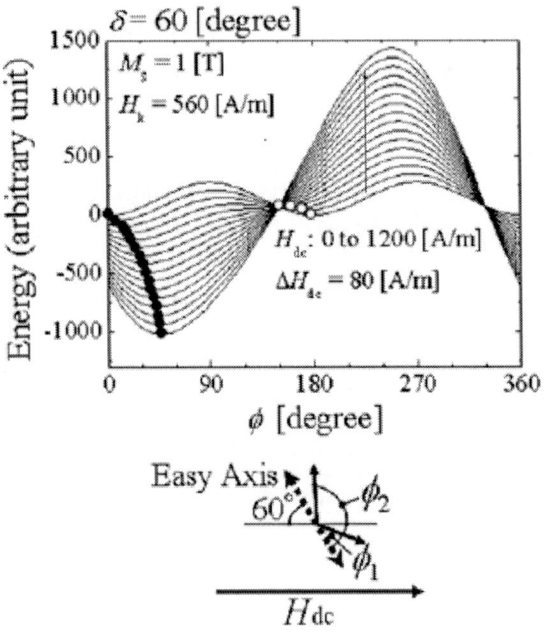

Figure 2.2. Variation of magnetic energy.

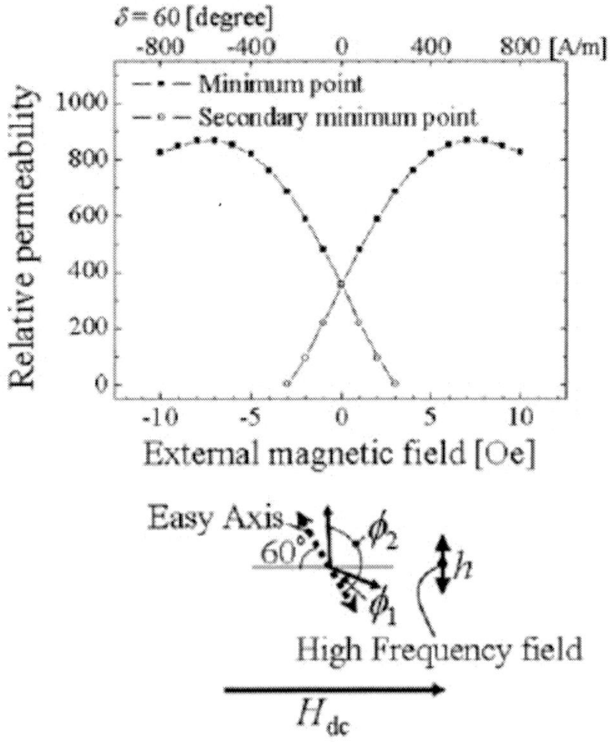

Figure 2.3. Calculate high-frequency permeability.

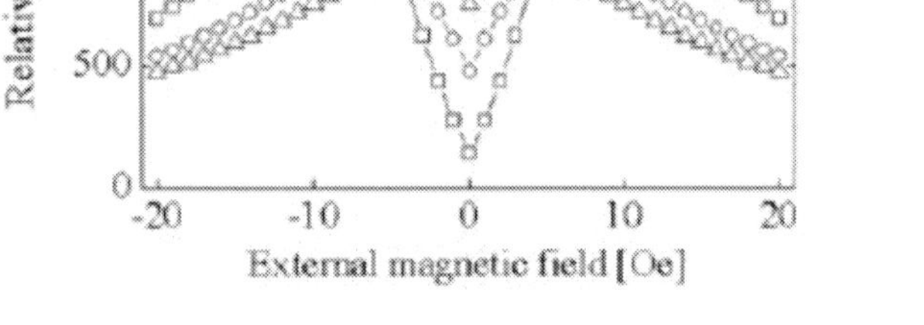

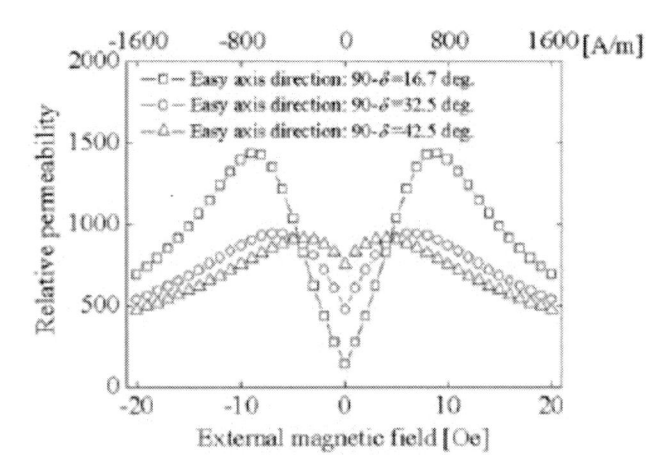

Figure 2.4. Calculated high-frequency permeability for 3 different easy axis.

From eq. (3) and eq. (4) we can calculate a high-frequency permeability both for steady state and for quasi-steady state as shown in Figure 2.3. The parameters used for the calculation are as follows: Ms ; 1T, H_k; 640 A/m, α ; 0.0139, γ; 2.21×10^5, and the frequency ; 500 MHz. The permeability in each state is quite different. The quasi-steady state has lower permeability than the steady state when an external magnetic field is applied. Figure 2.4. shows calculated results of high-frequency permeability for 3 different direction of easy axis as a function of external field. This result shows the permeability for steady state, and it is in good agreement with the measured result of Figure 1.5. As a result, the high-frequency permeability of magnetic thin film can be estimated by the theoretical model of magnetization rotation. Therefore both the direction of magnetic momentum in the magnetic film and the energy change during perturbation are the important parameters for discussing the high-frequency permeability.

3. MICRO-FABRICATED ELEMENT AND FABRICATION METHOD

The micro-fabricated element with the width of tens of microns is made by a photo-lithography method. There are some possible method of fabrication, such as wet etching, dry etching, ion-milling and lift-off. In case of magnetic element, which has close dimensions to the size of magnetic domain on the element, the magnetic property is strongly affected by domain wall movement. The needs for the fabrication method must be both a smooth outline of element and also without damage near the cutting-line of the element while micro-fabrication. Therefore the lift-off method by using negative tapered-edge type photo-resist is chosen. The lift-off process by using the negative tapered-edge type photo-resist makes the shape of edge of the element smooth and damage-less.

Figure 3.1. shows a fabrication process of making a rectangular shaped element. The RF-sputter and the magnetic field annealing is the same process as explained in section 1. The $Co_{85}Nb_{12}Zr_3$ film is a single layer, too. A Cu electrode is fabricated finally by using the lift-off method in order to measure the impedance of the element.

Figure 3.2. shows an edge of the fabricated rectangular magnetic strip. The thickness of the film was 1 μm and the width of the strip was 20 μm. A clear-shaped fringe without damage is obtained by the lift-off method.

Figure 3.3. shows the fabricated element. The magnetic strip is combined with coplanar Cu electrode. The electrode pads were designed for making contact with G-S-G type high-frequency wafer probe. The use of the wafer probe is indispensable for measurement of high-frequency impedance of the element. There are some special measurement systems which are needed to measure electric or magnetic property for the tens of micron sized element.

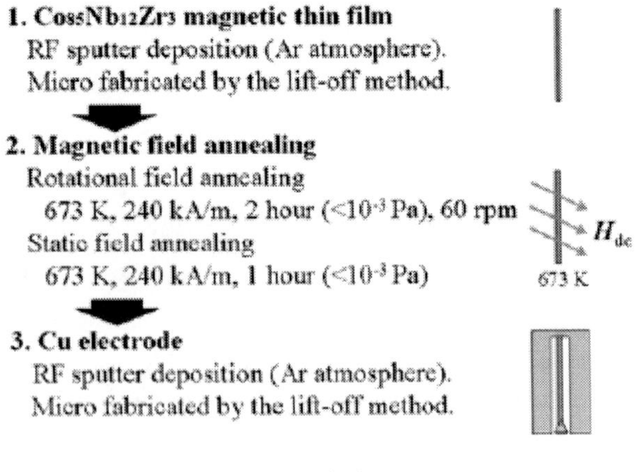

Figure 3.1. Fabrication process of rectangular shaped element.

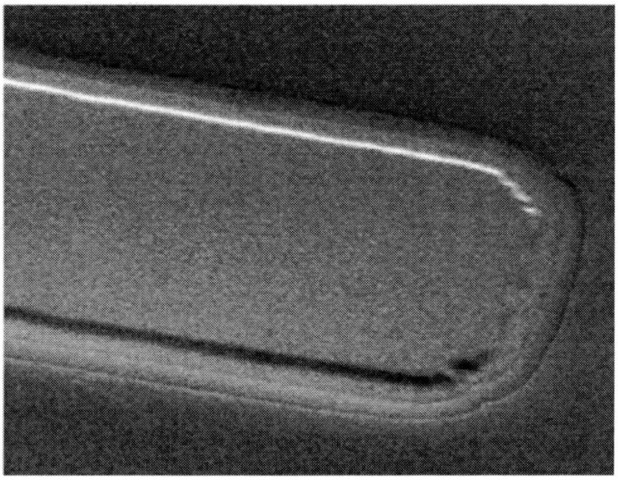

Figure 3.2. Edge of the fabricated rectangular magnetic strip.

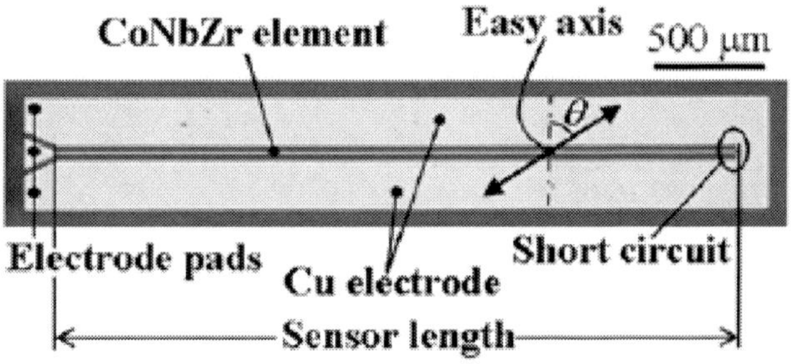

Figure 3.3. The fabricated element.

4. Measurement Method (Domain, High-Frequency Impedance and Permeability)

The hysteresis loop (MH loop) of the fabricated element was measured by VSM (PV-H300; Toei Scientific Industrial Co., Ltd.) along the longitudinal direction. An advantage of this apparatus is that it has an accurate sensor for the control of magnetic field, therefore a change of magnetization which occurs simultaneously with a change of magnetic domain can be measured in an accuracy of 1 A/m. (Figure 4.1)

The magnetic domain on the element was observed by a Kerr-effect microscope (BH-786iP; Neoarc Co., Ltd.). An in-plane domain image can be observed with the size of tens of microns elemnt. An impedance measurement fixture by using wafer probe with applying magnetic field by Helmholtz coil was made ourselves, which makes possible to observe magnetic domain simultaneously with impedance measurement with applying external field. (Figure 4.2)

The impedance of the element was measured by using a network analyzer based on S11 response with applying an external magnetic field by Helmholtz coil. The wafer probe contact was used to measure electric property of the element especially for high-frequency impedance. The calibration of measurement is possible to be carried out at the top of probe needle, therefore the high-frequency property of the element can be measured directly itself. The measurement system is shown in Figure 4.3. Several notices are needed for the design of shape of coplanar electrode, because of both a capacitance and an inductance induced by the element shape [4]. A measurement of temperature variation is possible by using this apparatus, by controlling the temperature of sample stage. The incident RF power to the element was -14 dBm in every measurement in this chapter.

The high-frequency permeability of a magnetic thin film with an in-plane area of several mm of diameter was measured by using an apparatus based on the shielded-loop method (PM-3000; Ryowa Electronics Co., Ltd.). (Figure 4.4)

Figure 4.1. VSM apparatus.

Figure 4.2. Kerr-effect microscope.

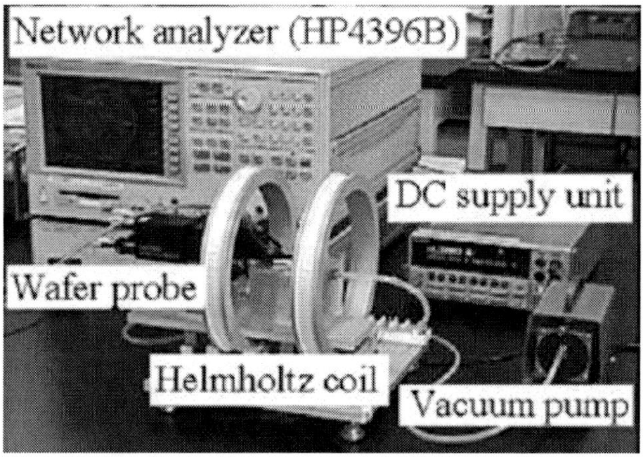

Figure 4.3. Impedance measurement system.

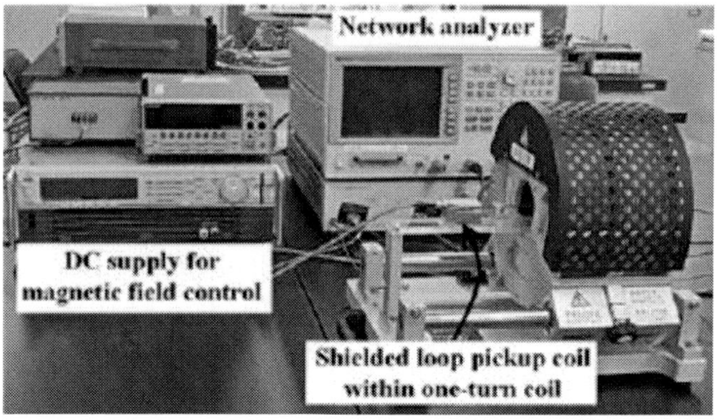

Figure 4.4. Measurement apparatus of high-frequency permeability.

5. Relationship between Easy Direction and Element Impedance

A variation of electrical impedance of a soft magnetic conductive element as a function of external magnetic field is called as the magnetoimpedance effect. The giant magnetoimpedance (GMI) sensor has an effect of a significant change in the impedance of a soft magnetic element, which is driven by a high frequency current, when it is placed in a static magnetic field [5-8].

The GMI effect is caused by a change of permeability induced by applying an external magnetic field. The change of permeability makes a change of current distribution in the sensor, which is well known as skin effect, and then makes the sensor impedance change. The change of permeability with applying the external field is caused by an interaction of magnetic anisotropy as well as magnetic domain structure with the external field.

In this section, an effect of in-plane direction of magnetic easy axis for the impedance of thin film GMI element is explained. The direction controlled in-plane easy axis makes an inclined Landau-Lifchitz-like domain structure on the micro-element. The in-plane inclined easy axis is made by the induction of magnetic anisotropy in the inclined direction of thin film strip by field annealing. Some technological benefits can be obtained by applying this method, such as bias reduction, high sensitivity, and wide linearity. A topic is that a stepped GMI property can be obtained in a certain domain inclination angle.

5.1. Preparation of Element

The element was made by the same process previously explained in the section 3. The dimensions of the element are ranging from 1000 μm to 2000 μm length, 10 μm to 100 μm width and 1 μm to 3.3 μm thickness. The direction of the magnetic anisotropy was controlled both by the direction of magnetic field while annealing and by the demagnetization force along the width direction of the element. In this section the magnetic field during annealing, 40 kA/m - 673 K, was oriented 1 degree relative to the width direction of the element. The inclination angle of domain is controlled by a cross-sectional aspect ratio of the element, because the demagnetization force is a function of the aspect ratio. The relationship between annealing field and internal field of the element with consideration of demagnetization force is schematically shown in Figure 5.1. In case of the annealing field, which is large enough to ignore the demagnetization force, the inclination angle between domain wall and annealing field is in good agreement (Figure 5.2). The direction of domain wall of 180 degrees is ordinary agreed with easy axis, result on which is obtained by magnetic energy analysis. Here, the measured anisotropy field, H_k, of the film made by the same process was 480A/m for a large disk-like shaped specimen. The method of impedance measurement was explained in section 4. The measurement frequency was 500 MHz or 300 MHz in the following data. The magnetic domain was observed by the Bitter-method, here.

5.2. Variation of Impedance Profile as a Function of Easy Axis

Figure 5.3. shows a schematic view of an inclined magnetic stripe domain of an element strip. The "domain wall angle" is defined in this figure. In this schematic, the domain is the inclined Landau-Lifshitz-like domain with closure domain. Figure 5.4. shows a typical change of the GMI profile as a function of the domain wall angle. An observed domain picture is shown on the left of each GMI profile. As increasing the domain wall angle, a magnetic field of maximum impedance decreases as shown in (a) and (b). When the domain wall angle is getting larger, the dip-point of GMI profile go up as shown in (c), eventually it turns to a profile with a maximum impedance at the zero field. In this figure the element thickness is 2.1 μm, width and the domain wall angle are (a)25 μm and 17.8 deg. , (b)15 μm and 42.9 deg. , (c)10 μm and 70.9 deg. Now we call a magnetic field strength of the maximum impedance as "peak field". Figure 5.5. shows the dependence of the peak field on the domain wall angle. It shows that the increment of the angle makes the decrement of the peak field. This phenomenon is beneficial because of a reduction of bias field. This figure also shows that the GMI profile changes from a profile with a dip-point at the zero magnetic field to a profile with one maximum impedance only at the zero field according to the inclination angle. This profile transformation occurs when the domain wall angle is nearly 70 degrees. A control of domain wall angle can maximizes a sensor gain, as shown in Figure 5.6.

The sensor gain is defined by a maximum value of $\Delta Z/\Delta H$ in the $Z(H)$ curve. A remarkable improvement of gain was obtained when the domain wall angle is in the range from 40 degree to 70 degrees. The parentheses in Figure 5.6. mean an estimated value of angle, because the domain wall could not be observed by bitter-method.

Figure 5.7. shows another beneficial property. We can get excellent linear property for wide range, more than 400 A/m (5 Oe). It is obtained in the region C of Figure 5.4. In Figure 5.7. the element thickness is 2.0 μm, width is 10 μm.

A stepped GMI property is obtained in the region B and right end part of the region A [9]. Figure 5.8. shows a typical profile of the stepped GMI property. The element impedance suddenly changes, more than 15 ohm, within one measurement step, 0.859 A/m (10.8 mOe). In this figure the element thickness is 2.7 μm, width is 20 μm.

Consideration was carried out based on both the bias susceptibility model of a single-domain magnetic thin-film with uniaxial anisotropy and observed domain structure with applying an external magnetic field. A qualitative explanation was obtained for the phenomena above mentioned [10-11], which will be discussed in section 6.

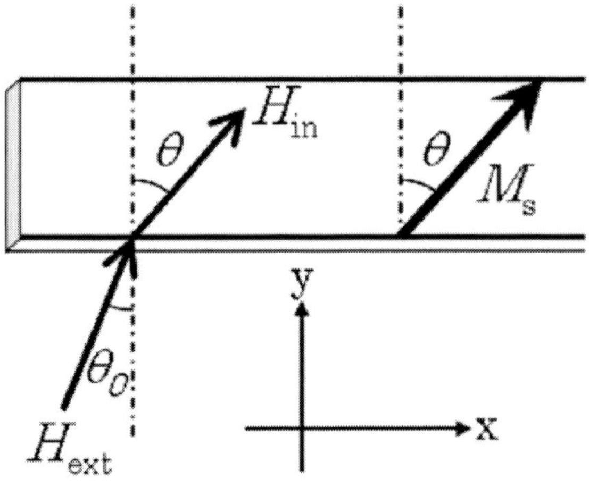

Figure 5.1. Relationship between annealing field and internal field.

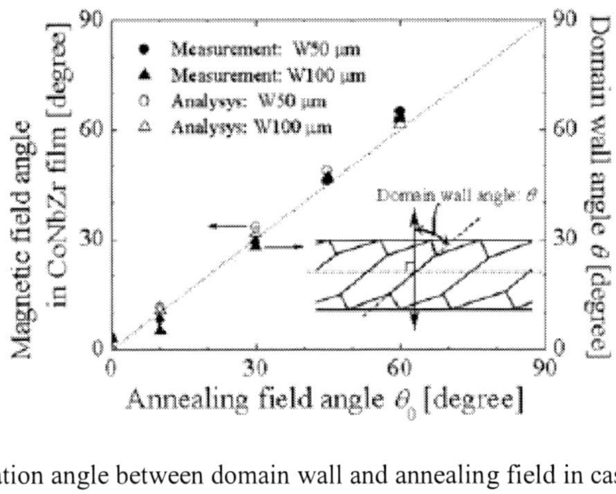

Figure 5.2. The inclination angle between domain wall and annealing field in case of ignoreable the demagnetization force.

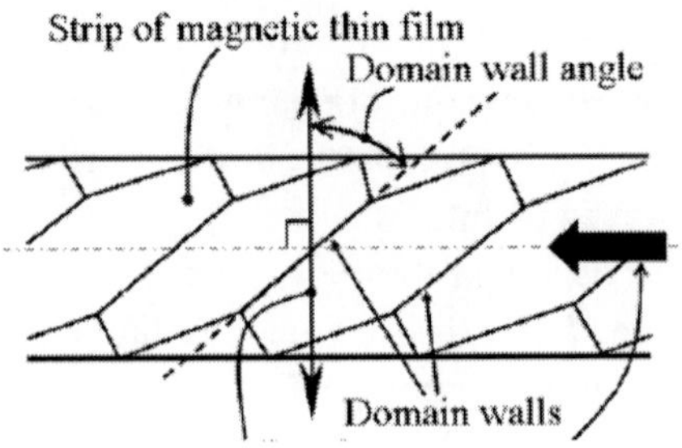

Figure 5.3. Schematic view of an inclined magnetic stripe domain.

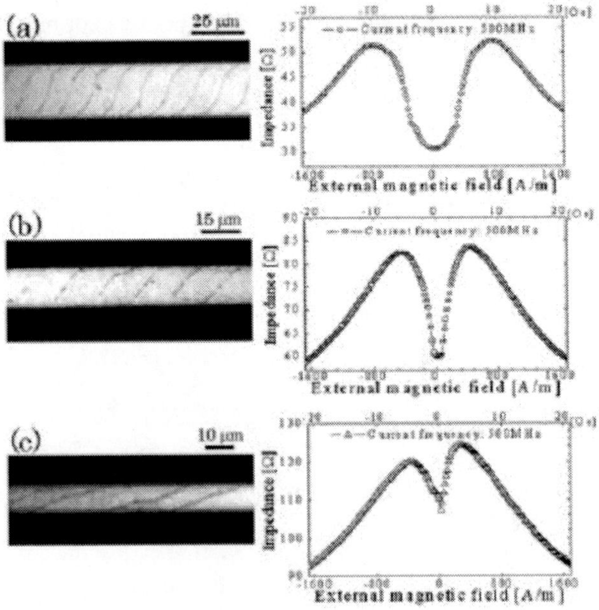

Figure 5.4. Typical change of the GMI profile as a function of the domain wall angle.

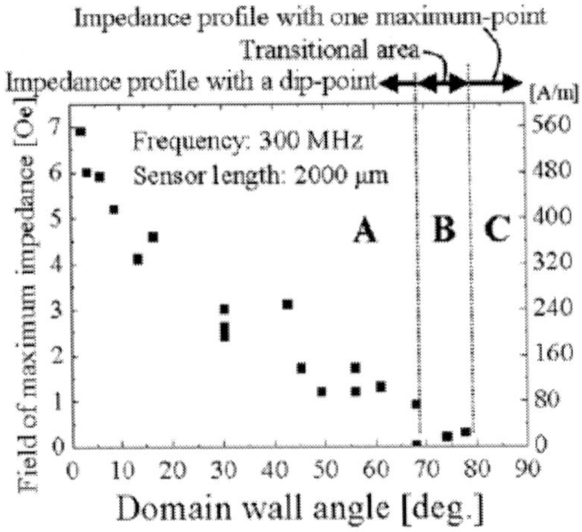

Figure 5.5. Dependence of the peak field on the domain wall angle.

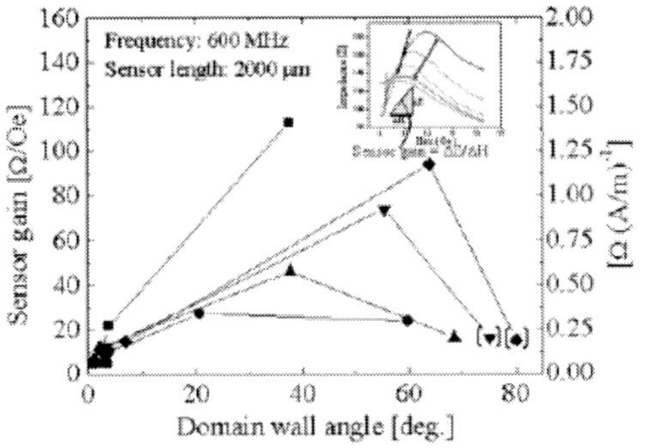

Figure 5.6. Sensor gain as a function of domain wall angle.

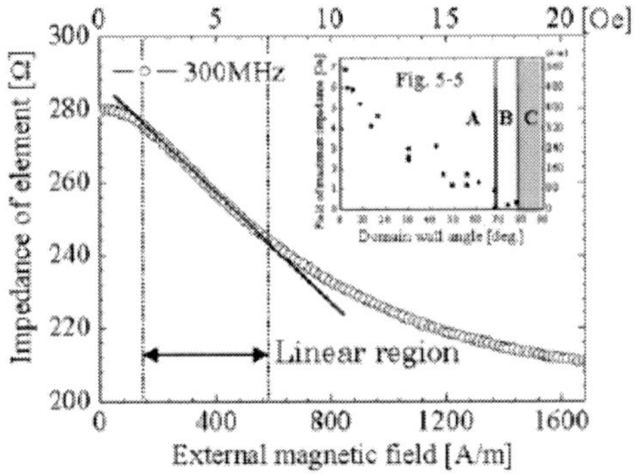

Figure 5.7. Linear property of the element.

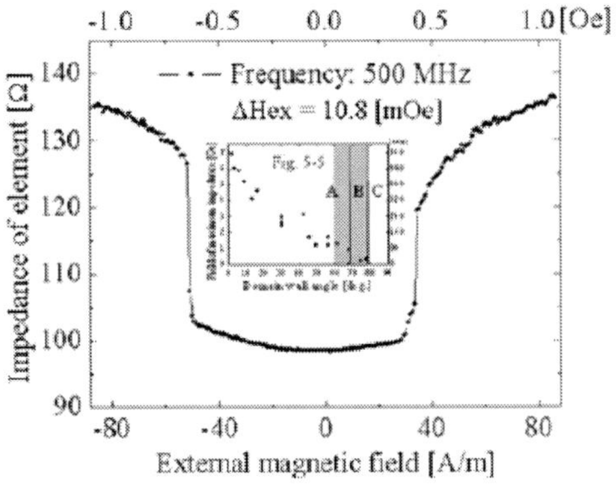

Figure 5.8. Typical profile of the stepped GMI property.

5.3. Conclusion of this Section

A method of making an inclined stripe domain on a rectangular element is explained. This method is based on applying a magnetic field during annealing for the element strip in an in-plane inclined direction. The inclination angle of the domain wall is controlled by both the cross-sectional aspect ratio of the element and the strength of applied field. It is also possible to control the inclination angle solely by direction of applied field. The control of inclination angle of domain wall is able to make the impedance property of GMI sensor beneficial, such as low bias, high gain, wide linear property. A stepped impedance property was also found in the region of property transformation from a dip-point profile to one maximum profile.

6. Domain Change and the Correlated Variation of Element Impedance

In the previous section, the variation of changing profile of the element impedance as a parameter of direction of easy axis was shown as a measurement result. In this section, a correlated domain change with the impedance change as a function of external magnetic field is shown. Two different pattern of domain change are shown as typical examples. One is the continuous profile of impedance, which has two peaks of impedance and a dip point in the vicinity of 0 A/m. Another is the stepped profile of impedance, which has abrupt jump of impedance and large hysteresis gap. A fundamental analysis, which explains an impedance change based on the domain change, is also carried out. In this chapter, the discussion is specialized in an element with inclined easy axis. So following two examples are the case with inclined easy axis.

6.1. Linear Profile with Two Peaks and a Dip-Point

Figure 6.1. shows a variation of inclined stripe domain as a function of external magnetic field in the case of the linear profile. As shown in Figure 6.1.(a), the impedance profile has two peaks and a dip-point. This element is a very high sensitivity element. The high sensitivity is achieved by controlling the domain-wall angle. In this case the angle is 52 degrees, and this angle is just below the occurence of the stepped property. Figure 6.1.(a) shows a variation of the element impedance in case of 500 MHz of the measurement frequency. The domains in match with the impedance are shown by caption numbers in Figure 6.1.(b). This domain was observed at the middle of the 2000 μm element. The variation of the linear impedance phenomenon is characterized as follows; the width of stripes of the domain changes as a function of external field in the range of magnetic field where the stripe domain is observed. The domains which have a magnetic moment in the direction of nearly parallel to the external field extend width of domains, and the domains with nearly anti-parallel moment narrow. This change of magnetic domain is executed with conserving the direction of domain wall. The directions of domain walls are almost in the direction of easy axis. The inclined domain disappears at a certain magnetic field where the impedance is still climbing up. A slight hysteresis of impedance, and also the hysteresis of domain change, is observed at the appearance and the disappearance of stripe domain. A theoretical investigation was carried out by using the bias susceptibility model of a single-domain magnetic thin-film with uniaxial anisotropy. This theoretical model predicts that contiguous inclined stripe domains on a 180 digrees wall will have different values of the high-frequency permeability. The discussion of bias susceptibility model was shown in the section 2. And the variation of the high-frequency permeability as a parameter of the direction of easy axis was shown in Figure 2.4. It is assumed that the total impedance of the element is calculated by a sum of the each impedance of contiguous domains in the following consideration. Figure 6.2. shows approximate results for the total sensor impedance obtained by adding the impedances of all the contiguous stripe domains. It explains qualitatively about the relationship between the magnetic domain and the element impedance. This result is merely a first step approximation, because the energy loss of domain-wall, the distribution of direction of magnetic moment in one domain area, the effect of closure domain and the effect of the demagnetization force in case of the longitudinal single domain (LSD) does not take into consideration. More accurate analysis will be a theme of future study.

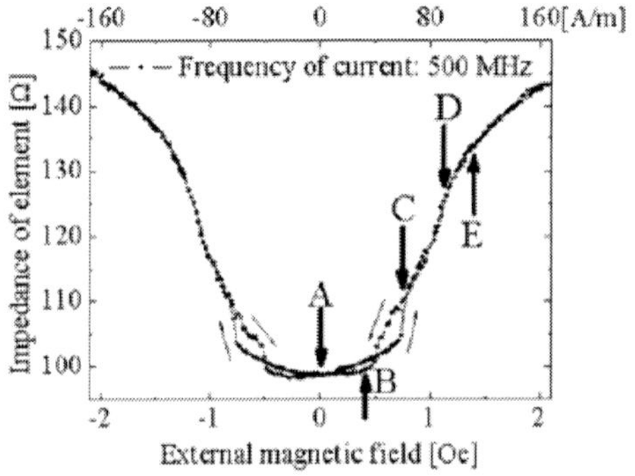

Figure 6.1.(a). Variation of impedance as a function of external magnetic field.

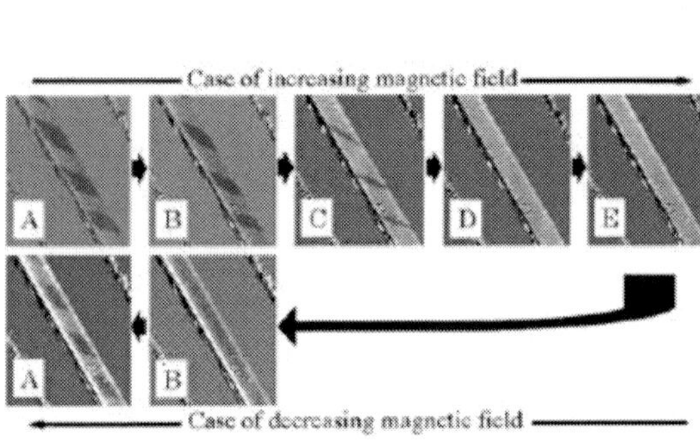

Figure 6.1.(b). Variation of inclined stripe domain as a function of external magnetic field.

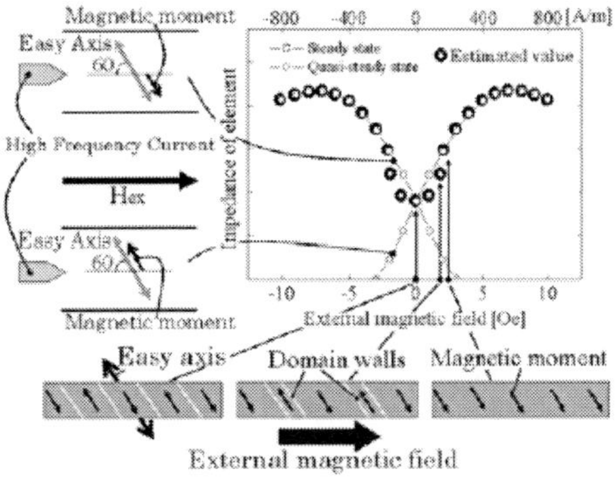

Figure 6.2. Approximate calculation of the element impedance.

6.2. Stepped Profile with Domain Change Phenomenon

Figure 6.3.(b) shows a variation of inclined stripe domain as a function of external magnetic field in the case of the stepped impedance element. Figure 6.3.(a) shows a variation of the element impedance in case of 50 MHz of the measurement frequency. The domains in match with the impedance are shown by caption numbers. The stepped impedance phenomenon is characterized as follows; the inclined stripe domain appears or disappears simultaneously with the abrupt step of impedance in a certain magnetic field. The width of stripes changes as a function of external field in the range of magnetic field where the stripe domain is observed. The domains which have a magnetic moment in the direction of nearly parallel to the external field extend width of domains, and the domains with nearly anti-parallel moment narrow. This change of magnetic domain is executed with conserving the direction of domain wall. The directions of domain walls are almost in the direction of easy axis.

A theoretical investigation was carried out by using the bias susceptibility model of a single-domain magnetic thin-film with uniaxial anisotropy, the same as the previous subsection. An approximate estimate shows that the sudden change of the domain leads to the impedance-step. In this estimation, it is assumed that the total impedance of the element is calculated by a sum of the each impedance of contiguous domains.

Figure 6.4. shows approximate results for the total sensor impedance obtained by adding the impedances of all the contiguous stripe domains. It explains qualitatively that the disappearance, also the appearance, of the inclined stripe domain leads to the impedance-step phenomenon for the GMI magnetic thin film. This result is a first step approximation too. The energy loss of domain-wall, the distribution of direction of magnetic momentum in one domain, the effect of closure domain, and the effect of the demagnetization force in case of the longitudinal single domain (LSD) must take into consideration for more accurate calculation.

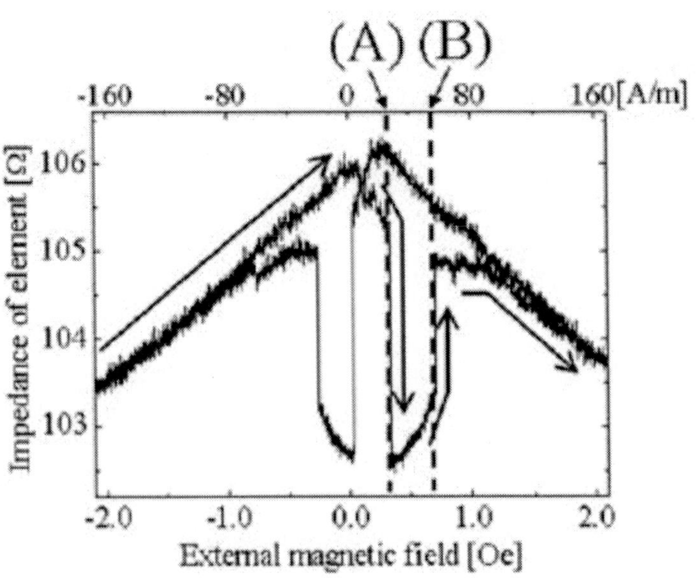

Figure 6.3.(a). Variation of the element impedance in the case of the stepped property.

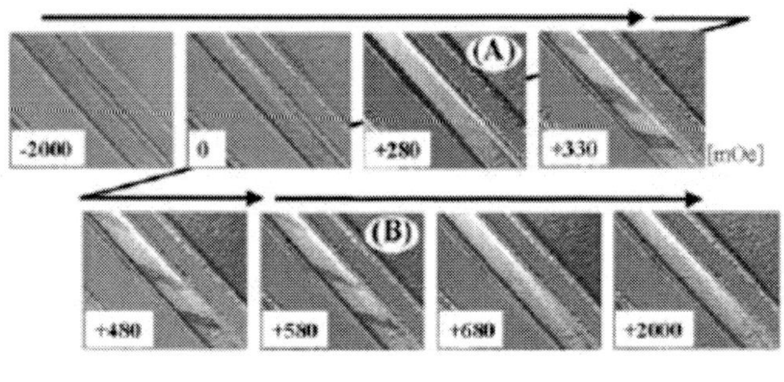

Figure 6.3.(b). Variation of magnetic domain in the case of the stepped impedance element.

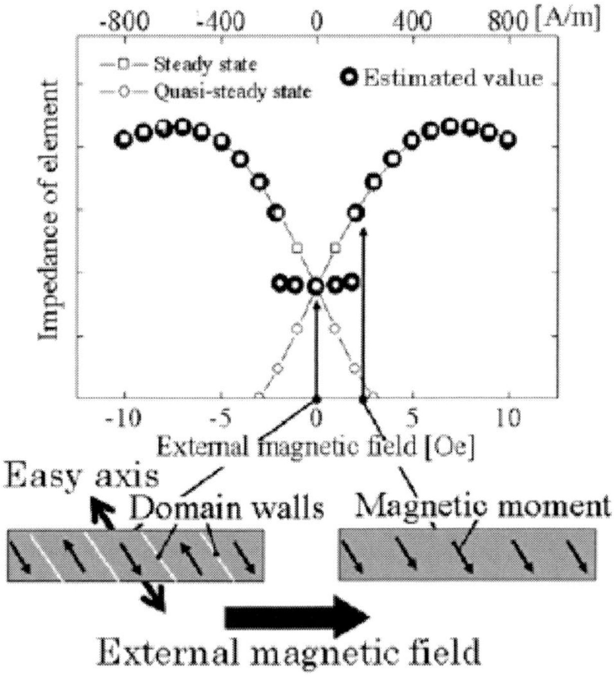

Figure 6.4. Approximate calculation of the impedance in the case of the stepped property.

7. Detailed Study of the Stepped Phenomenon

In this section, detailed study of the phenomenon of stepped impedance change is shown. A giant magnetoimpedance (GMI) element with a stepped impedance property was obtained for rectangular amorphous $Co_{85}Nb_{12}Zr_3$ soft magnetic thin film with an in-plane uniaxial easy axis in a direction of nearly 60 degrees relative to the short-side axis of the element [12]. The stepped change of impedance occurs simultaneously with the appearance or the disappearance of a certain magnetic domain [9].

A mechanism of the appearance or the disappearance of inclined stripe domain is discussed based on an analysis of magnetic domain energy. These extremely different domain structures is assumed as follows; the stripe domain which consists of contiguous anti-parallel domains with closure domains (the inclined Landau-Lifshitz-like domain (ILLD)), and the domain after disappearance of the stripe which consists of a single domain (longitudinal single domain (LSD)). This assumption is based on an experimentally observed domain structures. An actual variation of stepped impedance as a function of external magnetic field typically has a hysteretic property. The hysteresis of impedance means a hysteresis of the changing point of magnetic domain. In order to explain the hysteretic property, an asymmetrical threshold energy is introduced. Then the experimental result is compared with the theoretical result which is expected by the proposed analytical model.

7.1. Hysteresis of the Stepped Element

7.1.1. Hysteresis of Impedance and Magnetization

A Hysteresis of the stepped impedance element is discussed here. The profile of hysteresis of the impedance changes according to a certain rule. One of the parameter is element length, and the other is direction of easy axis. Firstly in this section, experimental results of the impedance hysteresis for stepped element will be shown, and then analytical consideration is explained.

The fabricated element is the same structure as shown in section 3. The dimensions of the sensor element are 3000 μm of length, 20 μm of width, 3.0 μm of thickness. The direction of the magnetic anisotropy was controlled by the direction of magnetic field. Here, the magnetic field during annealing, 240 kA/m, 673K, was oriented 65 degrees relative to the short-side axis of the element, and kept for 1 hour in vacuum atmosphere lower than 6.5×10^{-4} Pa.

Figure 7.1. shows an experimentally obtained MH loop of the element [13]. The MH loop was measured along the longitudinal direction of the element. The range of variation of magnetic field was controlled as follows; Figure 7.1.(a) shows a MH loop in case of applied field changes from -400 A/m to +400 A/m and then continuously from +400 A/m to -400 A/m, Figure 7.1.(b) shows a MH loop in case of applied field changes from +400 A/m to -24 A/m and then continuously from -24 A/m to +400 A/m, and Figure 7.1.(c) shows a case of applied field from -400 A/m to +31 A/m and then continuously from +31 A/m to -400 A/m. These figures show in the range of -160 A/m to +160 A/m as an expansion. In the range outside of it, the element is almost magnetically saturated and has constant value. The values of threshold fields of domain change are shown in Figure 7.1. The measurement accuracy is 1.6 A/m, because of the measurement interval was about 1.6 A/m and the accuracy of the field sensor of VSM is under 0.8 A/m. The thresholds appear at the same field values within measurement error, without relation of the range of magnetic field variation.

Figure 7.2. shows magnetic domains observed by Kerr-effect microscope in comparison with the MH loop measurement. There are three states of magnetic domains: (1) LSD with momentum of + direction. (2)-(4) ILLD. (5) LSD with momentum of – direction.

Based on the result of MH loop measurement (Figure 7.1) and the magnetic domain observation (Figure 7.2), it declares that there exists three different domain states as a function of external magnetic field, and these domain states individually make disconnected lines on the MH diagram.

As a result of this subsection, it declares that the stepped GMI element can have three different states of domain structure between -17 A/m to +19 A/m, and these states form disconnected lines on the MH diagram as shown in Figure 7.3. The existence of three stable states is expected to be explained by the relationship between the magnetic energy of ILLD, LSD domain state as a function of external field, and the threshold barrier which obstructs the change of domain.

Figure 7.4 shows the variation of element impedance as a function of external magnetic field. The range of variation of magnetic field was controlled as follows; Figure 7.4.(a) shows an impedance variation in case of applied field changing from -400 A/m to +400 A/m and then continuously from +400 A/m to -400 A/m, Figure 7.4.(b) shows an impedance variation in case of applied field changing from +400 A/m to –26 A/m and then continuously from -26 A/m to +400 A/m, and Figure 7.4.(c) shows a case of applied field from -400 A/m to +26 A/m and then continuously from +26 A/m to -400 A/m. The incident power from the network analyzer to the element was -14 dBm with the frequency of 300 MHz. The sweeping speed of the external field was 1.6 (A/m)/s. The measurement of impedance was carried out discretely with 0.16 (A/m)/step. The sampling rate of impedance measurement was 21 ms/point, and the measurement was carried out without using average-mode of the network analyzer. Therefore the measurement profile of Figure 7.4. includes measurement noise. But the noise level is much smaller than the range of impedance-step. The lower regions of impedance, which are divided by impedance-step, are in the state of ILLD domain structure. The higher regions of impedance are in the state of LSD domain structure. The Figure 7.4. shows in the range of -160 A/m to +160 A/m as an expansion, in order to coincide with the MH loop results of Figure 7.1.

Figure 7.5. shows a variation of element impedance as a function of external field, which is compared with MH diagram of the same element. This figure is a kind of state diagram for the element with three state domain structures. From this figure, it is possible that the three domain states can easily be detected by measuring the impedance of the element. The reason of occurrence of the stepped impedance is expected to be a change of high-frequency permeability and an existence of domain wall. The high-frequency permeability arises from magnetization rotation, and the degree of magnetization rotation is expected to be strongly affected by domain structure.

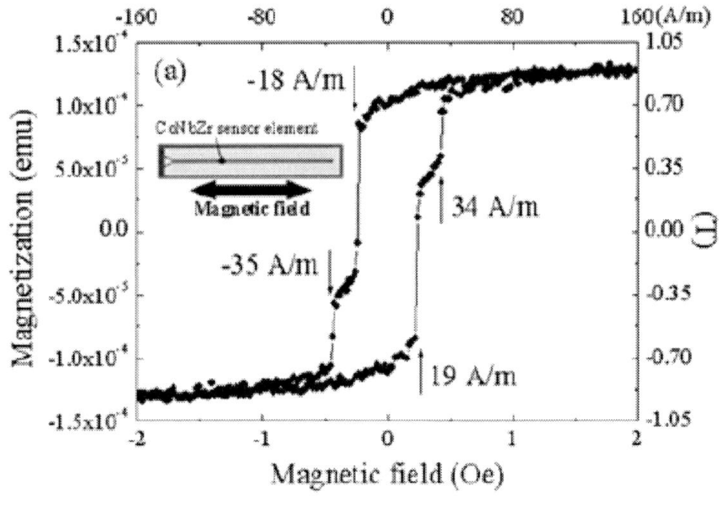

Figure 7.1.(a). Measured MH loop with the range between -400 A/m and +400 A/m.

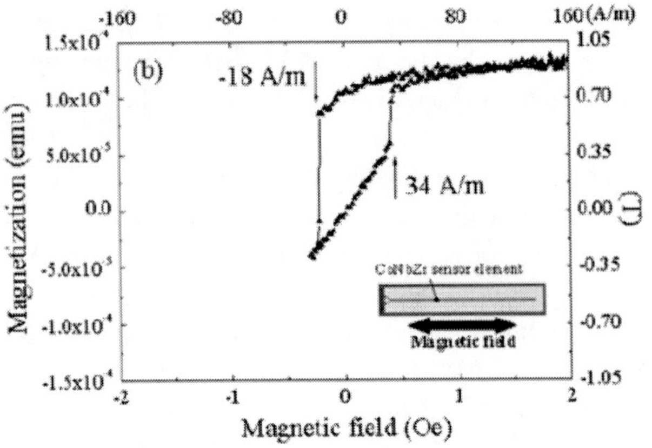

Figure 7.1.(b). Measured MH loop with the range between +400 A/m to -24 A/m.

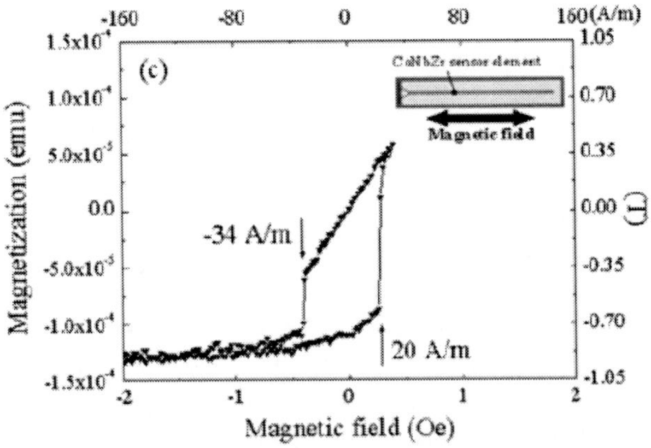

Figure 7.1.(c). Measured MH loop with the range between -400 A/m to +31 A/m.

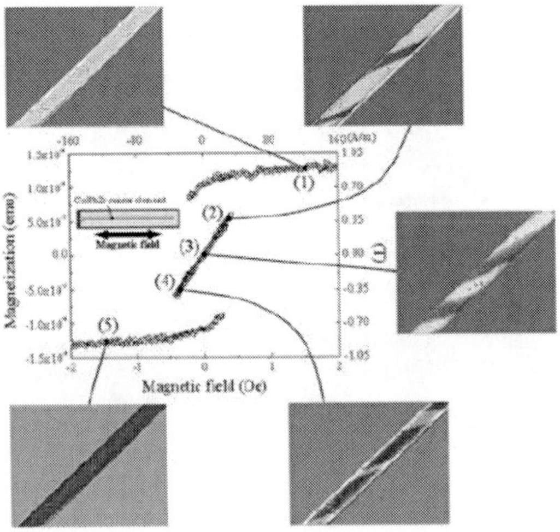

Figure 7.2. Magnetic domains observed by Kerr-effect microscope in comparison with the MH loop.

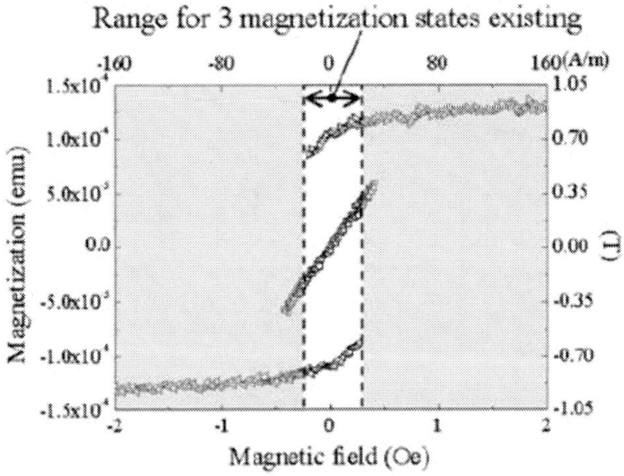

Figure 7.3. MH diagram which shows the three stable states.

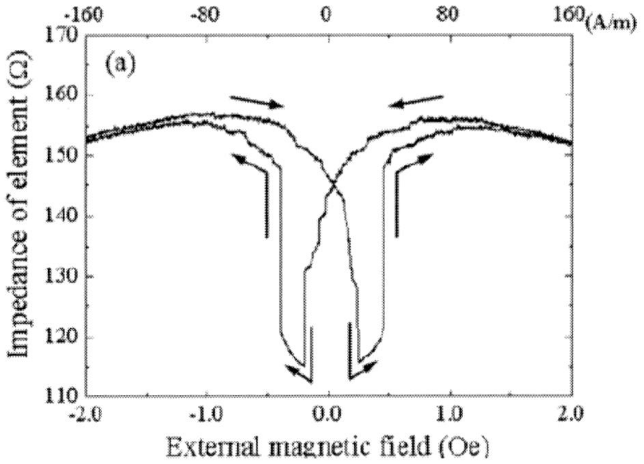

Figure 7.4.(a). Variation of element impedance with the range between -400 A/m to +400 A/m.

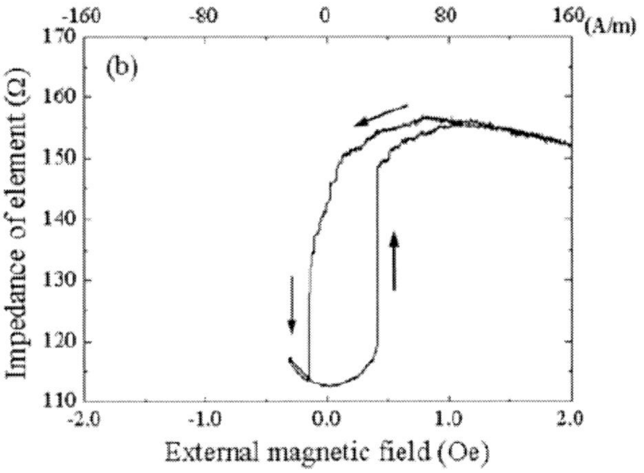

Figure 7.4.(b). Variation of element impedance with the range between +400 A/m to –26 A/m.

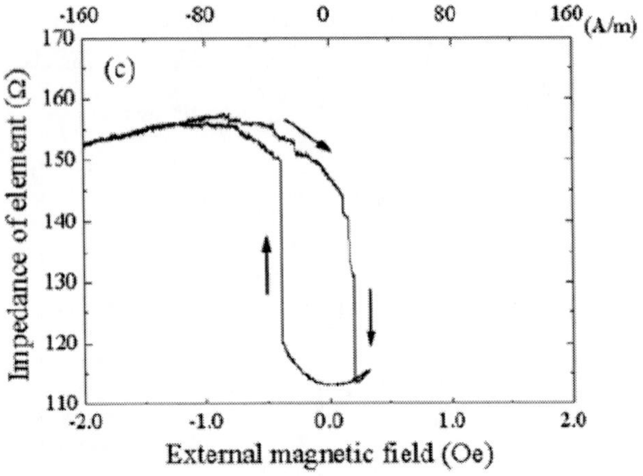

Figure 7.4.(c). Variation of element impedance with the range between -400 A/m to +26 A/m.

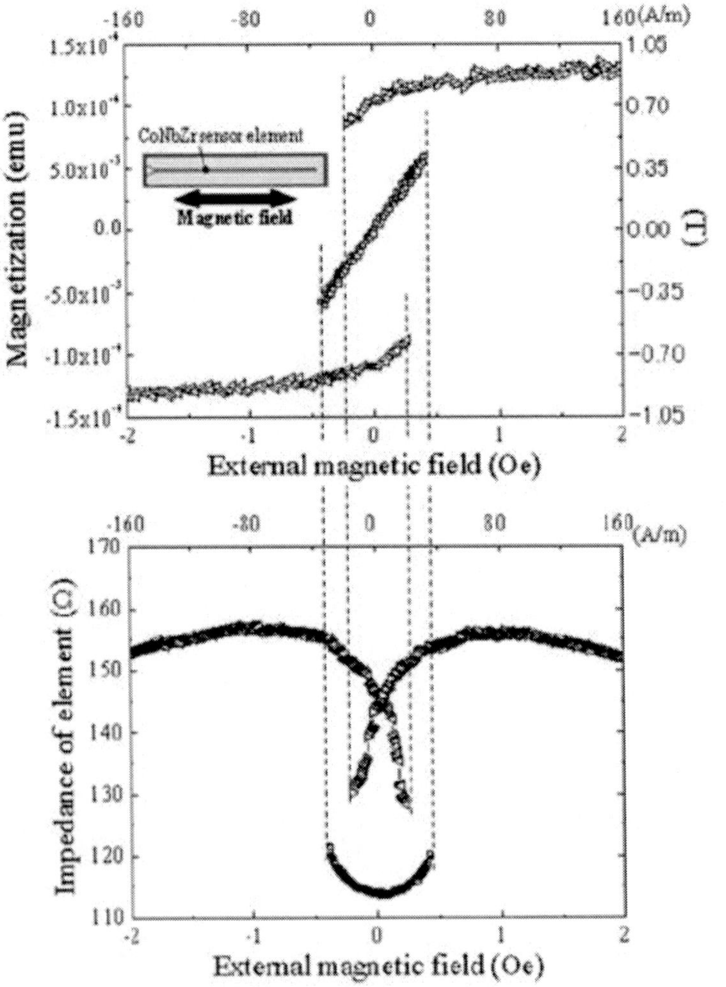

Figure 7.5. Variation of element impedance, which is compared with MH diagram of the same element.

7.1.2. Hysteresis as a Function of Element Length

The change of hysteresis is according to a certain rule. The following in this subsection is a result of variation of the profile of hysteresis as a function of element length. On the bottom of Figure 7.6. shows domain structures with different length of the element. The elements were fabricated on a same substrate, which means that they experienced a same fabrication process. The element length is, (a)1000 μm, (b)2000 μm, (c)3000 μm respectively. The width of element is 20 μm, and the thickness is 2.1 μm. The direction of the domain wall, as an average value, is (a)63 degrees, (b)62 degrees, (c)62 degrees respectively, relative to the short-side axis. In this case the direction of annealing field was 60 degrees. These domain structures were observed when the element had lower impedance, which means in the range of magnetic field between the step-down and step-up of the element impedance. The magnetic field of the domain observation is shown as an arrow on an impedance profile in the upper part of the figure. These domains are the in-plane inclined Landau-Lifshitz-like domain, and they disappear when the impedance steps up. The upper part of Figure 7.6. shows the variation of the element impedance as a function of an external magnetic field. The external field was varied from -173 A/m (-2.16 Oe) to +173 A/m (+2.16 Oe), and then continuously from +173 A/m (+2.16 Oe) to -173 A/m (-2.16 Oe). The incident power to the sensor was -14 dBm with the frequency of 50 MHz. The sweeping speed of the external field was 1.6 (A/m)/sec (20 mOe/sec). The measurement of impedance was carried out discretely with 0.16 (A/m)/step (2 mOe/step). The sampling rate of the impedance measurement was 21 msec/point, and the measurement was done without using average-mode of the network analyzer, this is the same measurement apparatus explained in section 4. These impedance profiles indicate a tendency that a lower impedance region, which exists between impedance step-down and step-up, narrows and splits when the element lengthening. The regions of the lower impedance overlap with each other in the case of Figure 7.6.(a), which indicates the impedance profile of 1000 μm of sensor length. Whereas the regions of the lower impedance sprits in the case of Figure 7.6.(c), which has 3000 μm of sensor length [14].

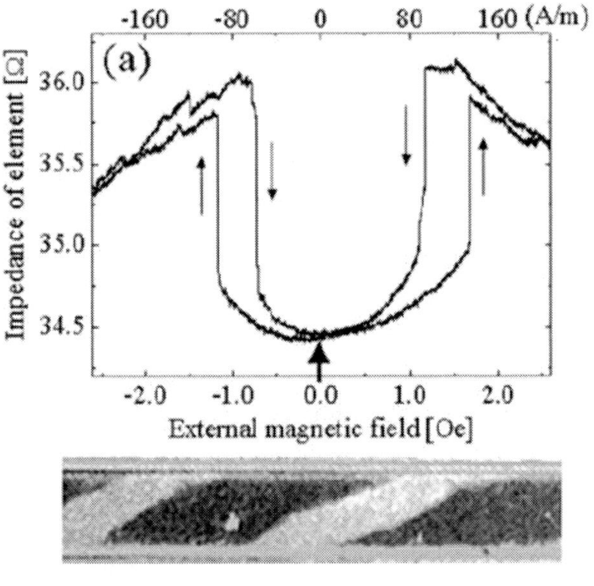

Figure 7.6.(a). Domain and impedance for the element with length of 1000 μm.

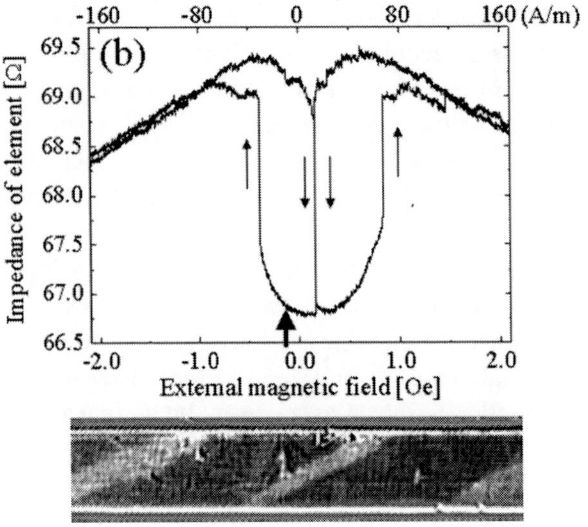

Figure 7.6.(b). Domain and impedance for the element with length of 2000 μm.

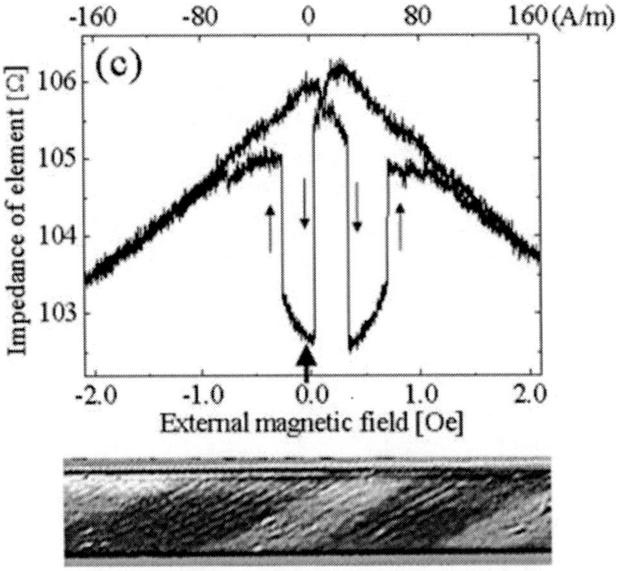

Figure 7.6.(c). Domain and impedance for the element with length of 3000 μm.

7.1.3. Hysteresis as a Function of Direction of Easy Axis

Figure 7.7. shows a view of the element which is used in this measurement. The element length is 3000 μm, the width is 20 μm, and the thickness is 2.0 μm. The electrode was fabricated in the middle area of the element in order to reduce an edge effect induced by a residual domain. The element impedance was measured along the length of 1000 μm in the middle area between the electrodes with the frequency of 500 MHz. The external field was varied from -320 A/m (-4.0 Oe) to +320 A/m (+4.0 Oe), and then continuously from +320 A/m (+4.0 Oe) to -320 A/m (-4.0 Oe).

Figure 7.8. shows a variation of element impedance as a function of external magnetic field. Figure 7.8(a) shows a case of easy axis in the direction of 55 degrees, Figure 7.8.(b) 65 degrees, and Figure 7.8.(c) 75 degrees. These easy axis directions were controlled by the direction of annealing field. These are the results of the same dimensions and different directions of easy axis. The measured profile of element impedance declares that the lower impedance region narrows and splits as the angle of easy axis increases. This tendency is the same as the previous experimental result, Figure 7.6, which is observed when the element length is changed.

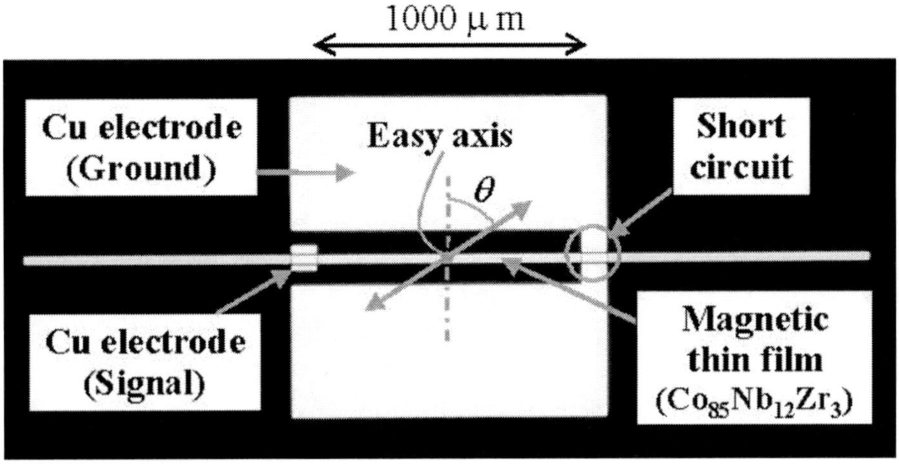

Figure 7.7. View of the element which is discussed in this subsection.

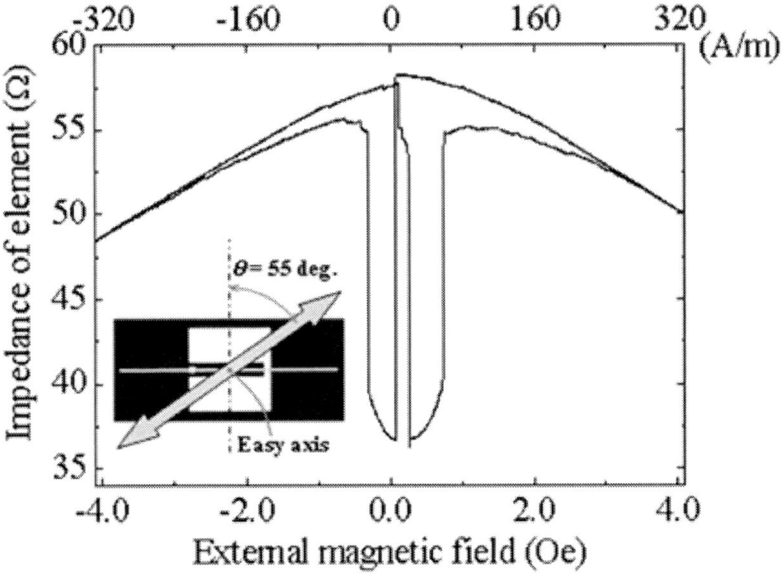

Figure 7.8(a). Variation of element impedance in case of easy axis 55 degrees.

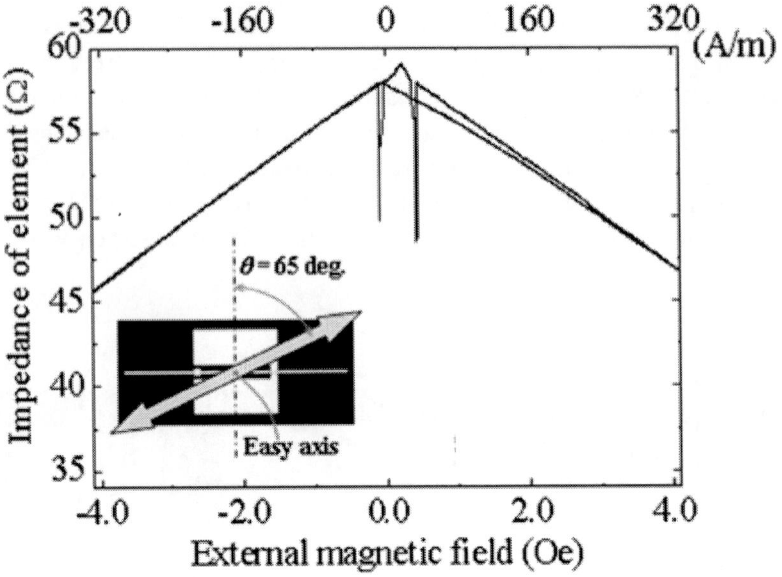

Figure 7.8.(b). Variation of element impedance in case of easy axis 65 degrees.

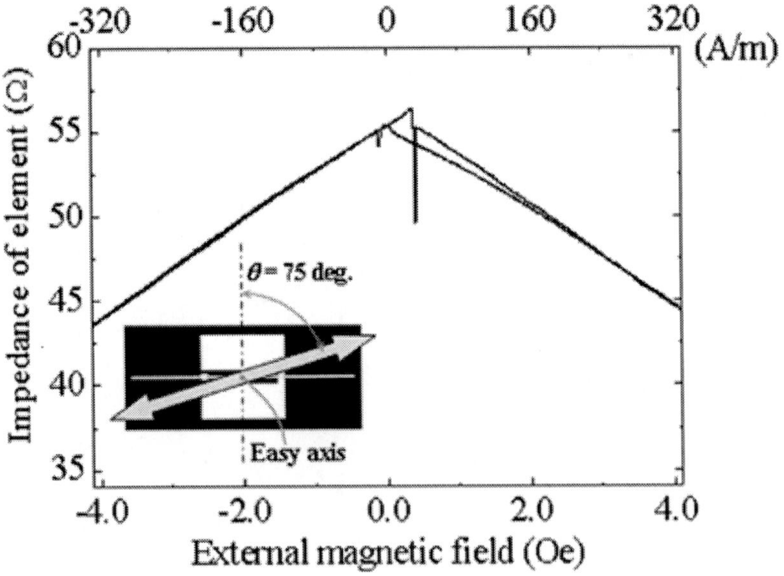

Figure 7.8.(c). Variation of element impedance in case of easy axis 75 degrees.

7.2. Analysis of Magnetic Energy

In this subsection, a mechanism of the appearance or the disappearance of inclined stripe domain is discussed based on an analysis of magnetic energy. It is assumed that these extremely different domain structures as follows; the stripe domain which consists of contiguous anti-parallel domains with closure domains, and the domain after disappearance of the stripe which consists of a single domain. This assumption is based on an experimentally observed domain structures. It is shown, as a result of this subsection, that the energy of these extremely different domain structures intersects each other in a certain external magnetic field in a certain range of in-plane angle of the easy axis direction.

7.2.1. Simulation model

Two simulation models for estimating the magnetic energy is discussed in this subsection. At first experimentally obtained change of magnetic domain is shown as a basis of our simulation model. Figure 7.9. shows a variation of inclined stripe domain as a function of external magnetic field. Figure 7.9. also shows a variation of the element impedance in case of 50 MHz. The domains in match with the impedance are shown by caption numbers. The stepped impedance phenomenon is characterized as follows; the inclined stripe domain appears or disappears simultaneously with the stepped impedance phenomenon in a certain magnetic field. The width of stripes changes as a function of external field in the range of magnetic field where the stripe domain is observed. The domains which have a magnetic moment in the direction of nearly parallel to the external field extend width of domains, and the domains with nearly anti-parallel moment narrow. This change of magnetic domain is executed with conserving the direction of domain wall. The directions of domain walls are almost in the direction of easy axis. Figure 7.10. shows models of magnetic domain which are applied to our simulation. Figure 7.10.(a) shows a model of domain in case of stripe domain, called as ILLD. An existence of closure domains is assumed in ILLD. Figure 7.10.(b) shows a model of domain in case of single domain, called as LSD. These models are 2-dimensional models with infinite length, therefore it means a neglect of longitudinal demagnetizing energy. The magnetic energy of ILLD is obtained by Zeeman energy, anisotropy energy and wall energy. The magnetostriction of amorphous $Co_{85}Nb_{12}Zr_3$ thin-film is small enough to negligible. The magnetic charge on the wall of closure domain is assumed to be neglected in this simulation. In this model, the anisotropy is uniaxial. The wall energy, γ, is estimated by eq. (1).

$$\gamma = 2\sqrt{A} \cdot \sin\xi \int_0^{\pi} \sqrt{g(\xi,\varphi)}\, d\varphi \quad [\mathrm{J/m^2}] \tag{1}$$

Where A is the exchange stiffness constant, assumed to be A=1.49×10^{-11} J/m in this simulation, and g is the anisotropy energy. The integral of the equation means an integration of anisotropy energy of the domain wall. The parameter φ is the azimuthal angle of magnetic moment based on the normal axis of the wall plane, and ξ is the angle of moment against the normal axis.

In case of 180 degrees wall, the wall energy γ_{180} is obtained as follows [15];

$$\gamma_{180} = 2\sqrt{A} \int_0^{\pi} \sqrt{K_u \cdot \sin^2\varphi}\, d\varphi = 4\sqrt{A \cdot K_u} \quad [\mathrm{J/m^2}] \tag{2}$$

Where K_u is the uniaxial anisotropy constant. In this simulation K_u is 260 J/m^3 and M_s is 0.93 T.

In case of closure domain wall, the wall energy γ_c is obtained as follows [15];

$$\gamma_c = 2\sqrt{A}\cdot\sin\xi\int_0^{\pi}\sqrt{g(\xi,\varphi)}\,d\varphi \quad [\mathrm{J/m^2}]$$

where $g(\xi,\varphi) = K_u\left\{1-\left[1-\sin^2\xi(1-\cos\varphi)\right]^2\right\}$ (3)

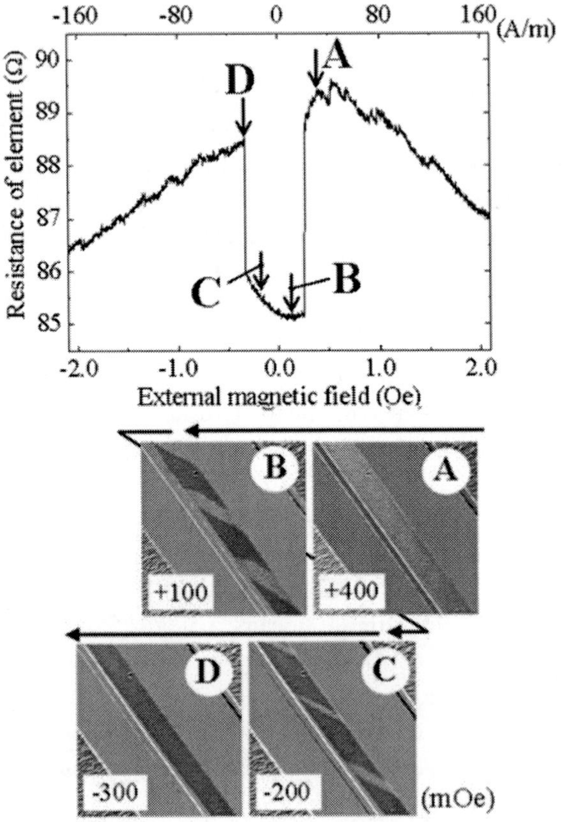

Figure 7.9. Variation of inclined stripe domain and impedance as a function of external magnetic field.

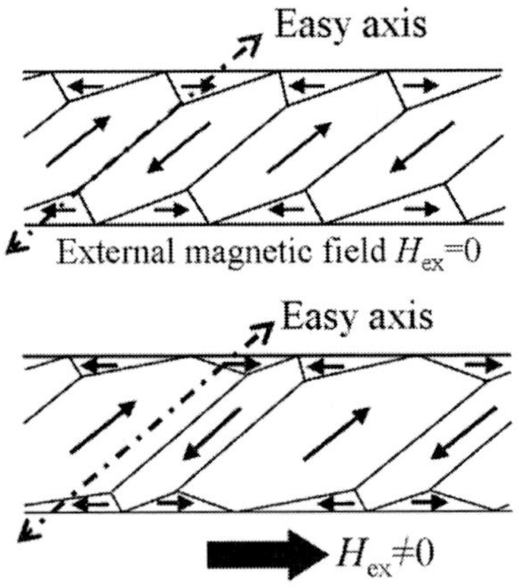

Figure 7.10. (a). Model of ILLD domain.

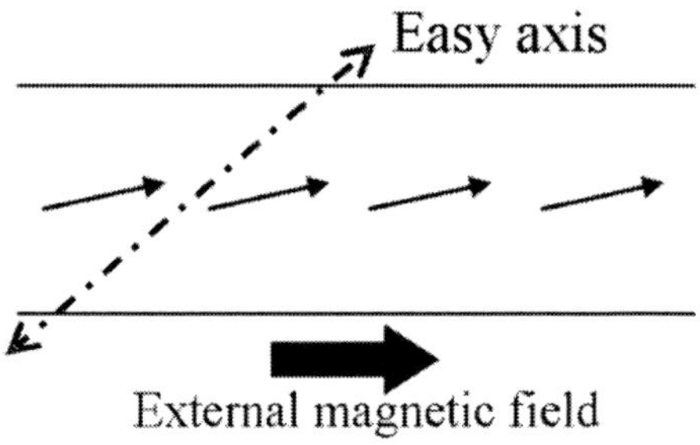

Figure 7.10.(b). Model of LSD domain.

In case of ILLD, the movement of domain walls are restricted as the observation in Figure 7.9, when the external magnetic field changes. It means that they move with conserving the direction of 180 degrees wall in the direction of easy axis. The ILLD consists of a Landau-Lifshitz-like configuration with closure domains at the edge [16]. It also resembles to the well-known sectional domain structure of stripe domain for perpendicular anisotropy thin-film [20-22]. But our target is in-plane domain structure for in-plane anisotropy thin-film with quite different dimensions, 20 μm of width. In case of LSD, the magnetic energy is obtained by Zeeman energy, magnetic static energy and anisotropy energy. In this case, the direction of magnetic moment is obtained as almost longitudinal. Our analysis shows that it is the direction of minimum energy, because of an effect of demagnetizing energy along the width direction. The demagnetizing energy is calculated only in the direction of width, because there is no demagnetizing field in the length direction with its infinite length.

7.2.2. Simulation Results and Discussion

Figure 7.11. shows a result of simulation [17]. This figure shows a relationship between external magnetic field and total magnetic energy. The relation of the stripe domain is plotted by solid circles, and the relation of the single domain is plotted by solid triangles. The direction of uniaxial easy axis against the short-side axis, θ, is (a)45°, (b)55°, (c)65° and (d)75°. Ratios of domain width for nearly parallel and nearly anti-parallel are indicated besides the plots. From these results, if the domain state is assumed to keep lower energy, three cases of transition are predicted. In case of (a), the stripe domain gradually disappears as increasing the external magnetic field, and then smoothly connected to the single domain. In case of (b), the stripe domain transforms to the single domain in spite of existing a certain width of anti-parallel magnetic domain. In case of (d), only the state of single domain exists. The previously reported experimental results show that the direction of easy axis in which the stepped phenomenon was observed was ranging from 55° to 75° [12]. On the other hand the range predicted by this simulation is ranging from 45° to 65°. This difference is assumed to be caused by a demagnetizing energy based on a longitudinal demagnetizing field in the LSD. The experimental results were obtained by 3-dimensional elements with the length ranging from 1000 μm to 3000 μm and the thickness from 2 μm to 4 μm, whereas the simulation model has an infinite length. The finite length brings an increment of magnetic energy for LSD, but there is slight difference for ILLD. If the longitudinal demagnetizing energy is taken into consideration, the simulated range of easy axis would be brought near to the experimental results.

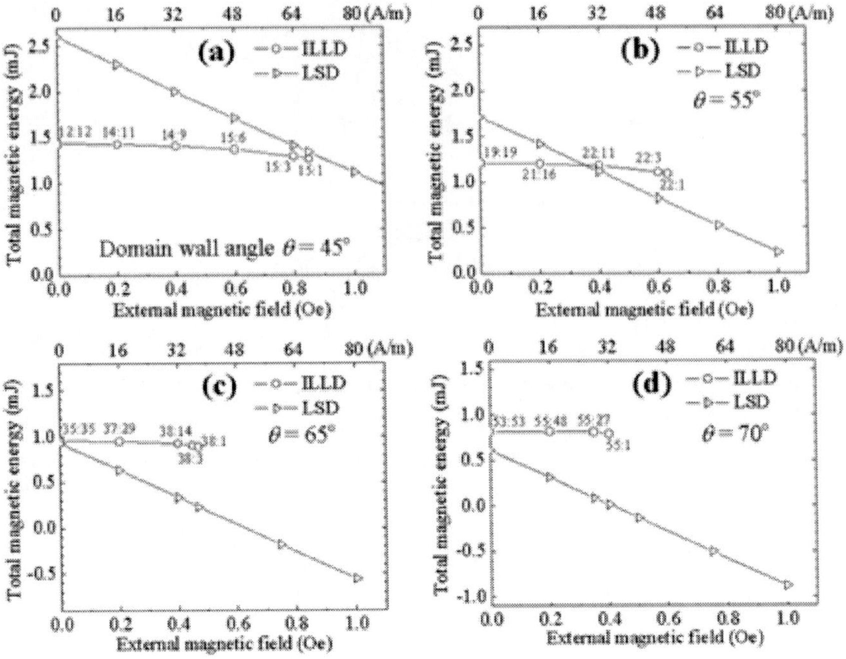

Figure 7.11. Relationship between external magnetic field and total magnetic energy.

7.2.3. Explanation of Hysteresis of Impedance

An actual variation of the stepped impedance as a function of external magnetic field typically has a hysteretic property, as shown in Figure 7.4, Figure 7.8, and Figure 7.9. The hysteresis of impedance means a hysteresis of the changing point of magnetic domain. In this subsection, in order to explain the hysteretic property, an asymmetrical threshold energies are introduced.

Figure 7.12. indicates the threshold energy levels added to the Figure 7.11.(b), which shows the magnetic energy of each domain structures as a function of external magnetic field in case of the direction of easy axis 55 degree. A dashed line in Figure 7.12. shows the threshold energy for the change from LSD to ILLD, and a one dot and dashed line shows the threshold energy in the opposite change. LSD changes to ILLD at a point where the energy of ILLD intersects with the dashed line. On the other hand, ILLD changes to LSD at a point where LSD intersects with the one dot and dashed line. In this discussion, the difference of energy between these two domain states is assumed to be a motive force to overcome the threshold energy. A schematic illustration of a variation of impedance when applying an alternating magnetic field is also shown in Figure 7.12.(b). In this figure, it is assumed that the impedance of the state of ILLD is lower than LSD. It is also assumed that the impedance of the state of LSD decreases as increasing the magnitude of external field, this tendency is predicted based on the bias-susceptibility theory for high frequency permeability. Figure 7.12.(a) shows the variation of magnetic domain and energy when the external magnetic field increases from -167 A/m (-2100 mOe) to +167 A/m (+2100 mOe). The schematic variation of impedance in match with Figure 7.12.(a) is shown by solid line in Figure 7.12.(b). Figure 7.13.(a) shows the variation in case of magnetic field decreases from +167 A/m (+2100 mOe) to -167 A/m (-2100 mOe), and the schematic variation of impedance in match with Figure 7.13.(a) is shown by dotted line in Figure 7.13.(b). In comparison with the measurement result, previously shown, the hysteretic variation of impedance is qualitatively explained.

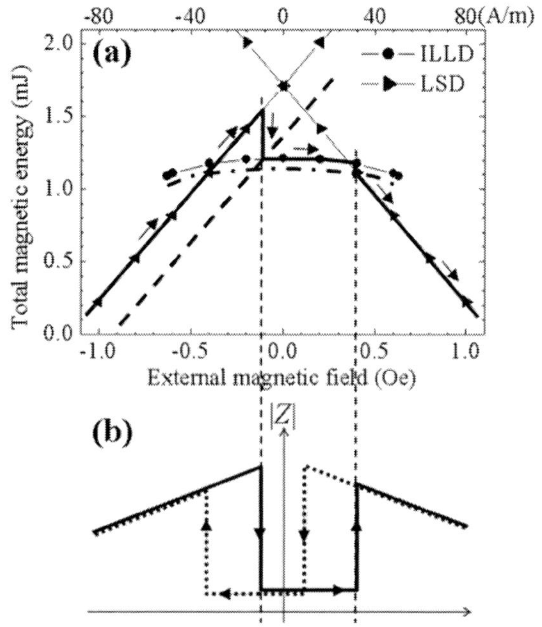

Figure 7.12. Variation of magnetic energy and impedance with consideration of hreshold energy levels (Increment of field).

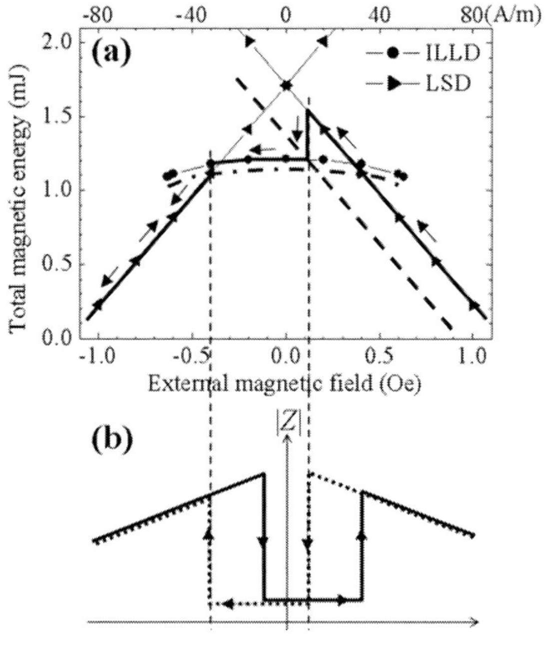

Figure 7.13. Variation of magnetic energy and impedance with consideration of threshold energy levels (decrement of field).

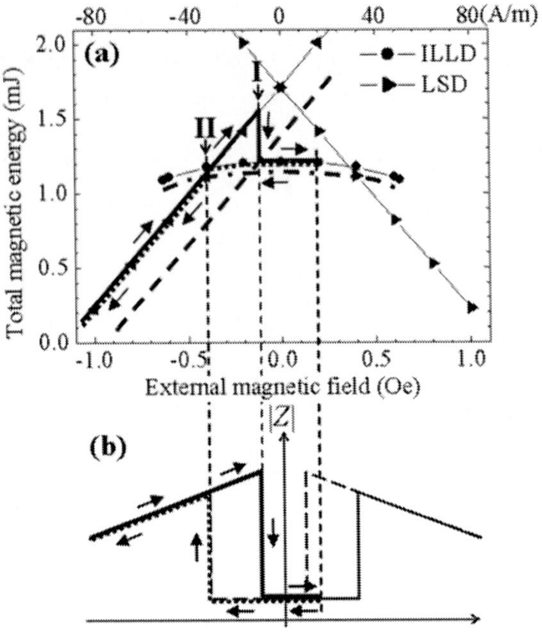

Figure 7.14. Variation of magnetic energy and impedance with consideration of threshold energy levels (Pivot variation of field).

In order to confirm the analytical model, which consists of intersected magnetic energy with asymmetrical threshold, a virtual experiment is carried out. The experiment is as follows; the sweeping direction of the external magnetic field was changed soon after the impedance step-down, which means the change of magnetic domain from LSD to ILLD happens. Based on our analytical model, the variation is predicted as shown in Figure 7.14. When the external magnetic field increases started from minus region of magnetic field, the initial state of LSD changes to ILLD after the magnetic energy of LSD exceeds the threshold energy at the point "I" in Figure 7.14.(a). Then, the changing direction of the external magnetic field switches to decrease. ILLD has been kept until the external field decreases to the point "II", where the magnetic energy of the ILLD exceeds the threshold energy to deform to LSD with opposite direction. The qualitative variation of impedance in match with Figure 7-14(a) is shown in Figure 7.14.(b). This theoretical prediction is confirmed experimentally by the Figure 7.4, which was previously shown as a result of impedance hysteresis [18].

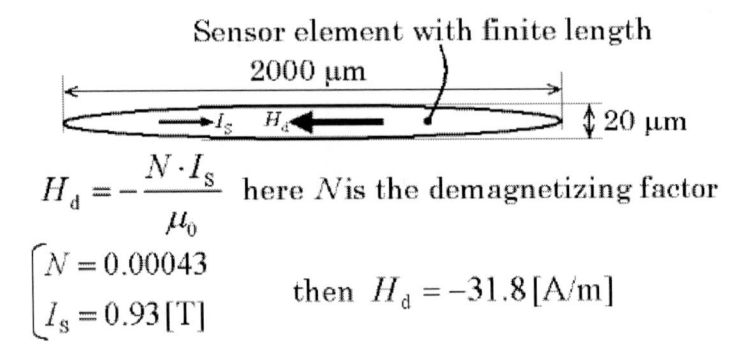

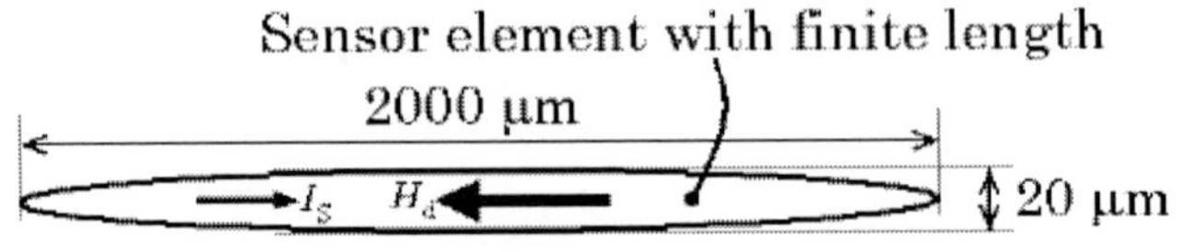

$$H_d = -\frac{N \cdot I_S}{\mu_0} \quad \text{here } N \text{ is the demagnetizing factor}$$

$$\begin{cases} N = 0.00043 \\ I_S = 0.93\,[\mathrm{T}] \end{cases} \quad \text{then } H_d = -31.8\,[\mathrm{A/m}]$$

Figure 7.15. Schematic figure of ellipsoidal magnetic body and demagnetizing field.

In order to take into consideration the effect of the sensor length, a demagnetizing energy is introduced in the energetic model. Figure 7.15. shows a schematic figure of ellipsoidal magnetic body and demagnetizing field in it. In this discussion, the demagnetizing energy is estimated based on ellipsoidal assumption. Based on the actual dimensions of the element in Figure 7.7.(b), 2000 μm of length, 20 μm of width, 2.1 μm of thickness, the demagnetizing field is assumed as H_d= -31.8 A/m, and which is 0.3 mJ of demagnetizing energy when applying to the 2-dimensional model.

Figure 7.16. shows an energy level of magnetic domains with consideration of the demagnetizing energy. The direction of easy axis is 60 degrees. The finite length of the element brings an increment of energy level especially for LSD, because the domain has a longitudinal magnetic moment. Figure 7.17. shows a schematic illustration of the impedance change based on the energy and threshold model. Figure 7.17.(a) shows the case of infinite length of element, whereas Figure 7.17.(b) shows the case of finite length with consideration of demagnetizing energy as shown in Figure 7.15. This variation of the element impedance, which is including the effect of the sensor length, resembles the profile of Figure 7.7.(c) and Figure 7.7.(b) with a slight difference, but a tendency of the profile is in good agreement. This result explains the variation of impedance profile as a function of element length.

The variation of the profile of hysteresis as a function direction of easy axis is discussed analytically. Based on the 2-dimensional analysis, a variation of magnetic energy as a parameter of the angle of easy axis is calculated, previously in this section. Figure 7.18. shows an analytical result of magnetic energy for ILLD and LSD for the easy axis of 55 degrees. A comparison with Figure 7.19, which is the result of the easy axis of 60 degrees, the vertical distance of energy level between ILLD and LSD widens. Therefore the inclement of the angle of easy axis makes the lower region of element impedance narrow and split, for the same reason in the case of decreasing the demagnetizing energy.

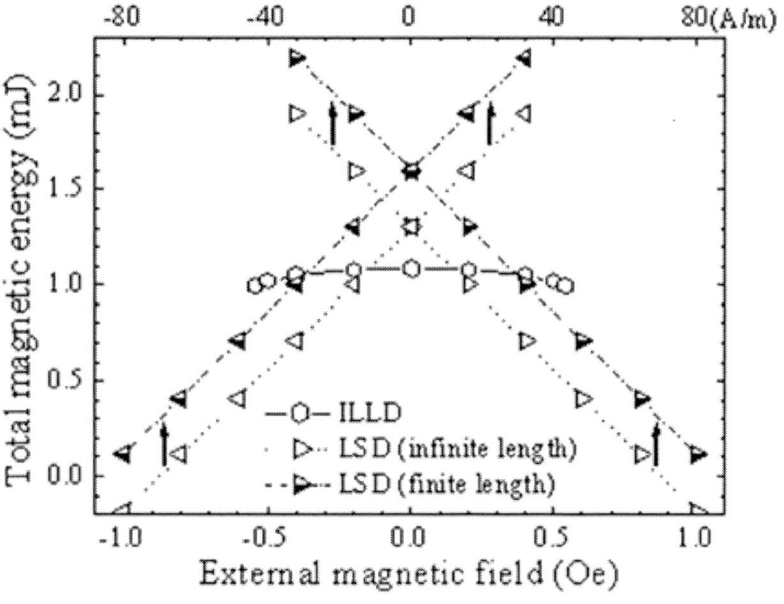

Figure 7.16. Energy level of magnetic domains with consideration of the demagnetizing energy.

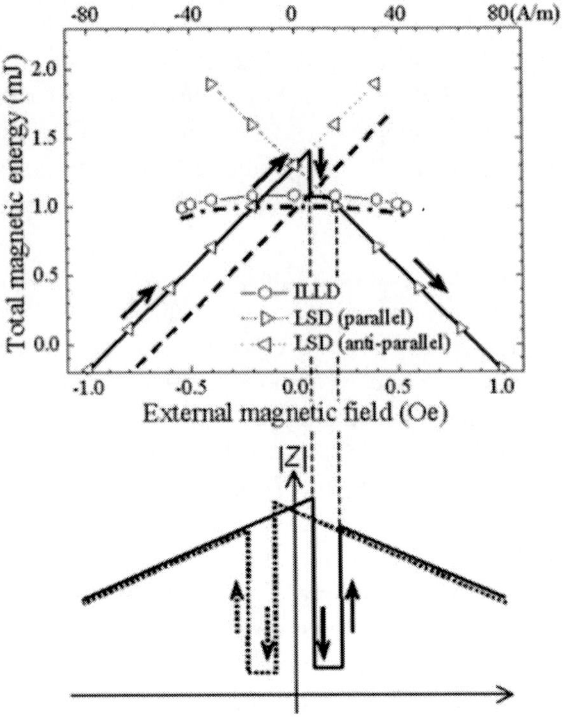

Figure 7.17.(a). Estimated variation of impedance in the case of infinite length of element.

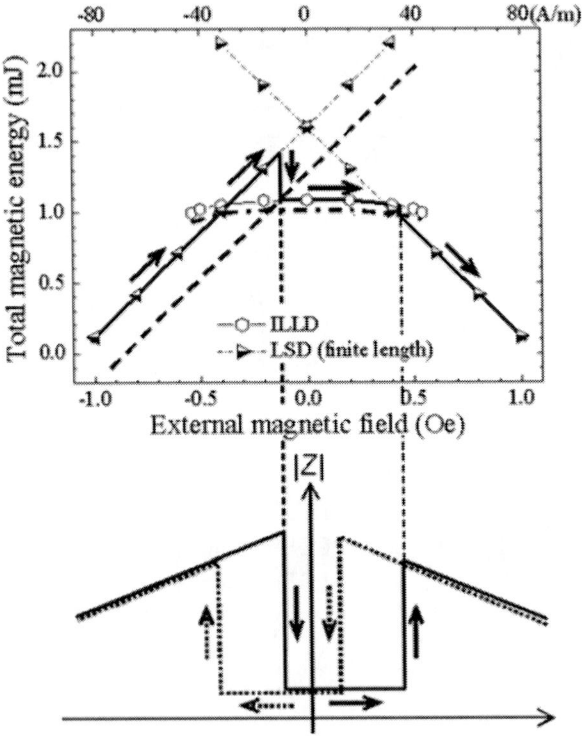

Figure 7.17.(b). Estimated variation of impedance in the case of finite length of element.

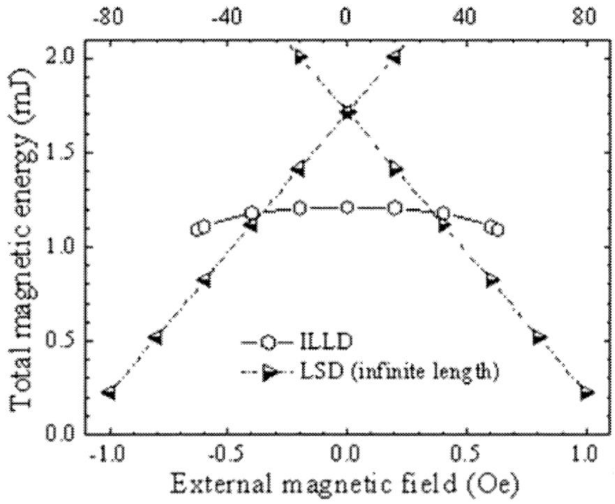

Figure 7.18. Analytical result of magnetic energy for the easy axis of 55 degrees.

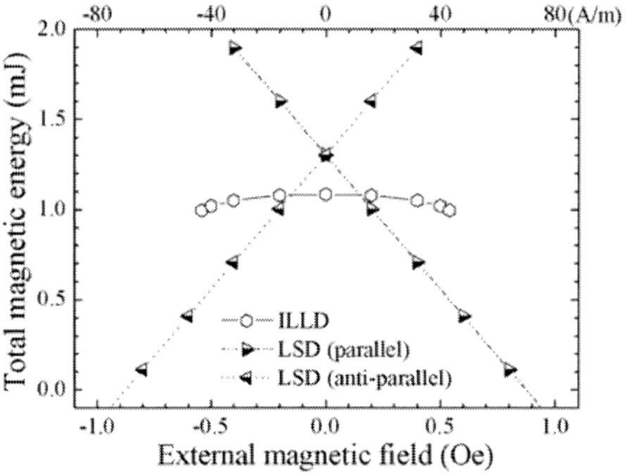

Figure 7.19. Analytical result of magnetic energy for the easy axis of 60 degrees.

7.3. Summary of this Section

A mechanism of the appearance or the disappearance of inclined stripe magnetic domain for stepped GMI phenomenon is discussed based on an analysis of magnetic domain energy. We assume these extremely different magnetic domains as stripe domain with closure domain (the inclined Landau-Lifshitz-like domain; ILLD) and single domain (the longitudinal single domain; LSD), based on experimentally observed domain structures. The result of analysis shows that the magnetic energy of these two phases intersects each other in a certain external magnetic field and in a certain direction of uniaxial easy axis, which is in the value near to the experimentally obtained. This means that the basis of stepped GMI of thin-film element is revealed as a structural change of magnetic domain based on magnetic energy.

It is also discussed that the change of impedance profile of the stepped GMI element as a function of length, and also as a function of easy axis direction. The discussion is based on the 2-dimensional analytical model for the domain energy, which consists of both the intersected magnetic energy and asymmetrical threshold energy. As a result, it is declared that both the demagnetizing energy induced by the finite length of the element and the direction of easy axis has the same effect for the change of impedance profile of the stepped GMI. This change of impedance profile is based on the relationship of magnetic energy between ILLD and LSD. Both the increment of element length and the increment of the angle of easy axis make the lower region of element impedance narrow and split. The experimentally obtained change of the impedance profile of the stepped GMI element can be explained based on the magnetic energy analysis, and the element with sprit profile of impedance is the element with three stable domain states.

In the previous study, an "non-linear GMI" which has a large jump of impedance with applying external field was reported in the case of plated magnetic film covered on a surface of non-magnetic wire with 100 μm diameter [19]. The thin film GMI element with a property of stepped-impedance, which consists of single-layered amorphous soft magnetic material with low-magnetostriction and also with directionally controlled uniaxial easy axis, was discussed in this subsection. The physical basis of the change of magnetic domain is possible to be discussed based on an analogy of the previous studies of sectional domain of the perpendicular or the canted anisotropy thin films. In these studies, the domain was called as the strong or the weak stripe domain [20]-[22]. These films with perpendicular anisotropy have a sectional Landau-Lifshitz-like domain in the case of low magnetic field, and change to in-plane single domain when a large magnetic field is applied. The points of difference between the perpendicular anisotropy thin film and our stepped GMI element are the existence in-plane domain structure induced by in-plane anisotropy, and the quite different dimensions such as 20 μm of in-plane width and thousands of μm of length. But a study with analogy of the sectional Landau-Lifshitz-like domain would be an important cutting point for the future study.

8. Dynamic Process of Domain Change for the Non-Linear Element

In this section, an induced pulse-voltage caused by a non-linear leap of magnetization in the domain change is discussed for the purpose of making dynamics of the domain change clear.

The results of measurement of the M-H loop for the stepped GMI element with resulting from a non-linear leap in magnetization is discussed, and then the results of detecting a pulse signal by using a planar coil fabricated with the stepped GMI element is shown. In this section, the domain change in the stepped GMI element is shown as a kind of Barkhausen jump with directionally different process of the domain-change [23].

8.1. Experimental Procedure

Figure 8.1. shows a view of fabricated stepped GMI element with planar coil. The dimensions of the $Co_{85}Nb_{12}Zr_3$ element were 5000 μm of length, 50 μm of width, 3.0 μm of thickness. The direction of the magnetic anisotropy was controlled by the direction of magnetic field during annealing.

In this study, 240 kA/m, 673 K, was oriented 65 degrees relative to the short-side axis of the element, and kept for 1 hour in vacuum atmosphere. The planar-coil, made by Cu thin film, was finally constructed by using the RF-sputter and the lift-off process with an insulating layer existing between the element and the planar-coil. The coil was wound 31 turns as a spiral configuration with 50μm of line width, 50μm of gap distance and 3.0 μm of thickness. The insulating layer was fabricated by using a high-temperature treated resist polymer, kept for 1 hour in 473 K air-atmosphere. The thickness of insulating layer was 3.2 μm.

The MH loop was measured by the VSM along the longitudinal direction of the element. The magnetic domain was observed by the Kerr-effect microscope. The induced pulse voltage in the planar-coil was measured by using both a 50Ω wafer-probe and high-speed oscilloscope with applying an alternating external magnetic field by Helmholtz coil. We used a 50Ω measurement system for the purpose of keeping the pulse wave form. A schematic of the measurement system is shown in Figure 8.2. The measured signal was amplified with a magnitude of 812 by the amplifier shown in the figure. The cutoff frequency of the amplifier was 15 MHz, so the measurement system could keep a waveform of 100 ns rectangular-pulse with slight deformation.

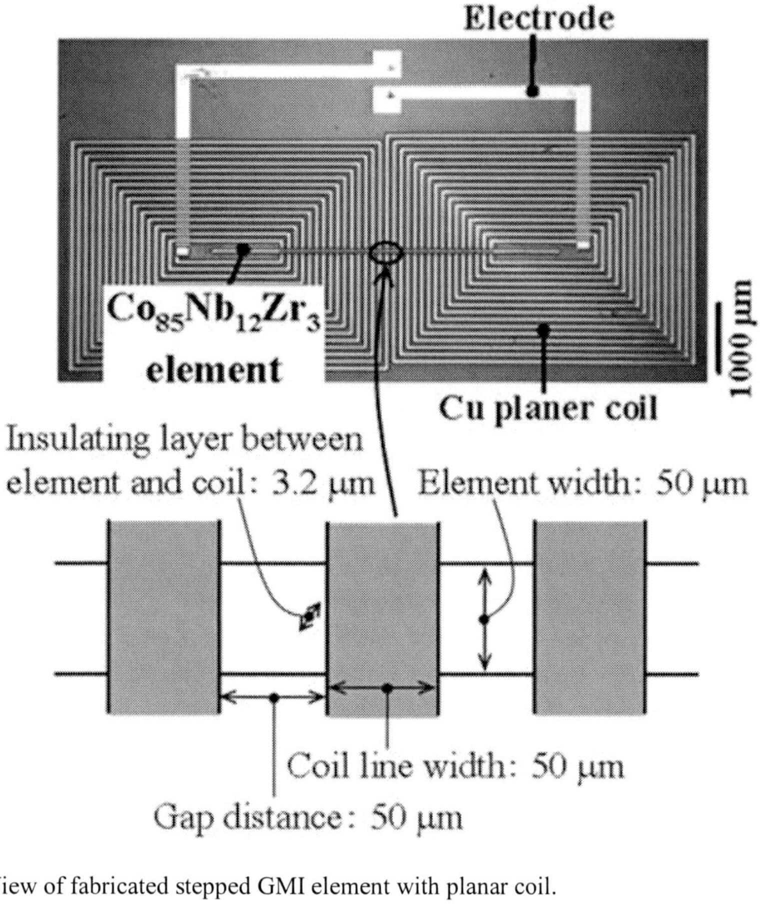

Figure 8.1. View of fabricated stepped GMI element with planar coil.

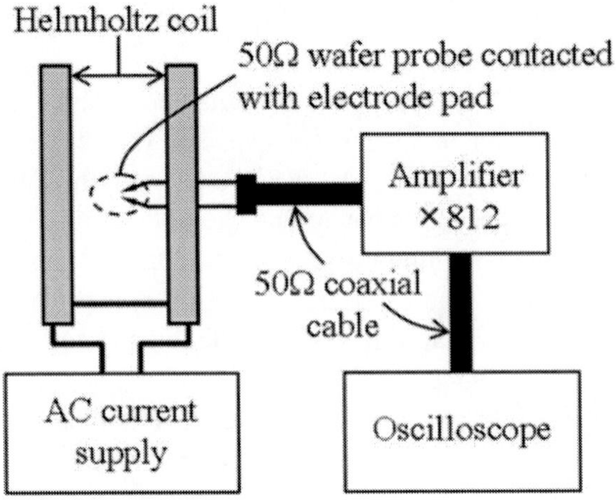

Figure 8.2. Schematic of the measurement system.

8.2. Experimental Results

Figure 8.3. shows the MH loop of the element. It also shows observed magnetic domains which are corresponding to the MH loop. The MH relation has some stepped curves of magnetization. The disconnections occur in the vicinity of 0 A/m and 40 A/m when the longitudinal external field increases, and also occur in the vicinity of 0 A/m and -40 A/m when the field decreases. The upper-curve is a state of longitudinal single domain (LSD) which is assumed to have a moment of positive longitudinal direction. The lower-curve is a state of LSD which is assumed to have a moment with negative direction. The middle-curve divided by two disconnections is a state of multi-domain (ILLD). In the state of ILLD, the width of the domains, which are shown as black and white stripes in this figure, change as a function of external field. This gradual change of domain width is a reason of the inclined line of the middle-curve of MH loop.

Figure 8.4. shows a measured signal of pulse voltage with application of 10 kHz alternating magnetic field with magnitude of 328 A/m in the longitudinal direction. This signal of pulse was measured on the electrode pads of the planar-coil. Two pulses were observed both in case of increasing of external field and in case of decreasing. The appearance of two pulses is explained by the result of MH loop of Figure 8.3. There are two disconnection of magnetization while increasing or decreasing of longitudinal magnetic field. One of them corresponds to the change from LSD to ILLD, the other correspond to the one in the opposite. Therefore the two pulses correspond to these leaps of magnetization. The polarity of pulse is also in match with an expected one. The difference of magnetic field between the occurrence of pulse and the occurrence of disconnections of magnetization is considered to be caused by a rate of changing field, therefore caused by frequency. The MH loop was measured in quasi-steady rate of changing, whereas the pulse measurement was done in 10 kHz. It is expected that the field, where the domain-change occurs, varies as a function of changing rate of magnetic field.

Figure 8.5. shows a pulse profile corresponding to the domain-change from LSD to ILLD for 100 Hz and 1 kHz, with applying 328 A/m of the alternating magnetic field. In this figure, correction of offset was not done.

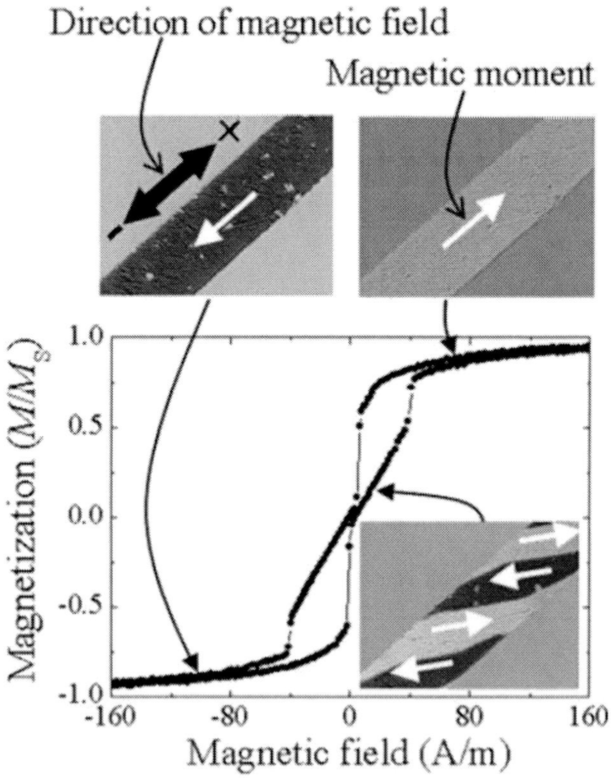

Figure 8.3. Observed magnetic domains and corresponding MH loop.

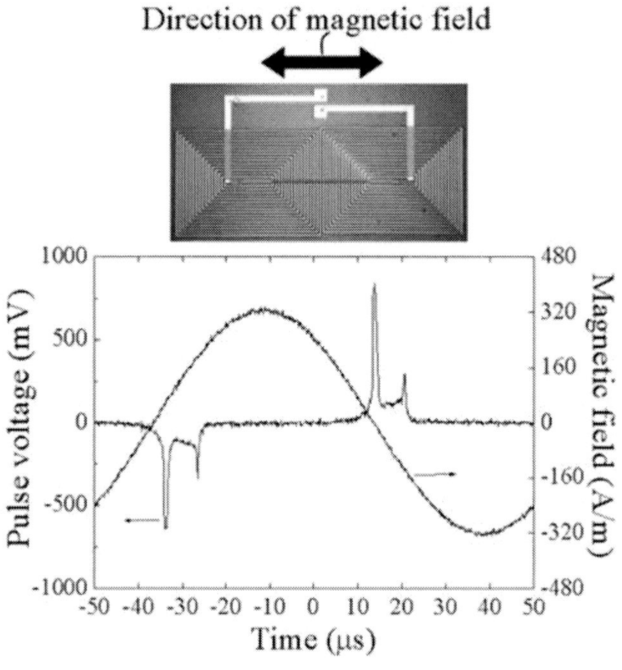

Figure 8.4. Measured signal of pulse voltage with application of 10 kHz alternating magnetic field.

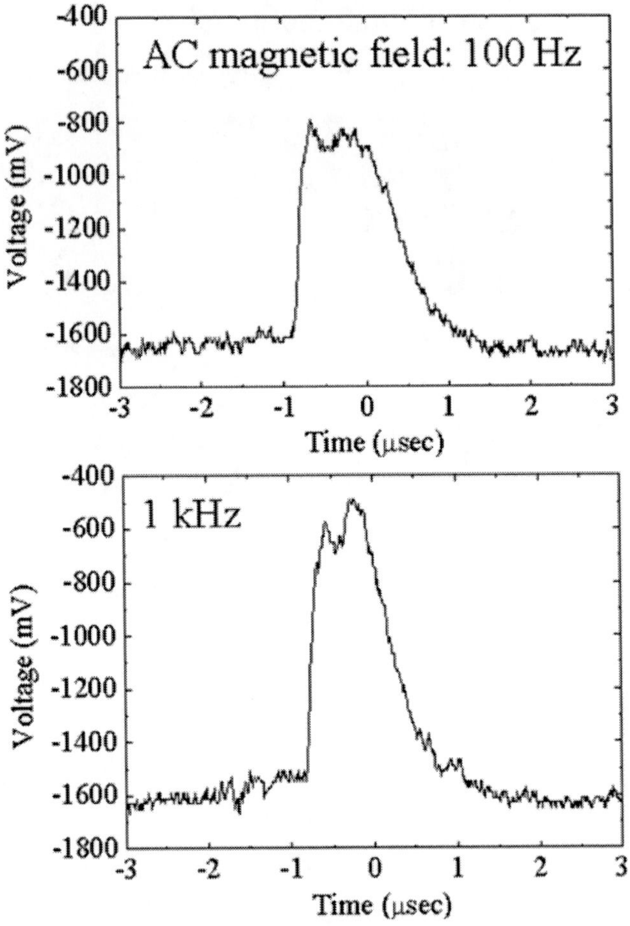

Figure 8.5. Pulse profile corresponding to the domain-change from LSD to ILLD for 100 Hz and 1 kHz.

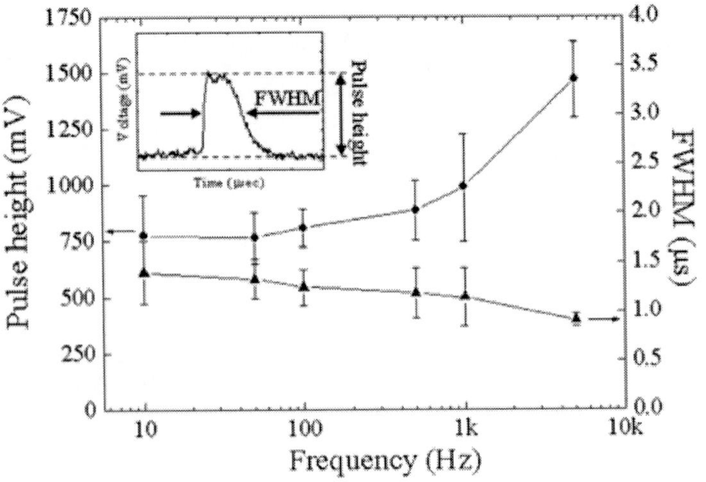

Figure 8.6. Variations in pulse height and FWHM as a function of frequency of alternating field for the domain-change from LSD to ILLD.

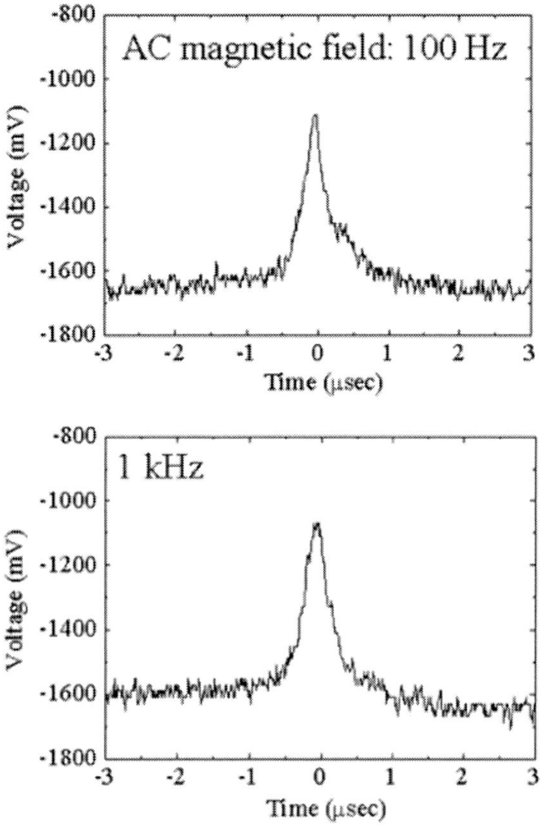

Figure 8.7. Pulse profile corresponding to the domain-change from ILLD to LSD for 100 Hz and 1 kHz.

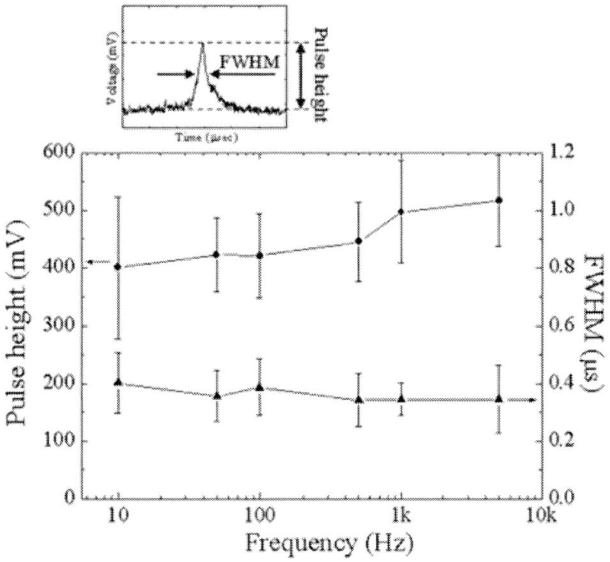

Figure 8.8. Variations in pulse height and FWHM as a function of frequency of alternating field for the domain-change from ILLD to LSD.

So the base line of the figure is not zero. From this figure the pulse height is about 800 mV, and the pulse width is about 1.2 μs of FWHM. Figure 8.6. shows variations in pulse height and FWHM as a function of frequency of alternating field for the domain-change from LSD to ILLD. The error bar is 1.96σ of 10 measurements. The pulse wave-form is kept as Figure 8.5. in the whole frequency range of Figure 8.6 measurement. The increment of pulse-height and the decrement of FWHM occur as the frequency increases.

Figure 8.7. and Figure 8.8. show the pulse profile and the variations in pulse height and FWHM for the domain-change from ILLD to LSD. The measurement condition was the same as Figure 8.5. and Figure 8.6. From Figure 8.7. the pulse has a peaked wave-form. The pulse-height is about 450 mV, and the FWHM is about 0.4 μs. There is a little change of pulse-height and FWHM as a function of frequency.

8.3. Summary and Discussions of this Section

Based on our measurement results, the phenomenon of domain-change is supposed to be a kind of Barkhausen jump with a certain propagation velocity. This resembles the phenomenon of magnetization reversal of amorphous wire which has constant velocity of domain-wall propagation. But a difference exists that our results are the case of domain-change between the single and the multi domain. The pulse wave-form is kept, such as pulse width, height, and FWHM, less than 5 kHz of alternating magnetic field. An enhancement of domain-propagation is considered to occur in higher frequency especially for the change from the single to the multi domain, because of the increment of pulse-height. The difference in wave-form of pulse between Figure 8.5. and Figure 8.7. is supposed to be showing a difference of dynamic process of magnetization, therefore it shows a difference of domain-change propagation.

9. Proposal of Application

9.1. High-Sensitivity Magnetic Sensor

The thin film magnetoimpedance sensor realizes very high sensitivity. An important technique for obtaining high sensitivity is the use of a high-frequency current applied to the sensor element. The maximum sensitivity is obtained at frequencies up to the ferromagnetic resonance of the sensor magnetic thin film; for example, hundreds of megahertz for amorphous CoNbZr films. It is important to design a high-frequency circuit for driving the sensor in order to achieve any practical applications such as nondestructive testing or biomedical applications. Here, a method of reflection signal measurement is introduced in order to measure alternating magnetic fields with very small magnitude. The measurement of a small alternating magnetic field, which is lower than 10^{-10} Tesla (T), was effected by combining reflection signal measurement with the carrier-suppressing method. Figure 9.1. shows an element which is used for this experiment. Figure 9.2. is the measured result of variation of impedance. The impedance property is linear and high sensitivity, which is achieved by using the control of inclined angle of magnetic domain. Figure 9.3. shows the high-frequency circuit used in the experiment. The points of this circuit are both the use of reflection signal from sensor element and the use of the carrier-suppress method. The signal reflection is a high-frequency phenomenon, and the reflection signal is extracted by the circulator device. The carrier-suppress method is a kind of noise suppression method by mixing an anti-phased carrier signal to the signal from the sensor. By using the carrier-suppress method, the signal with small amplitude and with different frequency from carrier signal is effectively detected. Figure 9.4. shows a measured magnetic signal by using the circuit previously shown. The alternating magnetic field of 1.7×10^{-8} T with the frequency of 500.025 kHz was detected. With consideration of noise level of -124 dBm, the sensitivity is achieved as 5.4×10^{-11} T/Hz$^{1/2}$.

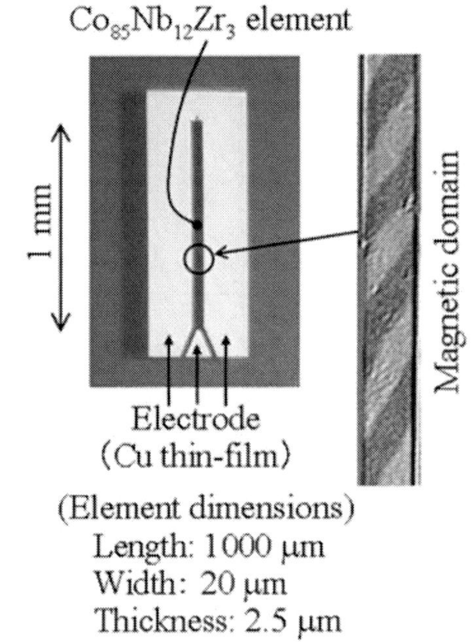

Figure 9.1. Element and domain which is used for the experiment in this subsection.

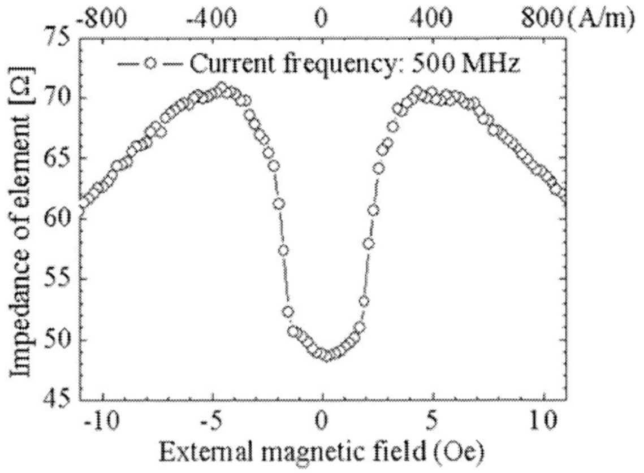

Figure 9.2. Measured result of variation of impedance.

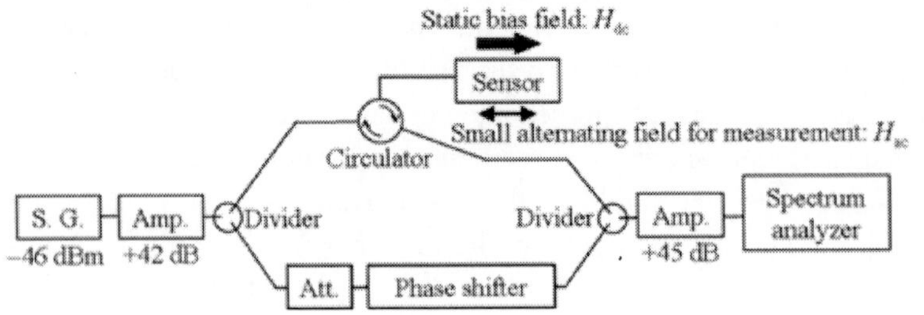

Figure 9.3. The high-frequency circuit used in the experiment.

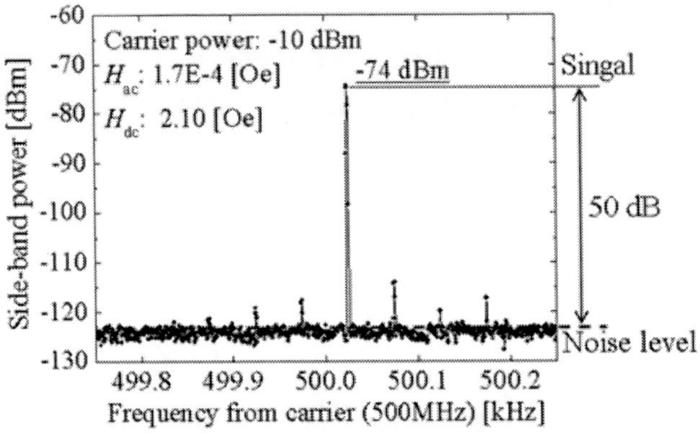

Figure 9.4. Measured magnetic signal by using the carrier-suppress circuit.

9.2. Magnetic Sensor by Using Stepped GMI Element

It was shown experimentally that the magnetic field in which the stepped impedance change occurs has an almost constant value with a standard deviation in the micro-Tesla (μT) range [25]. The magnetic field in which the stepped change occurs includes information on the external magnetic field. To realize a magnetic field sensor with an accuracy of nT/Hz1/2 by using this stepped phenomenon, a proposal was made by using differential circuit in combination with an alternating bias field in the kilohertz (kHz) frequency range [26]. The aim of the differential circuit was to make the driving circuit into a digital logic circuit, by using pulse signals timed to coincide with the stepped changes. The alternating bias field increases the sampling number, because of the stepped changes in the bias frequency. The aim of the alternating bias is to improve the sensor accuracy by using an averaging effect.

Figure 9.5. shows a variation of impedance of the element as a function of external magnetic field, which was used for this experiment. Figure 9.6. is a schematically explained procedure of the signal processing. The alternating bias field, which is large enough to make happen the stepped phenomenon and with lower frequency than the current for impedance measurement, is applied by a bias coil. The step-point of impedance was transformed to a pulse signal by using the differential circuit. In this schematic only the up-pulse, which was obtained for incremental step of impedance, is shown. The value of external magnetic field is measured through the calculation of $(H_{inc}+H_{dec})/2$, where the H_{inc} is the bias field at which the pulse occurs in case of the bias field increases and the H_{dec} is the one in case of the bias field decreases. This calculation also makes a compensation for temperature drift [27]. The algorithm of the signal processing is shown in Figure 9.7. Figure 9.8. shows a diagram of driving circuit. The function of the block 1 of the figure is both the measurement of element impedance and the application of bias field. The block 2 makes the differential pulse, and the block 3 makes the final output of measured magnetic field by using both the signal of bias field and the pulses where the step occurs. Figure 9.9. shows the measured signals. The lower signal was obtained at the point of the terminal of sensor element. The upper signal was obtained after the differentiation, which is the output of the block 2. Figure 9.10. is the result of magnetic field measurement. As a result, a high-linearity sensor without hysteresis was obtained, which had a linearity error of less than 0.5% in the range of ±100 μT. A measurement accuracy of 460 nT was achieved with a 20 Hz time constant of the output low pass filter (LPF).

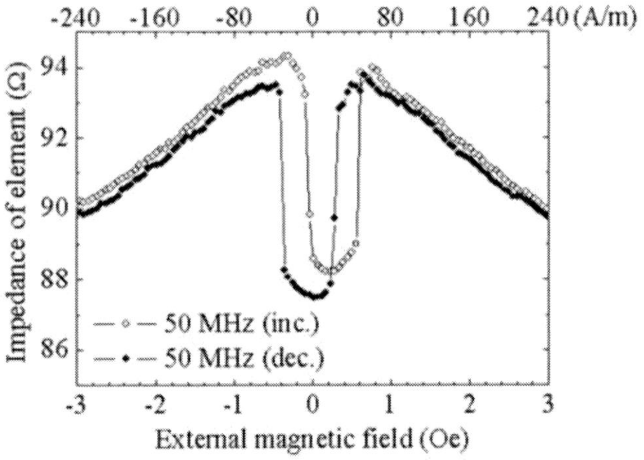

Figure 9.5. Variation of impedance of the elemen.

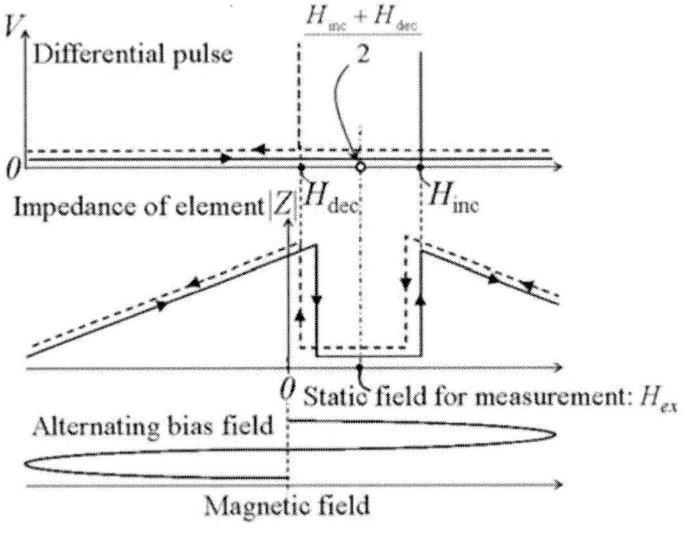

Figure 9.6. Schematically explained procedure of the signal processing.

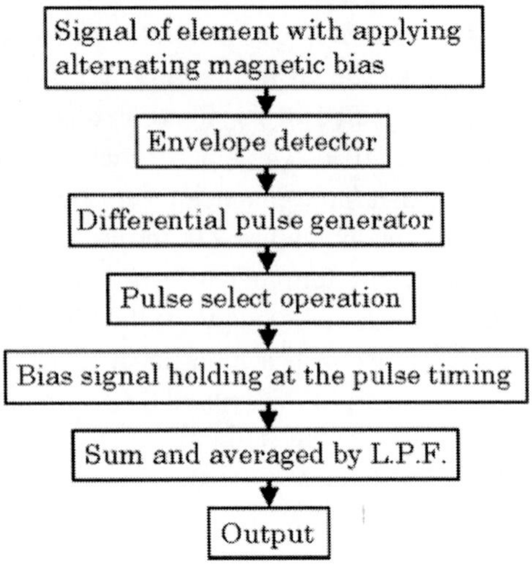

Figure 9.7. Algorithm of the signal processing.

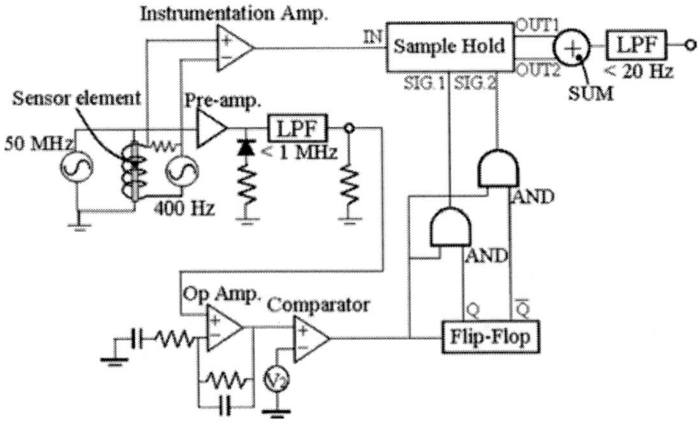

Figure 9.8. Diagram of driving circuit.

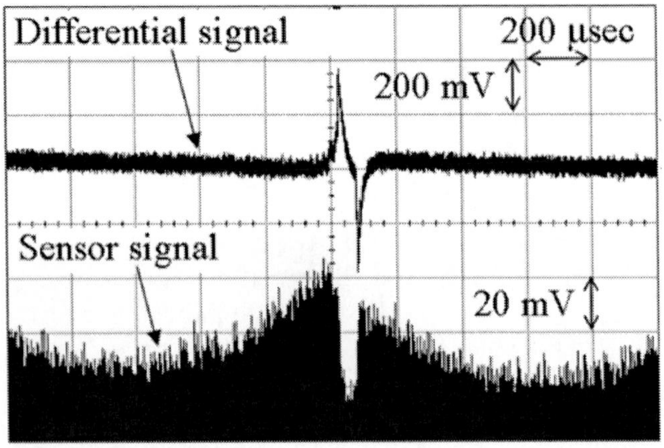

Figure 9.9. Experimentally measured signals.

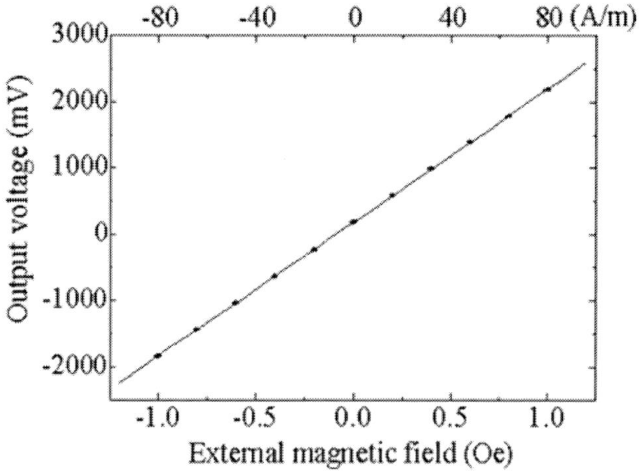

Figure 9.10. The result of magnetic field measurement.

9.3. Magnetic Sensor with Memory Function

Based on the results of section 7, we propose a new sensor which has a memory function [13]. An advantage of this sensor is a bipolar property of the memory function. The bipolar property means a sensor property which memorizes an event of application of both pole of magnetic field which is larger in absolute value than a certain threshold field. Figure 9.11. shows a schematic diagram that explains the function of bipolar memory sensor.

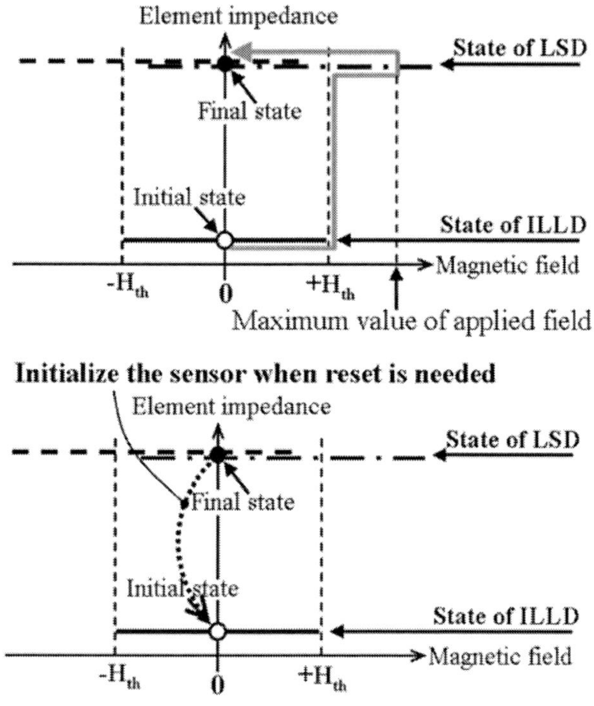

Figure 9.11. Schematic diagram that explains the function of bipolar memory sensor.

The sensor is made by the stepped GMI element with three stable domain states around zero magnetic field (Figure 7.5). The initial state of the sensor with bipolar memory function must be set in the ILLD domain state, which has a lower impedance value. When an external magnetic field with a magnitude larger than the threshold field is applied, the magnetic domain changes to LSD domain state which has higher value of impedance. The threshold field, H_{th}, is the magnetic field which makes the element domain change from ILLD to LSD. In the case of our stepped GMI element with three stable magnetic domains, the magnetic domain is maintained as LSD even after the applied field returns to zero value. This procedure of memory function also occurs even when the applied external field is a minus value. This proposed sensor memorizes an event of magnetic field application. It memorizes an event of applied magnetic field which is larger than a threshold value in absolute. This event of applied magnetic field, which causes a domain change, can easily be detected by impedance measurement of the sensor element.

9.4. Stability of the Step-Point

For the purpose of actual application, a stability of element property against the circumferential condition and the aging is important. An introduction of these data is carried out as a final of this subsection.

Firstly the stability against the circumferential condition is discussed. The following data are both the stability against the application of large magnetic field and the stability against the application of high and low temperature. The element, which used in this experiment, was the one with stepped property. Figure 9.12. shows the variation of element impedance as a function of external magnetic field. The minor-loop measurements of impedance variation are also shown. The magnetic field, where the impedance-step occurs, is stable less than 1.6 A/m (20 mOe) of standard deviation without effect of minor-loop range of magnetic field. The step-point is more stable for the up-step, which means the domain variation from ILLD to LSD, than the down-step, which means the one from LSD to ILLD.

Figure 9.13. shows the variation range of applied field in case of minor-loop measurement. Figure 9.13.(a) shows a range of variation between the started value 0 A/m and minus value of pivot-point. Figure 9.13.(b) shows a range of variation between the started value -173 A/m (-2.16 Oe) and a value of pivot-point and △(open-triangle). The following data are expressed by using a horizontal value as this pivot-point value.

Figure 9.14. shows a variation of the averaged value and the standard deviation of a certain step-point, which is the up-step in the minus region, after application of large magnetic field. The large magnetic field was applied in each direction of length, width and thickness. The strength of the applied field was 1.5 T during 1 minute. These measurements of step-point stability were measured after application of large field. The minor-loop was turned in 20 turns, therefore 20 data of step-point were obtained in each measurement point of this figure. The result shows that the application of large magnetic field makes very slight effect on the element property.

Figure 9.15. shows a variation of the averaged value and the standard deviation of the step-point, the same up-step in the minus region as the previous figure, after application of high and low temperature circumference. The applied temperature was 394K, 234K, 434K and 474K during 1 hour in air atmosphere, continually applied by using one element. The measurement of Figure 9.15. was carried out in room temperature, 295K±0.5, after the each heating treatment. The result shows that the step-point is stable less than 1.6 A/m (20 mOe) for the treatment with 394K and 234K, and also less than 3.2 A/m (40 mOe) for the treatment with 474K.

Figure 9.16. shows an aging variation of the stepped element. The measured element was kept in a room atmosphere in a box made by acrylic acid resin for 3 years. The element has been kept as bare CoNbZr without any protection layer. The result shows that the stepped property is kept in spite of the time period of 3 years.

These results would give us an impression of possibility of actual application of this element.

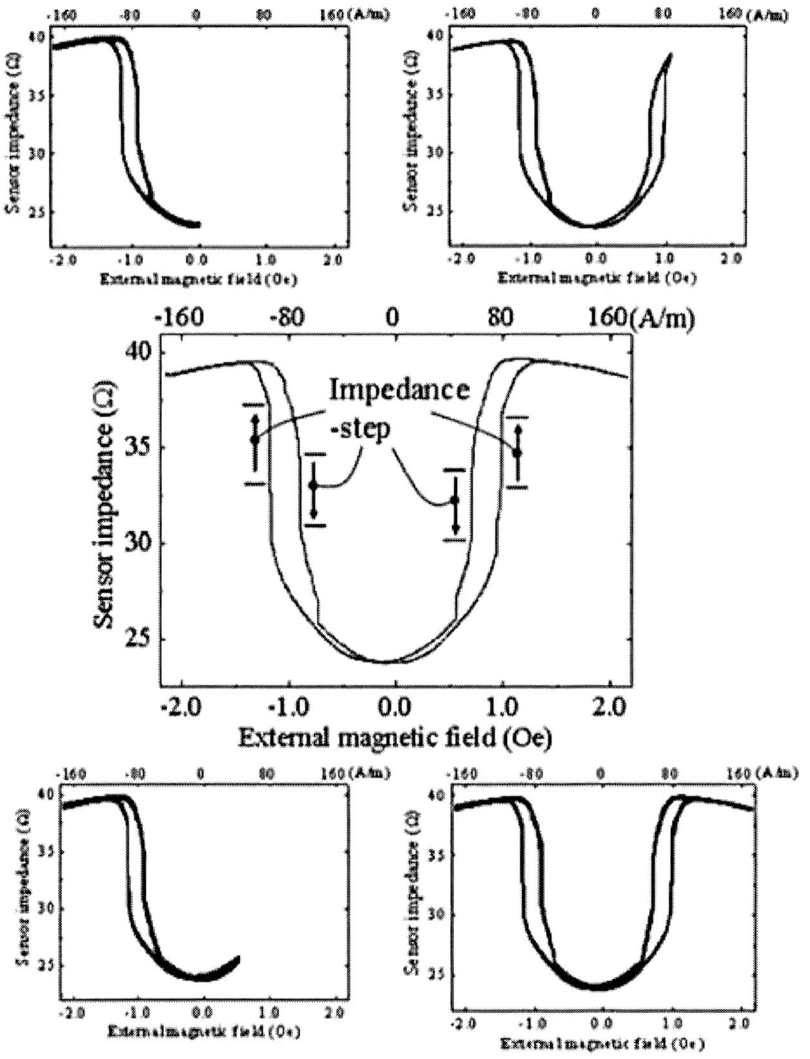

Figure 9.12. Variation of element impedance as a function of external magnetic field in case of minor-loop.

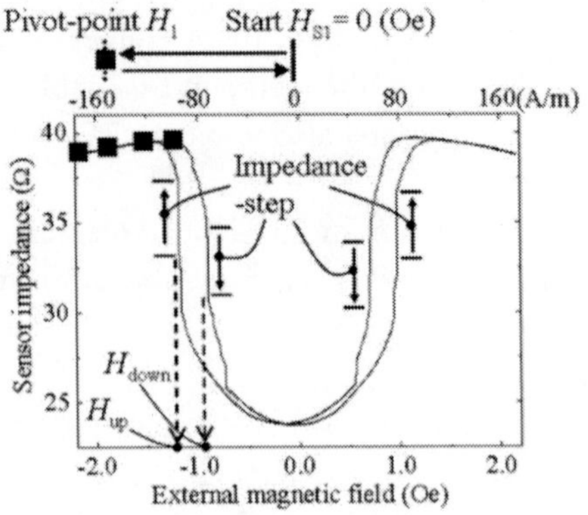

(a) Range of variation between the started value 0 A/m and minus value of pivot-point

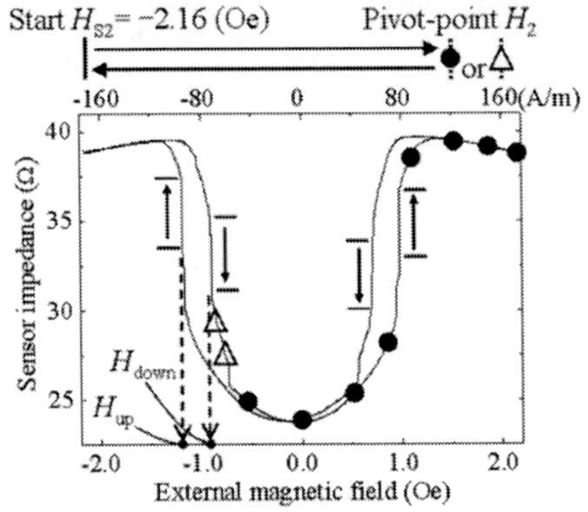

(b) Range of variation between the started value -173 A/m (-2.16 Oe) and a value of pivot-point

Figure 9.13. The variation range of applied field in case of minor-loop measurement.

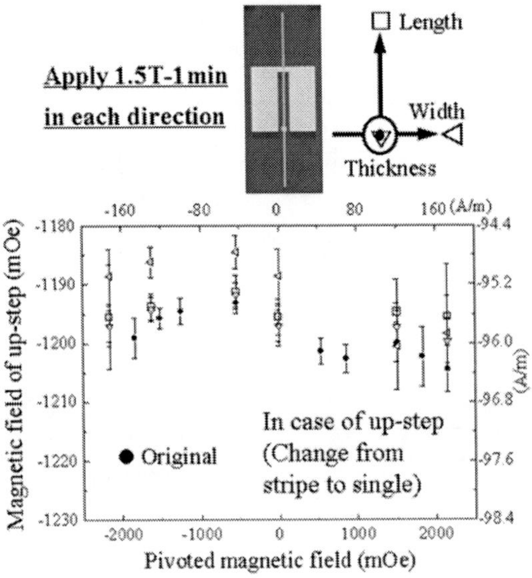

Figure 9.14. Variation of the averaged value and the standard deviation of the point of up-step after application of large magnetic field.

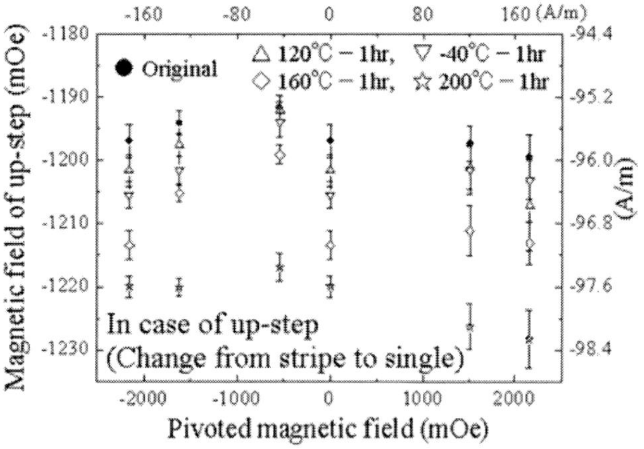

Figure 9.15. Variation of the averaged value and the standard deviation of the point of up-step after application of high and low temperature circumference.

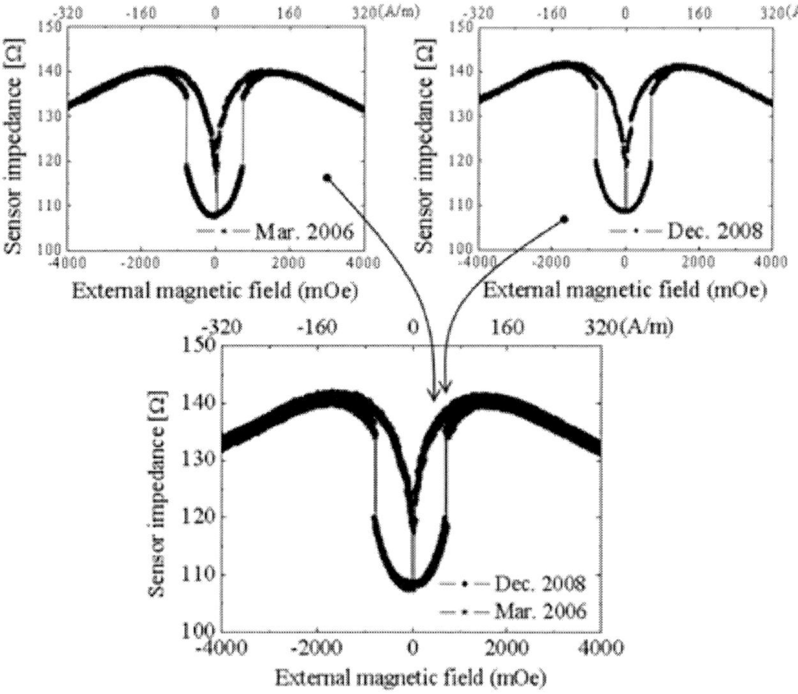

Figure 9.16. The aging variation of the stepped element.

10. Future Study of the Domain Controlled and the Stepped Element

Based on the experimental results and discussions of this chapter, a possibility for actual application of the element by using the control of domain structure is shown. In this section, an expected future study of the domain controlled and the stepped element is discussed.

Concerning about the sensor with memory function, which is shown in subsection 9-3, a desirable property must be achieved as a target of future study. Based on both the experimental and the theoretical results for the stepped GMI element with three stable states, we can predict a sensor with memory function which has a property as shown in Figure 10.1. This property must be desirable, because of the following reason.

A problem of the sensor with memory function, previously shown in the section 9, is an existence of step-down magnetic field of the element impedance. In the case of the sensor property as shown in Figure 9.11, the information of memorized event is easily erased by a spontaneous magnetic field.

As explained in section 9, the initial state for this sensor with memory function must be low impedance, i.e. multi domain (ILLD). When an external magnetic field larger than the threshold field is applied, the sensor impedance changes to high-level, therefore change to single domain (LSD), and keep this high impedance even after the external field returns to zero. This is the memory function which is proposed in section 9. But when an external field of in the vicinity of the minus threshold is applied, this state of high-impedance goes down to the state of low-impedance. This means that a spontaneous magnetic field with a certain value and direction makes the memorized information to be erased (Figure 10.2).

A proposal of sensor property in this section, as shown in Figure 10.1, can solve this problem. Once a sensor memorizes an event of field application, the state keeps high-impedance, because the domain changes only a directional swap between longitudinal and anti-longitudinal single domain (LSD). In this case both directions of longitudinal single domain have high-impedance.

An existence of the desirable property can be predicted from the previous discussion in this chapter. Figure 7.8(b) is an experimentally obtained impedance property which has three stable states, which is discussed in section 7. When an inclined angle of easy axis is set in a slight larger angle, the magnetic energy and expected impedance property is predicted as Figure 10.1. This prediction is confirmed experimentally as Figure 10.3, as a first step of this research. In Figure 10.3, element dimension is the same as Figure 7.8.(b), but different in the direction of magnetic easy axis. It is larger than 65 degrees. In Figure 10.3, the step-down of impedance disappeared, as predicted in Figure 10.1. Another important step of this research is to confirm an existence of multi domain state in the case of Figure 10.1. as a theme of the further experiment, and also realizing a reset-procedure from high to low impedance, that is to say realize a controlled change of domain from LSD to ILLD domain state. An initial step of research was carried out.

A partial application of magnetic field on the element can realizes a step-down of impedance. A pulsed magnetic field was partially applied to the element in the direction of width with a slight angle relative to the width. Figure 10.4. shows a result of step-down of impedance. After applying the pulsed field with zero external field, an abrupt step-down of impedance occurs. But the variation of impedance after the step-down occurrence was not an ideal. More investigation must be needed to realize this artificial step-down technique to be ideal. This technique would be a prospective candidate for a functional 3D material with controlling a property by using controlled permeability distribution derived from domain change. The study of the effect of magnetostriction is also important for the stepped phenomenon. It is not clear that this phenomenon is applicable for an element with magnetostriction. But possibility exists, and the experimental confirmation would be a future theme, too.

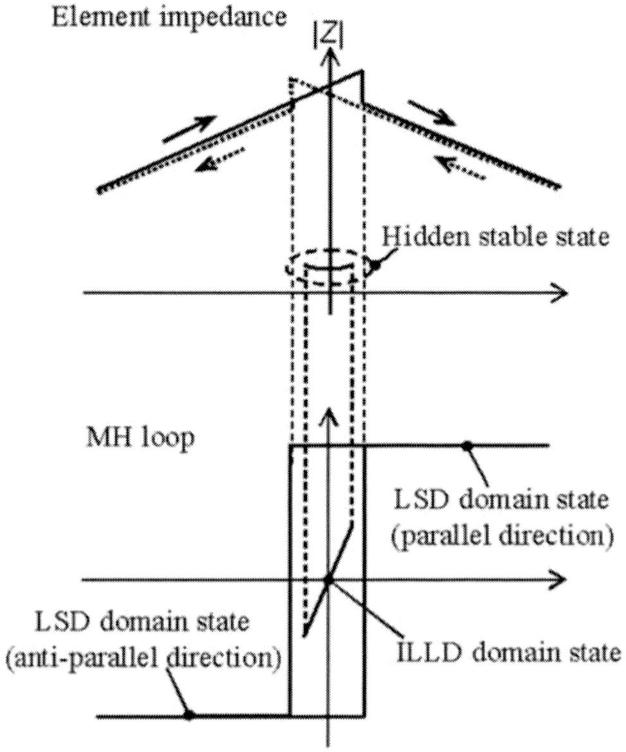

Figure 10.1. Predicted property of a sensor with memory function.

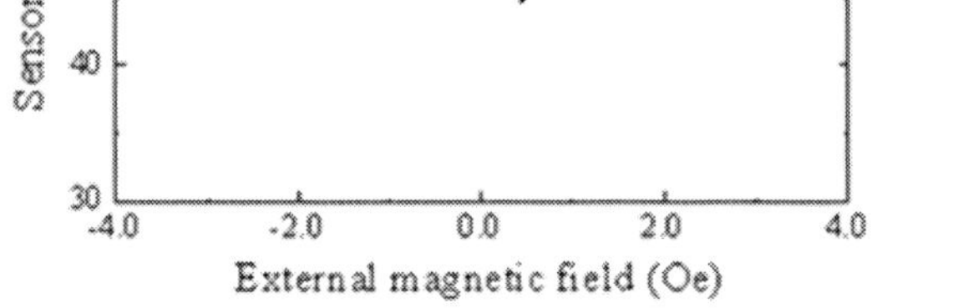

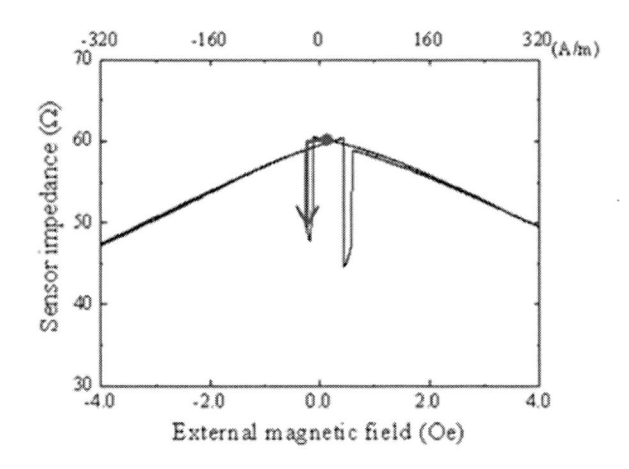

Figure 10.2. Mechanism of erasing the memorization by a spontaneous magnetic field.

Another important point of future study is that the control of step-point to occur in larger magnetic field. It is important for practical application. The reason is that in order to avoid influence of terrestrial magnetism, the step-point is preferable more than 800 A/m (10 Oe). It is not clear now that how to achieve this property. A possibility exists to control the step-point by control the magnetic property of the element and by control the dimensions and structure of the element, which means the control of demagnetization force and threshold energy.

In order to investigate to achieve this property, a domain simulator, which can simulate a domain structure and it's movement by external field, must be effective. A study, both actual experiment and virtual simulation, must be done with a wide range of magnetic property, with investigation of the element dimension and the element structure. The consideration of the variation of threshold energy which is needed to prevent the occurrence of domain change will be important. A development of suitable domain simulator and application for this study would be an effective way for the research in the next step.

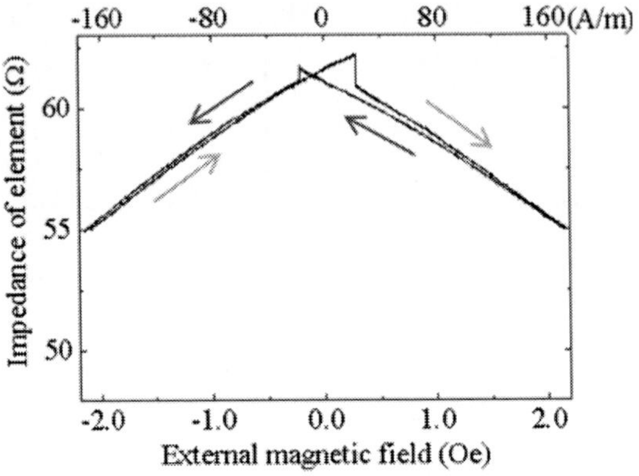

Figure 10.3. Experimentally obtained variation of impedance.

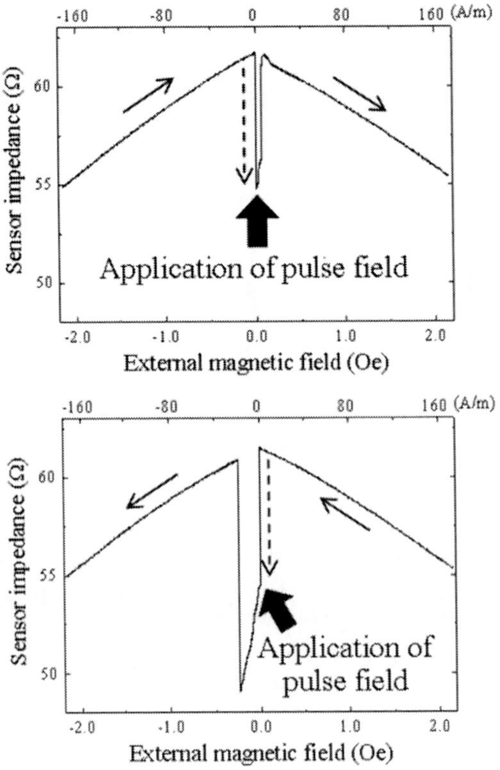

Figure 10.4. Result of step-down procedure and variation of impedance after that.

SUMMARY

In this chapter, a variation of magnetic domain and correlated magneto-impedance of a micro-fabricated rectangular-shaped magnetic thin film is discussed. The micro-fabricated magnetic element, which is made from an amorphous film with in-plane uniaxial anisotropy and low magnetostriction, the Landau-Lifshitz-like magnetic domain is observed. The in-plane direction of magnetic easy axis controls the angle of 180 degree wall of the Landau-Lifshitz-like magnetic domain.

The change of magnetic domain and the correlated high-frequency impedance occur as a function of external magnetic field. The variation of the domain structure and correlated magneto-impedance of the element is characterized by a direction of magnetic easy axis. In this chapter, a magnetic phenomenon and an analytical explanation was shown for the rectangular magnetic element with the width of some tens of microns. Firstly the phenomenon, which is based on the variation of magnetic domain and correlated magneto-impedance induced by external magnetic field as a parameter of the in-plane direction of uniaxial magnetic easy axis, was discussed. Both the high-sensitivity magnetic field sensor and the stepped magneto-impedance element were obtained. Secondary a proposal of application of these phenomenon were discussed. The future study of this element is also commented, finally.

ACKNOWLEDGMENTS

The authors would like to express gratitude to Prof. H. Wakiwaka, Shinsyu University, Prof. T. Matsuo, Kyoto University, Dr. S. Hashi, Tohoku University, Prof. M. Takezawa, Kyusyu Institute of Technology and Prof. H. Kikuchi, Iwate University for the useful discussions.

This work was supported by Nationwide Cooperative Research Projects by Research Institute of Electrical Communication, Tohoku University, Sendai, Japan.

REFERENCES

[1] T. Nakai, "*Study of Bias Reduction by Controlling in-plane direction of easy axis for the High-Frequency Carrier-Type Thin Film Magnetic Sensor*", Tohoku University, Sendai, Japan, 2005, Thesis.

[2] A. Hosono, Y. Shimada, J. Numazawa, Y. Yoneda. R*eport of Research Institute for Scientific Measurements,* Tohoku Univ. 39, 27 (1990) .

[3] D.O.Smith. *J. Appl. Phys.* 29, 264 (1958) .

[4] T. Nakai, H. Abe, M. Yamaguchi, S. Yabukami, H. Kikuchi, and K. I. Arai. *J. Magn. Soc. Jpn.*, vol. 27, 419 (2003).

[5] L. V. Panina, K. Mohri, K. Bushida, M. Noda. *J. Appl. Phys.* 76, 6198 (1994).

[6] K. V. Rao, F. B. Humphrey, J. L Costa-Krämer. *J. Appl. Phys.* 76, 6204 (1994).

[7] R. S. Beach , A. E. Berkowitz. *J. Appl. Phys*. 76, 6209 (1994).

[8] C. G. Kim, S. S. Yoon, K. J. Jang, C. O. Kim. *Appl. Phys. Lett.* 78, 778 (2001).

[9] T. Nakai, H. Abe, and K. I. Arai. *The Physics of Metals and Metallography*, vol. 101, S41-S44 (2006).
[10] T. Nakai, H. Abe, S. Yabukami, M. Yamaguchi and K. I. Arai. *J. Magn. Soc. Jpn.*, vol. 27, 832 (2003).
[11] T. Nakai, S. Yabukami and K. I. Arai. *J. Magn. Soc. Jpn.*, vol. 28, 128 (2004).
[12] T. Nakai, H. Abe, S. Yabukami, and K. I. Arai, *J. Magn. Magn. Mater.*, vol. 290-291, 1355 (2005).
[13] T. Nakai, K. Takada and K. Ishiyama, *IEEE Trans. Magn.*, 45, 10, 3499 (2009).
[14] T. Nakai, H. Abe and K. I. Arai. *J. Magn. Soc. Jpn.*, vol. 29, 747 (2005).
[15] A. Hubert, R. Schäfer, "Magnetic Domains", *Springer* (1998).
[16] L. D. Landau, E. Lifshitz, *Phys. Z. Sowjetunion*, 8, 153 (1935).
[17] T. Nakai, K. Ishiyama, and J. Yamasaki, *J. Appl. Phys.*, vol. 101, 09N106 (2007).
[18] T. Nakai, K. Ishiyama, and J. Yamasaki, *J. Magn. Magn. Mater.*, vol. 320, e958-e962 (2008).
[19] G. V. Kurlyandskaya, H. Yakabchuk, E. Kisker, N. G. Bebenin, H. Garcia-Miquel, M. Vazquez, and V. O. Vaskovskiy, *J. Appl. Pys.*, vol. 90, No. 12, 6280 (2001).
[20] A. Hubert, *IEEE Trans. Magn.* , MAG-21, No. 5, 1604 (1985).
[21] H. Aitlamine, M. Labrune, and I. B. Puchalska, *IEEE Trans. Magn.*, vol. 26, No. 1, 48 (1990).
[22] M. Labrune, J. Miltat, *J. Magn. Magn. Mater.* , 104-107, 241 (1992).
[23] T. Nakai, K. Takada, K. Ishiyama, *Journal of Physics: Conference Series*, 200, 042019 (2010).
[24] T. Nakai, M. Sendoh, S. Yabukami, K. Ishiyama and K. I. Arai. *J. Magn. Soc. Jpn.*, vol. 30, 550 (2006).
[25] T. Nakai, K. Takada, H. Komatsu and K. Ishiyama. *J. Magn. Soc. Jpn.*, vol. 32, 366 (2008).
[26] T. Nakai, K. Takada, H. Abe, N. Hoshi, H. Suzuki, K. Ishiyama and K. I. Arai. *J. Magn. Soc. Jpn.*, vol. 31, 216 (2007).
[27] T. Nakai, H. Abe, and K. I. Arai. *J. Magn. Soc. Jpn.*, vol. 29, 675 (2005).

In: Magnetic Thin Films
Editor: John P. Volkerts
ISBN: 978-1-61209-302-4

Chapter 6

MAGNETIC SWITCHING OF MAGNETIC NANOFILMS AND MANAGEMENT SPIN CURRENT BY PULSE LASER RADIATION

M. M. Krupa
Institute of Magnetism NASU, Vernadski st. 36b, Kyiv, Ukraine

1. INTRODUCTION

Spintronics belongs to one of the most quickly developing areas of science and technology; it is based on the control of the processes of transfer of spin current between the elements of electronic devices. Researches of a spin dependent transport and spin relaxation in solids, search for new materials with the high degree of an electron spin polarization and development of methods for active control of spin states in solid-state circuits constitute the main directions of the spintronics.

The spin current control involves availability of spin current injectors, control apparatuses of an electron spin orientation and elements of a spin current filtration. Magnetic materials with the high degree of spin polarization are used for obtaining effective injectors of the spin current. Such a high electron spin polarization at a room temperature is provided in magnetic semimetals, some magnetic Haussler alloys and magnetic semiconductors.

Heterogeneous magnetic nanostructures constitute the most perspective material for spintronics. This is primarily due to the fact that modern technology in microelectronics allows to get through the required elements of such structures and devices of spintronics. The study of such nanostructures allows to get a high magnetoresistance, establish basic regularities concerning the influence of the electronic structure and magnetic characteristics of component materials on the spin-polarized current [1-4], and also to reveal the effect of spin current-induced non-equilibrium magnetization of nonmagnetic nanolayers [5,7].

For creation of high-frequency elements of spintronics high-speed systems for controlling by a spatially localized magnetization are necessary. The solution of this problem with the help of conventional magnetic systems is rather hard. An interesting idea of the magnetization control in heterogeneous magnetic nanostructures was realized via spin current [8, 9]. A spin current in such structures becomes excited by the external electric field. It is well known [10-13] that photonic pressure of laser radiation due to the transmission of impulse of photons also allows electrons to excite an electric current in solids. Clearly, that effective excitation of current by means of laser radiation is possible to get only in small absorber materials or in the thin nanofilms.

In our works [14,15] it is shown that three-layer nanofilms of a magnetic-dielectric-nonmagnetic or magnetic-dielectric-magnetic by means of nanosecond and picosecond laser impulses is possible to excite a large spin current which creates non-equilibrium spin polarization in an output nanolayer to the direction of a laser beam. Such laser impulses allow not only to obtain a large spin current but also create in magnetic nanolayer a strong magnetic field due to the reverse effect of Faraday [16, 17, 18], that specifies on the prospect of their use in spintronics. Thus, as shown in works [19, 20], at large intensity of femtosecond laser impulses reversal time of magnetic nanofilms without the use of external magnetic-field does not exceed a few picoseconds.

In the presented work we have researched mechanisms of the laser-induced reversal magnetization in single-layered $Tb_{22}Co_5Fe_{73}$ and three-layered magnetic nanofilms $Tb_{19}Co_5Fe_{76}/Pr_6O_{11}/Tb_{22}Co_5Fe_{73}$ and $Co_{80}Fe_{20}/Pr_6O_{11}/Co_{30}Fe_{70}$ under nanosecond and picosecond laser pulse radiation. It is shown the role of the laser-induced spin current in magnetic transformation in the magnetic nanofilms and possibility of fast magnetic reversal in these nanostructures under ultra-short polarized laser pulses that can be used in the new spintronic devices.

2. Mechanisms of Magnetic Switching of Magnetic Films by Laser Radiation

Let us consider a three-layer film structure with two magnetic nanolayers divided by a thin non magnetic nanolayer. Magnetic reversal magnetic nanolayers at an irradiation of such structure laser radiation can be caused by thermal as well as non thermal influence of this radiation [21]. The thermal component of this impact is exhibited via the temperature dependence of the magnetization and magnetic anisotropy constants that lead to the change of spin states of the system. And perspective from the point of view of use in spitronics non thermal influence of laser radiation on magnetic properties of films is more efficient. The strongest not thermal influence on magnetic characteristics magnetic nanolayers is rendered by the local magnetic fields created by laser radiation at the expense of magnetooptical inverse Faraday effect [10] and injection in this layer of the big concentration nonequilibrium polarized electrons [11,12].

Heating of a single-layer magnetic nanofilm with non-absorptive protective layer and substrate exposed to nanosecond pulses can be estimated in the first approximation as

$$\Delta T \approx \frac{(1-R)[1-exp(-\alpha h_m)]I_0\tau_i}{c_m\rho_m h_m + 2c_1\rho_1 h_1 + \sqrt{K_2\tau_i}c_2\rho_2} \quad (1)$$

For picosecond pulses, we have

$$\Delta T < \frac{(1-R)[1-exp(-\alpha h_m)]I_0\tau_i}{c_m\rho_m h_m + 2c_1\rho_1\sqrt{K_1\tau_i}} \quad (2)$$

where I_0 and τ_i are the radiation intensity and duration of a laser pulse, respectively; α and R are the coefficients of absorption and reflection of laser radiation for a magnetic layer, respectively; h_m и h_1 are the thicknesses of the magnetic and protective layers, respectively; c_m, c_1, c_2, ρ_m , ρ_1 and ρ_2 are the specific heats and densities of the magnetic layer, protective layer and substrate, respectively; K_m, K_1 and K_2 are the thermal diffusivities of the magnetic layer, protective layer and substrate, respectively.

Using typical values for the parameters resulted higher: $\alpha=10^5\ cm^{-1}$, $R=0,5$, $h_m=20\ nm$, $h_1=40\ nm$, $c_m=0.5\ J/(g\ K)$, $c_1=1,1\ J/(g\ K)$, $c_2=1,0\ J/(g\ K)$, $\rho_m=7,9\ g/cm^3$, $\rho_1=4,0\ g/cm^3$, $\rho^2=2,2\ g/cm^3$, $K_m=0,15\ cm^2/s$, $K_1=0,05\ cm^2/s$, and $K_2=0.006\ cm^2/s$, we will get for $\Delta T\approx100°$ for a nanosecond pulse of duration $\tau_i=15\ ns$ and intensity $I_0=1MW/cm^2$ and $\Delta T<30°$ for a nanosecond pulse of duration $\tau_i=80\ ps$ and intensity $I_0=100\ MW/cm^2$.

In the case of thermal action, laser heating of nanolayers considerably reduces their coercive force. When the coercive force of a magnetic nanolayer becomes smaller than the oppositely directed external magnetic field H_0, reversal magnetization takes place in this layer. In magnetic films with perpendicular anisotropy (such is ferrimagnetic amorphous films TbCoFe) magnetization reversal may also occur in zero magnetic field under the action of demagnetizing fields H_{dm} emerging in the region of action of laser radiation. The demagnetizing field in the region of diameter d_0 and thickness h of a magnetic nanolayer heated to the Curie temperature (in TbCoFe the Curie point $T\sim500\ K$) can be estimated using the expression

$$H_{dm} \approx \frac{h}{d_0} H_a \quad (3)$$

where H_a is the coercive force of the magnetic nanolayer.

The time of laser action is determined by the spin-electron and spin-phonon relaxation times and lies on the nanosecond time scale. It is clear that a time scale is determined by spin-electron and spin-phonon relaxation times. Heating of magnetic nanolayers can cause a reduction of coercive forces and magnetic reversal in an external magnetic field or in a demagnetizing field. Such a mechanism of magnetic reversal is well-known and also is widely used for magneto-optical recording [22, 23] in ferrimagnetic films in which strong dependence coercive forces from temperature is observed.

In nonthermomagnetic mechanism of magnetic reversal of the magnetic nanolayers may cause their direct reversal in the magnetic field generated by laser radiation due to the inverse Faraday effect [16, 17]. Such a mechanism generates a magnetic field only when the circular polarization of laser radiation and a magnetic field is proportional to the square of the field intensity $\vec{E}$ of the light wave. The circular polarization of laser radiation due to this effect in the medium, a nonequilibrium magnetization $\vec{M}_F$ is directed along the propagation of the laser beam [16]

$$\overrightarrow{M_F} = \frac{\chi}{16\pi}\left[\vec{E}\times\vec{E^*}\right] \quad (4)$$

χ is the magneto-optical susceptibility of the medium.

In thin conducting magnetic films the value of the magnetization is induced by circularly polarized light wave can be found on the basis of a circular current induced by laser radiation [17]

$$\overrightarrow{M_F} = \frac{ie^3 N}{4(m_e^*)^2\omega^3}\left[\vec{E}\times\overrightarrow{E^*}\right] \quad (5)$$

The direction of the magnetization vector varies with the direction of rotation of the field vector of circularly polarized electromagnetic radiation on the opposite. Therefore, by using laser radiation with the right circular polarization or the left circular polarization we can obtain the magnetic field directed towards the laser beam or against him.

From (2) for the magnetic field H_F in the conductive magnetic film with a magnetic permeability μ circularly polarized laser light with the intensity of the radiation I we obtain the following formula

$$H_F = \frac{\left|\overrightarrow{M_F}\right|\mu}{4\pi\mu_0} = \frac{e^3\mu N E^2}{16\pi\mu_0 n(m_e^*)^2\omega^3} = \frac{e^3\mu N I}{8\pi\mu_0 c\varepsilon_0 n(m_e^*)^2\omega^3} \quad (6)$$

Where I and ω are the laser radiation intensity and frequency; c is a light speed; e, m_e^* and N are a charge, an effective masse and the average concentration of conductive electrons; ε and μ are an dielectric and magnetic film permeability.

The mentioned laser-induced internal magnetic field can reach large enough values $H_F > 10^5$–10^6 A/m for magnetic films with the magnetic permeability $\mu = 10^3$-10^4 under relatively low radiation intensity of picosecond pulses $I = 10^9$ W/cm^2 at the average concentration of conductive electrons $N \approx 10^{22}$ cm^{-3}. The characteristic relaxation time for the inverse nonlinear magnetooptical Faraday effect ranges from 10^{-13} to 10^{-14} s.

In multilayer magnetic structures a significant contribution to the variation in magnetooptical characteristics of the second and next magnetic nanolayer is made by a new physical mechanism of magnetic switching [8,9]. That mechanism is related to a laser-induced spin current because of a photon pressure [11-15]. The laser-induced injection of spin-polarized electron from the first magnetic layer produces a nonequilibrium magnetic field H_i in the next layer. This field consists of the self-field H_e of the electric current and magnetic field H_s related to the total magnetic moment of the injection spins $S = \gamma\sum_i s_i$ [8]

$\overrightarrow{H_i} = \overrightarrow{H_e} + \overrightarrow{H_s}$. These magnetic fields have different directions: the field produced by electric current H_e lies in the plane of the film, while the field of the total spin moment of the electrons injected from the first magnetic layer H_s is directed along the magnetization $\overrightarrow{M_1}$ of the first magnetic layer. In case of perpendicular anisotropy the first magnetic layer H_s is directed normally to the second magnetic layer. The results of the article [8] showed that the relation $H_e / H_s \sim d_0$ (d_0 is the diameter of the current conductor). In case of the plane anisotropy of magnetic layers it is necessary to use nanometer magnetic structures and a very strong current [9]. To estimate the magnitude of the magnetic field H_e and H_s, we will use the following expression for the density of injection current emerging under the action of photon pressure of laser radiation [11, 14]

$$j_s = -e\alpha(1-R)I\frac{n_0\tau_p\eta\xi}{m_e c} \tag{7}$$

Since the laser pulse duration τ_i is larger or comparable with the pulse τ_p and the spin relaxation time τ_s in the magnetic nanolayer, the internal magnetic field H_s can be described by the estimating expression [15,18]

$$H_s = l_s\alpha(1-R)I\mu_B\mu\frac{n_0\tau_p\tau_s\gamma\eta\xi}{2\mu_0 h_2 m_e c} \tag{8}$$

The magnetic field H_e can be described by the estimating expression

$$H_e = l_e\alpha(1-R)Ier\frac{n_0\tau_p\gamma\xi}{4\pi m_e c} \tag{9}$$

where I, α and R and also n_0 are an intensity, absorption and reflection coefficients, and also refractive index of laser radiation falling on the first magnetic layer, respectively; r is the radius of a laser beam, h_1 and m_e are the thickness of the first magnetic nanolayer and effective electron mass; c is a light speed; $\gamma<1$, $\eta<1$ and $\xi<1$ are coefficients characterizing a momentum transfer from photons to electrons in the first magnetic layer and the degree of an electron polarization and affectivity of the electron passage from the first into the second nanolayer; μ_B is the Bohr magneton; μ and μ_0 are the a magnetic and absolute magnetic permeability; l_s and l_e are proportionality constant. We will estimate the value of the magnetic field, created in the second magnetic nanolaer due to the injection of the polarized electrons. At $I=100\ MW/cm^2$, $\alpha=10^5\ cm^{-1}$, $R=0,5$, $r=10^{-6}\ m$, $\tau_s=10^{-10}$-$10^{-11}\ s$, $\tau_p=10^{-11}$-$10^{-12}\ s$, $\mu=10^4$, $\gamma=0,8$, $\eta=0,8$, $\xi=0,5$ we can obtain the value $H_s>10^6\ A/m$ and $H_e=10^4$-$10^5\ A/m$. Thus theoretical estimates have shown that the laser-induced spin-polarized current can causes the magnetic switching of magnetic layers with perpendicular and planar uniaxial magnetic anisotropy. Generally, because of absorption of laser radiation and heating magnetic layer the value of magnetic moment $\vec{M}$ in the area of irradiation will change, the dynamics of variation of magnetization in a magnetic nanolayer can be described with the help of the Landau–Lifshitz-Gilbert equation of the form

$$\frac{d\vec{M}}{dt} = -\gamma\left[\vec{M}\times\vec{H}_{eff}\right] + \frac{k}{M}\left[\vec{M}\times\frac{\partial\vec{M}}{\partial t}\right] \qquad (10)$$

Here γ is the gyromagnetic ratio, k· is the dimensionless damping constant, though $0 < k < 1$, t is a time. The effective magnetic field can be represented as the sum $\vec{H}_{eff} = \vec{H}_{ext} + \vec{H}_{an} + \vec{H}_F + \vec{H}_i + \vec{H}_{dm}$ where H_{ext} is the consisting of external magnetic field, H_{an} is coercive force of the magnetic layer, H_F is magnetic field produced by circularly polarized laser radiation owing to the inverse Faraday effect, $\vec{H}_i = \vec{H}_e + \vec{H}_s$ is magnetic field of the laser-injected spin-polarized electron current and H_{dm} is effective demagnetizing field. The equilibrium magnetic orientation corresponds to the condition $dM/dt=0$.

It follows from formula (6) that the sign reversal of the effective magnetic field at the instant of action of laser pulses may change the direction of the magnetization vector $\vec{M}$ in the irradiated magnetic nanolayer owing to any of the mechanisms described above. The reason for such magnetization reversal may be thermal heating in an external magnetic field or demagnetizing field, the magnetic field induced by circularly polarized laser radiation owing to the inverse magnetooptical Faraday effect, and the magnetic field of spin current produced in the second magnetic layer by polarized electrons injected from the first magnetic layer. Such a spin current emerges under the effect of photon pressure of laser radiation. Through variation of the experimental conditions, it is possible to make any of these mechanisms predominant and use it for information recording as well as for developing high-speed optoelectronic switches and other spintronics devices.

3. The Magneto-Optical Measurements and Results

In the present paper we have studied magnetic switching dynamics of magnetic nanolayers in the film with a single magnetic nanolayer Al_2O_3 /$Tb_{22}Co_5Fe_{73}$ /Al_2O_3 and the films with two magnetic nanolayers Al_2O_3 /$Tb_{22}Co_5Fe_{73}$ /Pr_6O_{11} /$Tb_{19}Co_5Fe_{76}$ /Al_2O_3 and Al_2O_3 /$Co_{80}Fe_{20}$/Pr_6O_{11}/$Co_{30}Fe_{70}$ /Al_2O_3 radiated by nanosecond (τ_i=*15 ns*) and picosecond ($\tau_i \approx$ *80 ps)* pulses of the Nd-YAG laser *(*λ*=1,06* μ*m)* under external magnetic field. Besides, we studied the influence of the laser pulse on the conductivity of the tunnel microcontacts $Tb_{22}Co_5Fe_{73}$/Pr_6O_{11}/$Tb_{19}Co_5Fe_{76}$ and $Co_{80}Fe_{20}$/Pr_6O_{11}/$Co_{30}Fe_{70}$.

The films are sprayed by a magnetron deposition technique on plates with sizes *10x14 mm* and discs with the diameter *110 mm* from optical fused quartz with the thickness *1,2 mm.* Thicknesses of magnetic nanolayers TbCoFe and CoFe constituted *20 nm.* For the barrier nanolayer Pr_6O_{11} and the cover layer Al_2O_3 that thickness constitutes *2-3 nm* and *40 nm*, respectively.

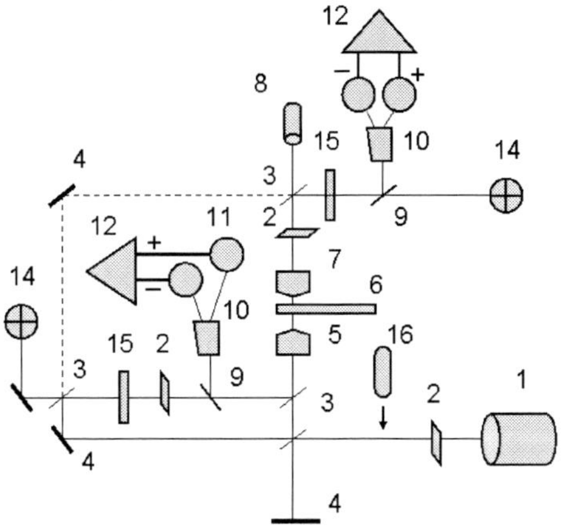

Figure 1. Experimental scheme of the optical investigation: Nd-YAG laser –1, polarizer –2, interference mirror –3, total reflection mirror –4, standard microscope objective –5, substrate with a film –6, special microscope objective –7, He-Ne laser, –8, semireflecting mirror –9, Senarmon prism –10, photodiode of reading –11, differential amplifier –12, light filter –13, photodiode with four active area –14, Babinet compensator –15.

The tunnel microcontacts with a conductive surface $S=20\ \mu^2$ are produced by a photolithography technique on the plates with sizes *10x14 mm*. The edge of plates through which the current was inputted to the tunnel contacts TbCoFe and CoFe is covered by platinum. The contact zone and conductive magnetic strips are also protected by the Al_2O_3 cover with thickness near 40 *nm*. Before measurements we carried out testing microcontacts with large and near tunnel resistances. The magnetic switching dynamics of the magnetic nanolayers in the films is studied with the help of the magnetooptical Kerr and Faraday effects on a polarization twisting of laser radiation. The scheme of corresponding optical researching is represented in Figure 1. The beam of the Nd-YAG laser ($\lambda=1{,}06\ \mu m$) with Gaussian energy distribution in its cross-section (TEM_{00}) passed through the polarizer 2 and through the *50 %* mirror 3, and then it was directed by *100 %* mirror 4 on the special microscope objective 5 with the numerical aperture *0.45*. This microscope objective focused laser radiation on a researched film through the substrate 6. Polarized radiation ($\lambda=630\ nm$) of the He-Ne laser 8 with Gaussian energy distribution in its cross-section (TEM_{00}) was focused by the microscope objective 7 with the numerical aperture *0.5* into that range of the film from its opposite side. Film reflected radiation of the Nd-YAG and He-Ne lasers with the help of interference mirrors 9 were directed on the polarization Senarmont prism 10, where they were separated into two beams and registered by reading photodiode 11. The electric signals from the reading photodiode are amplified by the differential amplifiers 12. Then these signals are registered by the double-beam oscilloscope. Reflected or transmitted through the film the laser beam can be directed to the polarization Senarmont prism by light filters 13. This light filter is used for registration of radiation of He-Ne or Nd-YAG lasers. The rotation of the laser radiation polarization plane upon reflection or transmission was measured with the help of the differential signal from photodiodes 11.The time resolution of the registration system constitutes *3 ns*. Signals input on the self-focusing microdrivers from photodiode 14 that provide constant focusing of the microscope objectives 5 and 7 on the film surface. By reposition the substrate with film we could direct the beam of the Nd-YAG laser on the film from the opposite side. For the research of the magnetic switching by circularly polarized laser pulses we introduced the Babinet compensator 15 into the system. For the time resolution enhancement of the system working with picosecond laser pulses the probing polarized beam of the Nd-YAG laser was formed by the system of 50 % mirrors 3 and 100 % mirrors 4. That beam with the controlled delay with respect to an excitation picosecond pulse could be focused into the researched area of the film structure both on the side of the excitation picosecond pulse and on the opposite film side. The rotation of the laser radiation polarization plane upon reflection or transmission through the film of such a probe beam was determined as changing of the signal amplitude registered by the reading photodiodes and oscilloscope. Such approach allows to study magnetic switching dynamics of magnetic nanolayers with temporary resolution up 5×10^{-11}s. The described scheme allows to study via changing of photoelectric signals (which are proportional to the Kerr and Faraday angles) the dynamic of magnetic switching of magnetic nanolayers in the films Al_2O_3 /$Tb_{22}Co_5Fe_{73}$ /Al_2O_3 and Al_2O_3 /$Tb_{22}Co_5Fe_{73}$ /Pr_6O_{11} /$Tb_{19}Co_5Fe_{76}$ /Al_2O_3 and also registration recording (magnetic switching) information in investigated magnetic nanolayers. For this purposes we used films on substrates in the form of optical discs. The Nd-YAG laser worked on repetition frequency *20 Hz*. The dynamic of magnetic switching of magnetic nanolayers in the films Al_2O_3 /$Co_{80}Fe_{20}$/Pr_6O_{11}/$Co_{30}Fe_{70}$ /Al_2O_3 under the laser pulses was studied on the variation of current transmitted through the tunnel contact $Co_{80}Fe_{20}$/Pr_6O_{11}/$Co_{30}Fe_{70}$. Amorphousness ferromagnetic films amorphous ferrimagnetic films of alloys of transition and rare-earth metals may operate as effective spin-current injectors with magnetic parameters that can be controlled by ultrashort laser pulses. The high energy of the perpendicular anisotropy and considerable coercive force [22-24] ensure a long lifetime for the magnetization state in such materials, which is close to saturation magnetization even in the zero external magnetic field. It should result in the high degree of an electron polarization. The strong temperature dependence of the coercive force and the low Curie temperature (near 300 K) make it possible not only to use such films for optical information recording, but also to develop optoelectronic microelements of spintronics on the basis of these films. The ferrimagnetic amorphous films $Tb_{22}Co_5Fe_{73}$ and $Tb_{19}Co_5Fe_{76}$ have a composition which is close the composition in compensation point $Tb_{22}Fe_{78}$, that provides high energy of the perpendicular anisotropy. A small distinction in composition (cobalt addition reduces egging and, virtually, does not influence magnetic characteristics of the layers) results in a constitutive difference in coercive force values [24]. At $T=300\ K$ the coercive forces of the layers $Tb_{22}Co_5Fe_{73}$ and $Tb_{19}Co_5Fe_{76}$ constitutes $H_1\approx 3x10^5\ A/m$ and $H_2\approx 1,2x10^5\ A/m$, respectively (Figure 2). The ferromagnetic films CoFe in the magnetized state also have a high degree of electron polarization [25]. In ours samples the substrate at film deposition of the nanolayers $Co_{30}Fe_{70}$ and $Co_{80}Fe_{20}$ was under the external magnetostatic field, that provides production of the films with a small angular dispersion ($\Delta\alpha \approx 3^0$) of planar single-axis magnetic anisotropy. The coercive force of the ferromagnetic nanolayers $Co_{30}Fe_{70}$ and $Co_{80}Fe_{20}$ differs appreciably because of distinctions of compositions in $Co_{30}Fe_{70}$ ($H'_1\approx 300\ A/m$) and $Co_{80}Fe_{20}$ ($H'_2\approx 800\ A/m$) (Figure 2). As a result we obtained the tunnel microcontact $Co_{80}Fe_{20}$/Pr_6O_{11}/$Co_{30}Fe_{70}$ with a small dispersion of magnetization axis. That allowed a magnetic switching of low coercive nanolayers $Tb_{19}Co_5Fe_{76}$ and $Co_{30}Fe_{70}$ by the external magnetic field without changing the magnetization direction of high coercive nanolayers $Tb_{22}Co_5Fe_{73}$ and $Co_{80}Fe_{20}$. Magnetic field direction coincided with an easy magnetization axis and its value varied from 0 to $8x10^5\ A/m$. The results of our research have shown that different mechanisms of the laser-induced magnetic switching of the magnetic nanolayers depend on the magnetic structure of the nanofilm and laser pulse duration and polarization. For single-layered magnetic films $Al_2O_3/Tb_{22}Co_5Fe_{73}/Al_2O$ the magnetic switching behavior of the nanolayer $Tb_{22}Co_5Fe_{73}$ under nanosecond laser pulses is in well agreement with the known thermomagnetic mechanisms of an information recording [22]. During radiation of the nanosecond Nd-YAG laser pulses the photoelectric signal I_R for radiation of a He–Ne laser reflected from the $Tb_{22}Co_5Fe_{73}$, which is proportional to the Kerr angle, varies even in zero magnetic field (Figure 3). First, the value of I_R changes at the trailing edge of a nanosecond laser pulse; upon an increase in the radiation intensity of this pulse, these values are shifted on the time axis towards the maximal value, and ΔI_R increases . In the external magnetic field antiparallel to the initial magnetization of the $Tb_{22}Co_5Fe_{73}$ layer, the value of I_R passes through zero and becomes negative, indicating magnetization reversal in the $Tb_{22}Co_5Fe_{73}$ layer in the region exposed to the nanosecond laser pulse. For a very high radiation intensity (close to the film disintegration threshold), we can observe in the nanosecond laser pulse magnetization reversal in $Tb_{22}Co_5Fe_{73}$ nanolayer even in the zero magnetic field. The characteristic time of changing of the value I_R(He-Ne) on the initial stage $\tau \geq 10^{-8}$s well conforms with the heat time for the film $Tb_{22}Co_5Fe_{73}$ and the temperature variation of the Kerr angle. The magnetic switching (transition via zero the signaled I_R(He-Ne) occurs faster than $\tau \leq 3\times10^{-9}$s.

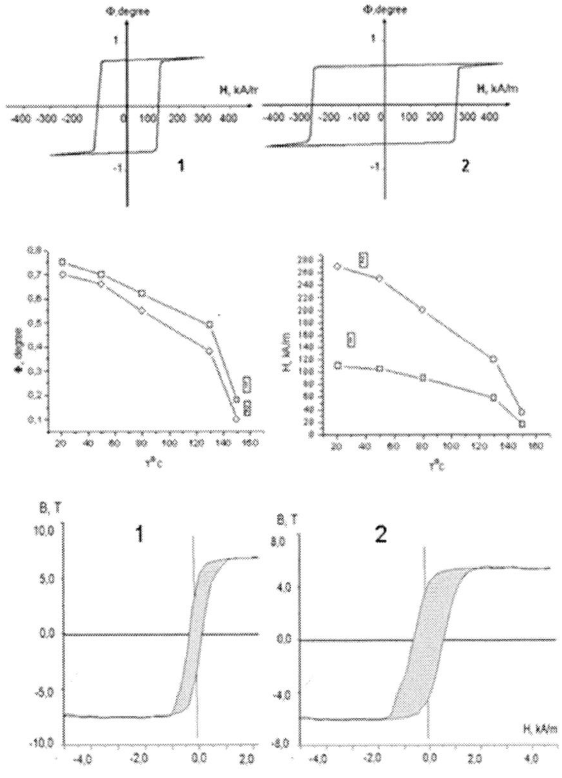

Figure 2. The curves of change of the Kerr angle (Φ), of the coercetive force (H) in nanolayers $Tb_{19}Co_5Fe_{76}$ (1) and $Tb_{22}Co_5Fe_{73}$ (2) and the B-H curves of nanolayers $Co_{30}Fe_{70}$ (1) and $Co_{80}Fe_{20}$ (2).

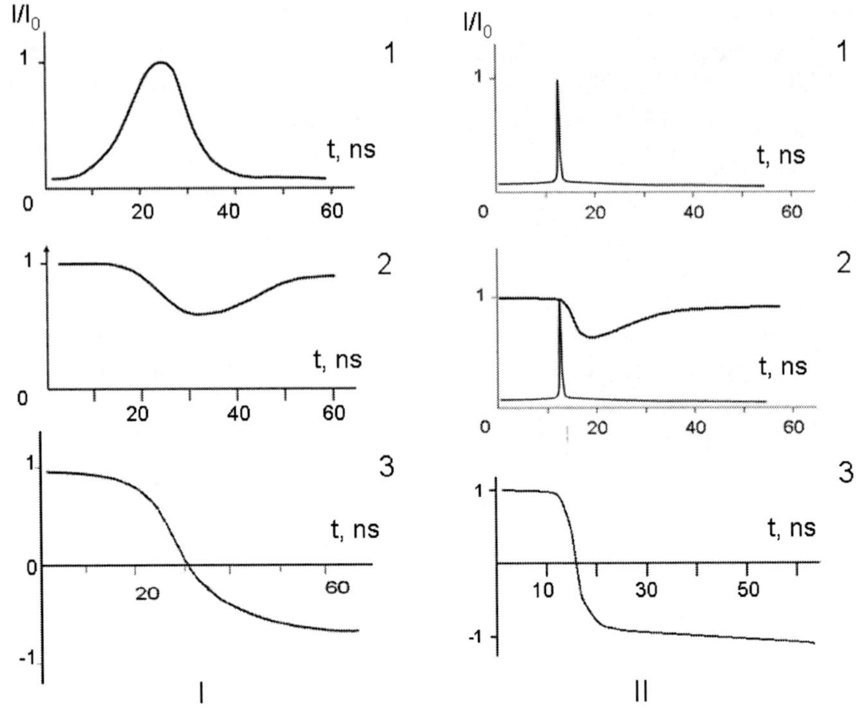

Figure 3. The change of photo-electric signal I_R of the reflected from film $Al_2O_3/Tb_{22}Co_5Fe_{73}/Al_2O_3$ radiation of He-Ne laser under the action nanosecond (τ_i=15 ns-I) and picosecond (τ_i=80 ps-II) pulse of Nd-YAG laser: 1- pulse of Nd-YAG laser, 2 and 3 - radiation of He-Ne laser; I-2 – linear polarization of radiation of Nd-YAG laser, I=0,2 MW/cm^2, external magnetic field H=0; I-3 –circular polarization of radiation of Nd-YAG laser, I=1 MW/cm^2 , external magnetic field H=3 kA/m; II-2 – linear polarization of radiation of Nd-YAG laser, I=300 MW/cm^2, external magnetic field H=0; II-3 – circular polarization of radiation of Nd-YAG laser, I=300 MW/cm^2, external magnetic field H=0.

After irradiation of the rotating substrate with an $Al_2O_3/Tb_{22}Co_5Fe_{73}/Al_2O_3$ film with a uniformly magnetized $Tb_{22}Co_5Fe_{73}$ nanolayer (in an external magnetic field antiparallel to the initial magnetization of the $Tb_{22}Co_5Fe_{73}$ layer) by a sequence of nanosecond pulses of the same power, the film acquires a sequence of point regions with magnetization opposite to the initial magnetization. This sequence of points can be read out using a He–Ne laser and a narrowband high-sensitivity amplifier if the substrate rotates with an elevated stable velocity.

The described results agree well with thermomagnetic switching behavior for which a laser radiation absorption results in the strong diminishing of the size of the coercive force of the nanolayer $Tb_{22}Co_5Fe_{73}$ and reorientation of a magnetic moment under the external magnetic field or a demagnetizing field, which emerges at film overheating in the radiated area above the Curie temperature (see the formula 1 and 3). Estimation according to formula 3 the demagnetizing field in the nanolayer $Tb_{22}Co_5Fe_{73}$ heated above the Curie temperature T_C gives the value $H_{dm} \approx 10^3$–$10^4 A/m$. After cooling of the nanolayer $Tb_{22}Co_5Fe_{73}$ (for which $T_c < 450$ K) the heated area is magnetized in the opposite direction of its initial magnetization. The magnetic switching of the nanolayer $Tb_{22}Co_5Fe_{73}$ in the films $Al_2O_3/Tb_{22}Co_5Fe_{73}/Al_2O_3$ can be produced under picosecond laser pulses not only on the thermomagnetic mechanism, but also under the effective internal magnetic field of the inverse magneto-optical Faraday effect generating by laser circularly polarized radiation. The laser-induced internal magnetic field (of the inverse magneto-optical Faraday effect) formed in the film $Al_2O_3/Tb_{22}Co_5Fe_{73}/Al_2O_3$ under circularly polarized picosecond pulses can be enough for magnetic switching of the nanolayer $Tb_{22}Co_5Fe_{73}$ even in the absence of the external magnetic field. The above estimation according to formula 6 shows that at the intensity of radiation of $I=10^9$ W/cm^2 the value of the magnetic field of the inverse Faraday effect for our magnetic films can achieve the large values of $H_F>10^5–10^6$ A/m, that considerably exceeds the value of the coercive force in these films. If the direction of the magnetic field of the inverse Faraday effect is opposite to the magnetization of a nanolayer $Tb_{22}Co_5Fe_{73}$ and its strength exceeds the coercive force, the magnetic switching occurs. Respectively, the photoelectric signal I_R(He-Ne) reflected from the nanofilm radiation of the Nd-YAG laser with the intensity I_{c0}(Nd-YAG) $\approx 10^9$ W/cm^2 passes through a zero and becomes negative (Figure 3). The magnetic switching of the nanolayer $Tb_{22}Co_5Fe_{73}$ results in a significant change of the photoelectric signal amplitude I_R(Nd-YAG) of the probe linear polarized picosecond pulse of the Nd-YAG laser (Figure 4). Dependence of the amplitude of this probe pulse on the time of its delay in relation to the excitation picosecond pulse also passes through a zero, and becomes negative. For the opposite circular polarization or linear polarization of picosecond Nd-YAG laser radiation the magnetic switching of the nanolayer $Tb_{22}Co_5Fe_{73}$ is not observed without an external magnetic field, even when the intensity of radiation becomes two times bigger. The analysis of the results of these measuring shows that the time of the magnetic switching of nanolayer $Tb_{22}Co_5Fe_{73}$ at an irradiation a circularly polarized picosecond laser pulse is smaller than $\tau \leq 10^{-10}$ s. The beginning of magnetization reversal of the $Tb_{22}Co_5Fe_{73}$ nanolayer by a picosecond laser pulse with circular polarization practically coincides in all cases with the instant of action of this pulse and has peculiar features. Depending on the direction of magnetization of the $Tb_{22}Co_5Fe_{73}$ nanolayer (antiparallel to the laser beam or parallel to it), magnetization reversal occurs for the right or the opposite left circular polarization of laser radiation.

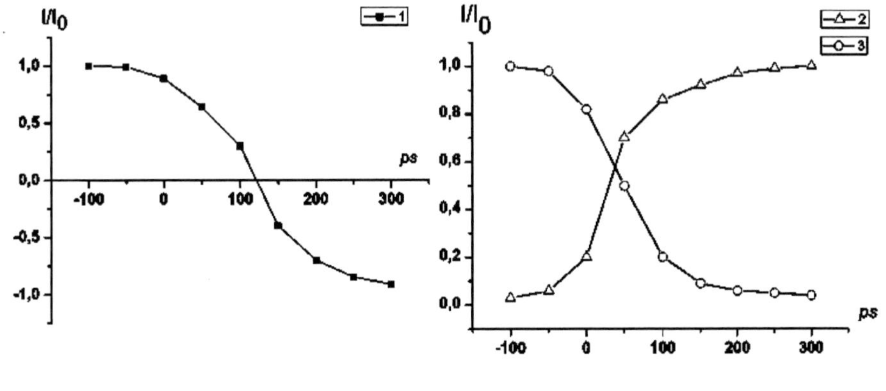

Figure 4. Change in the amplitude of the linearly polarized probe picosecond pulse of Nd-YAG depending on the time delay of this pulse with respect to the circularly polarized picosecond laser pulse causes the magnetization reversal of nanolayer $Tb_{22}Co_5Fe_{73}$ (1) and nanolayer $Tb_{19}Co_5Fe_{76}$ (2 and 3): 1 – probe laser pulse is reflected from the film $Al_2O_3/Tb_{22}Co_5Fe_{73}/Al_2O_3$; 2 and 3 –probe laser pulse passes through the film $Al_2O_3/Tb_{22}Co_5Fe_{73}\uparrow/Pr_6O_{11}/Tb_{19}Co_5Fe_{76}\downarrow/Al_2O_3$ (2) and $Al_2O_3/Tb_{22}Co_5Fe_{73}\uparrow/Pr_6O_{11}/Tb_{19}Co_5Fe_{76}\uparrow/Al_2O_3$ (3).

If magnetization reversal occurs for the right circular polarization of radiation, no magnetization reversal in a zero magnetic field is observed for the left circular polarization. If we reverse the direction of magnetization of the $Tb_{22}Co_5Fe_{73}$ nanolauer with the help of an external magnetic field the magnetic switching will be observed in the film for the left circular polarization of laser radiation. In the case of linear polarization of radiation of the same power of laser radiation, the signal intensity I_R(Nd-YAG) decreases slightly, but returns to its initial value at the end of the laser pulse. Irradiation by a sequence of circularly polarized picosecond pulses of a rotating substrate with an $Al_2O_3/Tb_{22}Co_5Fe_{73}/Al_2O_3$ film produces a sequence of point regions in it, which is detected with the help of a He–Ne laser. All these facts lead to the conclusion that irradiation of $Al_2O_3/Tb_{22}Co_5Fe_{73}/Al_2O_3$ films by picosecond laser pulses with circular polarization of radiation lead to the magnetic switching in the $Tb_{22}Co_5Fe_{73}$ nanolayer by the magnetic field of the laser pulse owing to the inverse Faraday effect. Therefore, a high-speed magnetic switching of magnetic nanolayers without an external magnetic field can be realized on the basis of the inverse magneto-optic Faraday effect with the help of irradiation by circularly polarized picosecond pulses. Heating of the $Tb_{22}Co_5Fe_{73}$ nanolayer by laser radiation facilitates such reversal magnetization owing to the decrease in of the coercive force. In the $Al_2O_3/Tb_{22}Co_5Fe_{73}/Pr_6O_{11}/Tb_{19}Co_5Fe_{76}/Al_2O_3$ film with two magnetic nanolayers $Tb_{22}Co_5Fe_{73}$ and $Tb_{19}Co_5Fe_{76}$, magnetization reversal at the thermomagnetic mechanism (presence of the external magnetic field directed in the opposite direction of magnetization in a magnetic nanolayer) occurs in the same way as in the $Al_2O_3/Tb_{22}Co_5Fe_{73}/Al_2O_3$ film. At the set value of the external magnetic field and set value of intensity of radiation in a nanosecond or picosecond pulse of Nd-YAG laser we can get magnetization reversal of one nanolayer $Tb_{19}Co_5Fe_{76}$ with a minor coercive force or for a higher intensity of laser radiation of two magnetic nanolayers $Tb_{22}Co_5Fe_{73}$ and $Tb_{19}Co_5Fe_{76}$ simultaneously. Without the external magnetic field, magnetization reversal of the first magnetic nanolayer (relative to the incident pulsed laser radiation) occurs in the $Al_2O_3/Tb_{22}Co_5Fe_{73}/Pr_6O_{11}/Tb_{19}Co_5Fe_{76}/Al_2O_3$ film in the same way as in the $Al_2O_3/Tb_{22}Co_5Fe_{73}/Al_2O_3$ film. Magnetization reversal of the second magnetic nanolayer depends not only on the pulse duration and the polarization of laser radiation, but also on the state of magnetization in both magnetic layers. In the film $Al_2O_3/Tb_{22}Co_5Fe_{73}\uparrow/Pr_6O_{11}/Tb_{19}Co_5Fe_{76}\uparrow/Al_2O_3$ with parallel magnetized magnetic nanolayers the laser-induced switching by circularly polarized picosecond pulses occurs in the result of the inverse Faraday effect. Depending on the laser radiation intensity magnetic switching can be realized only for one $Tb_{19}Co_5Fe_{76}$ nanolayer with a less coercive force, or simultaneous magnetic switching of two $Tb_{19}Co_5Fe_{76}$ and $Tb_{22}Co_5Fe_{73}$ nanolayers. At magnetic switching of the nanolayer $Tb_{19}Co_5Fe_{76}$ the film in the area of action of laser radiation passes from the state $Al_2O_3/Tb_{22}Co_5Fe_{73}\uparrow/Pr_6O_{11}/Tb_{19}Co_5Fe_{76}\uparrow/Al_2O_3$ with parallel magnetized magnetic nanolayers to the state $Al_2O_3/Tb_{22}Co_5Fe_{73}\uparrow/Pr_6O_{11}/Tb_{19}Co_5Fe_{76}\downarrow/Al_2O_3$ with antiparallel magnetized nanollayers. At magnetic switching of two $Tb_{19}Co_5Fe_{76}$ and $Tb_{22}Co_5Fe_{73}$ nanolayers the film in the area of action of laser radiation passes from the state $Al_2O_3/Tb_{22}Co_5Fe_{73}\uparrow/Pr_6O_{11}/Tb_{19}Co_5Fe_{76}\uparrow/Al_2O_3$ with parallel magnetized magnetic nanolayers to the state $Al_2O_3/Tb_{22}Co_5Fe_{73}\downarrow/Pr_6O_{11}/Tb_{19}Co_5Fe_{76}\downarrow/Al_2O_3$ with parallel magnetized magnetic nanolayers, but their magnetization is directed to the opposite side. Such switching is registered by the photoelectric signal I_T(He-Ne) for the radiation of the He-Ne laser passed through the film (Figure 5).

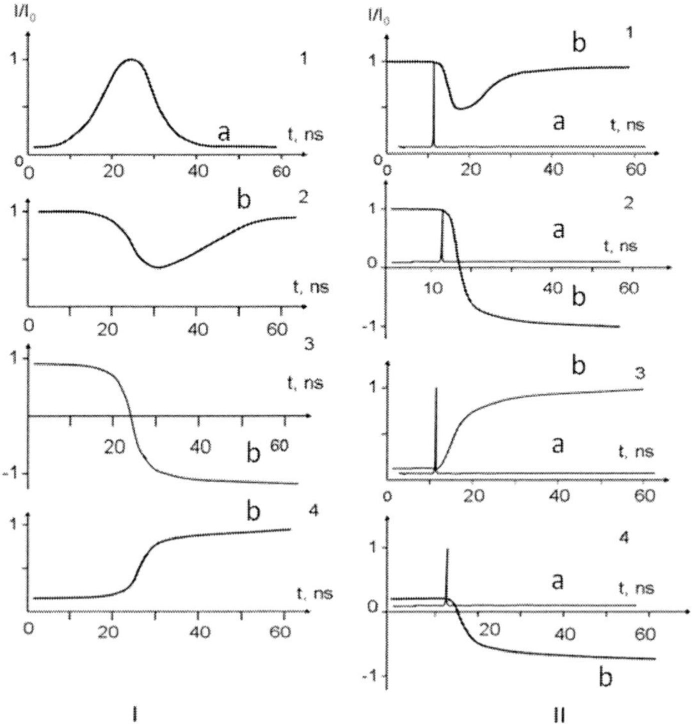

Figure 5. The change of the photoelectric signal of the helium-neon laser (b) is reflected from the layer $Tb_{19}Co_5Fe_{76}$ (I_R–I-2, I-3, II-1, II-2) and transmitted through film $Al_2O_3/Tb_{22}Co_5Fe_{73}\uparrow/Pr_6O_{11}/\downarrow Tb_{19}Co_5Fe_{76}/Al_2O_3$ (I_T–I-4, II-3, II-4) (2) upon irradiation this film with Nd-YAG laser pulses (a): I-2 and II-1 – linearly polarized laser pulse incident from the side of nanolayer $Tb_{19}Co_5Fe_{76}$; I-3, I-4 and II-2 –linearly polarization pulse incident from the side of nanolayer $Tb_{22}Co_5Fe_{73}$; II-3 and II-4 – right-hand circularly polarized pulse incident from the side of nanolayer $Tb_{22}Co_5Fe_{73}$, I=300 MW/cm^2; II-4 – left-hand circularly polarized pulse incident from the side of nanolayer $Tb_{22}Co_5Fe_{73}$, I=800 MW/cm^2.

At magnetic switching of one nanolayer the signal I_T(He-Ne) is decreased nearly twice, and at magnetic switching of two nanolayers it becomes negative. Temporal characteristics of such magnetic switching are close to dynamic behavior of the magnetic switching in the film $Al_2O_3/Tb_{22}Co_5Fe_{73}/Al_2O_3$ differing only in a power-level of the laser radiation.

The laser-induced magnetic switching without an external magnetic field in the film with antiparallel magnetized nanollayers $Al_2O_3/Tb_{22}Co_5Fe_{73}\uparrow/Pr_6O_{11}/Tb_{19}Co_5Fe_{76}\downarrow/Al_2O_3$ depends not only on the pulse duration and polarization but on the incidence direction of laser pulses in respect of the film. At the irradiation of this film from the side of the nanolayer $Tb_{19}Co_5Fe_{76}$ by the linearly polarized laser radiation we did not succeed in getting magnetic switching without an external magnetic field even for the nanolayer $Tb_{19}Co_5Fe_{76}$. The laser irradiation of the film $Al_2O_3/Tb_{22}Co_5Fe_{73}\uparrow/Pr_6O_{11}/Tb_{19}Co_5Fe_{76}\downarrow/Al_2O_3$ from the side of the nanolayer $Tb_{19}Co_5Fe_{76}$ by circularly polarized picosecond pulses results in magnetic switching of the input nanolayer $Tb_{19}Co_5Fe_{76}$ or the output nanolayer $Tb_{22}Co_5Fe_{73}$ (without an external magnetic field) depending on the direction of rotation of the plane of polarization of laser radiation.

At the right-handed laser polarization when the laser-induced internal magnetic field (the inverse magneto-optic Faraday effect) is directed along the magnetization of the output nanolayer $Tb_{22}Co_5Fe_{73}$ the magnetic switching only of the input nanolayer $Tb_{19}Co_5Fe_{76}$ occurs. As a result of such magnetic switching the film in the area of action of laser radiation passes from the state $Al_2O_3/Tb_{22}Co_5Fe_{73}\uparrow/Pr_6O_{11}/Tb_{19}Co_5Fe_{76}\downarrow/Al_2O_3$ with antiparallel magnetized nanollayers to the state $Al_2O_3/Tb_{22}Co_5Fe_{73}\uparrow/Pr_6O_{11}/Tb_{19}Co_5Fe_{76}\uparrow/Al_2O_3$ with parallel magnetized magnetic nanolayers. Such magnetic switching takes place at the intensity I_{c1} of radiation of Nd-YAG laser. It is accompanied by increasing of the photoelectric signal I_T of the He-Ne laser radiation transmitted through the film at a eventually constant signal I_R(He-Ne) related to the nanolayer $Tb_{22}Co_5Fe_{73}$ (Figure 5).

At the left-handed polarization of laser radiation, when the magnetic field of inverse Faraday effect is directed along the magnetization of the input nanolayer $Tb_{19}Co_5Fe_{76}$, the magnetic switching of the output nanolayer $Tb_{22}Co_5Fe_{73}$ occurs at a more high radiation intensity - I_{c2} (Nd-YAG) $\approx$ 2 I_{c1}(Nd-YAG). As a result the film in the area of action of laser radiation passes from the state $Al_2O_3/Tb_{22}Co_5Fe_{73}\uparrow/Pr_6O_{11}/Tb_{19}Co_5Fe_{76}\downarrow/Al_2O_3$ with the antiparallel magnetized nanollayers to the state $Al_2O_3/Tb_{22}Co_5Fe_{73}\downarrow/Pr_6O_{11}/Tb_{19}Co_5Fe_{76}\downarrow/Al_2O_3$ with the layers magnetized in the direction of magnetization of the input nanolayer $Tb_{19}Co_5Fe_{76}$. The appropriate photoelectric signal I_T of the He-Ne laser radiation transmitted through the film becomes negative and the signal I_R(He-Ne) of reflection from the nanolayer $Tb_{22}Co_5Fe_{73}$ nearly does not change (Figure 5).

For the laser radiation from the side of the nanolayer $Tb_{22}Co_5Fe_{73}$ the magnetic switching of the nanolayer $Tb_{19}Co_5Fe_{76}$ can occur, even for linearly polarized picosecond and nanosecond pulses without magnetic switching of the nanolayer $Tb_{22}Co_5Fe_{73}$. Such magnetic switching takes place at the radiation intensity I_{nl} (Nd-YAG) and I_{pl}(Nd-YAG). In this case the magnetic switching into the state $Al_2O_3/Tb_{22}Co_5Fe_{73}\uparrow/Pr_6O_{11}/Tb_{19}Co_5Fe_{76}\uparrow/Al_2O_3$ occurs in the film areas subjected to the irradiation by the Nd-YAG laser and it is accompanied by the increase of the photoelectric signal I_T(He-Ne) of the He-Ne laser radiation transmitted through the film. The value of the signal I_R(He-Ne) (radiation of the He-Ne laser reflected from the nanolayer $Tb_{22}Co_5Fe_{73}$) does not virtually change.

The laser irradiation of the film $Al_2O_3/Tb_{22}Co_5Fe_{73}\uparrow/Pr_6O_{11}/Tb_{19}Co_5Fe_{76}\downarrow/Al_2O_3$ from the side of the nanolayer $Tb_{22}Co_5Fe_{73}$ by circularly polarized picosecond pulses results in magnetic switching of the output nanolayer $Tb_{19}Co_5Fe_{76}$ or the input nanolayer $Tb_{22}Co_5Fe_{73}$ (without external magnetic field) depending on the direction of rotation of the polarization plane of laser radiation. At the right-handed laser polarization when the laser-induced internal magnetic field (the inverse magneto-optic Faraday effect) is directed along the magnetization of the input nanolayer $Tb_{22}Co_5Fe_{73}$ a magnetic switching of only the output nanolayer $Tb_{19}Co_5Fe_{76}$ occurs. As a result of such magnetic switching this film passes from the state $Al_2O_3/Tb_{22}Co_5Fe_{73}\uparrow/Pr_6O_{11}/Tb_{19}Co_5Fe_{76}\downarrow/Al_2O_3$ to the state $Al_2O_3/Tb_{22}Co_5Fe_{73}\uparrow/Pr_6O_{11}/Tb_{19}Co_5Fe_{76}\uparrow/Al_2O_3$.

Such magnetic switching takes place at the radiation intensity I_{c3}(Nd-YAG). The value of this intensity is smaller than for the case of the linear polarization of radiation of laser picosecond pulses and smaller as well for the case of irradiation of this film by circularly polarized picosecond pulses from the side of the nanolayer $Tb_{19}Co_5Fe_{76}$: I_{c3}(Nd-YAG)< I_{pl}(Nd-YAG and I_{c3}(Nd-YAG)< I_{c1}(Nd-YAG). It is accompanied by increasing of the photoelectric signal I_T(He-Ne) of the He-Ne laser radiation transmitted through the film at an eventually constant signal I_R(He-Ne) related to the nanolayer $Tb_{22}Co_5Fe_{73}$.

At the left-handed laser polarization of laser radiation when the magnetic field of inverse Faraday effect is directed along the magnetization of the output nanolayer $Tb_{19}Co_5Fe_{76}$ the magnetic switching of the input nanolayer $Tb_{22}Co_5Fe_{73}$ occurs at a more high radiation intensity I_{c4} (Nd-YAG) ≈2 I_{c3}(Nd-YAG. As a result, the film in the area of action of laser radiation passes from the state $Al_2O_3/Tb_{22}Co_5Fe_{73}\uparrow/Pr_6O_{11}/Tb_{19}Co_5Fe_{76}\downarrow/Al_2O_3$ with parallel magnetized magnetic nanolayers to the state $Al_2O_3/Tb_{22}Co_5Fe_{73}\downarrow/Pr_6O_{11}/Tb_{19}Co_5Fe_{76}\downarrow/Al_2O_3$ with the layers magnetized along the direction of magnetization of the input nanolayer $Tb_{19}Co_5Fe_{76}$. The appropriate photoelectric signal I_T(He-Ne) of the He-Ne laser radiation transmitted through the film becomes negative and the signal I_R(He-Ne) of reflection from the nanolayer $Tb_{22}Co_5Fe_{73}$ nearly does not change.

The dynamics of the laser-induced magnetic switching in the film $Al_2O_3/Tb_{22}Co_5Fe_{73}\uparrow/Pr_6O_{11}/Tb_{19}Co_5Fe_{76}\downarrow/Al_2O_3$ under the circularly polarized picosecond laser pulses from the side of the nanolayer $Tb_{22}Co_5Fe_{73}$ are similar to the case of the laser radiation from the side of the nanolayer $Tb_{19}Co_5Fe_{76}$ differing only by a laser power.

The results of the investigations of the films $Al_2O_3/Tb_{22}Co_5Fe_{73}\uparrow/Pr_6O_{11}/Tb_{19}Co_5Fe_{76}\downarrow/Al_2O_3$ with antiparallel nanolayer magnetizations show that a new physical mechanism of magnetic switching occurs in them. That mechanism is related to laser-induced spin current caused by a photon pressure. The laser-induced injection of spin-polarized electron from the layer $Tb_{22}Co_5Fe_{73}$ through the tunnel barrier results in the nonequilibrium magnetic field H_i. As it was shown above this field consists of the self-field H_e of the electric current and magnetic field H_s related to the total magnetic moment of the injection spins $\vec{H_i} = \vec{H_e} + \vec{H_s}$. The self-field of the electric current H_e is directed to the film plane and the field of the total spin moment of injecting electrons is perpendicular to the film plane H_s. Our films are characterized by a large perpendicular anisotropy. Therefore, the effect of the magnetic field H_e on magnetic switching of the nanolayers is minimal. The conducted estimation of the field (8) size shows that a spin current can result in magnetic switching of the nanolayer $Tb_{19}Co_5Fe_{76}$. The results of experimental researches of magnetic switching of the film $Al_2O_3/Tb_{22}Co_5Fe_{73}\uparrow/Pr_6O_{11}/Tb_{19}Co_5Fe_{76}\downarrow/Al_2O_3$ at its radiation by nanosecond and picosecond laser pulses certify this conclusion well.

The investigation of the tunnel microcontacts $Co_{80}Fe_{20}/Pr_6O_{11}/Co_{30}Fe_{70}$ has shown, that the laser-induced switching can be realized not only for the case of perpendicular but also uniaxial planar magnetic anisotropy of the magnetic nanolayers. However, in the latter case the magnetic switching technique is very limited. As our research showed, the magnetic reversal of one of the electrodes changes the value of the tunnel resistance in microcontacts $Co_{80}Fe_{20}/Pr_6O_{11}/Co_{30}Fe_{70}$ almost in two times at the temperature $T=80\ K$ and in one and a half times at the temperature $T=300K$. In microcontacts $Tb_{22}Co_5Fe_{73}/Pr_6O_{11}/Tb_{19}Co_5Fe_{76}$ the similar changes of resistance reach large values (Figure 6).

If to determine the value of tunnel magnetoresistance (TMR) as [26]

$$TMR = (R_{max} - R_{min}) / R_{min} \tag{11}$$

where R_{max} and R_{min} are the maximal and minimum values of resistance, we will get for microcontacts $Tb_{22}Co_5Fe_{73}/Pr_6O_{11}/Tb_{19}Co_5Fe_{76}$ at *T=300 K* value of TMR>70% and TMR>240% at *T=80 K*. For microcontacts $Co_{80}Fe_{20}/Pr_6O_{11}/Co_{30}Fe_{70}$ values of TMR will be smaller and reach the values of TMR ≈25% at *T=300 K* and TMR≈100% at *T=80 K*.

The research of influencing of laser radiation on resistance of microcontacts $Co_{80}Fe_{20}/Pr_6O_{11}/Co_{30}Fe_{70}$ at the absence of magnetic field showed that at small intensity of laser radiation the resistance of microcontacts falls at the moment of action of laser pulse, but after finishing of the pulse it returns back to the practically initial value. At a large intensity of laser radiation the resistance of the microcontact $Co_{80}Fe_{20}/Pr_6O_{11}/Co_{30}Fe_{70}$ changes after the end of the laser pulse and this change depends on the state of magnetization of magnetic layers, on the intensity, and even on the incidence direction of laser pulses on the contact.

In the external magnetic field directed towards the magnetization of the low-coercitive nanolayer $Co_{30}Fe_{70}$, the resistance of the microcontacts $Co_{80}Fe_{20}\uparrow/Pr_6O_{11}/\uparrow Co_{30}Fe_{70}$ after an radiation a powerful laser pulse increases on the value ΔR_0. The resistance of the microcontacts $Co_{80}Fe_{20}\uparrow/Pr_6O_{11}/\downarrow Co_{30}Fe_{70}$ with antiparallel magnetizations of the nanolayers after an irradiation in the same magnetic field decrease on the value $\Delta R \approx \Delta R_0$. Such a change of resistance of the microcontacts $Co_{80}Fe_{20}/Pr_6O_{11}/Co_{30}Fe_{70}$ is conditioned by the magnetic switching of nanolayer $Co_{30}Fe_{70}$. The obtained change of resistance of contacts $Co_{80}Fe_{20}/Pr_6O_{11}/Co_{30}Fe_{70}$ ΔR_0 on a value is close to the similar change of resistances that we observed during the magnetic switching of the same contacts only by the external magnetic field.

In the microcontacts $Co_{80}Fe_{20}\uparrow/Pr_6O_{11}/\uparrow Co_{30}Fe_{70}$ with parallel magnetizations of the nanolayers in the absence of the magnetic field it is impossible to get the change of resistance after radiation of a laser pulse. For such magnetic switching in the microcontact $Co_{80}Fe_{20}\uparrow/Pr_6O_{11}/\uparrow Co_{30}Fe_{70}$ it is necessary to select laser radiation intensity and external magnetic field carefully.

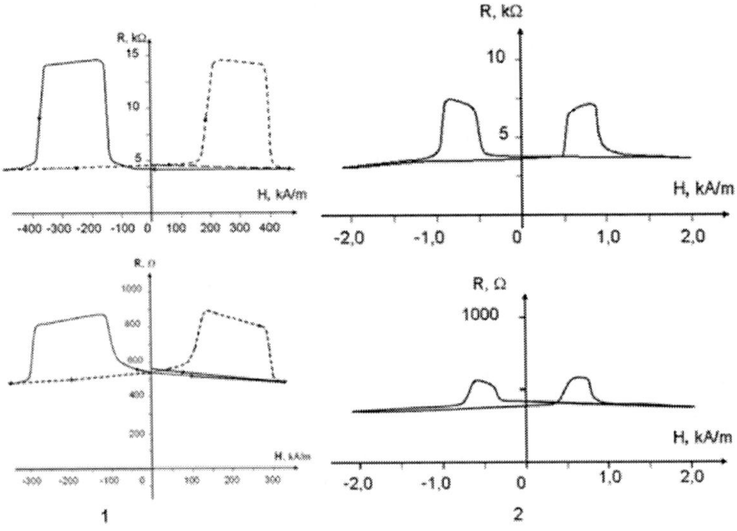

Figure 6. The change of resistance of tunnel contact $Tb_{22}Co_5Fe_{73}/Pr_6O_{11}/Tb_{19}Co_5Fe_{76}$ (1) and $Co_{80}Fe_{20}/Pr_6O_{11}/Co_{30}Fe_{70}$ (2) at their magnetic switching of an external magnetic field at T=300K (down) and T=80K (up).

In the microcontacts $Co_{80}Fe_{20}\uparrow/Pr_6O_{11}/\downarrow Co_{30}Fe_{70}$ with antiparallel magnetizations of the nanolayers it is possible to get the change of resistance even in default of the magnetic field at an radiation of the nanosecond and picosecond pulses of lasers. The change of resistance in such microcontacts occurs only for the laser pulses directed from the side of the nanolayer $Co_{80}Fe_{20}$. For the laser pulses directed from the side of the nanolayer $Co_{30}Fe_{70}$ the change of resistance does not take place. At a small intensity of radiation directed to the microcontact $Co_{80}Fe_{20}\uparrow/Pr_6O_{11}/\downarrow Co_{30}Fe_{70}$ from the side of the nanolayer $Co_{80}Fe_{20}$ its resistance in the moment of action of laser pulse diminishes in one or two times but later it returns to the initial value. At a large laser radiation the microcontact resistance in the laser pulse end becomes smaller than its initial value R_1 on the amount $\Delta R_1 \leq \Delta R_0$ (Figure 7), that denotes the magnetic switching of the nanolayer $Co_{30}Fe_{70}$ and the passage of the microcontact from the state $Co_{80}Fe_{20}\uparrow/Pr_6O_{11}/\downarrow Co_{30}Fe_{70}$ to the state $Co_{80}Fe_{20}\uparrow/Pr_6O_{11}/\uparrow Co_{30}Fe_{70}$.

The obtained results show that the magnetic switching of the low-coercive nanolayer $Co_{30}Fe_{70}$ without the external magnetic field takes place under the action of the internal effective magnetic field $\overrightarrow{H_i} = \overrightarrow{H_e} + \overrightarrow{H_s}$ laser-induced spin injection through the tunnel barrier $Pr_6O_{11.}$ The magnetic layers in the microcontacts $Co_{80}Fe_{20}/Pr_6O_{11}/Co_{30}Fe_{70}$ have a planar single-axis magnetic anisotropy, therefore the self-field H_e of the electric current and the magnetic field H_s related to the total magnetic moment of the injection spins have the direction in the plane of nanolayer $Co_{30}Fe_{70}$. In this case the value of the internal effective magnetic field is the sum of these two fields $H_i = k_sH_s + k_eHe$, where где k_s и k_e are constants.

Conducted by us the theoretical estimation (see formulas 8 and 9) shows that the values of such magnetic fields for the case of the microcontacts $Co_{80}Fe_{20}/Pr_6O_{11}/Co_{30}Fe_{70}$ have one order of magnitude. Total internal effective magnetic field H_i is more the coercive force of nanolauer $Co_{30}Fe_{70}$, that results in his magnetic switching at the irradiation microcontacts $Co_{80}Fe_{20}/Pr_6O_{11}/Co_{30}Fe_{70}$ by laser pulses from the side of nanolayer $Co_{80}Fe_{20}$.

The large variety of physical mechanisms of magnetic switching of magnetic nanolayers in the tunnel contacts $Tb_{19}Co_5Fe_{76}/Pr_6O_{11}/Tb_{22}Co_5Fe_{73}$ result in wide capabilities for their control by the short laser pulses. The known dependences of the tunnel resistance of such microcontacts on external magnetic field and thickness of the tunnel nanolayer [4] showed that that tunnel structures have well perspective for spintronic.

The laser-induced magnetic switching in the microcontacts $Tb_{22}Co_5Fe_{73}\uparrow/Pr_6O_{11}/\uparrow Tb_{19}Co_5Fe_{76}$ with parallel magnetizations of nanolayers is related to thermomagnetic and magneto-optic mechanism of the inverse Faraday effect at circularly polarized laser pulses. It is impossible to get the change of resistance in such microcontacts $Tb_{22}Co_5Fe_{73}\uparrow/Pr_6O_{11}/\uparrow Tb_{19}Co_5Fe_{76}$ after an irradiation the lasers pulses with linear polarization of radiation at the absence of the magnetic field.

The laser-induced thermomagnetic switching of the low-coercive nanolayer $Tb_{19}Co_5Fe_{76}$ occurs when external magnetic field is directed antiparallel to the magnetization of this layer. At the small laser radiation intensity the resistance of the microcontact $Tb_{22}Co_5Fe_{73}\uparrow/Pr_6O_{11}/\uparrow Tb_{19}Co_5Fe_{76}$ is decreased in some times in the initial time and subsequently it reverts to its initial value R_2. At greater laser radiation intensity the microcontact resistance decreases also in the initial time of the laser pulse, but at some laser radiation intensity I_1 after laser pulse end the microcontact resistance is increased on the value ΔR_2 (Figure 7).

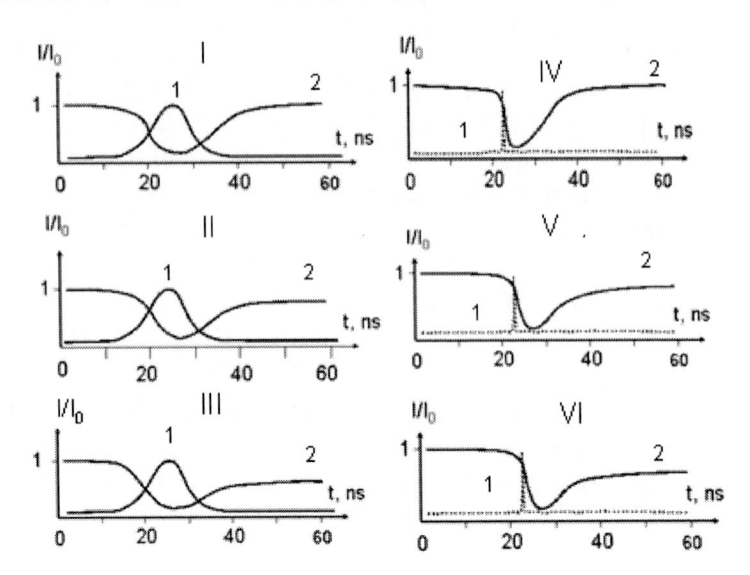

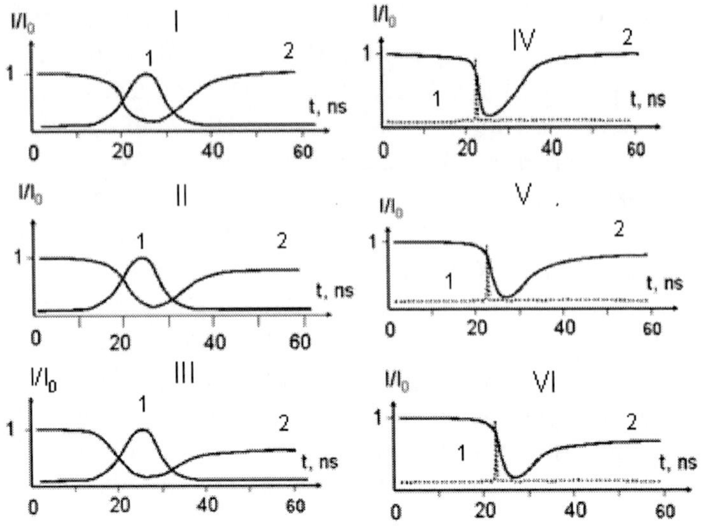

Figure 7. The change of resistance (curve 2) of the contact $Tb_{22}Co_5Fe_{73}/Pr_6O_{11}/Tb_{19}Co_5Fe_{76}$ (III, IV, VI) and $Co_{80}Fe_{20}/Pr_6O_{11}/Co_{30}Fe_{70}$ (I, II, V) at an radiation without external magnetic field by linearly polarized pulse of Nd-YAG laser (curve 1) from the side of low-coercive nanolayers ($Tb_{19}Co_5Fe_{76}$ and $Co_{30}Fe_{70}$–I, IV) and high-coercivity nanolayers ($Tb_{22}Co_5Fe_{73}$ and $Co_{80}Fe_{20}$–I,II, III, V, VI): I –I_i=10 MW/cm^2, T = 80 K ; II –I_i=10 MW/cm^2, T = 80 K; III –I_i=5 MW/cm^2, T = 80K; IV –I_i=300 MW/cm^2, T = 80K; V –I_i=1 GW/cm^2, T = 80K; VI –I_i=300 MW/cm^2, T = 80K.

Increasing of the resistance in the result of the thermomagnetic switching of the low-coercive nanolayer $Tb_{19}Co_5Fe_{76}$ is near to its change in external field $\Delta R_2 \approx \Delta R_0$. At yet greater laser radiation intensity $I_2 > I_1$ when external magnetic field is directed antiparallel to the nanolauers magnetization the resistance of the microcontact $Tb_{22}Co_5Fe_{73}\uparrow/Pr_6O_{11}/\uparrow Tb_{19}Co_5Fe_{76}$ is decreased but after end of laser pulse again goes back to the primary value R_2. Reason of such change conductivity of contact is transition from the state $Tb_{22}Co_5Fe_{73}\uparrow/Pr_6O_{11}/Tb_{19}Co_5Fe_{76}\uparrow$ with parallel magnetized magnetic nanolayers in the state $Tb_{22}Co_5Fe_{73}\downarrow/Pr_6O_{11}/$ $Tb_{19}Co_5Fe_{76}\downarrow$ also with parallel magnetized magnetic nanolayers in the direction of initial magnetized of nanolayer $Tb_{22}Co_5Fe_{73}$.

Under picosecond circularly polarized laser pulses occurs the change of resistance in the microcontact $Tb_{22}Co_5Fe_{73}\uparrow/Pr_6O_{11}/\uparrow Tb_{19}Co_5Fe_{76}$ without the external magnetic field. The change of resistance was observed for laser pulses directed from the side of the nanolayer $Tb_{19}Co_5Fe_{76}$. At the right-handed laser polarization when the laser-induced internal magnetic field (the inverse magneto-optic Faraday effect) is directed antiparallel to the magnetization of the this nanolayer was increased the microcontact resistance on the value ΔR_3 (Figure 8). The magnitude ΔR_3 is some less the appropriate resistance change under external magnetic field ($\Delta R_3 < \Delta R_2$). The change of resistance did not occur in the microcontact $Tb_{22}Co_5Fe_{73}\uparrow/Pr_6O_{11}/\uparrow Tb_{19}Co_5Fe_{76}$ under picosecond laser pulses with the left-handed circular polarization of radiation when external magnetic field is absent.

Increasing of the resistance in the result of the thermomagnetic switching of the low-coercive nanolayer $Tb_{19}Co_5Fe_{76}$ is near to its change in external field $\Delta R_2 \approx \Delta R_0$. At yet greater laser radiation intensity $I_2 > I_1$ when external magnetic field is directed antiparallel to the nanolauers magnetization the resistance of the microcontact $Tb_{22}Co_5Fe_{73}\uparrow/$ $Pr_6O_{11}/\uparrow Tb_{19}Co_5Fe_{76}$ is decreased but after end of laser pulse again goes back to the primary value R_2. Reason of such change conductivity of contact is transition from the state $Tb_{22}Co_5Fe_{73}\uparrow/Pr_6O_{11}/Tb_{19}Co_5Fe_{76}\uparrow$ with parallel magnetized magnetic nanolayers in the state $Tb_{22}Co_5Fe_{73}\downarrow/Pr_6O_{11}/$ $Tb_{19}Co_5Fe_{76}\downarrow$ also with parallel magnetized magnetic nanolayers in the direction of initial magnetized of nanolayer $Tb_{22}Co_5Fe_{73}$.

Under picosecond circularly polarized laser pulses occurs the change of resistance in the microcontact $Tb_{22}Co_5Fe_{73}\uparrow/Pr_6O_{11}/$ $\uparrow Tb_{19}Co_5Fe_{76}$ without the external magnetic field. The change of resistance was observed for laser pulses directed from the side of the nanolayer $Tb_{19}Co_5Fe_{76}$. At the right-handed laser polarization when the laser-induced internal magnetic field (the inverse magneto-optic Faraday effect) is directed antiparallel to the magnetization of the this nanolayer was increased the microcontact resistance on the value ΔR_3. The magnitude ΔR_3 is some less the appropriate resistance change under external magnetic field (ΔR_3 $<\Delta R_2$). The change of resistance did not occur in the microcontact $Tb_{22}Co_5Fe_{73}\uparrow/Pr_6O_{11}/\uparrow Tb_{19}Co_5Fe_{76}$ under picosecond laser pulses with the left-handed circular polarization of radiation when external magnetic field is absent.

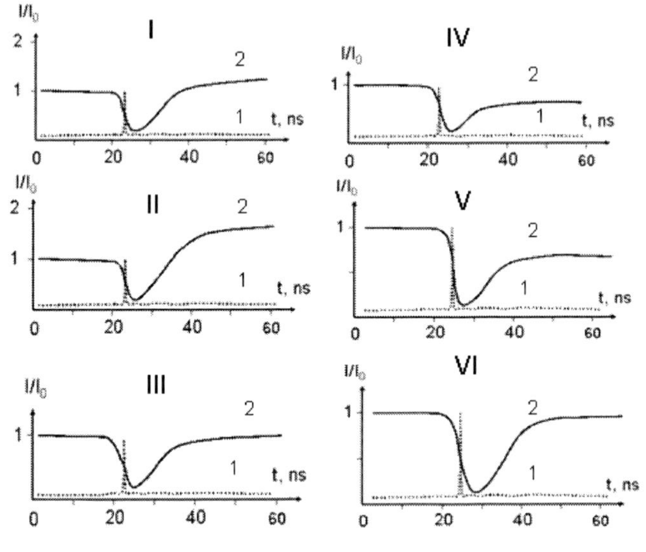

Figure 8. The change of resistance (curve 2) of the contact $Tb_{22}Co_5Fe_{73}\uparrow/Pr_6O_{11}/\uparrow Tb_{19}Co_5Fe_{76}$ (I-III) and $Tb_{22}Co_5Fe_{73}\uparrow/Pr_6O_{11}/\downarrow Tb_{19}Co_5Fe_{76}$ (IV-VI) at an radiation from the side of layer of $Tb_{19}Co_5Fe_{76}$ by the circular polarized picosecond pulse of Nd-YAG laser (curve 1) with different direction of rotation (I-II – right-hand, III-V – left-hand, VI – right-hand) of plane of polarization: I – radiation intensity I =1 GW/cm^{2,} T = 300 K; II - I =3 GW/cm^2, T = 80 K; III –I=1 GW/cm^2, T = 300 K; IV – I=1 GW/cm^2, T = 300 K; V – I=3 GW/cm^2, T = 80 K; VI – I=8 GW/cm^2.

The change of resistance was observed in the microcontact $Tb_{22}Co_5Fe_{73}\uparrow/$ $Pr_6O_{11}/\downarrow Tb_{19}Co_5Fe_{76}$ under laser radiation dependences on a pulse polarization and the direction of incidence of the laser pulse on the microcontact. For linearly polarized pulsed laser radiation directed from the side of the low-coercive nanolayer $Tb_{19}Co_5Fe_7$ the change of resistance without external magnetic field does not occur. For the same pulses directed from the side of the highly-coercive nanolayer $Tb_{22}Co_5Fe_{73}$ was observed the resistance reduction of the microcontact $Tb_{22}Co_5Fe_{73}\uparrow/Pr_6O_{11}/\downarrow Tb_{19}Co_5Fe_{76}$ on the value ΔR_4 without external magnetic field. The value ΔR_4 approximately equal to the value ΔR_3. Such changes of resistance of contacts $Tb_{22}Co_5Fe_{73}$↑/Pr_6O_{11}/↓$Tb_{19}Co_5Fe_{76}$ after the irradiation the pulses of lasers are caused magnetic switching nanolayer $Tb_{19}Co_5Fe_{76}$ and passes from the state $Tb_{22}Co_5Fe_{73}$↑/Pr_6O_{11}/↓$Tb_{19}Co_5Fe_{76}$ with antiparallel magnetized nanolayers in the state $Tb_{22}Co_5Fe_{73}$↑/Pr_6O_{11}/↑$Tb_{19}Co_5Fe_{76}$ with parallel magnetized magnetic nanolayers. The magnetic switching of the low-coercive nanolayer $Tb_{19}Co_5Fe_{76}$ without the external magnetic field takes place under the action of the internal effective magnetic field H_s related to the total magnetic moment of the injection through the tunnel barrier Pr_6O_{11} polarized electrons.

For the circular polarization of picosecond laser pulses which induce the magnetic field antiparallel to the initial magnetization of the nanolayer $Tb_{19}Co_5Fe_{76}$ the resistance of microcontact $Tb_{22}Co_5Fe_{73}$↑/Pr_6O_{11}/↓$Tb_{19}Co_5Fe_{76}$ R_4 decreases on the value $\Delta R_5 \approx \Delta R_4$ without external magnetic field at the irradiation of this contact from any direction. At the irradiation of this contact from the side of the highly-coercive nanolayer $Tb_{22}Co_5Fe_{73}$ the his resistance reduction on the value ΔR_5 takes place at some laser radiation intensity I_3. At the irradiation of this contact from the side of the low-coercive nanolayer $Tb_{19}Co_5Fe_7$ the his resistance reduction on the value ΔR_5 takes place at some laser radiation intensity $I_4 > I_3$. Such changes of resistance of contacts $Tb_{22}Co_5Fe_{73}$↑/Pr_6O_{11}/↓$Tb_{19}Co_5Fe_{76}$ after the irradiation the circular polarization laser pulses are caused magnetic switching nanolayer $Tb_{19}Co_5Fe_{76}$ of the laser-induced magnetic field (the inverse magneto-optic Faraday effect) and passes from the state $Tb_{22}Co_5Fe_{73}$↑/Pr_6O_{11}/↓$Tb_{19}Co_5Fe_{76}$ with antiparallel magnetized nanolayers in the state $Tb_{22}Co_5Fe_{73}$↑/Pr_6O_{11}/↑$Tb_{19}Co_5Fe_{76}$ with parallel magnetized magnetic nanolayers.

At the opposite rotation of polarization plane of circular polarization picosecond laser pulses the change of resistance of microcontact $Tb_{22}Co_5Fe_{73}$↑/Pr_6O_{11}/↓$Tb_{19}Co_5Fe_{76}$ takes place at more high intensity of laser radiation. At the irradiation of this contact from the side of the nanolayer $Tb_{22}Co_5Fe_{73}$ was increased this resistance on the value $\Delta R_6 \approx \Delta R_5$ at radiation intensity $I_5 > I_4$. At the irradiation of this contact from the side of the nanolayer $Tb_{19}Co_5Fe_7$ was increased this resistance on the value $\Delta R_6 \approx \Delta R_5$ at radiation intensity $I_6 > I_5$. Such changes of resistance are it is related to the transition of microcontacts from the state $Tb_{22}Co_5Fe_{73}$↑/Pr_6O_{11}/↓$Tb_{19}Co_5Fe_{76}$ with antiparallel magnetized nanolayers in the state $Tb_{22}Co_5Fe_{73}$↓/Pr_6O_{11}/↓$Tb_{19}Co_5Fe_{76}$ with parallel magnetized magnetic nanolayers in the direction of initial magnetized of nanolayer $Tb_{19}Co_5Fe_{76}$.

The difference in intensities of radiation of the circular polarization picosecond laser pulses inducing a change conductivity of contacts $Tb_{22}Co_5Fe_{73}/Pr_6O_{11}/Tb_{19}Co_5Fe_{76}$ explained those, that the magnetic switching of the input magnetic nanolayer takes place only the magnetic field of inverse Faraday effect H_F, and the magnetic switching of the output magnetic nanolayer the sum of this field and field H_s related to the total magnetic moment of the injection spins.

Conclusions

Thus the obtained results have shown that magnetic switching effects in the nanolayers under nanosecond and picosecond laser pulses can be used for creation elements and systems of high-speed controlling of spin currents with the response time $\tau \leq 10^{-10} s$. For effective work of such optical controlled spin elements it is necessary not only to develop a design and to choose a material with given characteristics but to set duration, intensity and polarization of controlling laser radiation.

Thus in the tunnel microcontact $Co_{80}Fe_{20}\uparrow/Pr_6O_{11}/\uparrow Co_{30}Fe_{70}$ with parallel magnetized nanolayers, which have the planar uniaxial anisotropy, the change of resistance under laser pulses is possible only under external magnetic field. In this case increasing of the microcontact resistance occurs via laser-induced thermomagnetic switching of the low-coercive nanolayer $Co_{30}Fe_{70}$ and passage of the microcontact in the state $Co_{80}Fe_{20}\uparrow/ Pr_6O_{11}/\downarrow Co_{30}Fe_{70}$. For the opposite magnetization of nanolayers in the given microcontact $Co_{80}Fe_{20}\uparrow/Pr_6O_{11}/\downarrow Co_{30}Fe_{70}$ the change of resistance can be realized in the low-resistance state $Co_{80}Fe_{20}\uparrow/Pr_6O_{11}/\uparrow Co_{30}Fe_{70}$ without external magnetic field under laser pulse radiation from the side of the nanolayer $Co_{80}Fe_{20}$. In this case magnetic switching of the nanolayer $Co_{30}Fe_{70}$ occurs under action of internal nonequilibrium magnetic field in result of laser-induced spin injection in this nanolayer from the nanolayer $Co_{80}Fe_{20}$ under the action of photon drag effect.

More effective control by spin current under nanosecond and picosecond laser pulses can be realized in the microcontact with perpendicular magnetic anisotropy of the nanolayers. In the under study by us tunnel microcontacts $Tb_{22}Co_5Fe_{73}$ / Pr_6O_{11}/ $Tb_{19}Co_5Fe_{76}$ with perpendicular magnetic anisotropy of magnetic nanolayers it is possible to get the increase and decrease of conductivity without the external magnetic field. The increase of conductivity in this microcontacts related to the magnetic switching of nanolayer $Tb_{19}Co_5Fe_{76}$ and transition of microcontacts from the state $Tb_{22}Co_5Fe_{73}\uparrow/Pr_6O_{11}/\downarrow Tb_{19}Co_5Fe_{76}$ with antiparallel magnetized nanolayers in the state $Tb_{22}Co_5Fe_{73}\uparrow/Pr_6O_{11}/\uparrow Tb_{19}Co_5Fe_{76}$ with parallel magnetized magnetic nanolayers it is possible to get due to the irradiation unpolarized radiation or linearly polarized radiation of nanosecond or picosecond laser pulses from the side of nanolayer $Tb_{22}Co_5Fe_{73}$ or due to the irradiation the circular polarization picosecond laser pulses from any direction. The decrease of conductivity related to the reverse transition of microcontacts from the state $Tb_{22}Co_5Fe_{73}\uparrow/Pr_6O_{11}/\uparrow Tb_{19}Co_5Fe_{76}$ with parallel magnetized nanolayers in the state $Tb_{22}Co_5Fe_{73}\uparrow/Pr_6O_{11}/\downarrow Tb_{19}Co_5Fe_{76}$ with antiparallel magnetized magnetic nanolayers without the external magnetic field it is possible to get due to the irradiation the circular polarization picosecond laser pulses. The magnetic switching of nanolayers with perpendicular magnetic anisotropy without the external magnetic field under the action only of laser radiation takes place under the action of the magnetic field of the laser-injected spins or magnetic field created the circular polarization picosecond laser pulses due to the reverse Faraday effect.

The relative resistance change in the microcontacts $Co_{80}Fe_{20}\uparrow/Pr_6O_{11}/\uparrow Co_{30}Fe_{70}$ at the laser induced thermomagnetic switching achieves the values $\Delta R/R > 0,1$ at $T=300\ K$ and $\Delta R/R > 0,6$ at $T=80\ K$. In the microcontacts $Tb_{22}Co_5Fe_{73}\uparrow/Pr_6O_{11}/\uparrow Tb_{19}Co_5Fe_{76}$ $\Delta R/R > 0,5$ at $T=300\ K$ and $\Delta R/R_0 >1,4$ at $T=80\ K$. The laser-induced resistance switching of the microcontacts without external magnetic field is accompanied by some less relative resistance changes. Thus for the microcontacts $Co_{80}Fe_{20}\uparrow/Pr_6O_{11}/\downarrow Co_{30}Fe_{70}$ $\Delta R/R > 0,08$ at $T=300\ K$ and $\Delta R/R_1 \geq 0,28$ at $T=80\ K$; for the microcontacts $Tb_{22}Co_5Fe_{73}\uparrow/Pr_6O_{11}/\uparrow Tb_{19}Co_5Fe_{76}$ $\Delta R/R_0 \geq 0,35$ at $T=300\ K$ and $\Delta R_3/R_0 > 0,8$ at $T=80\ K$. For similar laser-induced resistance switching in the microcontacts $Tb_{22}Co_5Fe_{73}\uparrow/Pr_6O_{11}/\downarrow Tb_{19}Co_5Fe_{76}$ the relative resistance changes achieve the values $\Delta R/R= 0,4$ at $T=300\ K$ and $\Delta R_3/R_0= 0,9$ at $T=80\ K$.

Thus received results denote on prospect of use of nanosecond and picosecond laser pulses for high-speed spin current control in spintronic elements.

REFERENCES

Albert, F. J., Buhrman, R. A., Katine, J. A., Myers, E. B.,. Ralph, D. C., *Phys. Rev. Lett. 84, 3149* (*2000*)

Ashkin, A., *Scientific American*. 226, 63 (1972)

Askaryan, G. A., Rabionovych, M. C., Smirnova, A. D., and Studenov, V. B., *JETF Letters*, 5, 116 (1967)

Bauer, G. E. W., Brataas, A., Nazarov, Yu, V., *Phys. Rev. Lett.* 84, 2481 (2000)

Bobo, J. F., Cibert, J., Luders, U.,. *C.R.Physique*, 6, 977 (2005)

Chantrell, R., Hinzke, D., Itoh, A., Kalashnikova, A. M., Kimel, A. V., Kirilyuk, A.,

Chen, P., Kotissek, Ph., Moser, J., Sadowski, J., Wegscheider, W., Weiss, D.,

Dirks, A. G., Leamy, H. J., *J. Applied Phys.* 50, 2871 (1979).

Djayaprawira, D. D., Yuasa, S., *J. Phys. D: Applied Physics*., 40, R337 (2007)

Fabian, J., Sarma, S, Zutic, I. *Rev. Mod. Phys*., 76, 323 (2004)

Goff, J. E., Schaich, W. L., *Phys. Rev. B*, 61, 10471 (2000)

Hansteen, F., Itoh, A., Kimel, A. V., Kirilyuk, A., Rasing, Th., Stanciu, C. D.,

Hansteen, F., Kimel, A. V., Kirilyuk, A., Pisarev, R. V., Rasing, Th., *J. Phys. Condens. Matter.* 19, 043201 (2007)

Hertel, R., *JMMM.* L1, 303 (2006)

Hull, G. F., Nuchols, E. F., *Phys. Rev*., N 17. p. 26 (1903)

Inokyti, C., Komori, M., Nukata, T., Tsutsumi, K., Sakyrai, I., *IEEE Trans. Magnetic.* 20,1042 (1984)

Julliere, M., *Phys. Letter*. 54A, 225 (1975).

Karaseva, V. Yu., Krupa, M. M., Kuzmak, O. M., *Surface*, №11, 92 (2001)

Korostil, A. M., Krupa, M. M., *Inter. J. Modern Physics* B 21, 2339 (2007)

Krupa, M. M., *JETF*, 120, 1268 (2001)

Krupa, M. M., *JETF*, 132, 782 (2007)

Krupa, M. M., *JETF Letters*, 87, 635 (2008)

Krupa, M. M., *JETF*, 135, 981 (2009)

Landau, L. D., Lifshitz, E. M., Theoretics Physics. *Electrodynamics of Continuous Media Phys*, (Fiz.-Mat. Izd., Moscow, 1984.

Merservey, R., Tedrov, P. M., *Phys. Rep*. 238, 175 (1994).

Nowak, U., Rasing, Th., Tsukamoto, A., Vahaplar, K., *Phys. Rev. Lett.* 103, 117201 (2009)

Slonczewski, J. C., *JMMM*, 159, 1191 (1996).

Tsukamoto, A., *Phys. Rev. Lett.* 99, 047601 (2007)

Zenger, M., *Phys. Rev.*, B 74, 241302 R (2006)

In: Magnetic Thin Films
Editor: John P. Volkerts

ISBN: 978-1-61209-302-4

Chapter 7

RF – SPUTTERED NANOCRYSTALLINE CU-ZN FERRITE THIN FILMS

R. Singh*[*] *and M. Sultan

School of Physics, University of Hyderabad, Central University P. O.
Hyderabad -500046, Andhra Pradesh, India

ABSTRACT

The ferrites are ceramic magnetic materials consisting of the entire family of iron oxides including spinels, garnets, hexaferrites, and orthoferrites. These materials have significant potential in applications ranging from millimeter wave integrated circuitry to transformer cores and magnetic recording.

The ferrite thin films have application in microwave devices, micro-inductors, micro-transformers, magnetic recording, gas sensors, catalysts and as shield for the electromagnetic interference. The RF– magnetron sputtering is a useful technique for depositing dense and homogeneous thin films of insulating ferrite compounds. We have used RF-magnetron sputtering to deposit Cu , Zn and CuZn spinel ferrite thin films on glass substrates at room temperature in oxygen (O_2), argon (Ar) and mixture of ($Ar+O_2$) environment.

A detailed study of the effect of process gas environment and pressure on the crystal structure, magnetic and optical properties of these ferrite thin films is undertaken. The XRD and AFM studies confirm the nanocrystalline nature of the as deposited films. The observed changes in the thin film crystal structure, magnetization and optical properties as a function of deposition conditions are attributed to the random distribution of the cations among the tetrahedral A-sites and octahedral B-sites during the deposition process.

The presence of multivalent cations with differing sites preferences gives rise to structural and magnetic disorder. The change in disorder leads to the change in the properties of thin films.

[*] E-mail: rssp@uohyd.ernet.in . Ph.: +91 40 2313 4321, Fax: +91 40 2301 0227

1. INTRODUCTION

Ferrites include the entire family of iron oxides i.e. spinels, garnets, hexaferrites, and orthoferrites. These are an important class of technological ceramic magnetic materials that have been recognized to have significant potential in various applications such as millimeter wave integrated circuitry, transformer cores, magnetic recording, ferrofluids etc [1-4]. At microwave frequencies ferrite materials have low losses, high resistivity and strong magnetic coupling, making them irreplaceable constituents in microwave device technology such as isolators, circulators and phase shifters. The technological importance of ferrites increased continuously as many discoveries required the use of magnetic materials. The progress in electronics was accompanied by the development of computing, transportation and non-volatile data storage technologies. The reduction of the size of the magnetic materials to dimensions comparable to those of the atoms and molecules is required for small electronic devices. In the recent years, nanoscale ferrites have found enormous potential applications in medicine and life sciences. The synthesis of ferrite nanoparticles with controllable dimensionality and tailorable magnetic properties along with the understanding of the structure-property correlations have become one of the topics of fundamental scientific importance [5]. Ferrite thin films have also attracted much attention in recent years for applications such as high density magnetic recording and magneto-optical recording media because of their unique physical properties such as high Curie temperature, large magnetic anisotropy, moderate magnetization, high corrosion resistance, excellent chemical and mechanical stability and large Kerr and Faraday rotations. They can provide unique circuit functions that cannot be produced by other materials [6] Magnetic soft ferrite thin films with high resistivity and ac permeability were advanced for applications to multilayer chip inductors, micro-transformers, magnetic recording and high frequency switched mode power supply [7]. As fillers in composites, ferrites can improve the modulus of the polymer matrix and provide additional functionality such as electromagnetic interference (EMI) shielding [8].

1.1. Spinel Ferrites:

The spinel ferrites are mixed oxides with general formula AB_2O_4, where A is a metal ion with +2 valence and B is the Fe^{+3} ion. They have a spinel-type structure similar to that of the mineral "spinel", $MgAl_2O_4$. Substitution of A by divalent cation M (M: Mn, Co, Ni, Cu, Zn, Mg) and B by Fe forms ferrite MFe_2O_4 ($MO \cdot Fe_2O_3$). The unit cell of spinel ferrite contains eight molecules (8 formula units MFe_2O_4).

$8\ MFe_2O_4 = (8M + 16Fe)$ cations + 32 Oxygen anions = 56 ions

The interstitial sites are schematically represented in Figures (1.4). There are 64 tetrahedral and 32 octahedral possible positions for cations in the unit cell. In order to achieve a charge balance of the ions, the interstitial voids will be only partially occupied by positive ions. Therefore, in stoichiometric spinels, only one-eighth (8/64) of the tetrahedral sites and one-half (16/32) of the octahedral sites are occupied by metal ions in such a manner as to minimize the total energy of the system. The distribution of cations that can be influenced by several factors such as the chemical composition (ionic radius, electronic configuration and electrostatic energy), method of preparation and preparation conditions [4].

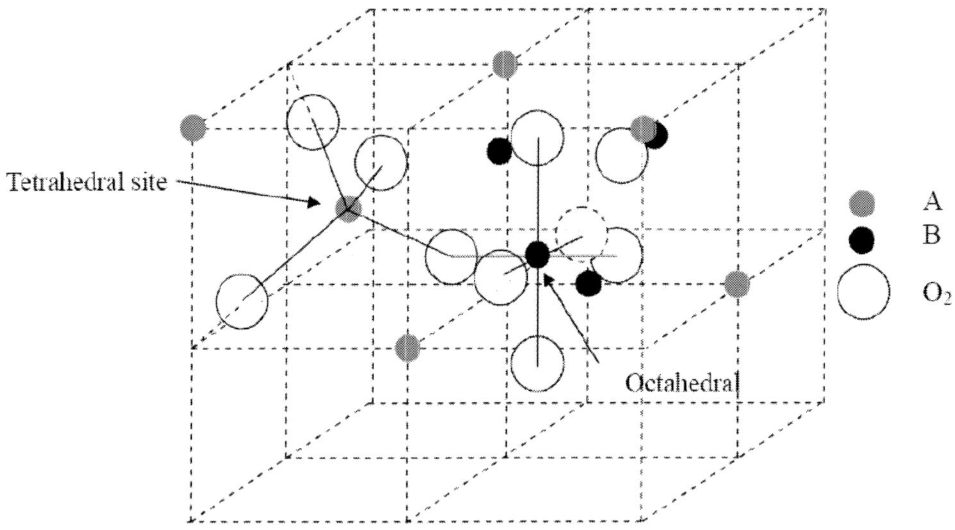

Figure 1. Schematic drawing of the spinel unit cell structure.

Depending on the cation distribution over the different crystallographic sites, the spinel compounds can be generally classified into two categories: normal and inverse spinels. Normal spinels have the general formula $(A)_t[B_2]_oO_4$ and contain all the trivalent metal ions $[B^{3+}]$ in the octahedral sites (o), whereas all the divalent metal ions (A^{2+}) reside in the tetrahedral sites (t). Transition metal ferrites, such as $ZnFe_2O_4$ and $CdFe_2O_4$, assume the normal spinel structure. In the inverse spinels, the divalent cations (A^{2+}) and half of the trivalent cations (B^{3+}) occupy the octahedral sites, whereas the other half of the B^{3+} metal ions lie in the tetrahedral sites. Most of the first series transition metal ferrites, such as $NiFe_2O_4$, $CoFe_2O_4$, Fe_3O_4, and $CuFe_2O_4$, are inverse spinels; their site occupancy can be described by the general formula $(B)_t[AB]_oO_4$. Some other spinel compounds posses the divalent and the trivalent metal ions randomly distributed over both the tetrahedral and the octahedral interstices. These compounds are named mixed spinels and their composition is best represented by the general formula $(A_{1-d}B_d)_t[A_dB_{2-d}]_oO_4$, where the inversion parameter, d, denotes the fraction of the trivalent cations (B^{3+}) residing in the tetrahedral sites. Mixed spinels occur in a wide range of compositions and are usually considered as intermediate compounds between the two extreme cases, i.e., normal and inverse spinels [9]. The mixed spinel structure include $MgFe_2O_4$ and $MnFe_2O_4$. Usually, d varies between 0 (for normal spinel compounds) and 1 (for inverse spinels); the inversion parameter can also take intermediate values ($0 < d < 1$) and the corresponding ferrites adopt a mixed spinel structure characterized by a random occupancy of the tetrahedral and octahedral sites by the divalent (A^{2+}) and trivalent (B^{3+}) cations. Nanoparticles and thin films of inverse spinels such as $CuFe_2O_4$ and normal spinel such as $ZnFe_2O_4$ also adopt a mixed spinel structure .

The exchange interaction between A and B sites is negative and the strongest among the cations so that the net magnetization comes from the difference in magnetic moment between A and B sites. The Fe cations are the sole source of magnetic moment and can be found on any of two crystallographically different sites: tetrahedral and octahedral sites [10]. Given the complexity of the crystal structures of cubic spinel ferrites, the structural disorder, ranging from cation disorder to grain boundaries, has a significant effect on the electronic and magnetic properties. Moreover, these ferrites have open crystal structures so that diffusion of metal cations readily occurs.

1.2. Thin Films of Cu-Zn Ferrites:

Thin films of spinel ferrites MFe_2O_4 (M = Mn, Cu, Zn, Ni, Co, Cd, etc.) exhibit excellent chemical stability and high corrosion resistivity and seem to be applicable as recording media and microwave absorbing layers [11]. Thin films of copper ferrites are of great interest. There have been a number of studies on Cu-ferrite thin films with the objective of enhancing magnetization for applications deposited either by RF-sputtering [12-19] or Pulsed laser deposition [20, 21]. In these studies higher magnetization values were achieved either by substrate heating or annealing the films at high temperatures or by applying high power and high working gas pressure for sputtering the films. Zuo et al. [21] predicted that a high magnetization may be achieved by increasing the number of Cu^{2+} ions on A-sublattice. Sultan and Singh [22] have reported high magnetization of $CuFe_2O_4$ nanocrystalline thin films prepared by RF-sputtering in Argon gas environment. It is considered that the preparation of the $ZnFe_2O_4$ thin film by the sputtering method, which involves very *rapid cooling* of vapor to form the solid state phase, causes the random distribution of Zn^{2+} and Fe^{3+} ions in the spinel structure. In such a situation, Fe^{3+} ions occupy both octahedral and tetrahedral sites, and the strong superexchange interaction among them gives rise to ferrimagnetic properties accompanied with high magnetization [23]. There are few reports on thin films of $ZnFe_2O_4$ prepared either by RF-sputtering [24-25] or pulsed laser deposition (PLD) [26-30]. Epitaxial thin films of spinel $ZnFe_2O_4$ were grown along the (111) direction using laser-molecular beam epitaxy (laser-MBE) on an Al_2O_3 (0001) single crystal substrate [31]. Recently Sultan and Singh [32,33] reported the properties of nanocrystalline thin films of $ZnFe_2O_4$ deposited by RF-sputtering. This chapter contains the experimental results on the growth, structure, magnetic and optical properties of ferrospinel $Cu_{1-x}Zn_xFe_2O_4$ (x = 0, 0.4 and 1) thin films deposited by rf-magnetron sputtering. The RF-Magnetron sputtering is a useful technique for getting very high deposition rate and homogeneous films with unique properties achieved by controlling the sputtering parameters . The main objective is to study the influence of deposition environment on the structure and film growth as well as on the magnetic and optical properties of the sputtered films. These studies would be helpful in finding out the role of Cu^{2+} and Cu^{1+} ions in controlling the magnetic and optical properties of these ferrites. Bulk $Cu_{1-X}Zn_XFe_2O_4$ ferrites show that highest magnetization of ~330 emu/cc for the composition $Cu_{0.6}Zn_{0.4}Fe_2O_4$ [34]. The synthesis and properties of copper zinc ferrite thin films of this composition are reported in by Sultan and Singh in a recent study [35].

2. Experimental Methods:

2.1. Thin Film Deposition by RF-Magnetron Sputtering:

RF-Sputtering has become one of the most versatile techniques in thin film technology for preparing thin solid films of almost any material. The main advantages of this technique are (a) high uniformity of thickness of the deposited films, (b) good adhesion to the substrate, (c) better reproducibility of films, (d) ability of the deposit to maintain the stoichiometry of the original target composition, and (e) relative simplicity of film thickness control. In the present work ferrite films were deposited by employing rf-magnetron sputtering system schematically shown in figure 2. The deposition equipment is composed of two 2-inch planar high performance water-cooled magnetrons, a stainless steel chamber, high vacuum system, switch cabinet, a heated rotatable substrate holder etc. The equipment has provision of cooling water supply to cathode, power supply and turbo pump. The base pressure of 10^{-6} Torr could be achieved in the vacuum chamber. The material to be deposited (target) acts as the cathode and is connected to a negative voltage RF power supply. The substrates are placed exactly below the target which makes 10° with the normal to the substrate. The pure Argon (Ar), pure oxygen (O_2) and mixture of ($Ar+O_2$) were used as sputtering gases at pressure of 5, 8, 12 and 15 mTorr. The films were deposited on glass substrates mounted on a non-rotating, unheated platform placed at a distance of ~ 60 mm below the target inside the vacuum chamber. For all the films the deposition time was kept 60 minutes. The films were deposited using rf-power of 100 W at various working gas pressure of Ar, O_2 and $Ar+O_2$ from a 2-inch diameter polycrystalline sintered targets. The targets were synthesised by ceramic method using high purity starting materials with final sintering at 900 °C for 12 hrs.

2.2. Characterization of Thin Films:

The structural characterization of the films was done using X-ray diffraction (XRD). The information about microstructure, surface morphology, chemical composition, grain size and surface roughness were obtained from atomic force microscope (AFM), scanning electron microscope (SEM) and energy dispersive X-ray analysis (EDAX) studies. The magnetization measurements were carried out using a vibrating sample magnetometer (VSM) (Lakeshore 7400). The thickness of the films was measured using a surface stylus profilometer (Ambios tech model XP-1).The optical transmission spectra of the deposited thin films were recorded at RT using a JASCO V570 spectrophotometer in the wavelength range 190-2700nm.

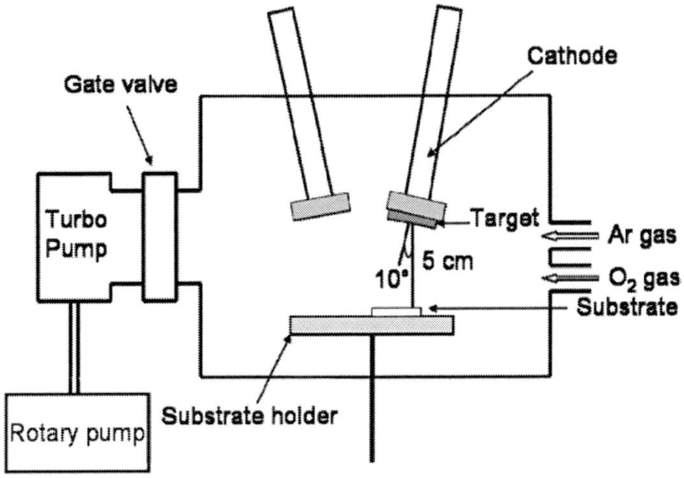

Figure 2. Schematic diagram of the sputtering system with magnetron[33].

3. Copper Ferrite ($CuFe_2O_4$) Thin Films:

In this section, we present the results on the structure, magnetic and optical properties of Cu-ferrite thin films deposited under various process gas environment. The films were deposited in pure Ar and pure O_2 at different working pressure. Then the films were deposited in a mixture of Ar and O_2 to understand the effect of oxidizing environment on their properties.

3.1. Effect of Ar-O_2 Pressure on Structure and Properties:

3.1.1. Structural Analysis (XRD):

Figure 3. shows the XRD patterns for the as-deposited $CuFe_2O_4$ films. All the peaks belong to cubic spinel structure without any secondary phase. The diffraction (220) lines indicate the textured growth in the case of film deposited in Ar. With increase in oxygen content the intensity of (400) peak increases at the expense of (220) peak intensity. Ohnishi and Teranishi [36] reported that for spinel–ferrite the intensity ratios of planes I220/I400, I440/I400 and I422/I400 are considered to be sensitive to the cation distribution parameter (d). Our XRD data shows a strong dependence of these intensity ratios on the deposition environment.

The higher values of these ratios for the as-deposited films in pure Ar indicate higher inversion parameter. The average crystallite size (D) of the films was estimated from XRD data using the Scherrer formula [37] $D = 0.9\lambda / \beta\cos\theta$, where λ is the wavelength of the used X-rays, β is the broadening of diffraction line measured at the half maximum intensity in radians and θ is the angle of diffraction. The values of the crystallite sizes are estimated to be about 15±2 nm for the film deposited in Ar and decreases with increasing oxygen partial pressure to be about 10 ± 2 nm for the film prepared in pure oxygen.

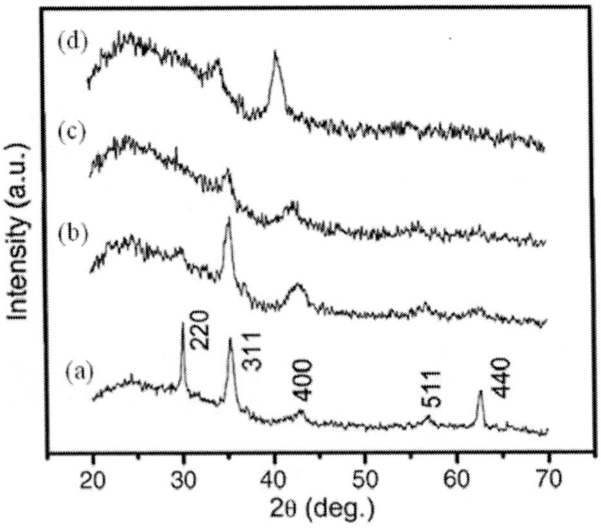

Figure 3. XRD patterns of as-sputtered Cu-ferrite films prepared under (Ar+O2) at 12 mTorr in (a) 100% Ar, (b) 75% Ar+25% O2, (c) 50% Ar+50% O2, and (d) 100% O2.

3.1.2. Surface Morphology (AFM):

Figure 4. shows the AFM images of the films. The film sputtered in Ar shows dense formation of oblong crystallites with partial alignment in one direction. The shape of grains becomes Y-shaped and their density decreases with increase in O_2 content. The average grain sizes and root mean square roughness (rms) values decreases with oxygen content (Table 1).

The average composition of the films was determined by EDAX studies using systematic data collected from the centre and edges of the films. The relative elemental concentration was used to estimate the atomic composition of the as-deposited films by considering their values normalized to the nominal composition of the target. The composition of the films was found to be close to the composition of the target.

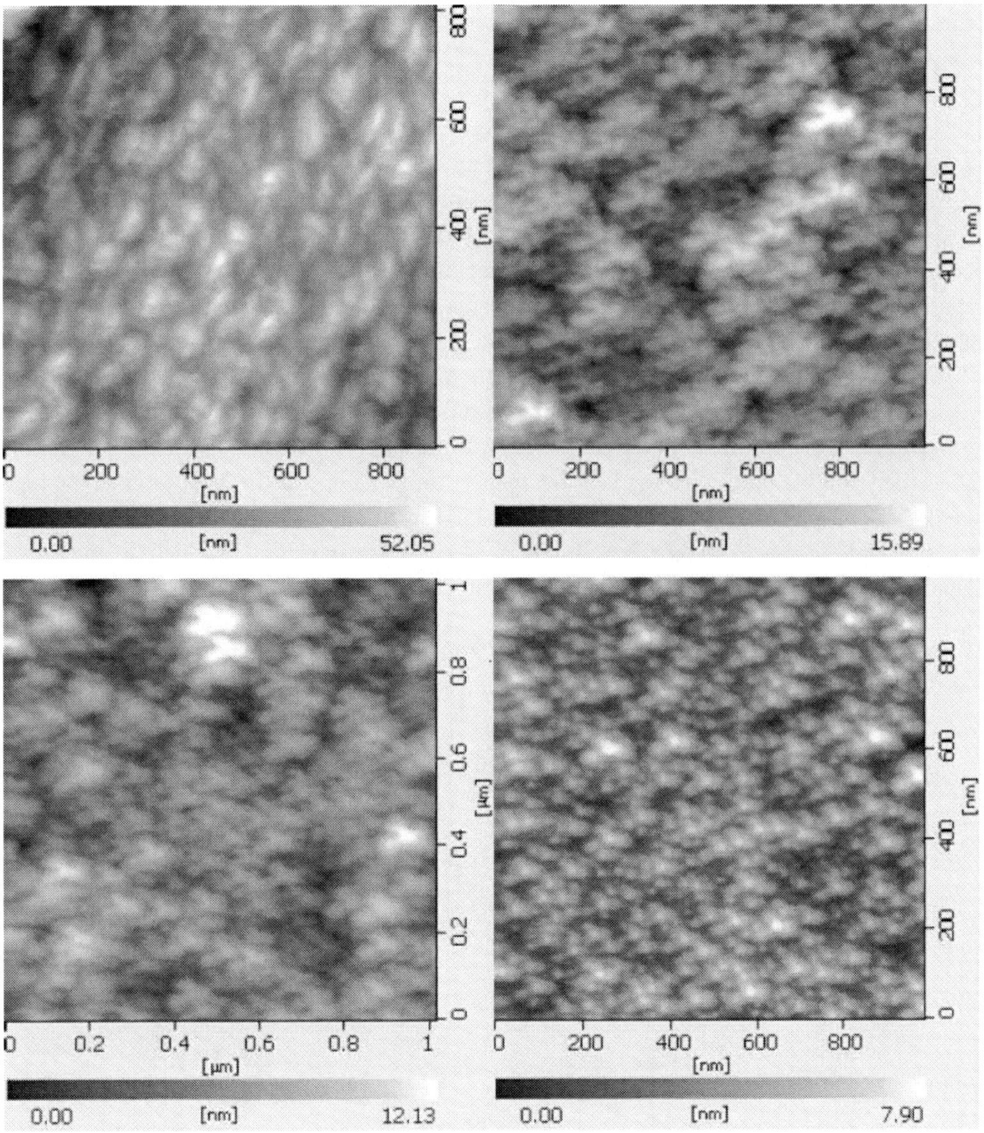

Figure 4. AFM images of as-sputtered Cu-ferrite films prepared under (Ar+O2) at 12 mTorr. from top left 100% Ar ,(b) 75% Ar+25% O2 (c) 50% Ar+50% O2 and (d) 100% O2.

3.2. Magnetization Studies (M-H plots):

Figure 5. shows the M–H plots at RT for the Cu-ferrite films deposited under Ar gas environment with external field parallel ($H_{\parallel}$) and perpendicular ($H_{\perp}$) to the film planes. The estimated magnetic parameters are summarized in Table 1. The films show large magnetization values and exhibit ferrimagnetic behavior with strong magnetic anisotropy. The saturation magnetization (M_S) for bulk Cu-ferrite target is found to be 170 emu/cm^3 . The magnetization value of the as-deposited film strongly depends on Ar gas pressure. The in-plane magnetization $M_{S\parallel}$ values of the as-deposited films are 120, 142, 264 and 247 emu/cm3 at 5, 8, 12 and 15 mTorr of pure Ar gas respectively. Since $M_{S\parallel}$ value is found to be maximum for films deposited under 12 mTorr, the effect of O_2 addition was therefore studied by maintaining the ($Ar+O_2$) pressure at this level. The magnetization values and coercivity (H_C) as a function of oxygen content are listed in Table 1.

The estimated deposition rate is ~58 Å /min for the films prepared under pure Ar and it decreases to ~17 Å /min in pure oxygen at pressure of 12 mTorr. The high deposition rate in pure Ar gas environment can lead to freezing of some Cu-ions on tetrahed

ral sites and equivalent number of Fe ions on octahedral sites during the deposition process. Furthermore, the deposition in reducing (argon) atmosphere may lead to the formation of Cu^+ ion, having larger ionic radius than the Cu^{2+} ion. The Cu^+ ions prefer occupation of the smaller four-coordinated tetrahedral site in the spinel structure and displace Fe^{3+} cations to occupy the octahedral sublattice. This will increase octahedral Fe^{3+} ion concentration in expense of Cu^{2+} ions which in turn causes ferrimagnetic behavior with large magnetization. For the films sputtered under Ar atmosphere, the estimated magnetic moment is 2.1 μ_β per formula unit, whereas it is 1.5 μ_β per formula unit for the bulk target. The higher value of M_S can therefore arise due to ordering of Cu^+ to tetrahedral A-sites. The decrease in M_S value with increase in oxygen content in ($Ar+O_2$) mixture is ascribed to decrease in growth rate and Cu^+ concentration which allow the cations to take up their preferable sites.

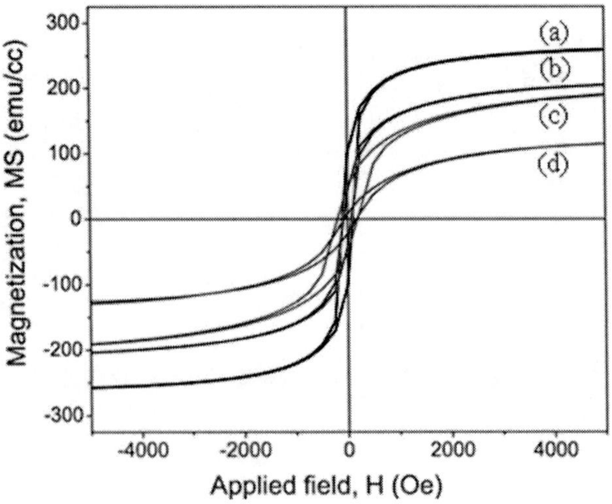

Figure 5. M-H curves of as-sputtered Cu-ferrite films prepared under (Ar+O2) at 12 mTorr (a) 100% Ar, (b) 75% Ar + 25%O2, (c) 50% Ar + 50%O2, (d) 100% O2.

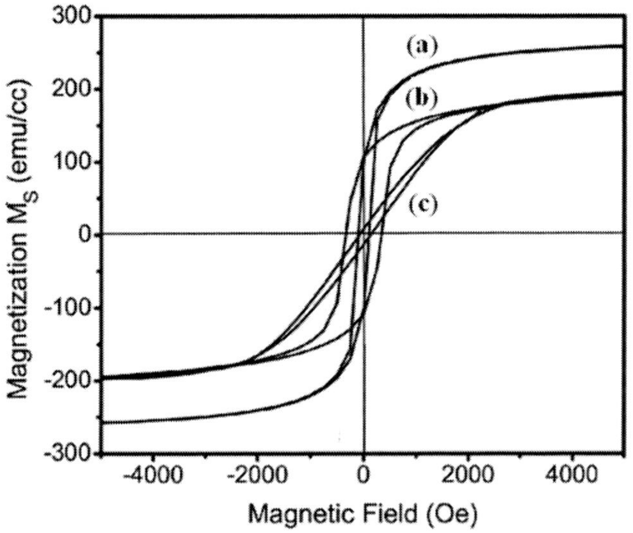

Figure 6. M–H plots at RT for (a) as-sputtered CuFe2O4 film (b) film annealed at 500 °C for 3 h with field parallel to film surface and (c) as-sputtered CuFe2O4 film with field normal to film surface (Ar pressure 12 mTorr).

The formation of Fe^{2+} cannot be ruled out in the spinel structure where a fraction of Fe^{3+} will be replaced by Fe^{2+} ions. So the general cation distribution for copper ferrite can be represented as

$$((Cu^{2++}Cu^{+})_d Fe^{3+}{}_{1-d})_{tetra}[Cu^{2+}{}_{1-d}\,(Fe^{2+}Fe^{3+})_{1+d}]_{octa}O_4{}^{2-}.$$

The strong magnetic anisotropy of the films may be due to weak coupling between Cu^{2+} on the A-sites and Fe^{3+} on B-sites as a result of ordering of the Cu^{+} ions to tetrahedral A-sites.

The films sputtered in Ar environment were annealed in air at 500 °C for 3 h followed by slow cooling. This lowered the M_S value by about 30% of the as-deposited film (Figure 6). The coercivity however increases to a value of 330 Oe. The annealing removes the randomness of A-site and B-site cations and also oxidizes Cu^{+} to Cu^{2+} leading to lower magnetization value.

Table 1. Structural and magnetic parameters of as-sputtered Cu-ferrite films prepared under ($Ar + O_2$) at 12 mTorr

($Ar+O_2$) (%)	Thickness (nm)	Grain size (nm)	RMS (nm)	$M_{S\parallel}$ (emu/cc)	$M_{S\perp}$ (emu/cc)	$H_{C\parallel}$ (Oe)	$H_{C\perp}$ (Oe)
	±5%						
100 + 0	350	126	3.7	264	188	100	107
75 + 25	210	112	2.6	212	150	77	110
50 + 50	130	98	2.2	202	110	196	210
0 + 100	100	68	1.3	118	58	125	150

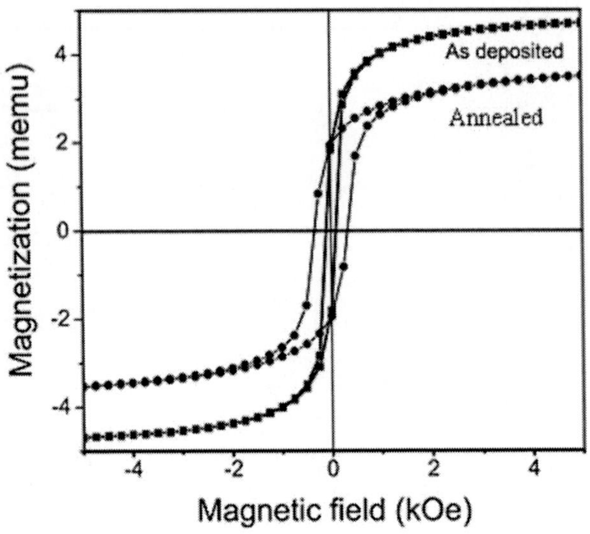

Figure 7. M–H plots at RT for as-sputtered CuFe2O4 film (P(Ar) = 12 mTorr) and film annealed at 500 °C for 3 h with field parallel to film surfaces.

Figure 8 shows room temperature hysteresis curves of the film magnetic moment versus static applied field in both the in-plane and perpendicular to plane static field configurations. The in-plane and perpendicular-to-plane intrinsic coercivities are 100 and 132 Oe, respectively. The saturation flux density $4\pi M_S$ of the film is obtained as the value of applied field H required saturating the sample in the perpendicular-to-plane configuration. This saturation field is 2.5–3 kOe. With the perpendicular to-plane demagnetizing factor equal to 4π for a thin film, $4\pi M_S$ is 2.5–3 kG. Magnetization as a function of the external magnetic field, *M(H)*, are shown at 300 and 80 K for $CuFe_2O_4$ thin film in figure 7. The magnetization at 10 kOe is as high as 264 emu/cc at 300 K, and reaches 300 emu/cc at 80 K. The magnetic hysteresis loops are clearly observed in the low-field ranges. The hysteresis loop at 80 K (H_C = 132 Oe) is larger than that measured at 300 K (H_C = 100 Oe). The higher value of H_C at lower temperature can be assigned to the large disorder in the system.

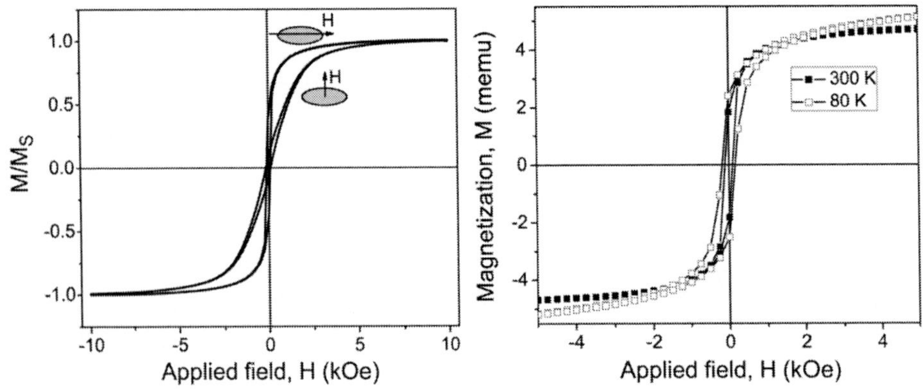

Figure 8. M–H plots (Left) recorded at RT with applied field in plane and out of plane configurations (Right) recorded at 300 and 80 K for as-sputtered CuFe2O4 film (P(Ar) = 12 mTorr) with field parallel to film surfaces.

3.3. Optical Properties:

The variation of (Ar/O_2) ratio influences the optical properties of the ferrite films. Figure 9. shows the transmittance spectra measured at RT for the as-grown $CuFe_2O_4$ films deposited in 12 mTorr of Ar, O_2 and mixture of ($Ar+O_2$) gas pressure.

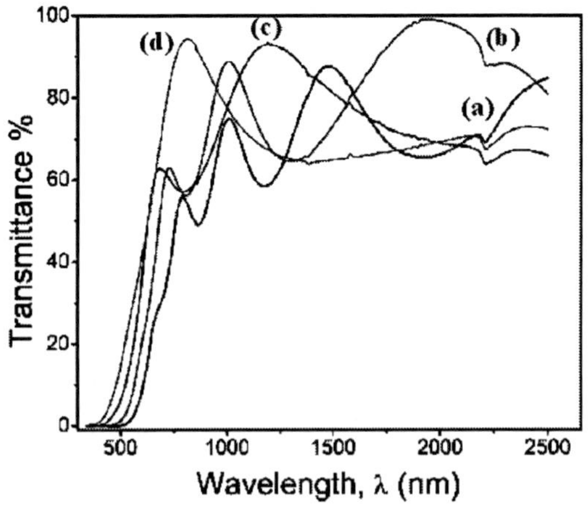

Figure 9. Optical transmittance of as-sputtered Cu-ferrite films prepared at 12 mTorr where (a) is at 100% Ar atmosphere, (b) at 75% Ar and 25%O_2, (c) at 50% Ar and 50% O2 and (d) at 100% O_2.

The transmittance spectra show that all the films have sharp absorption edges which vary between 570 and 700 nm with increase in Ar percentage. The Ar-deposited films show less transmittance due to the higher thickness. With increase in the O_2 percentage the film thickness decreases, absorption edge shifts towards lower wavelengths and transmittance is enhanced. This can also be due to the decrease of average grain size and formation of pores in the films. The compositional dependence of transmittance for films having same thickness shows that copper ferrite films have less transmittance. Zinc ferrite films show higher transmittance and blue shift.

4. ZINC FERRITE ($ZnFe_2O_4$) THIN FILMS:

This section presents the results on films of zinc ferrite deposited by RF-Magnetron sputtering from a sintered target of $ZnFe_2O_4$ on glass substrates. The rf-power used was 100W. The films were sputtered under argon, oxygen and (argon + oxygen) mixture at various pressures. The effect of annealing at temperatures from 200 to 500 °C on structural, magnetic and optical properties of the films was investigated.

4.1. Effect of Working Gas Pressure on Structure and Properties:

4.1.1. Structural Analysis (XRD)

Figure 10. shows the XRD patterns of the $ZnFe_2O_4$ thin films sputtered under pure Ar and oxygen at various working pressure. The patterns were indexed to the cubic spinel structure belonging to Zn-ferrites. Three main peaks were observed with orientations along the (311), (400) and (511) directions in all the films. Using Bragg's law, the interplanar lattice spacing, $d_{(311)}$, was calculated. The lattice parameters, *a*, calculated by using $a = \sqrt{11}d_{311}$ are listed in table 5.2. The lattice parameter is higher for the film deposited at lower pressure and decreases with increase in pressure. At lower pressure (8 mTorr), the diffraction peaks are broad and slightly shifted towards lower angle as compared to the spectra for the bulk Zn-ferrite powder sample. The shift is more for the films prepared under Ar gas. This can be attributed to the internal stress caused by the trapped Ar atoms into the film and its consequent effect on the interaction between the grains and the substrate. The XRD peak shift for the films deposited under oxygen is lower and it decreases with the increase in the oxygen pressure. The XRD spectra peaks for the films deposited at 31 mTorr are very close to the corresponding peak positions of the bulk Zn-ferrite powder indicating considerably lower stress in these films. The average crystallite size *(D)* of the films was estimated from XRD data using the Scherrer formula are listed in table 2. The crystallite size is smaller for the films deposited at 8 mTorr pressure of argon or oxygen. There is a slight increase in crystallite size with increase in oxygen pressure.

4.1.2. Film Surface Morphology (AFM)

The surface morphology of the Zn-ferrite thin films was analyzed using the non-contact mode of AFM. Figure 11. displays AFM images of the Zn-ferrite films grown on glass substrates at 8mTorr of Ar and 8–31mTorr of O_2. It can be seen that the topography of the ferrite films depends on the sputtering gas type and pressure. The film deposited under Ar exhibits a dense microstructure compared with those sputtered in the O_2 environment. The average grain sizes and surface roughness (root mean square (RMS)) values are listed in table.2.

The increase in roughness is consistent with the increase in crystallinity with oxygen pressure during deposition. This result is supported by XRD patterns as shown in Figure 10. The film deposited under Ar shows higher RMS values compared with that sputtered at the same pressure in oxygen environment.

The average RMS roughness increased from 1.7 to 2.7 nm as the oxygen pressure increases. This is due to variation in the distribution of incident angles and energy of the depositing atoms at low and high deposition pressures [38]. At low deposition pressure few collisions occur within the plasma; therefore incident angles tend to be close to normal and bombarding energy tends to be high.

The high bombarding energy leads to a high adatom mobility (atoms present at the surface of the layer) and thus, to a higher film density. However, the increase in oxygen pressure results in focusing the particles ejected from the target on a small area, which leads to an increase in the number of collisions occurring in the space between the target and the substrate (plasma) leading to an increase in the incident angles of the particles. This causes a shadowing effect leading to the formation of clusters and voids within the growing layer [39, 40]. This makes the films porous.

The increase in the surface roughness could be attributed to the formation of clusters of particles with voids in between due to the variation in the distribution of incident angles and energy of the depositing atoms at low and high gas pressures. These results support the optical refractive index variation with oxygen pressure.

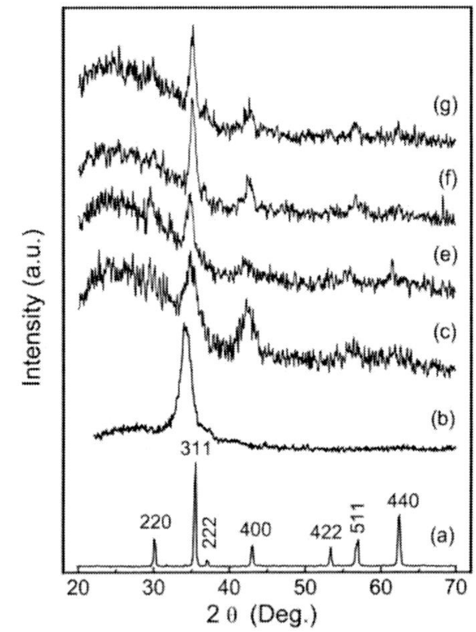

Figure 10. The XRD patterns of $ZnFe_2O_4$ (a), powder sample and thin films deposited at a pressure of (b) 8mTorr Ar, (c) 8mTorr O_2, (d) 18mTorr O_2, (e) 27mTorr O_2 and (f) 31mTorr O_2.

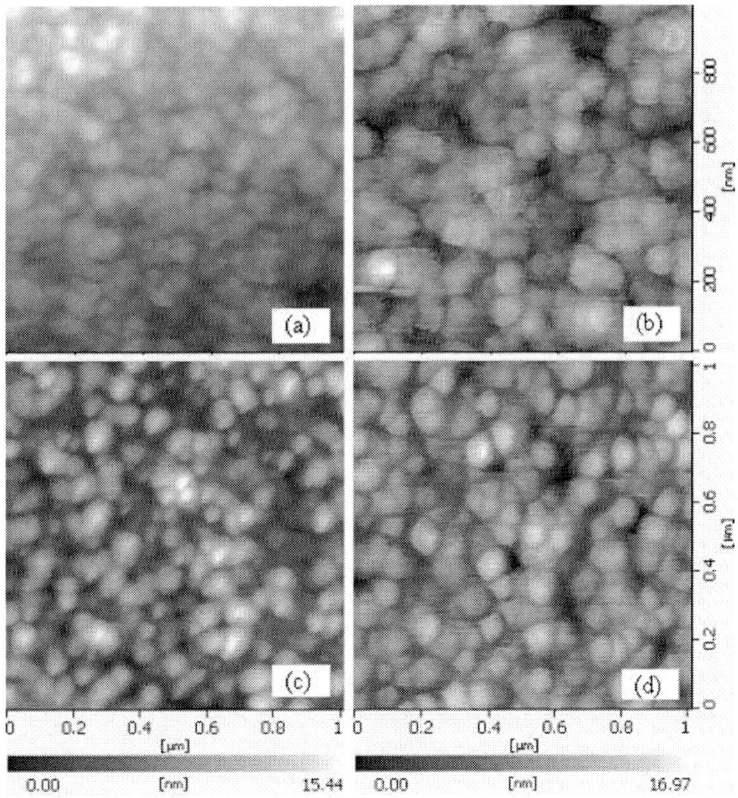

Figure 11. AFM images of $ZnFe_2O_4$ thin films deposited at a pressure of (*a*) 8mTorr Ar, (*b*) 8mTorr O_2, (*c*) 27mTorr O_2 and (*d*) 31mTorr O_2.

Table 2. Structural parameters of the films deposited under Ar and O_2 environments. t_1 and t_2 refer to film thicknesses as estimated from the profilometer and the optical transmission spectra, respectively

Sample	P (mT)	RMS (nm)	Grain Size (nm)	Crystallite Size (nm)	Lattice Parameter (Å)	t_1 (nm)	t_2 (nm)
		± 0.1	± 5	± 2	± 0.002	± 15	± 15
ZF-Ar8	8	4.2	55	11	8.68	420	450
ZF-O8	8	1.7	70	11	8.55	300	320
ZF-O18	18	2.3	66	13	8.53	280	265
ZF-O27	27	2.7	60	15	8.46	240	250
ZF-O31	31	2.6	72	15	8.45	200	230

4.2. Magnetic Properties (M-H plots):

Figure 12. shows the plots of magnetization versus applied magnetic field curves at RT for the zinc ferrite films deposited at 8 mTorr Ar and 8, 18, 27 and 31mTorr oxygen gas pressure. The as-deposited $ZnFe_2O_4$ thin films show large magnetization values compared to that of the bulk sample and exhibit ferrimagnetic behavior at room temperature. The magnetization value of the argon-deposited film is lower value compared to that prepared under oxygen atmosphere.

Significant increase in M with increase in O_2 pressure to reach maximum value of 230 emu/cc (42 emu/g) obtained for the film deposited at 27 mTorr. The value of 42 emu/g for the as-deposited film prepared under normal conditions in the present study is the highest so far reported in the literature. The increase in the magnetization of $ZnFe_2O_4$ has been attributed to oxygen vacancies and disorder [41], surface effects [42], as well as to cation random distribution of magnetic iron Fe^{3+} ions and diamagnetic Zn^{2+} ions among the interstitial octahedral (*B*) and tetrahedral (*A*) sites of the spinel lattice of $ZnFe_2O_4$ [43]. The occupancy of Fe^{3+} ions on both *A*- and *B*-sites leads to strong negative superexchange interaction between Fe^{3+} ions with large magnetization values. During the film deposition, the oxygen vacancy and the disorder on the nanocrystal surface can contribute to cation redistribution on *A* and *B* sites leading to high magnetization. The oxygen working pressure of around 27 mTorr appears to be critical as it may provides sufficient time for the nanocrystals to settle down on the substrate in a preferred direction to give maximum magnetization.

Figure 13. shows typical magnetization curves of as-sputtered $ZnFe_2O_4$ thin films at 27 mTorr of oxygen pressure for parallel ($H_{\parallel}$) and perpendicular ($H_{\perp}$) orientations of external magnetic field with respect to the film plane. One observes strong dependence of magnetization (M) on the orientation of H. For in-plane magnetization, the saturation takes place at field of about 2 kOe, which is smaller than the saturation field for transverse magnetization. The M_S per unit volume of the as-deposited ZFO films is 230 emu/cc. A well-defined hysteresis is observed with coercivity H_C value of 50 Oe. Such an M_S versus H behavior is an indicative of the expected ferromagnetic ordering at room temperature of $ZnFe_2O_4$ thin films. Thus the magnetization curves of Figure 24. for Zn-ferrite films are totally different from the corresponding magnetization for the bulk ferromagnetic counterparts.

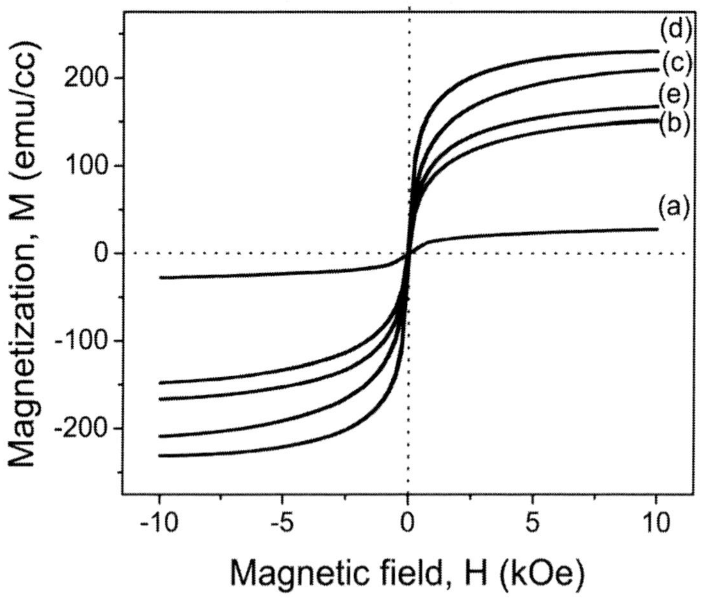

Figure 12. Effect of oxygen gas pressure on M-H loops at RT for ZnFe2O4 thin films prepared under (a) 8mTorr Ar, (b) 8mTorr O2, (c) 18mTorr O2, (d) 27mTorr O2 and (e) 31mTorr O2 gas pressure.

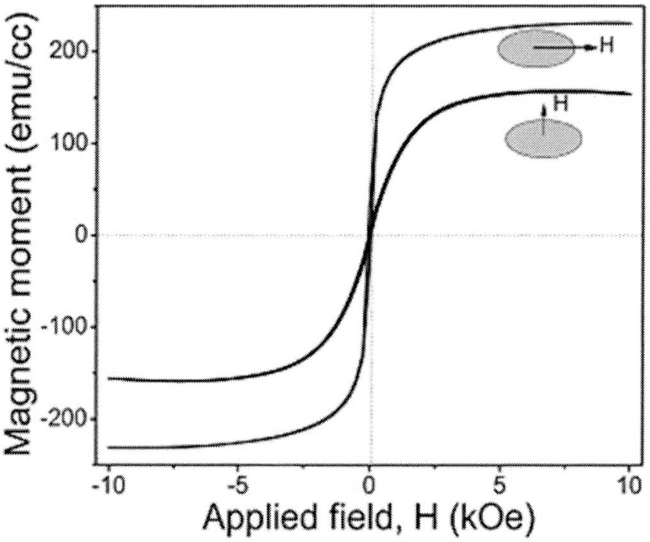

Figure 13. Angular dependent M-H curves of as-sputtered $ZnFe_2O_4$ thin films at 27 mTorr of oxygen pressure.

Magnetizations as a function of the external magnetic field, *M(H)*, are shown at 80 K for $ZnFe_2O_4$ thin film in figure 14. The magnetization at 10 kOe is as high as 230 emu/cc (42 emu g^{-1}) at 300 K and reaches 350 emu/cc (65 emu g^{-1}) at 80 K. The magnetic hysteresis loops are clearly observed in the low-field ranges. The hysteresis loop at 80 K (330 Oe) is much larger than that at 300 K (55 Oe). The higher value of H_C at lower temperature can be assigned to the large disorder in the system.

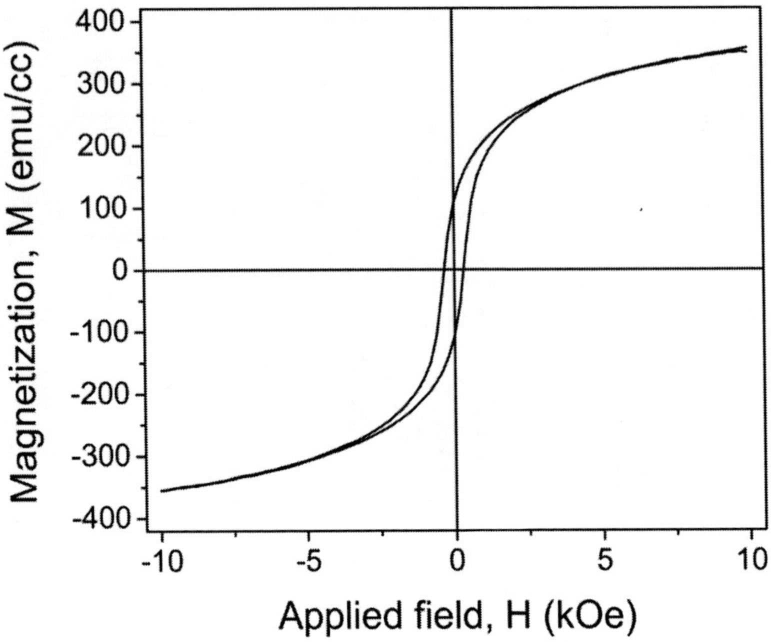

Figure 14. Magnetization as a function of the external magnetic field for $ZnFe_2O_4$ thin at 80K.

4.3. Effect of Heat Treatment on Structure and Properties:

4.3.1. Structural Studies:

The key inference is that the structural, optical and magnetic properties are influenced by the annealing of the Zn-ferrite films. The M_S decreases with increase in annealing temperature which is assigned to the redistribution of the cations among A- and B-sites where they take up their preferable sites as in the bulk sample.

The X-ray diffraction spectra for the as-deposited film and the film annealed at 500 °C are displayed in Figure 15. The XRD patterns confirm the presence of spinel phase without any impurity phase for both the films. The lines can be indexed to the characteristic interplanar spacing (220), (311), (400), (422), (511) and (440) of the spinel structure with cubic symmetry.

The average grain size was estimated from the width of the (311) reflection of the XRD patterns by using the Scherrer's formula. The values of average grain size are 50 and 60 nm for the as-deposited and annealed film respectively. (After subtracting the broadening effects due to strain and experimental broadenings)

To confirm the chemical composition of the as-deposited and the heat treated $ZnFe_2O_4$ films, EDAX spectra were recorded at a number of positions of the film surface. The chemical signatures obtained are identical within experimental accuracy (5%) and only Zn, Fe, and O elements are observed with the expected stoichiometric proportions of $ZnFe_2O_4$ (Zn/Fe = ½). This further confirms the formation of ferrite crystallites in the as-synthesized sample. XRD and EDAX studies demonstrate that the as-sputtered thin films are nanocrystalline $ZnFe_2O_4$ ferrites.

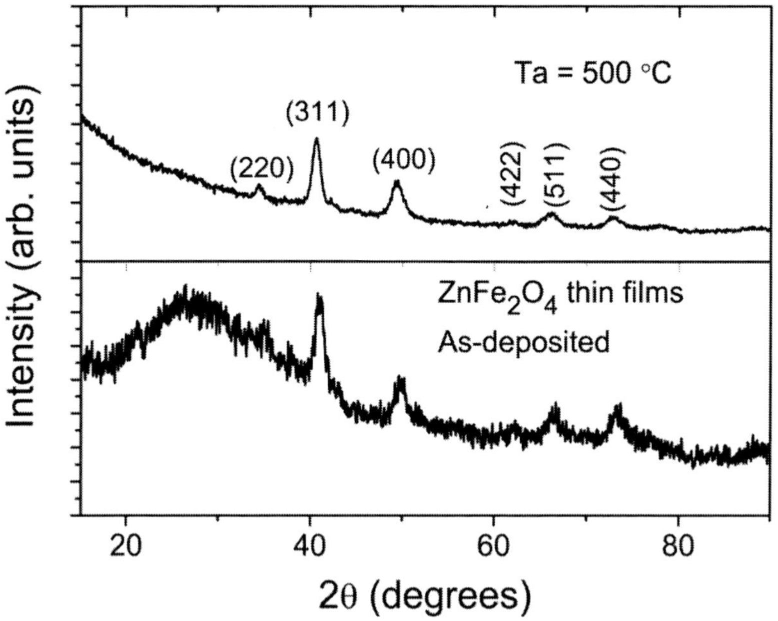

Figure 15. X-ray diffraction pattern of the as-deposited and annealed films of $ZnFe_2O_4$.

4.3.2. Surface Morphology (AFM Studies):

The influence of heat treatment on the surface morphology of the film was studied by atomic force microscopy (AFM). The surface morphology of the film is smooth with no cracks and defects, as shown in Figure 16. The film exhibited a dense microstructure and the grain size was found to increases from 80 to 120 nm as the annealing temperature is increased from 200 to 500 °C (Table 3). The surface morphology of the deposited film is related to the surface kinetic energy and depends on the processing parameters such as annealing temperature, duration and environment [44].

Surface roughness (RMS) was improved with increase in annealing temperature. The average surface roughness is around 2 nm for the film annealed at 500 °C. Higher annealing temperature improved, the structural quality of the $ZnFe_2O_4$ thin films. The kinetic energy of the sputtered atoms increases with increase in annealing temperature resulting in structural improvement of the deposited $ZnFe_2O_4$ thin films.

Table 3. Effect of annealing on the AFM and magnetization parameters

$ZnFe_2O_4$ thin film	⟨Grain size⟩ (nm)	⟨RMS⟩ (nm)	$M_{S\parallel}$ (emu/cc)	$H_{C\parallel}$ (Oe)	$\Delta H_{\parallel}$ (Oe)	$H_{R\parallel}$ (Oe)
As-deposited	80	2.4	230	55	630	2235
Ta = 200 °C	100	2.2	254	45	551	2221
Ta = 300 °C	110	2.2	220	25	360	2180
Ta = 400 °C	105	2.1	157	14	200	2430
Ta = 500 °C	120	2.0	70	3	210	2587

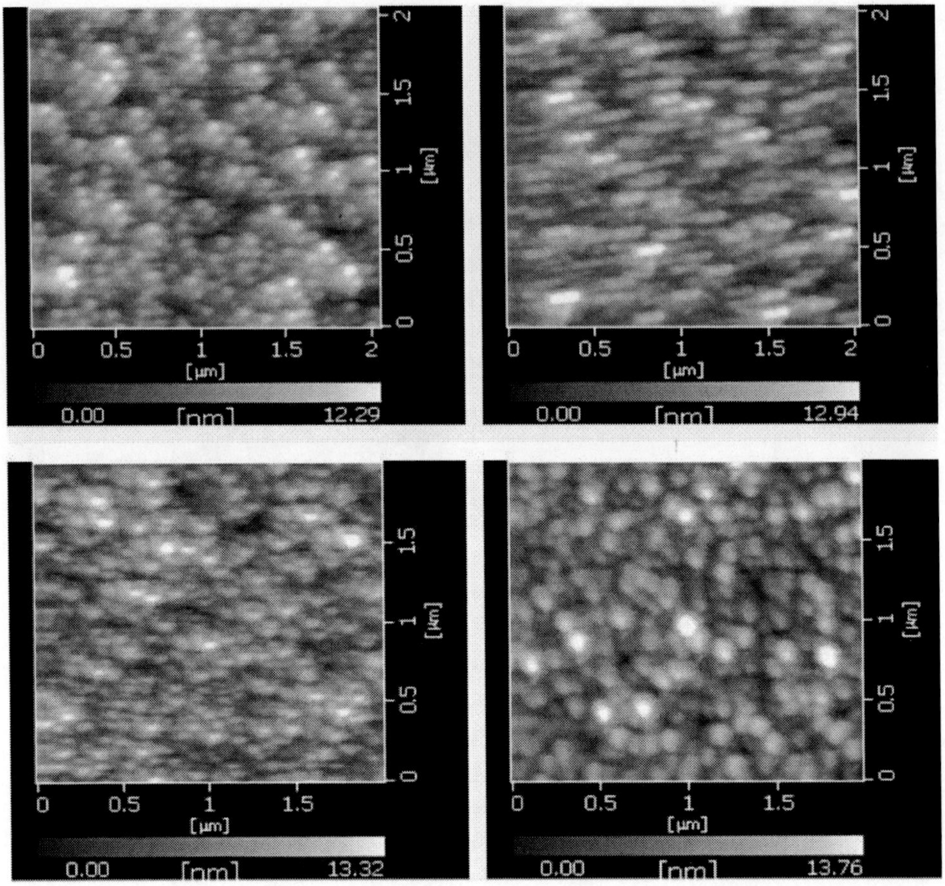

Figure 16. AFM images of $ZnFe_2O_4$ thin films prepared under 31 mTorr Oxygen pressure from top left clockwise (a) As deposited (b) annealed for 3 hrs at 200°C (c) 300 °C and (d) 500 °C.

4.3.2. Magnetization Studies:

Magnetization vs magnetic field plots of $ZnFe_2O_4$ films deposited in pure O_2 and annealed in air for 3 h at temperatures in the range 200–500 °C are presented in Figure 17. It is observed that as annealing temperature increases the saturation magnetization (M_S) of the films increases and reaches the highest value of 254 emu/cc at Ta = 200 °C followed by decreasing trend with further increase in annealing temperature (Table 3) .

According to XRD observations, the as-deposited film is composed of nanocrystalline particles in amorphous background. AFM images also show the nanocrystalline nature of the films. These nanocrystals align in one direction at 200 °C. The nanocrystals merge forming bigger crystals as the annealing temperature is increased to 300 °C. The TEM observations on the films prepared by similar method by Nakashima et al. [45] show that the film is composed of crystalline nanoparticles dispersed in an amorphous matrix, and the amorphous phase is converted into the crystalline phase upon annealing at 300 °C. The enhancement of magnetization is presumably due to precipitation of a disordered $ZnFe_2O_4$ crystalline phase from the amorphous phase viz. the volume fraction of the ferrimagnetic $ZnFe_2O_4$ phase is increased by annealing the film at 200 and 300 °C.

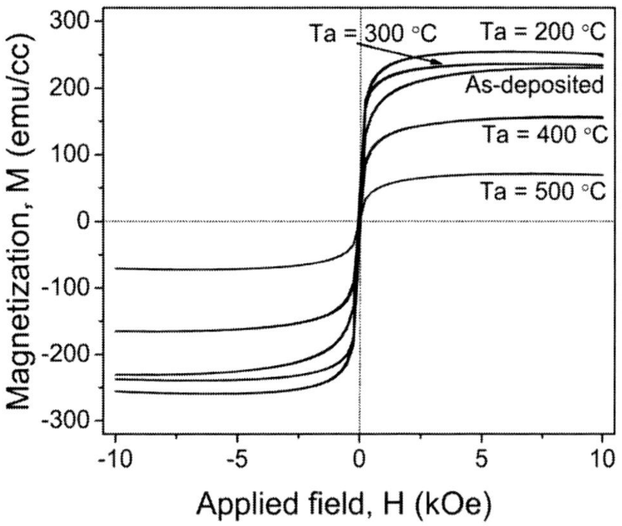

Figure 17. Magnetization curves of $ZnFe_2O_4$ thin films annealed in air for 3 hrs at temperatures in the range 200–500 °C.

The thermal annealing of the film at Ta > 300 °C decreases M_S and H_C values drastically (figure 17 and table 3). This reduction can be due to the redistribution of the cations among A- and B-sites where they take up their preferential sites as in the paramagnetic bulk sample and consequently reduce the disorder in the system.

4.3.3. Optical Properties:

Figure 18. shows the optical transmission spectra, as a function of photon energy in the wavelength range 350–2500 nm, of $ZnFe_2O_4$ thin films deposited on glass substrates in argon and oxygen at room temperature. The films have strong band edge absorption in the wavelength region of less than 600 nm. The fringes in the transmission spectra result from the interference of the incident light at the air–film, film–substrate and substrate–air interfaces. The optical transmittance of the films prepared in O_2 is more than 80% above the band edge and higher than the film deposited in argon. Zhi-hao *et al.* [46] reported a band edge of 700 nm for $ZnFe_2O_4$ nanoparticles and Wu *et al.* [47] reported 650 nm for the $ZnFe_2O_4$ film prepared by spray pyrolysis. The optical constants of the films such as refractive index (n), extinction coefficient (k) and absorption coefficient (α) estimated from the transmission spectra using the envelope method proposed by Manifacier *et al.* [48] and developed by Swanepoel [49] are reported elsewhere [33]. Figure 19. shows the RT transmittance spectra for the as-deposited and annealed films. The transmittance spectra show that all the films have sharp absorption edges at ~ 550 nm. The films exhibit high transmittance above the absorption edge. Post-deposition annealing in air influences the optical properties of $ZnFe_2O_4$ films. The film annealed at 200 °C has absorption edge at lower wavelength and enhanced transmittance which could be due to Burstein–Moss band-filling. However, the annealing above 200 °C decreases the optical transmittance and shifts the absorption edge towards higher wavelength. This can be due to the increase in grain size in the film and a change in the nature and strength of the interaction potentials between defects and host materials, which increase the tailing of the absorption edge.

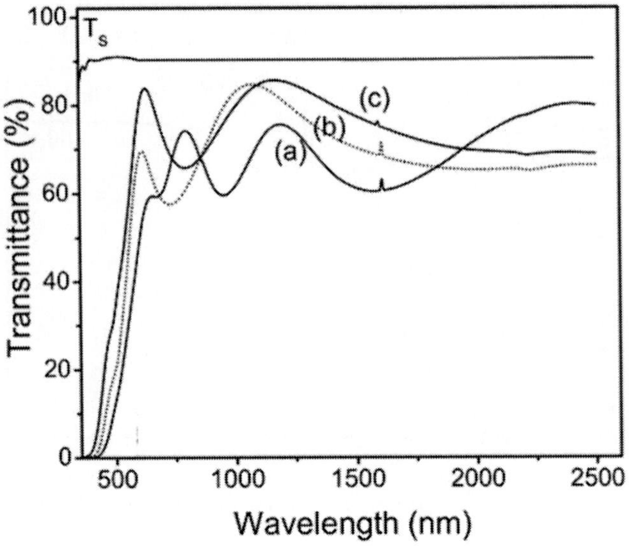

Figure 18. Spectral transmittance of Zn-ferrite thin films prepared under (a) 8 mT of Ar (b) 8 mT of O_2, (c) 31 mT of O_2. The inset shows the envelope for the data for film prepared under O_2 at 8 mT. The curve T_s is for glass substrate.

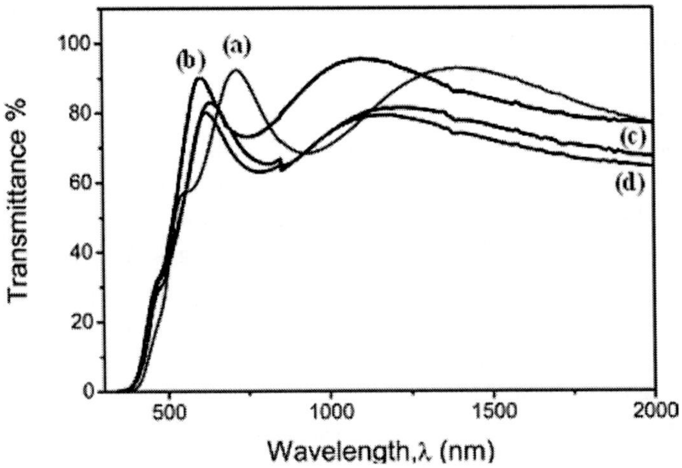

Figure 19. Optical transmittance spectra of the as-grown and annealed $ZnFe_2O_4$ films (a) as deposited, (b) annealed at 200°C, (c) annealed at 300°C, (d) annealed at 500°C.

5. Copper-Zinc Ferrite ($Cu_{0.6}Zn_{0.4}Fe_2O_4$) Thin Films:

This section includes the study of the structure, morphology, magnetic and optical properties of $Cu_{0.6}Zn_{0.4}Fe_2O_4$ thin films. This composition was chosen because bulk ferrite of this composition has the highest magnetization among all the compositions of Cu-Zn ferrite. The films were deposited from a sintered target of the above composition onto glass substrate using rf-magnetron sputtering system operating at 100 W of rf-power in argon and oxygen environment.

5.1. Effect of Ar Gas Pressure on Structure and Properties:

5.1.1. Structural Analysis (XRD)

Figure 20. shows the XRD patterns of the as-sputtered thin films at various pressure of pure Ar gas. All the peaks belong to spinel structure with cubic symmetry without any secondary or impurity phases. It can be seen that the peak broadening decreases with increase in Ar pressure. The average crystallite size increases from 5 to 21 nm with increase in Ar pressure from 5 to 15 mTorr. With increase in Ar content the relative intensity ratios of planes I_{220}/I_{400}, I_{440}/I_{400} and I_{422}/I_{400} increases confirming the increase in disorder parameter and consequently to the increase in magnetization . The diffraction data were fitted with a Lorentzian function to determine the location *(2θ)* of each peak. Using Bragg's law, the interplanar lattice spacing d_{hkl} and the cell parameter a were calculated. The variation of the average lattice spacing, i.e. the slight shift of the (hkl) peak position, in the direction normal to the plane of the film gives the measure of strain in the film; either compressive or tensile [50]. A characteristic shift towards lower angle compared to that of ideal crystal indicates lattice expansion. The films deposited at lower pressure show higher cell parameter (8.5Å) compared to the bulk due to the strain introduced in the film during deposition. The origin of the strain in the sputtered films may be related to several factors, including voids, argon inclusions and film substrate mismatch [51]. The strain reduces with increase in Ar pressure and the cell parameter approaches the bulk value of 8.41 Å. With increase in Ar gas pressure the phase collision increases which consequently reduces the kinetic energy of sputtered neutral atoms leading to decrease in strain [40].

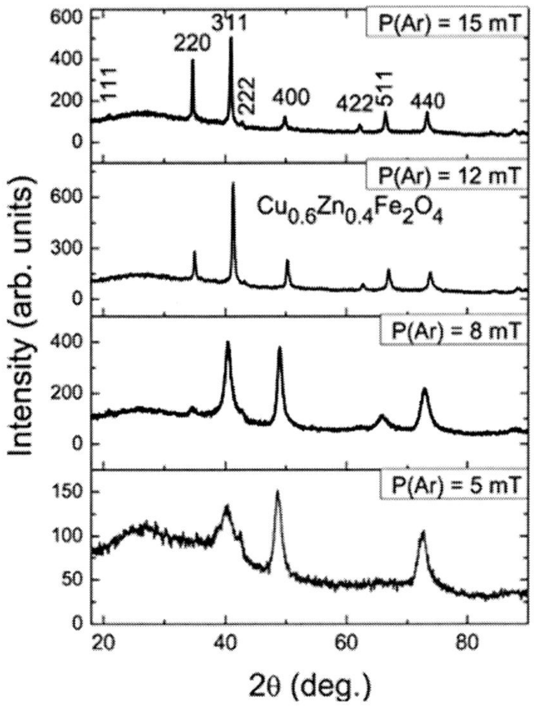

Figure 20. The XRD patterns of the as deposited $Cu_{0.6}Zn_{0.4}Fe_2O_4$ thin films under different Ar gas pressure.

5.1.2. Surface Morphology (AFM)

Figure 21. shows the AFM images of the films deposited at 5 and 12 mTorr of Ar pressure. The average particle size ⟨D⟩ and RMS values for various films are listed in table 4. The increase in particle size and decrease in (RMS) with increase in Ar gas pressure could be attributed to shadowing effects. There is a variation in the distribution of incident angles and energy of the depositing atoms at low and high deposition pressures. At low deposition pressure few collisions occur within the plasma; therefore incident angles tend to be close to normal and bombarding energy tends to be high. The high bombarding energy leads to a high variation in surface roughness. However, the increase in Ar gas pressure increases the number of collisions occurring in the plasma and consequently to an increase in the incident angles of the particles which leads to smoothening of the films. The results presented here are in good agreement with the model proposed in references [38-40].

5.1.3. Magnetization Studies (M-H Plots):

Figure 22. shows the in-plane M-H plots at RT for the films deposited at various Ar gas pressure. The magnetization (M_s) increases from 126 to 132 emu/cc as the pressure increases from 5 to 8 mTorr followed by sharp increase to 283 emu/cc at 12 mTorr. Further increase in Ar pressure to 15 mTorr leads to an increase in M_S value to 295 emu/cc.

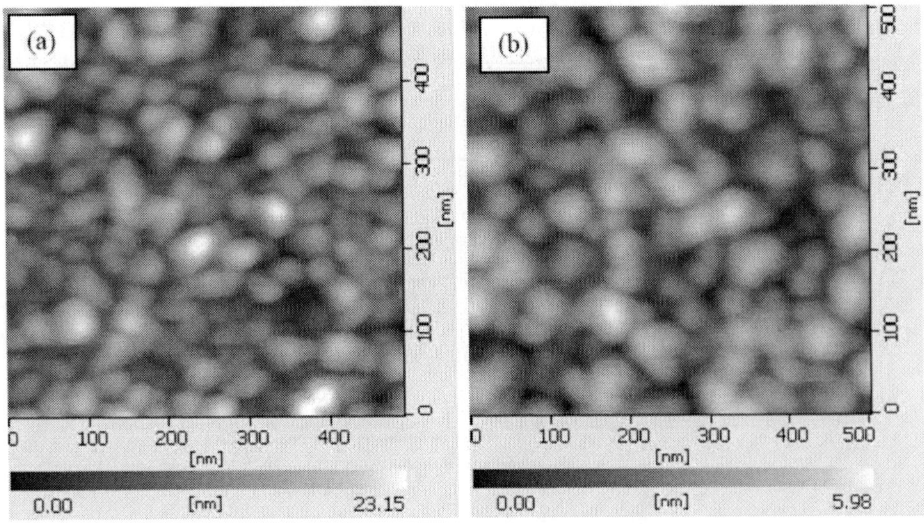

Figure 21. AFM images of the as-deposited $Cu_{0.6}Zn_{0.4}Fe_2O_4$ thin films at (a) 5 mTorr (b) and 12 mT.

Table 4. AFM and magnetic parameters of $Cu_{0.6}Zn_{0.4}Fe_2O_4$ thin films deposited under various Ar gas pressure

Sample Name	⟨D⟩ (nm)	RMS (nm)	M_S (emu/cc)	H_C (Oe)	M_R/M_S %
CZF-Ar5	33	3.8	126	170	25
CZF-Ar8	40	3	132	100	15
CZF-Ar12	43	1.15	283	90	32
CZF-Ar15	55	2.1	295	67	17

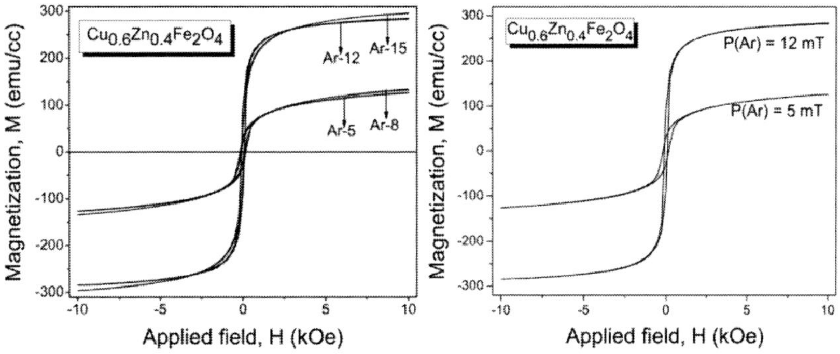

Figure 22. Effect of Ar pressure on M-H curves of $Cu_{0.6}Zn_{0.4}Fe_2O_4$ film.

The change in crystal structure and magnetization can be explained in view of preferential occupation of A and B –sites in spinel structure. The high deposition rate in pure Ar environment can lead to freezing of some Cu-ions on tetrahedral A-sites and equivalent number of Fe ions on octahedral B-sites during the deposition process. Furthermore, the deposition in reducing (argon) atmosphere may lead to the formation of Cu^+ ion, having larger ionic radius than the Cu^{2+} ion. The Cu^+ ions prefer occupation of the smaller four-coordinated A-site in the spinel structure and displace Fe^{3+} cations to occupy the B- sites. The formation of Fe^{2+} cannot be ruled out in the spinel structure where a fraction of Fe^{3+} will be replaced by Fe^{2+} ions. So the general cation distribution can be represented as

$$((Cu^{2+}Cu^{+})_d\, Zn^{2+}Fe^{3+}{}_{1-d})_{tetra}[Cu^{2+}{}_{1-d}(Fe^{2+}Fe^{3+})_{1+d}]_{octa}O_4{}^{2-}$$

where d is the inversion parameter. The occupancy of the nonmagnetic Cu^+ and Zn^{2+} ions on A-sites dilutes the magnetic moment on A-sites leading to large difference between magnetization on B and A sites causing a strong negative super-exchange interaction between Fe^{3+} ions on both sites which in turn cause ferrimagnetic behavior with large magnetization [32]. The value of coercivity, H_C decreases from 170 to 67 Oe with increase in Ar pressure from 5 – 15 mTorr. This can be correlated to the crystallite size and the presence of significant inhomogeneities. The smaller crystallites in the case of thin films deposited under low Ar pressure has larger volume of the boundaries and consequently more energy is required to rotate the grains by the external field and thus leading to higher coercivity.

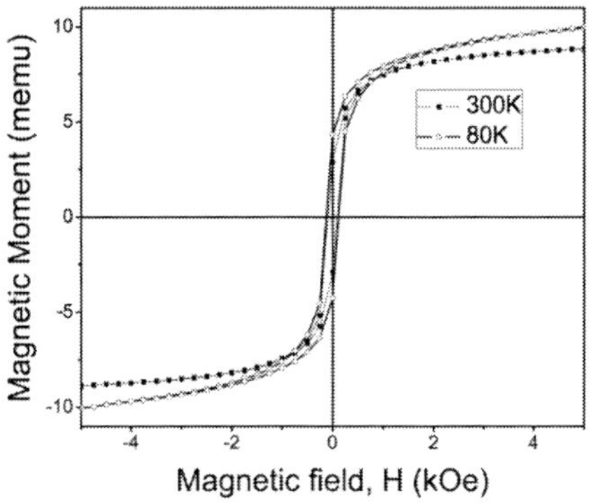

Figure 23. M-H plots for $Cu_{0.6}Zn_{0.4}Fe_2O_4$ thin film deposited at 12 mT measured at 300 and 80K.

Figure 23. shows the M-H plots for $Cu_{0.6}Zn_{0.4}Fe_2O_4$ thin film deposited at 12mT of Ar pressure measured at 300 and 80K. It can be seen that the saturation magnetization and coercivity are slightly higher at 80 K.

5.2. Effect of O_2 Gas Pressure on Structure and Properties:

5.2.1. Surface Morphology

The surface morphology of the $Cu_{0.6}Zn_{0.4}Fe_2O_4$ thin films was analyzed by atomic force microscope (AFM) using non-contact mode with amplitude modulation. The scan area was 1μm x 1μm (Figure 24).

The films exhibited a dense microstructure and the grain size was found to decrease with the increase in O_2 pressure. The average grain size for the films deposited at 5 and 12 mTorr of O_2 are 50 and 30 nm respectively.

The surface morphology of the films was smooth with no cracks and defects and the average surface roughness was 1.5 nm for films deposited under O_2 environment. There is no significant difference in the surface roughness for the deposited films in O_2 environment as compared to that deposited in Ar atmosphere. The narrow distribution of the grains size is observed for these films.

5.2.2. Magnetization Studies (M-H Plots)

The M-H plots at RT for the films deposited at 5, 8, 12 and 15 mTorr of O_2 gas pressure provide magnetization (M_s) values of the films are around 150 emu/cc for various oxygen gas pressures. The magnetization is not sensitive to oxygen gas pressure in these films unlike Cu and Zn ferrite thin films.

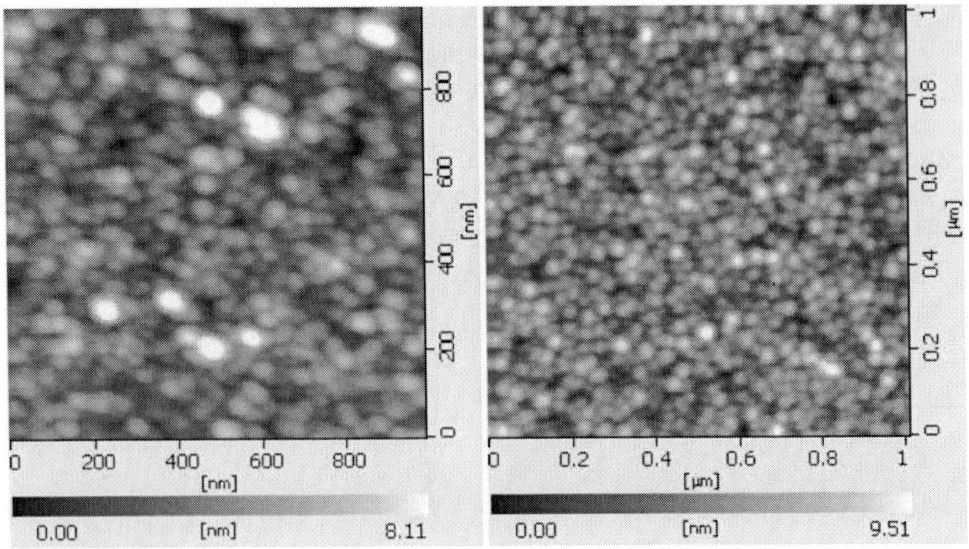

Figure 24. AFM images of $Cu_{0.6}Zn_{0.4}Fe_2O_4$ thin films deposited under 5 mTorr (left) and 12 mTorr (right) of Oxygen gas pressure.

5.3. Effect of (Ar+O_2) Gas Mixture Pressure on Structure and Properties:

5.3.1. Structural Analysis (XRD)

Figure 25. shows the effect of O_2 content in (Ar+O_2) mixture on the XRD patterns of as-deposited $Cu_{0.6}Zn_{0.4}Fe_2O_4$ thin films at 12 mTorr of gas pressure. The XRD data for the as-deposited films reveal the nanocrystalline nature of the films and confirm the spinel structure with cubic symmetry having the stoichiometry of the ceramic target (metallic cation/oxygen anion ratio equal to ~ 3/4). These results were also confirmed by EDAX measurements.

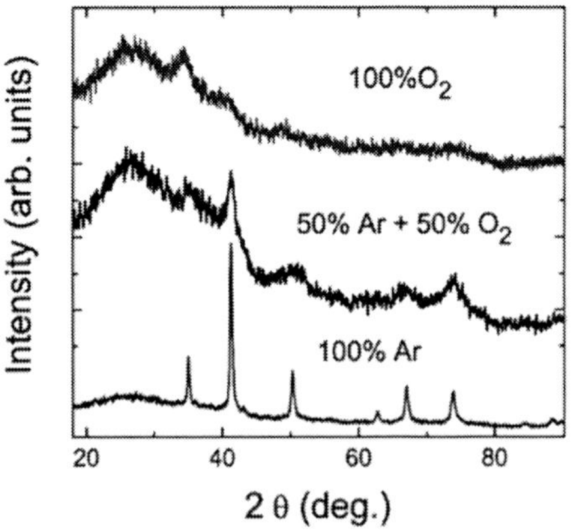

Figure 25. Effect of Ar-O_2 content on XRD patterns of $Cu_{0.6}Zn_{0.4}Fe_2O_4$ thin films deposited at 12 mTorr.

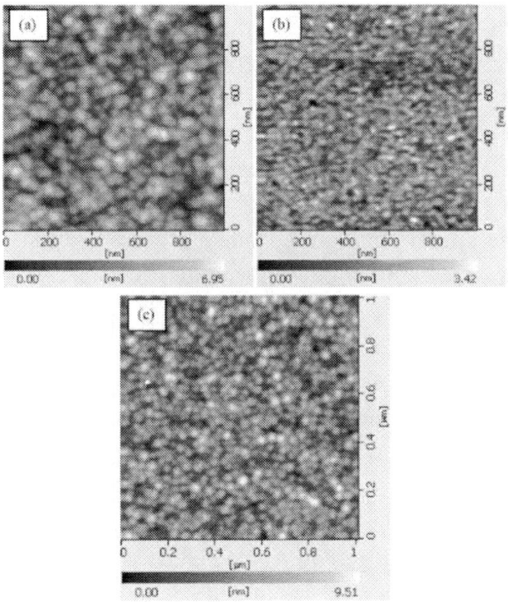

Figure 26. AFM images of $Cu_{0.6}Zn_{0.4}Fe_2O_4$ thin films deposited in 12 mTorr of (a) Ar (b) 50% Ar + 50% O_2 (c) O_2 gas pressure.

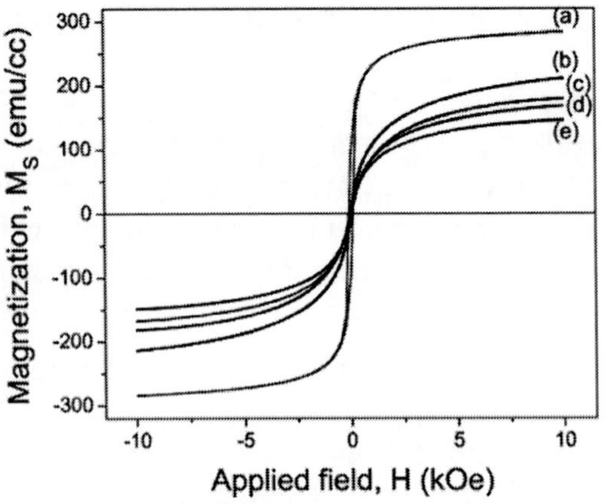

Figure 27. M-H curves measured at RT of $Cu_{0.6}Zn_{0.4}Fe_2O_4$ thin films deposited in 12 mTorr of (a) 100% Ar, (b) 90% Ar + 10% O_2, (c)75% Ar + 25% O_2, (d) 50% Ar + 50% O_2 and (e) 100% O_2 gas pressure.

5.3.2. Surface Morphology (AFM)

Figure 26. shows the AFM images of $Cu_{0.6}Zn_{0.4}Fe_2O_4$ thin films deposited at 12 mTorr of (a) Ar (b) 50% Ar + 50% O_2 (c) O_2 gas pressure. AFM data analysis reveals that the films are made up of grains with a mean diameter centered on 60, 20 and 30 nm for films deposited at 12 mTorr of Ar, Ar+O_2 and O_2 respectively. The corresponding average RMS values are 1.15, 0.8 and 1.5 nm. The AFM images of the films show a very narrow grain size distribution.

5.3.3 Magnetization Studies (M-H Plots)

Figure 27. shows the plots of magnetization versus applied magnetic field curves at RT for $Cu_{0.6}Zn_{0.4}Fe_2O_4$ thin films deposited in 12 mTorr of (Ar + O_2) gas pressure. The magnetization decreases with increasing O_2 content. The high deposition rate in pure Ar environment can lead to freezing of some Cu-ions on tetrahedral A-sites and equivalent number of Fe ions on octahedral B-sites during the deposition process. Furthermore, the deposition in reducing (argon) atmosphere may lead to the formation of Cu^+ ion, having larger ionic radius than the Cu^{2+} ion. The Cu^+ ions prefer occupation of the smaller four-coordinated A-site in the spinel structure and displace Fe^{3+} cations to occupy the B- sites . The formation of Fe^{2+} cannot be ruled out in the spinel structure where a fraction of Fe^{3+} will be replaced by Fe^{2+} ions.

5.3.4. Optical Properties

Figure 28. shows the transmittance spectra measured at RT for the as-grown $Cu_{0.6}Zn_{0.4}Fe_2O_4$ films deposited in 12 mTorr of (AR+ O_2) gas pressure. The transmittance spectra show that all films have sharp absorption edge at about 550 nm. The films exhibit high transmittance above the absorption edge. The variation of Ar-O_2 ratio influences the optical properties of the ferrite films. The Ar-deposited films show less transmittance due to the higher thickness. With increase in the O_2 percentage the film thickness decreases and consequently gives rise to shifting of the absorption edge towards lower wavelengths with enhancement in transmittance. This can also be due to the decrease of grain size and increase in porosity of the film.

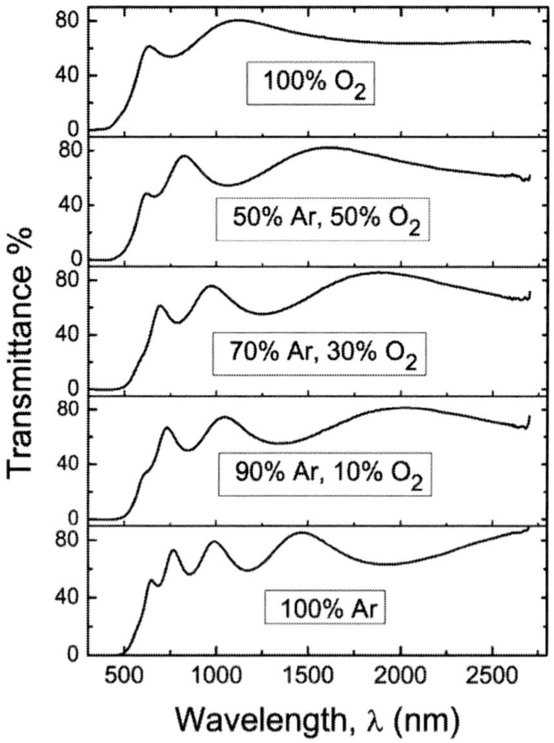

Figure 28. Optical transmittance spectra of the as-grown $Cu_{0.6}Zn_{0.4}Fe_2O_4$ films at 12 mTorr of Ar, O_2 and Ar+O_2.

CONCLUSION

The thin films of $CuFe_2O_4$, $ZnFe_2O_4$ and $Cu_{0.6}Zn_{0.4}Fe_2O_4$ were deposited by RF – sputtering with RF power of 100 W in pure Ar, O_2 and mixture of (Ar+O_2) at various process gas pressure. The detailed study of crystal structure, magnetic and optical properties of the as-deposited and heat treated thin films of these ferrites were carried out. The XRD and AFM studies of the as-deposited films indicate nanocrystalline cubic spinel structure. The crystal structure and properties of the films depend on the process gas environment composition and pressure. The heat treatment of the films affects the properties markedly. The present study shows that the changes in the film properties with working gas environment and heat treatment are due the presence of multivalent copper and iron ions with differing site preferences.

REFERENCES

[1] Wohlfarth, E. P.*Handbook of Magnetic Materials*, , North-Holland,1982.3.

[2] Willard, M. A.;Kurihara, L. K ; Carpenter, E. E. ; Calvin S.; Harris, V. G. *International Materials Reviews* 2004, 49,125–170.

[3] Cullity B. D.; Graham, C. D Introduction to magnetic materials,2009, 2nd Ed. IEEE Press. USA

[4] Lue, J. T. Physical Properties of Nanomaterials, in *Encyclopedia of Nanoscience and Nanotechnology* Ed.by H. S. Nalwa, American Scientific ,2007, Vol. X, 1 –46.

[5] Suzuki, Y. Annu. *Rev. Mater. Res*. 2001, 31 ,265–89.

[6] Zaquine, I.; Benazizi H. and Mage, J. C. *J. Appl. Phys*. 1988, 64, 5822–5824.

[7] Celozzi, S. ; Araneo R.;Lovat, G . Electromagnetic Shielding, Wiley, New Jersey ,2008.

[8] Sickafus, K. E.; Wills J. M.; Grimes, *N. W. J. Am. Ceram. Soc*. 1999, 82, 3279–92.

[9] Willard, M. A.;Nakamura, Y.; Laughlin D. E. ; McHenry, M. E. *J. Am. Ceram. Soc*.1999, 82, 3342–46.

[10] Miclea, C. ; Tanasoiu, C.; Miclea, C. F.; Gheorghiu, A.; Tanasoiu, V..*J. Magn. Magn. Mater*. 2005, 290– 291 ,1506–1509.

[11] Srinivasan, G. ; Rao, B. U. M. ; Zhao,J.; Seehra, *M*. S. *Appl. Phys. Lett*. 1991 ,59, 372-374.

[12] lbrahim, M. M.; Seehra M. S.; Srinivasan,*G. J. Appl. Phys*. 1994 75, 6822.

[13] Baubet, C.; Tailhades, Ph.; Bonningue, C.; Rousset, A; Simsa, *Z. J. Phys. Chem. Solids* 2000, 61,863–867.

[14] Desai, M.; Prasad, S.; Venkataramani, N.; Samajdar, I.; Nigam, A. K.;Krishnan, *R.J. Magn. Magn. Mat*er. 2002, 246,266–269.

[15] Desai, M.; Prasad, S.; Venkataramani, N.; Samajdar, I.; Nigam A. K.; Krishnan, *R. J. Appl. Phys*. 2002, 91,2220- 2227.

[16] Kulkarni, P. D.; Desai, M.; Venkataramani, N.; Prasad S.; Krishan, *R. J. Magn. Magn. Mater*. 2004,272–276 ,e793–e794.

[17] Mugnier, E. ;Pasquet, I.; Barnabe,′ A.; Presmanes, L.; Bonningue C; Tailhades, *P. Thin Solid Films* 2005,493, 49–53.

[18] Ahmad, M.; Desai M.; Khatirkar, *R. J. Appl. Phys*. 2008, 103, 013903.

[19] Yang, A. ; Zuo, X.; Chen, L.; Chen, Z.; Vittoria, C.; Harris,*V.G. J. Appl. Phys*. 2005,97, 10G107.

[20] Zuo, X. ;Yang, A.; Vittoria, C.; Harris, *V. G. J. Appl. Phys*. 2006, 99, 08M909.

[21] Sultan , M.; Singh,R. *Mater. Lett*. 2009, 63,1764.

[22] Tanaka, K. ; Nakashima, S.; Fujita K.; Hirao, *K. J. Phys.: Condens. Matter* 2003 ,15, L469.

[23] Nakashima, S.; Fujita, K.; Tanaka, K.; Hirao, K.; Yamamoto T.; Tanaka, I.*Phys. Rev. B* 2007 75,174443.

[24] Nakashima, S. ;Fujita, K.; Nakao, A.; Tanaka, K.; Shimotsuma, Y.; Miura K.; Hirao, K..*Appl. Phys. A* 2009 ,94, 83–88.

[25] Wakiya, N. ;Muraoka, K.; Kadowaki, T.; Kiguchi, T.; Mizutani, N.; Suzuki H.; Shinozaki, K.. *J. Magn. Magn. Mater*. 2007 ,310 2546–2548.

[26] Ogale S. B. ; Nawathey, *R.J. Appl. Phys*. 1989 ,65 ,1367-1369.

[27] Sorescu, M. ;Diamandescu, L.; Swaminathan, R.; McHenry M. E,; Feder, *M. J. Appl. Phys*. 2005, 97, 10G105.

[28] Bohra, M.; Prasad, S.; Kumar, N.; Misra, D. S.; Sahoo, S.; C. Venkataramani N.; Krishnan, R. *Appl. Phys. Lett*. 2006 ,88, 262506.

[29] Chen, Y. F. ; Spoddig D.; Ziese, *M. J. Phys. D: Appl. Phys*. 2008 ,41, 205004.

[30] Yahiro, H.; Tanaka, H.; Yamamoto Y.; Kawai, *T. Solid State Comm*. 2002 ,123 ,535–538.

[31] Sultan ,M; Singh, *R. J. Appl. Phys.* 2009 ,105, 07A512.
[32] Sultan ,M; Singh, *R. J. Phys. D: Appl. Phys.* 2009 ,42 ,115306.
[33] Sultan ,M; Singh, *R. Solid State Physics* (India) 2005 ,50, 691-2.
[34] Sultan ,M; Singh, *R. J. Appl. Phys.* 2010 , 107, 09A510.
[35] Ohnishi, H.; Teranishi, *T. J. Phys. Soc. Jpn.* 1961 ,16, 35–43.
[36] Cullity B. D.; Stock, S. R. Elements of X-Ray Diffraction, Prentice Hall, NJ, 2001.
[37] Capdeville, S.; Alphonse, P.; Bonningue, C.; Presmanes, L.; Tailhades, *P. J. Appl. Phys.* 2004,96 , 6142
[38] Sandu, I. ;Presmanes, L.; Alphonse ,P.; Tailhades, *P. Thin Solid Films* 2006, 495 ,130.
[39] Oudrhiri-Hassani, F.; Presmanes, L.; Barnabe, A.; Tailhades, *P. Appl. Surf. Sci.* 2008, 254, 5796.
[40] Goya ,G. F.; Rechenberg, *H. R. J. Magn. Magn. Mater.* 1999 ,196–197 ,191.
[41] Kodama, R. H. *J. Magn. Magn. Mater.* 200, (1999) 359.
[42] Ammar,S.; Jouini,N.; Fievet,F.; Beji,Z.; Smiri, L.; Moline, P.; Danot, M.; Greneche,J.-*M. J. Phys.: Condens. Matter* 2006,18, 9055.
[43] Choi, C-H.; Kim, S-H. *J. Crystal Growth* 2005 ,283 ,170–179.
[44] Nakashima, S.; Fujita, K.; Tanaka, K.; Hirao, K.; Yamamoto, T.;Tanaka, I. *J. Magn. Magn. Mater.* 2007, 310, 2543–2545.
[45] Zhi-hao, Y.; Wei, Y.; Jun-Hui, J.; Li-de, Z. *Chin. Phys. Lett.* 1998, 15,535.
[46] Wu, Z.; Okuya, M.; Kaneko, S. *Thin Solid Films* 2001, 385,1090.
[47] Manifacier, J. C.; Gasiot J; Fillard,J.P. *J. Phys. E: Sci. Instrum.* 1976, 9,1002.
[48] Swanepoel, *R.J. Phys. E: Sci. Instrum.* 1983, 16,1214.
[49] Ahmad ,M.; Desai, *M. J. Magn. Magn. Mater.* 2008, 320,L74.
[50] Yang, A.; Chen, Z.; Zuo, X.; Arena ,*D. Appl. Phys. Lett.*2005,. 86, 252510.

Reviewed by Prof. T.K.Dey,Thermophysical Laboratory,
Cryogenic Engineering Center,Indian Institute of Technology
Kharagpur, West Bengal,India

In: Magnetic Thin Films
Editor: John P. Volkerts

ISBN: 978-1-61209-302-4

Chapter 8

$L1_0$ FePt (001) Thin Films for Perpendicular Magnetic Recording Media

Y.F. Ding, X. Zhao and E. Liu
School of Mechanical and Aerospace Engineering, Nanyang Technological University
50 Nanyang Avenue, Singapore 639798, Singapore

Abstract

Chemically ordered $L1_0$ (CuAu-I type structure) FePt thin films with a FCT (face centered tetragonal) structure possess a high K_u (magnetic anisotropy constant) value (>10^7 erg/cm^3) that allows them to have very small thermally stable magnetic grains (~ 2.6 nm) and makes them the most promising candidates for recording media having an ultra-high areal density. Usually, as-deposited FePt alloy thin films are a disordered FCC (face centered cubic) phase and tend to show a (111) texture.

A high temperature (above 550 °C) post annealing is needed to form an ordered $L1_0$ phase. In order to develop FePt films as perpendicular magnetic recording media, $L1_0$ FePt ordering temperature has to be reduced and the easy axis of the films should be perpendicular to the film surfaces. Moreover, the grain size and exchange coupling between grains should be decreased. Thus, this chapter reviews two main issues for $L1_0$ FePt thin films to be successful perpendicular magnetic recording (PMR) media: (1) how to lower $L1_0$ FePt ordering temperature while getting crystallographic texture controlled at the same time, and (2) how to reduce grain size and inter-granular exchange coupling. The first issue is related to the study of the intrinsic relationships between the chemical ordering, lattice constant, and anisotropy energy of FePt films. It is demonstrated that residual stress can assist to expand the a-axis and shrink the c-axis of FePt films and thus favors the chemical ordering of the films at relatively low temperatures. The second issue covers the study of the structural and magnetic properties of FePt films grown on a CrX alloy underlayer (X=Mo, Ru, Ti, or W). It is demonstrated that the texture of FePt films strongly depends on the texture of CrX underlayers, for which Cr (200) texture and substrate temperature (T_s) are the key parameters for the growth of the FePt (001) films.

To decrease the grain size and maintain the (001) preferred orientation in FePt films, the effects of carbon additive on the structural and magnetic properties of FePt films are also discussed. A good FePt (001) texture can be maintained even with C content up to 20 vol.% in FePt films by a fine control of sputtering deposition process. It is also highlighted that a small amount of C doping can slightly enhance the ordering degree, while a large amount of C doping can prohibit the ordering degree. Well-defined grain boundaries and relatively uniform grains with a mean diameter of about 5.6 nm and a standard deviation of about 1.6 nm can be achieved by doping 20 vol.% C in FePt films.

1. Background

The storage density of hard disk drives increased by over 100% per annum during the late 1990s and reached 100 Gb/in^2 in 2002 as shown in Figure 1. The increases in data density, data rate, and other performance metrics have generally been achieved by scaling to make the read-write head smaller in size, the media thinner in thickness and higher in coercivity, and the head-medium spacing smaller. The performance of a thin film medium is limited by the noise originated from the granular microstructure of the film, so there has been a trend to decrease the grain size of the recording layer. However, for Co-based media, when the grain size is reduced to below 7 nm, the thermal energy stored in the grains is sufficient to allow their magnetization to reverse spontaneously, with a consequent loss of the recorded signals (so-called superparamagnetism).

In 2005, CoCrPt-Oxide based perpendicular magnetic recording (PMR) media were successfully implemented into hard drives by Seagate Technology. After that, almost all of the hard drive companies started to implement PMR media into their products. The capacity of CoCrPt-Oxide based PMR media already reached 500 Gb/in^2 in 2010.

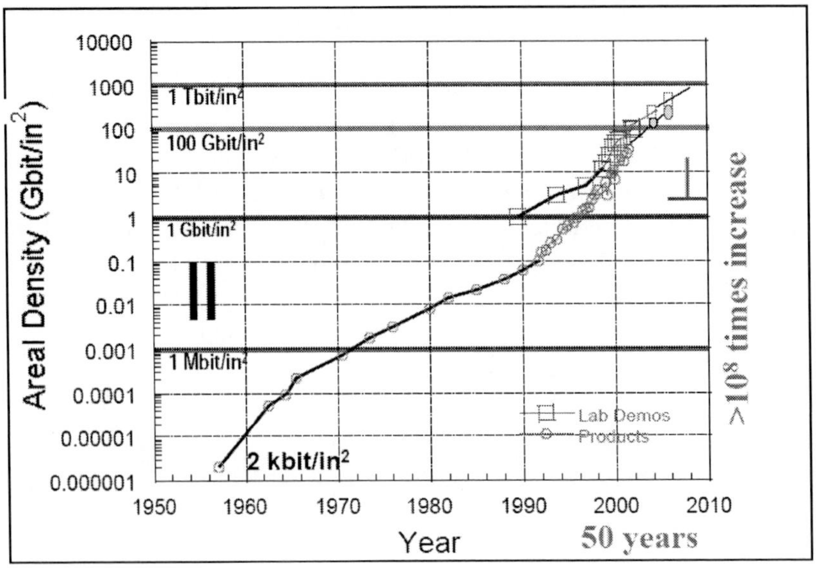

Figure 1. Evolution of recording density of magnetic media (From Weller D., IEEE Distinguished Talk (2005)).

However, it was expected that CoCrPt-Oxide based PMR media would reach their recording areal density limit at around 1 Tb/in^2 in 2013. To keep the current recording density growth rate (~ 40%/year), an alternative is needed to achieve ultra-high data storage and a new recording medium with high K_u is needed to achieve 1 Tb/in^2 and beyond.

Table 1 shows a compilation of candidate materials for perpendicular recording media. The last three columns show minimal thermally stable single grain diameters D_p calculated based on the stability criterion outlined with $r_K \sim ln(\tau f_0) \sim 50$, for (a) cylindrical grains of constant height $\delta = 10$ nm, (b) cubic grains with $D = \delta$ and (c) spherical grains.

Here, $H_K = 2K_1/M_S$: anisotropy field; K_1: intrinsic first order anisotropy constant; D_P: minimal stable grain size ($\tau = 10$ years; $T = 350$ K); T_C: Curie temperature; $^{(a)}$: cylindrical grains of constant height $\delta = 10$ nm; $^{(b)}$: cubic grains; and $^{(c)}$: spherical grains. Among the candidate materials shown in Table 1, the chemically ordered FePt with $L1_0$ (CuAu-I) structure (FCC) possesses very high magnetic anisotropy constant ($6.6\text{-}10\times10^7$ erg/cm^3), large Ms, high corrosion resistance, relatively low T_c and small thermally stable grain size, all of which make it the most promising candidate for ultra-high areal density (>1000 Gb/in^2) recording media. $L1_0$ FePt is preferred due to the poor corrosion resistance of rare-earth alloys.

Table 1. Magnetic properties and theoretical minimal grain diameters of various perpendicular media candidates. The minimal physical grain diameter is estimated using maximum demagnetization 4πMS and constant thickness of δ = 10 nm (a), cubic grains Dp (b), and spherical grains (c). (From Weller D., Lu B., and Kryder M., IEEE Distinguished Talk, (2005))

alloy system	Material	K_1 (10^7erg/cm^3)	M_S (emu/cm^3)	H_K (kOe)	T_C(K)	$D_p^{(a)}$ (nm)	$D_p^{(b)}$ (nm)	$D_p^{(c)}$ (nm)
	$CoCr_{20}Pt_{15}$	0.25	330	15.2	--	15.5	12.4	15.4
Co-alloys	Co_3Pt	2.0	1100	36.4	--	6.4	6.9	8.5
	$(CoCr)_3Pt$	0.39	410	19.0		12.4	10.6	13.2
	$CoPt_3$	0.5	300	33.3	600	9.0	8.6	10.7
CoX/Pt(Pd)	Co2/Pt9	1	360	55.6	500	6.1	6.7	8.3
multilayers	Co2/Pd9	0.6	360	33.3	500	8.4	8.2	10.2
	FePd	1.8	1100	32.7	760	7.3	7.5	9.3
$L1_0$	FePt	7	1140	122.8	750	2.4	3.6	4.4
phases	CoPt	4.9	800	122.5	840	2.8	3.9	4.9
	MnAl	1.7	560	60.7	650	4.9	5.7	7.1
Rare-earth	$Fe_{14}Nd_2B$	4.6	1270	72.4	585	3.4	4.5	5.5
transition m.	$SmCo_5$	20	910	219.8	1000	1.9	3.0	3.8

2. PHYSICAL PROPERTIES OF FePt

The properties of FePt alloy systems were first studied systematically in 1907. A transformation between ordered and disordered phases was observed in the equiatomic composition range (the transformation temperature is about 1300 °C), which was confirmed by measurements of crystal structure and magnetic, electrical and mechanical properties. Kussman and Rittberg found three stable crystal structures existing in the Fe-Pt system, i.e., $FePt_3$, FePt and Fe_3Pt as illustrated in the phase diagram in Figure 2. The phases and properties of these have been documented by Hansen and Bozorth.

As shown in Figure 2, a disordered γ-phase exists above the transformation temperature (1300 °C), which has a FCC crystal structure. In 1941, Lipson H, Schoenburg D, and Rittberg G.V reported that the ordered r_1-phase of the FePt alloy has a CuAu-I type structure with lattice parameters $a = 3.838$ Å and $c = 3.751$ Å with the structure shown in Figure 3.

The magnetic properties of Fe-Pt alloys have been studied since 1930's. Fallot determined that an equiatomic alloy is a ferromagnet with a Curie temperature of 670K. Kussman and Rittberg found that the saturation magnetization is greater for a disordered alloy than for an ordered alloy. The FePt $L1_0$ alloy has uniaxial magnetocrystalline anisotropy and K_1 has been measured as about 7.0×10^6 J/m^3 for bulk alloy. A similar value, $K_1 = 6\times10^6$ J/m^3, was measured for thin film. In comparison, the disordered alloy has cubic anisotropy and $K_1 = 6\times10^3$ J/m^3 and the ordered alloy has a saturation magnetization of 1150 emu/cm^3 at 298 K. The critical diameter for a single domain FePt particle is around 600 nm. The thickness of a domain wall in the FePt bulk alloy is around 3.9 nm.

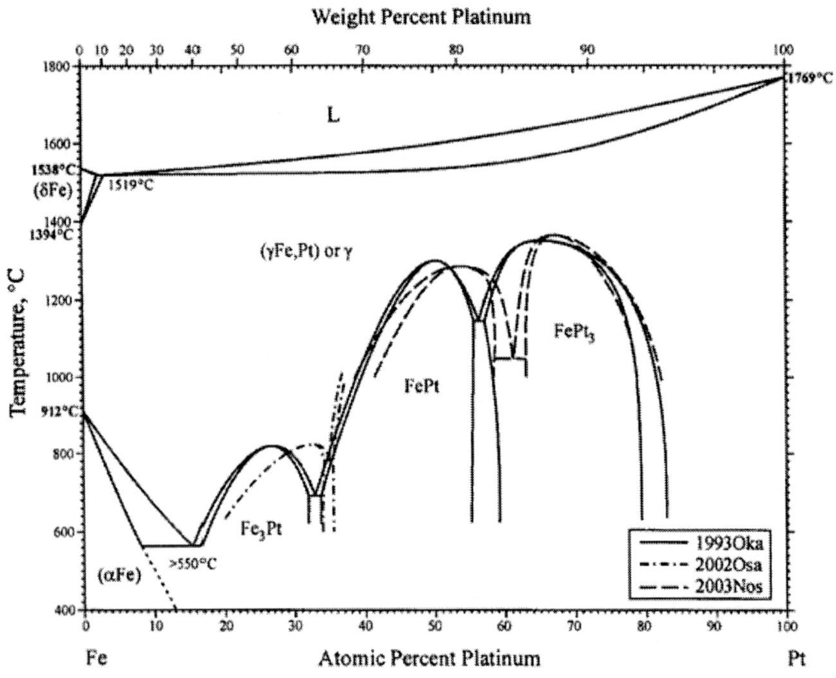

Figure 2. Phase diagram for iron-platinum alloys.

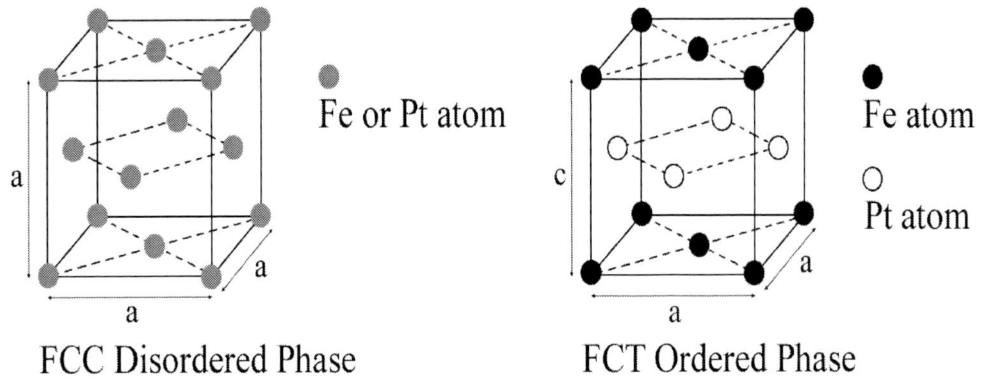

Figure 3. Crystal structures of disordered and ordered equiatomic FePt alloy.

3. Challenges for FePt Recording Media

Usually, as-deposited FePt alloy thin films are a disordered FCC phase and tend to show a (111) texture. A high temperature (above 550 °C) post annealing is needed to form an ordered $L1_0$ phase. In order to develop FePt films as perpendicular magnetic recording media, the $L1_0$ FePt ordering temperature has to be reduced and the easy axis of the films should be perpendicular to the film surface. Moreover, the grain size and exchange coupling between grains should be decreased. Thus, this chapter reviews two main issues: (1) how to lower the $L1_0$ FePt ordering temperature while getting the crystallographic texture controlled at the same time and (2) how to decrease the grain size and inter-granular exchange coupling.

In order to develop FePt as recording media, the $L1_0$ FePt ordering temperature has to be reduced and the easy axis should be either perpendicular or parallel to the thin film plane according to the recording mode.

3.1. Fabrication Methods of $L1_0$ FePt (001) Thin Films

Many researchers made much effort to produce $L1_0$ ordered FePt films with (001) preferred orientation by a variety of methods. So far, the methods for producing $L1_0$ ordered FePt films with (001) preferred orientation can be roughly sorted as follows:

(a) Molecular beam epitaxial (MBE) deposition of FePt films on MgO (100) single crystal substrates at elevated temperatures. Lipson H, Schoenburg D, and Rittberg G.V reported formation of c-axis oriented epitaxial FePt with perpendicular magnetic anisotropy by annealing (001) oriented multiplayer precursors and obtained magnetic properties far superior to those of the as-deposited films or comparable disordered alloys. This processing, while resulting in films with desirable properties, is somewhat complicated, involving both multiplayer deposition and post annealing. Simplification of this was achieved by using MBE to directly synthesize c-axis oriented ordered FePt from element sources.

(b) Co-sputtering deposition of FePt films on single crystal MgO (200) substrates at elevated temperatures. Airson B.M, Visokay M.R, Marinero. E.E, Sinclair R, and Clemens B.M. reported a further simplification of processing by directly synthesizing an epitaxial c-axis oriented FePt film by co-sputtering Fe and Pt on a single crystal MgO (200) substrate at a high deposition temperature (≥400 °C), with resulting properties comparable to those reported by the previously mentioned researchers. As of today, this method has been widely used to study the structure and magnetic properties of FePt films for fundamental purpose, but the use of crystalline MgO substrates are usually not accepted by industry.

(c) Deposition of FePt films on MgO or Ag underlayers. Hsu Y.N reported that a thick (175-nm) Ag underlayer on silicon substrate was found not only to induce epitaxial growth of FePt films, but also to reduce the FePt ordering temperature. In 2003, Kang K, Zhang Z. G, Papusoi C, and Suzuki T obtained (001)-textured nanogranular FePt composite films on amorphous quartz substrates using an MgO seedlayer and an Ag underlayer prior to the FePt/Ag multilayers followed by annealing at 550 °C for 1 h. They found that a 20 nm thick Ag underlayer played a significant role in improving (001) texture and also in decreasing the FCT FePt ordering temperature.

(d) Non-epitaxial methods with Fe/Pt, FePt/B_2O_3, FePt/SiO_2, Fe/Pt/C, Fe/Pt/Ag: Zeng H. and co-authors reported that highly oriented, FCT ordered FePt thin films can be grown non-epitaxially by directly depositing $(Fe/Pt)_n$ multilayers on thermally oxidized Si substrates followed by annealing. Although this process results in films with desired magnetic properties, it involves both multilayer deposition and long time post annealing at elevated temperatures (≥550 °C). The orientation mechanism of this method is still not quite clear. Luo C.P. and co-authors obtained nanostructured FePt:B_2O_3 and FePt:SiO_2 thin films with perpendicular magnetic anisotropy by annealing FePt/B_2O_3 and FePt/SiO_2 multilayers. This method needs rf-sputtering and long time post annealing at high temperature, which is very challenging for industry.

3.2. Methods for Lowering Ordering Temperature of $L1_0$ FePt Thin Films

As discussed above, it is desirable to find a way to produce $L1_0$ FePt magnetic thin films at low temperature. For this purpose, several attempts were made to fabricate the $L1_0$ FePt films at low temperatures by MBE and multilayer films by sputtering. However, both methods require a rigorous control for film processing and are slow fabrication processes. The development of a cheaper and rapid low-temperature processing of $L1_0$-ordered thin films by sputtering process is more desirable for successful industrial application. One method of fabricating $L1_0$-ordered FePt films at low temperature is to grow a $L1_0$ film in situ on a heated substrate. For example, Suzuki T., Harada K., Honda N., and Ouchi K. reported an in situ growth of an $L1_0$ FePt film under a high Ar pressure at a substrate temperature of 400 °C. Takahashi Y.K, Ohnuma M., and Hono K. reported that a fully ordered $L1_0$ FePt film can be growth in situ on a heated substrate only at 300 °C. Hsu Y.N also reported the fabrication of an epitaxial $L1_0$ FePt sputtered film using an Ag underlayer. These in-situ annealing processes are among the most promising methods for preparing $L1_0$-ordered FePt films. Another interesting approach is to reduce the temperature for ordering by adding ternary elements to the FePt binary alloy. For example, Seong-Rae Lee reported that Zr-doped FePt alloy films showed accelerated ordering transformation kinetics accompanying smaller grains then pure FePt alloy, Tomoyuki Maeda found that the addition of Cu to an FePt alloy film is an effective approach for reducing the ordering temperature of FePt. Takahashi Y.K., Ohnuma M., and Hono K. reported that the temperature that A1 phase orders to the $L1_0$ structure decreased from 500 to 400 °C by alloying 4 at% Cu. Ravelosona D., Chappert C., and Mathet V. reported that the long-range order parameter S of sputtered FePt (001) films may be improved by using post growth with He ion irradiation.

3.3. Methods for Controlling Grain Size of FePt Thin Films

In order to control grain size of magnetic particles as well as to achieve magnetic isolation among grains, various substitutions of nonmagnetic elements (in multiplayer fashion, introduction of buffer layer, and/or exchange-decoupling layer between FePt layer and soft-magnetic underlayer) have been investigated in FePt systems. These substitutions obviously determine the mechanisms of grain growth, inter-grain interactions.

Most of the studies on earlier systems are followed by annealing the (001) oriented multilayer precursors and obtaining the magnetic properties superior to those of the as-deposited films. Although this process may result in some desired magnetic properties, it involves both multilayer deposition and long time post-deposition annealing. Moreover, this method has disadvantages as a manufacturing process because such a high temperature post annealing can cause irreversible changes in substrates for magnetic media and enhance particle growth considerably. Ko H.S, Perumal A, and Shin S.C reported that the ordering and grain growth kinetics of FePt films could be finely controlled by C doping. Despite C doping, materials such as AlO_x, Al_2O_3, AlN, TaN, Ag, W and Ti have also been widely studied and shown to be promising in controlling FePt grain size in ordered films.

Other studies involving FePt with Ni, B, BN as additives have demonstrated that these can reduce magnetic coupling. Oxide addition to FePt can provide a better grain size control and reduce magnetic coupling in the FePt film. Luo C.P and Sellmyer D.J reported well-separated grains with strong perpendicular orientation in nanostructured FePt/B_2O_3 multilayers while FePt/SiO_2 multilayers showed a limited grain size of about 10 nm but poor orientation. HfO_2 and MnO additives were studied by C.L. Platt et.al, where the HfO_2 additive reduced the grain size with a random texture. MnO additive affected neither grain size nor $L1_0$ ordering of annealed FePt films. Suzuki T. and Ouchi K. reported that the MgO additive reduced the domain size of FePt film. In 2008, Ding Y.F., Chen J.S., Lim B.C., Hu J.F., Liu B., and Ju G. reported that the FePt:TiO_2 (001) film with about 5 nm grain size, adjustable coercivity and reduced exchange coupling was obtained by doping 20 vol.% TiO_2 into FePt film.

3.4. Self-organized Magnetic Arrays

In 2000, Sun S.H. and co-works reported the formation of FePt nano-particles with self-organized-magnetic-arrays (SOMA). SOMA shows promising to fabricate the FePt nanoparticles by nucleation and growth processes separated from each other during chemical synthesis. This allows for the formation of nearly mono-disperse magnetic nanoparticles, e.g., FePt and Co. These nanoparticles are encapsulated in a surfactant molecule shell, which prevents agglomeration in solution as well as during subsequent self-assembly and drying processes. So-formed SOMA media may serve as (i) conventional media with reduced dispersions, (ii) bit-patterned media with bit-transitions defined by rows of particles, and (iii) single-particle-per-bit recording media scaled by (ii).

In this last scenario, the ultimate areal density capability of SOMA media is governed by the minimal thermally stable size and center-to-center spacing of neighboring particles. Under most optimistic conditions, i.e. assuming that writability, signal retrieval, bit-addressability and spacing issues can be solved, areal densities of up to 40 Tb/in^2 may be reached.

However, to use such SOMA media for either conventional (longitudinal or perpendicular) or bit-patterned media, a number of obstacles have to be overcome. First, while self-organization and size distribution of better than 5% have been found in as-deposited FePt SOMA structures, annealing, as required to convert FePt material into a desired magnetically hard FCT ($L1_0$) phase, has been found to destroy the array order and size distribution via metallurgical sintering processes.

Proposals to avoid sintering include synthesis of core-shell structures with harder shell (diffusion barrier) and enhanced adhesion of the particles to the substrate surface via chemical modification of substrates as well as particle legend shell. It has been shown that by sandwiching FePt nanoparticle layers between carbon films, sintering can be avoided even after heating to 700 °C. Second, all FePt SOMA structures reported to-date have more or less random easy-axis orientation, which is not acceptable for future high density media as it will result in extra media noise and reduce the signal strength. This appears to be one of the most difficult unsolved problems today. Attempts to anneal SOMA structures in large fields or deposit them in the presence of a magnetic field so far, to our knowledge, have failed to produce oriented magnetic structures. Third, FePt SOMA nanoparticles today are approximately spherical (truncated octahedral before anneal).

The surfactant shell coating is between 1-2 nm thick (oleic acid, oleylamine). Scaling under these conditions is much more unfavorable for cylindrical grains as advocated in conjunction with PMR or HAMR media. Finally, self-organization is limited to dimensions of order one to ten micrometer. To overcome these limitations two-step processes involving lithographic pre-patterning of substrates and subsequent nanoparticle assembly on smaller scales of 100 - 10 nm have been proposed. This new concept could be used to push out the superparamagnetic limit. In rotating storage one would align the lithographically pre-patterned regions into a circumferential structure. For alternative probe storage devices one could choose an x-y pattern, while both topographic and chemical pre-patterning schemes have been proposed.

3.5. Nanoclusters

Another approach to controlling of FePt grain size is nano-cluster. In 2003, Stappert and co-workers reported a technique to prepare FePt nanoparticles from a gas-phase that aimed to obtain L1$_0$ ordered FePt nanoparticles by thermal sintering in the gas-phase. The particles in a size range of 3 - 20 nm are generated via inert-gas condensation using DC-sputtering at very high pressures in a millibar range. This method shows to be promising for controlling of FePt grain size.

However, to use such media for either conventional (longitudinal or perpendicular) or bit-patterned media, similar obstacles as SOMA media have to be overcome.

4. Effect of Lattice Mismatch on Chemical Ordering of Epitaxial L10 FePt Thin Films

The growth of magnetic layers can be controlled by appropriate underlayers, and then the magnetocrystalline anisotropy of the films can be optimized. A lattice mismatch at the interface between FePt film and underlayer may expand the a axis and shrink the c axis of the FePt film and thus favors the ordering at low temperatures.

Relationship between the lattice mismatch and the chemical ordering of the FePt films and their magnetic anisotropic constant were identified with 30 nm thick Pt, Cr, $Cr_{95}Mo_5$ and $Cr_{90}Mo_{10}$ intermediate layers being used to adjust the lattice mismatch.

The sample structure is shown in Figure 4. The calculated lattice constant values of the intermediate layers grown on both MgO and glass substrates are listed in Table 2. The lattice constants of the intermediate layers grown on the MgO substrates are only slightly larger than those grown on the glass substrates, which illustrates that a 30 nm thick intermediate layer is enough for relaxing the strains from the MgO substrates.

FePt (5~60 nm)
Pt, $Cr_{95}Mo_5$, $Cr_{90}Mo_{10}$ (30 nm)
MgO (200)

Figure 4. Illustration of sample structure. (From Ding Y.F., Ph.D. Thesis, Nanyang Technological University, Singapore, 2006)

The calculated results are plotted in Figure 5 with respect to different buffer layers. As can be seen, the *c* holds the minimum while the *a* holds the maximum when the $Cr_{95}Mo_5$ underlayer is used. The results also show that the *c* axis of the FePt films has been indeed shrunken and *a* axis has been expanded with different buffer layers.

Table 2. Lattice constants of intermediate layers, epitaxial relationships, and lattice mismatches. (From Ding Y.F., Chen J.S., Liu E., Sun C.J., and Chow G.M., J. Appl. Phys. 97, 10H303 (2005))

Sample	Intermediate layer	a_1 (Å)	a_2 (Å)	Epitaxial relationship	ε (%)
A	Pt	3.928	3.923	MgO (100)<001>$\parallel$ Pt (100)<001>$\parallel$ FePt (001)<100>	2.23
B	Cr	2.884	2.878	MgO (100)<001>$\parallel$ Cr (100)<110>$\parallel$ FePt (001)<100>	5.88
C	$Cr_{95}Mo_5$	2.898	2.891	MgO (100)<001>$\parallel$ Cr (100)<110>$\parallel$ FePt (001)<100>	6.33
D	$Cr_{90}Mo_{10}$	2.915	2.911	MgO (100)<001>$\parallel$ Cr (100)<110>$\parallel$ FePt (001)<100>	6.89
E	MgO	4.211		MgO (100)<001>$\parallel$ FePt (001)<100>	8.86

a_1: Lattice constant of intermediate layer (grown on MgO)
a_2: Lattice constant of intermediate layer (grown on glass)
ε: Lattice mismatch

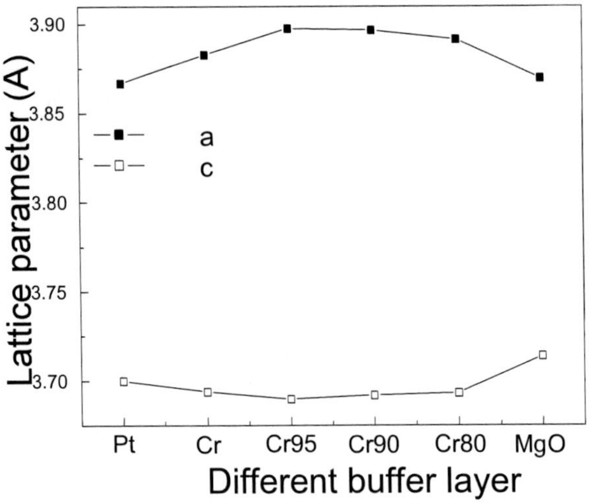

0

Figure 5. Lattice constants of FePt films with respect to different buffer layers (From Ding Y.F., Ph.D. Thesis, Nanyang Technological University, Singapore, 2006).

5. FePt Films Grown on Chromium Intermetallic Underlayers

In the last section, a strain induced ordering mechanism was demonstrated, and a critical lattice mismatch around 6.33% was found to be the most suitable for improving the chemical ordering and keeping the epitaxial stress in FePt films. However, the work introduced in the last section was carried out on MgO single crystal substrates. As far as the industrial application is concerned, it is essential to deposit magnetic recording media on commercial disk substrates, such as glass disks, Al alloy disks, etc. Thus, to apply the proposed idea to actual applications, it is necessary to develop CrMo(200) underlayers on glass and/or Al alloy substrates. It is well known that Cr based alloy (CrX) (X is Mo, Ru, W, V, or Ti, etc.) underlayers and/or seed layers have been widely used in commercial longitudinal media. Effort was made to deposit FePt films on CrMo seeded glass substrates at low temperatures, but the crystallographic orientation of the FePt films was random. Thus, the development of $Cr_{1-x}X_x$ (200) underlayers on glass substrates is of both scientific and technological interests. In this section, the structural and magnetic properties of FePt films deposited on $Cr_{1-x}Mo_x$ underlayers at relatively low temperatures are mainly introduced. A brief comparison of the FePt films grown on the CrX (X = Mo, Ru, W, Ti) underlayers is also introduced. Finally, the study of the initial layers for the FePt films is addressed. The samples used for this study have a configuration of Glass / CrX (X = Mo, Ru, W, Ti) / Pt (4 nm) / FePt (Figure 6).

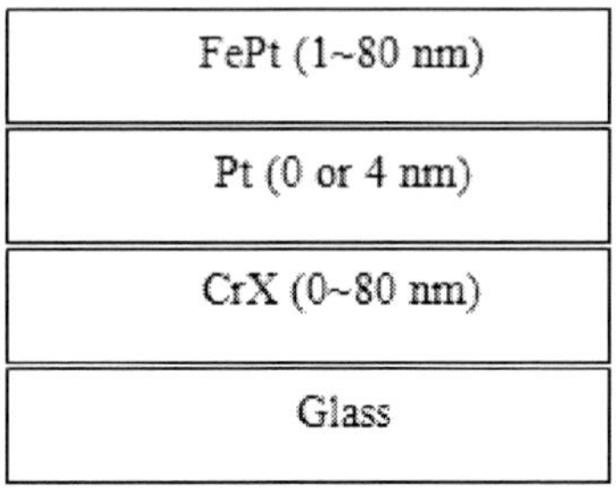

Figure 6. Illustration of sample configuration. (From Ding Y.F., Ph.D. Thesis, Nanyang Technological University, Singapore, 2006)

5.1. Influence of Mo Content

The deposition temperature for all the films was fixed at 350 °C and the thicknesses of the FePt, Pt and $Cr_{1-x}Mo_x$ layers were fixed at about 20, 4 and 30 nm, respectively. The Mo content in the underlayers varies from 0 to 20 vol.%.

Figure 7 shows the XRD spectra of the FePt films deposited on the CrMo underlayers containing various contents of Mo. For the FePt film grown on the pure Cr underlayer, in addition to an intense (001) peak, a FePt (111) peak is also observed [Figure 7(a)], which indicates that the (001) preferred orientation of the film is still not strong. When the $Cr_{90}Mo_{10}$ underlayer is used, the FePt (111) peak disappears, which implies an improved (001) preferred orientation [Figure 7(b)].

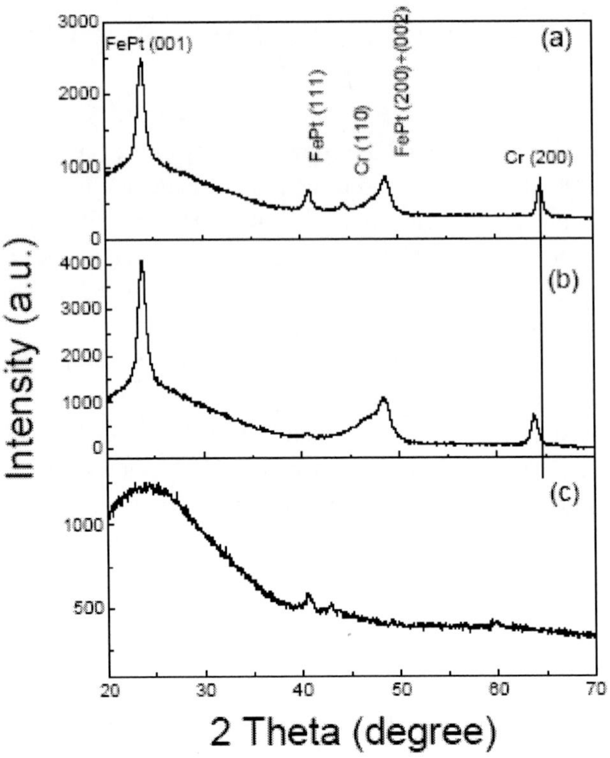

Figure 7. XRD spectra of FePt films deposited at 350 °C on (a) Cr, (b) $Cr_{90}Mo_{10}$, and (c) $Cr_{80}Mo_{20}$ underlayers (From Ding Y.F., Ph.D. Thesis, Nanyang Technological University, Singapore, 2006).

While for the FePt film grown on the $Cr_{80}Mo_{20}$ underlayer, the FePt (001) peak disappears and only very weak FePt (111) peak appears [Figure 7(c)]. It is expected that a (001) preferred FePt film can hetero-epitaxially grow on a Cr (200) underlayer with a crystallographic relationship of Cr (200)[110] || FePt (001)[100]. Thus, the texture evolution of the FePt films should be mainly governed by the texture of the $Cr_{1-x}Mo_x$ underlayers in this study.

For the pure Cr underlayer, both the Cr (110) and Cr (200) peaks appear [Figure 7(a)], while for the $Cr_{90}Mo_{10}$ film only the Cr (200) peak appears [Figure 7(b)]. In addition, the Cr (200) peak shifts from 64.5^o to 63.86^o, which corresponds to the expansion of the lattice constant *a* when Mo is doped into the Cr underlayer. However, only a weak Cr (110) for the $Cr_{80}Mo_{20}$ layer appears under this deposition condition [Figure 7(c)].

5.2. Effect of Substrate Temperature

In this section, the thicknesses of the $Cr_{90}Mo_{10}$ underlayers, Pt buffer layers and FePt films are about 20, 4 and 20 nm, respectively. The deposition temperature for the $Cr_{90}Mo_{10}$ underlayers is 350 °C and those for the Pt and FePt layers vary from 150 to 350 °C. The effect of the substrate temperature on the structural and magnetic properties of the FePt films is discussed.

Figure 8 shows the XRD spectra of the FePt films deposited on the $Cr_{90}Mo_{10}$ underlayers at various substrate temperatures. For the FePt film grown on the $Cr_{90}Mo_{10}$ underlayer at 150 °C, a FePt (200) peak dominates. A weak FePt (001) superlattice peak observed at 200 °C implies that the chemical ordering has occurred. With an increase of the substrate temperature, the intensity of the FePt (001) peak increases, while the FePt (111) peak weakens and eventually disappears at 350 °C, which implies an improved (001) preferred orientation. In addition, the peak between 45° and 50° shifts to a higher angle with increased substrate temperature due to the increase of the ordering degree.

The selected area electron diffraction (SAED) patterns of the FePt films produced at different substrate temperatures are shown in Figure 9. As the substrate temperature is 150 °C, only some fundamental diffraction rings appear. At 200 °C, weak (110) reflections confirm the occurrence of the chemical ordering and a low degree of chemical ordering of the FePt film. At 300 °C, a weak (001) reflection indicates the presence of (100) and/or (010) oriented grains. The (001) reflection is not observed from the FePt films grown at 150 and 200 °C, because of a low ordering degree and/or a weaker (001) reflection compared to some other reflections. At 350 °C, the (001) reflection disappears again, which may be due to the improved FePt (001) preferred orientation.

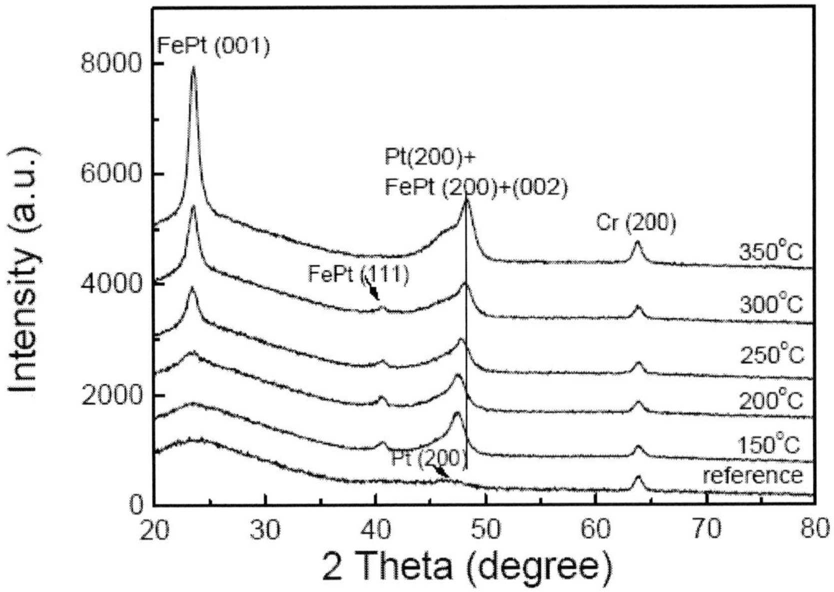

Figure 8. XRD spectra of FePt films deposited on $Cr_{90}Mo_{10}$ underlayers at various substrate temperatures (150 ~ 350 °C). The reference spectrum is from a 4 nm thick Pt film deposited on a $Cr_{90}Mo_{10}$ underlayer at 350 °C (From Ding Y.F., Ph.D. Thesis, Nanyang Technological University, Singapore, 2006).

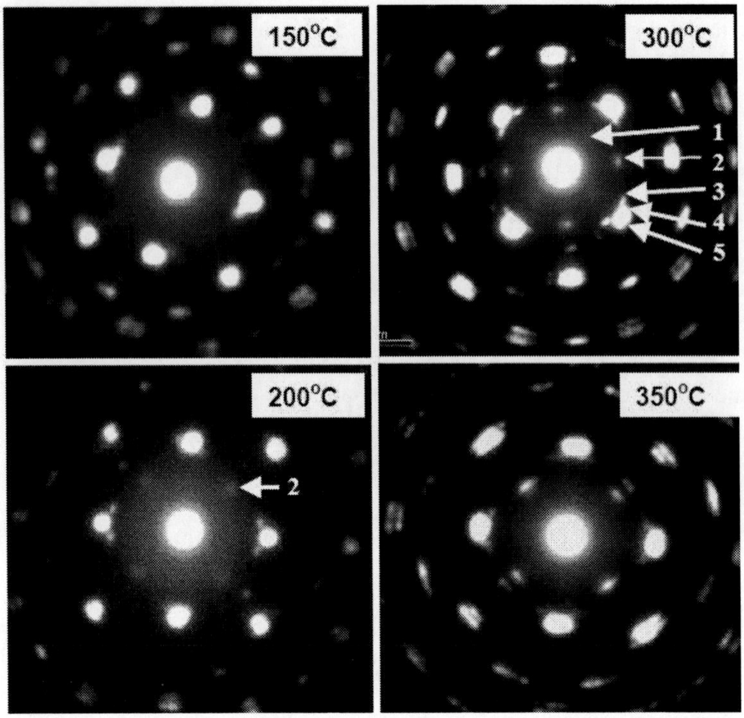

Figure 9. SAED patterns of FePt films deposited on $Cr_{90}Mo_{10}$ underlayers at 150, 200, 300 and 350 °C, respectively. [Note: 1: FePt (001); 2: FePt (110); 3: FePt (111); 4: Cr (110); and 5: FePt (200), (002)] (From Ding Y.F., Ph.D. Thesis, Nanyang Technological University, Singapore, 2006).

5.3. Effect of Underlayer

In this series of experiments, the deposition temperature for all the films was fixed at 350 °C and the thicknesses of the FePt, Pt and $Cr_{90}X_{10}$ (X= Mo, Ru, Ti, or W) layers are fixed at about 20, 4 and 30 nm, respectively.

Figure 10 shows the XRD spectra of the FePt films grown on the different underlayers. The inserted spectra are the rocking curves of the FePt (001) peaks. As can be seen from the graphs, the FePt films grown on the $Cr_{90}Ru_{10}$, $Cr_{90}Mo_{10}$ and $Cr_{90}W_{10}$ (200) underlayers show a (001) preferred orientation due to the epitaxial growth, while the FePt film grown on the $Cr_{90}Ti_{10}$ underlayer shows a (111) preferred orientation. The inserted rocking curves of the FePt (001) peaks reveal that the FWHM of the FePt film grown on the $Cr_{90}Ru_{10}$ underlayer is the narrowest among all the samples, depicting a good FePt (001) texture.

According to the proposed epitaxial relationship between Cr (200) and FePt (001) films, the FePt films grown on the CrX (200) underlayers should have a (001) preferred orientation, and the texture quality of the FePt films mainly depends on the texture quality of the underlayers.

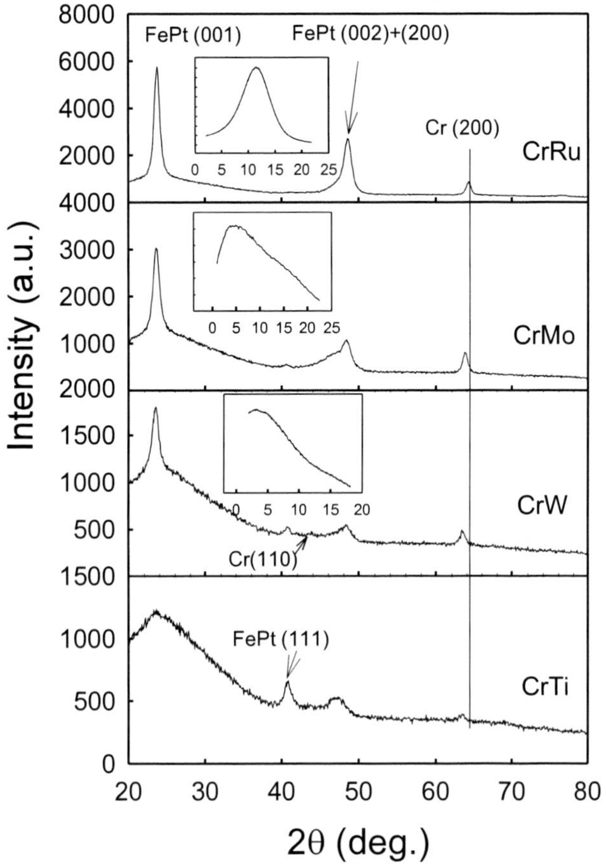

Figure 10. XRD spectra of about 20 nm thick FePt films grown on about 20 nm thick CrX underlayers (From Ding Y.F., Ph.D. Thesis, Nanyang Technological University, Singapore, 2006).

However, the FePt film grown on the $Cr_{90}Ti_{10}$ (200) underlayer shows a (111) preferred orientation. To understand the mechanism, the atomic radii of different additives, lattice constants of different underlayers and FePt films, FWHM of Cr (200) peak, and lattice mismatches between underlayers and FePt layers are analyzed and summarized in Table 3. The lattice parameter of the Cr intermediate layers increases with the increase of atomic radius of X, which appears to follow the Vegard's law. The lattice mismatch between FePt (001) and CrX (200) increases with the increase of lattice parameters of CrX. The lattice constant *c* of the FePt films increase with increased lattice mismatch, indicating a decreased ordering degree. This result is consistent with our previous findings, i.e., the critical lattice mismatch which is the most favorable for chemical ordering, is about 6.33%. With a further increase in lattice mismatch, the chemical ordering degree decreases. It is found that the residual stress decreased with increased lattice mismatch from about 6 to 8%. Thus, a possible reason for the FePt film grown on the $Cr_{90}Ti_{10}$ (200) underlayer showing a (111) preferred orientation is due to the released residual epitaxial stress caused by a larger lattice mismatch.

Table 3. Structural properties of both underlayers and FePt films. (From Ding Y.F., Ph.D. Thesis, Nanyang Technological University, Singapore, 2006)

Underlayer (CrX)	Atomic Radius of X (Å)	CrX $\sqrt{2}\,a$ (Å)	Cr (200) FWHM (°)	FePt c (Å)	Mismatch* (%)
CrRu	1.34	4.09	0.7121	3.73	6.5
CrMo	1.39	4.12	0.7549	3.76	7.1
CrW	1.41	4.14	0.8917	3.78	7.6
CrTi	1.47	4.15	1.3769		7.9

* Refers to the lattice mismatchs between FePt (001)[100]||CrX (200)[110].

Both the in-pane and the out-of-plane hysteresis loops are shown in Figure 11. The FePt films grown on the $Cr_{90}Ru_{10}$, $Cr_{90}Mo_{10}$ and $Cr_{90}W_{10}$ underlayers show a perpendicular magnetic anisotropy, while the FePt films grown on the $Cr_{90}Ti_{10}$ underlayer show a longitudinal anisotropy. The easy axis variation is consistent with that reflected by the XRD spectra.

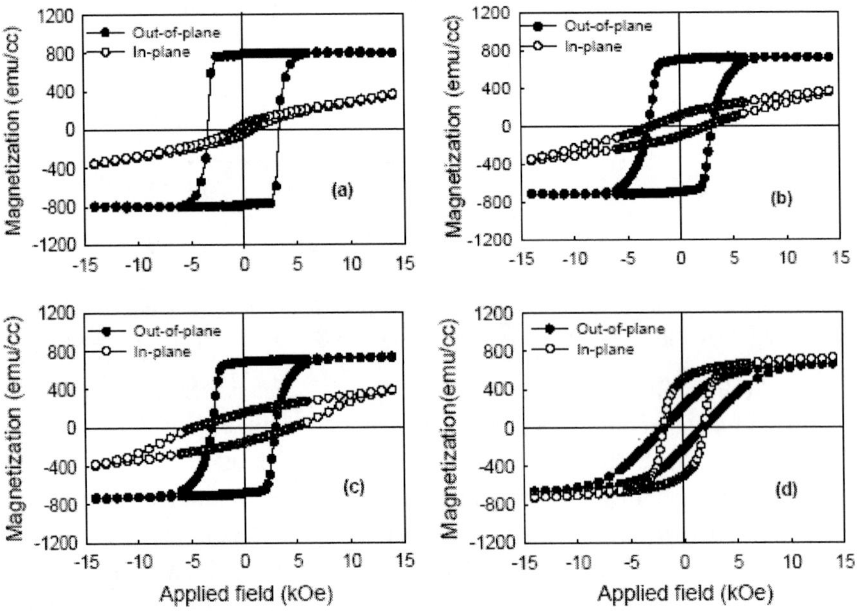

Figure 11. Hysteresis loops of FePt films grown on (a) $Cr_{90}Ru_{10}$, (b) $Cr_{90}Mo_{10}$, (c) $Cr_{90}W_{10}$, and (d) $Cr_{90}Ti_{10}$ underlayers (From Ding Y.F., Ph.D. Thesis, Nanyang Technological University, Singapore, 2006).

6. FePt:C Based Magnetic Recording Media

This section introduces the effect of C additive on the microstructure and magnetic properties of the FePt films. The main purpose here is to control the FePt (001) grain size at relatively low substrate temperatures. A direct synthesis of the ordered FePt:C (001) nanocomposite films with ultra small and uniform grain size, adjustable out-of-plane coercivity and narrow *c*-axis distribution on glass substrates by magnetron sputtering at 350 °C is introduced. Figure 12 shows the configurations of the FePt:C systems.

In this section, the thicknesses of the $Cr_{90}Ru_{10}$ underlayers, Pt buffer layers and FePt:C films are about 30, 2 and 10 nm, respectively, with the C content in the FePt:C films varied from 0 to about 30 vol.%. The deposition temperature for all the layers is 350 °C. The microstructural properties of the FePt:C films are discussed. Figure 13 shows the XRD spectra of the FePt:C films with various C volume fractions deposited at 350 °C. From the XRD spectra, one can see that all the peaks are attributed to the $Cr_{90}Ru_{10}$ underlayers and FePt films, while no peaks are from carbide phases, which indicates that most C remains as a pure element matrix. With increasing C content from 0 to 30 vol.%, the FePt (001) peak locations remain almost unchanged, illustrating that the lattice constant *c* has no change. The FePt (002)+(200) peaks of the samples doped with 5, 10 and 15 vol.% C locate at a slightly higher angle than that of the pure FePt film, indicating an increase of the FePt (002) content. With further increase of the C content from 15 to 30 vol.%, the FePt (002)+(200) peak shifts to a lower angle, indicating an increase of the FCC FePt (200) content.

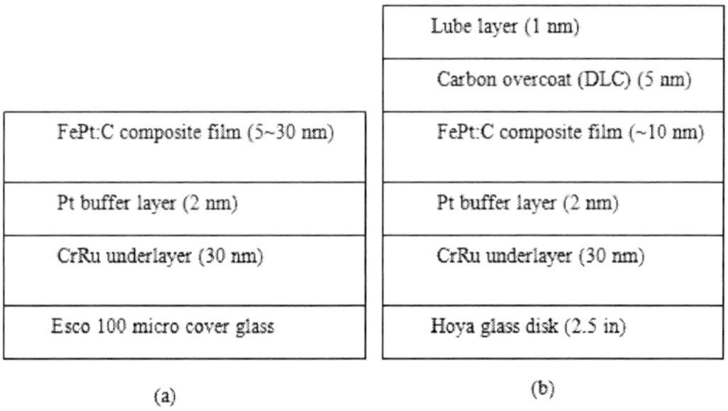

Figure 12. Configurations of (a) FePt:C films and (b) media (From Ding Y.F., Ph.D. Thesis, Nanyang Technological University, Singapore, 2006).

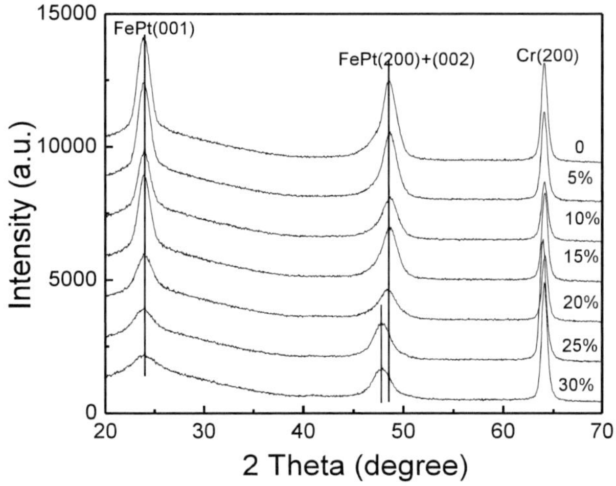

Figure 13. XRD spectra of FePt films with different C volume fractions at 350 °C (From Ding Y.F., Chen J.S., Liu E., Lim B.C., Hu J.F., and Liu B., Thin Solid Films 517, 2638 (2009)).

The XRD results depict that the FePt films can keep a (001) preferred orientation after incorporating C up to 20 vol.%, which is different from those doped with Ag or Cu. In the previous discussion, the easy axis of FePt films grown on CrRu or CrMo underlayers changed from perpendicular to longitudinal by Ag and Cu doping, for which the suppressed mobility of Fe and Pt adatoms by Cu and Ag doping was believed to be a main reason. The atomic radii of Cu and Ag atoms are comparable with those of Fe and Pt atoms, and thus some Fe and Pt atoms have to overcome the energy barriers caused by the Cu and Ag atoms before reaching their minimum energy sites. While for the C additive, the atomic radius of carbon atom (0.077 nm) is much smaller than those of Pt (0.138 nm) and Fe (0.126 nm) atoms. Thus, it is very likely that the C atoms can diffuse through interstitial sites as in the case of CoCr based alloys and the mobilities of the Fe and Pt adatoms are not restricted so much. Thus, most of the Fe and Pt adatoms can move to their minimum energy sites and align with the atomic arrangement of the underlayer. The chemical ordering degree first increases with increasing C content from 0 to 15 vol.%, and then decreases with further increasing C volume fraction. This signifies that a small amount of C doping can slightly enhance the ordering degree, while a large amount of C doping can decrease the ordering degree. These results are similar to those reported previously. The variation of the ordering degree of the FePt films with increased C content may be explained as follows. When the C content is small, C atoms may diffuse through FePt interstitial sites during growth; the diffusion of C atoms may provide a driving force for FePt phase transformation from FCC to FCT. As the C content is large, the epitaxial growth of the FePt on the Pt buffer layer may be suppressed; the FePt/Pt interface may be destroyed by the C adatoms; the lattice mismatch induced epitaxial stress may be released; and thus the FePt films have a lower ordering degree. The diffusion of the C during deposition has been verified by transmission electron microscopy (TEM) images and will be discussed in the following section. The microstructural properties of the FePt:C films are also investigated by TEM. Figure 14(a) and (b) show the TEM plane-views of the FePt:C films with 10 and 20 vol.% C, respectively. The pure FePt films grown on $Cr_{90}Ru_{10}$ underlayers show a continuous microstructure. When 10 vol.% C is doped, a granular structured film is formed though some grains aggregate together and not all grain boundaries can be distinguished. The grain size roughly follows a Gaussian distribution with a mean diameter of about 9.9 nm and a standard deviation of about 2.3 nm. As the C content increases to 20 vol.%, the TEM image shows well-defined grain boundaries and relatively uniform grain sizes. The mean grain size and standard deviation are about 5.6 and 1.6 nm, respectively. In the both cases, the grain size distribution of the FePt:C films is relatively narrow and encouraging with reference to the results reported previously. Figures 15-17 show both the low magnification and the high-resolution cross-sectional images of the FePt:C media doped with 10, 20 and 30 vol.% C, respectively. All the images in Figures 15-17 reveal that the films have three layers, according to different contrast. The top FePt:C layers show a granular microstructure with clear grain boundaries, confirming that the FePt:C films have well-defined grain boundaries and uniform grains, while the $Cr_{90}Ru_{10}$ and Pt layers show continuous microstructures. In all the images, the middle continuous layers are about 7 nm thick which is much thicker than 2 nm from the design, while the top FePt:C layers are thinner than the designed thickness. It has been reported that the Cr diffusion could be effectively blocked by a thin Pt buffer layer. Thus, the middle continuous layers consist of a 2 nm thick Pt layer and a 5 nm thick FePt continuous layer. This may be explained as follows. It is well known that an epitaxial stress is very strong when a film is very thin. Thus, at the initial growth stage of the FePt:C film, a strong epitaxial stress governs the thin film growth mechanism: i.e. the Fe and Pt adatoms can follow the atomic arrangement of the Pt layer to form a continuous microstructure, and then the C adatoms may diffuse onto the surface of the FePt layer at the same time. With the increase of the FePt:C film thickness, the epitaxial stress is released. More C atoms on the FePt film surface may inhibit the mobility of the Fe and Pt adatoms and reduce the surface bonding of the FePt film. Thus, some Fe and Pt adatoms cannot reach their epitaxial sites, instead forming three-dimensional islands. The ball-like shape of the FePt grains should be mainly governed by the film surface tension. That the FePt grains can have a (001) preferred orientation is mainly due to the epitaxial growth as observed in high-resolution TEM images [Figures 15(b), 16(b) and 17(b)]. A proposed two-step growth mode of the FePt:C films is shown in Figure 18. There exists a light gray layer (~ 2 nm) between the middle continuous and top granular layers as seen in the low magnification images [Figures 15(a), 16(a) and 17(a)]. This layer has been generally considered a "initial layer" caused by inter-diffusion between two layers and/or by elastic strain due to a lattice misfit between two layers in Co-based recording media. In this section, the continuous lattice images from the Pt to the FePt continuous layers and further to the FePt:C granular layers are observed [Figures 15(b), 16(b), and 17(b)], which indicate the epitaxial growth between Pt and FePt/FePt:C layers. In addition, some amorphous areas are also observed between Pt/FePt and FePt:C layers [Figures 15(b), and 16(b),], which should be formed by carbon. Thus, the regions with the light contrast between the continuous FePt layer and overlying granular FePt grains should be mainly due to the carbon. For the FePt:C film doped with 30 vol.% C, the high resolution image does not show a clear continuous lattice from the Pt/FePt layer to the FePt:C layer, but a rather poor texture that is believed to be caused by the C additive [marked with circle in Figure 17(b)].

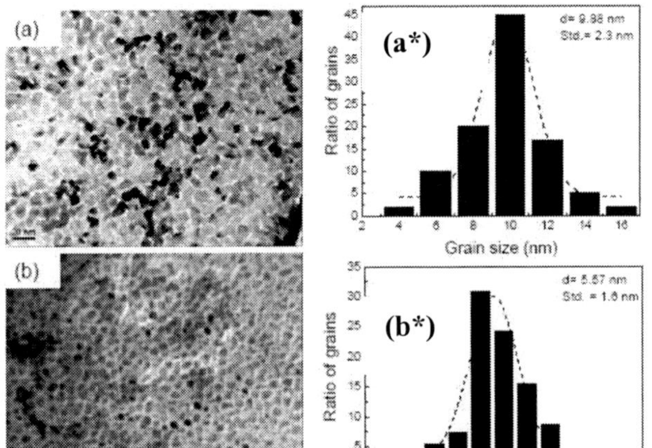

0

Figure 14. TEM plane views of FePt:C films with (a) 10 and (b) 20 vol.% C doping. (a)* and (b)* are grain size distributions of (a) and (b), respectively. (From Ding Y.F., Chen J.S., Liu E., Lim B.C., Hu J.F., and Liu B., Thin Solid Films 517, 2638 (2009)).

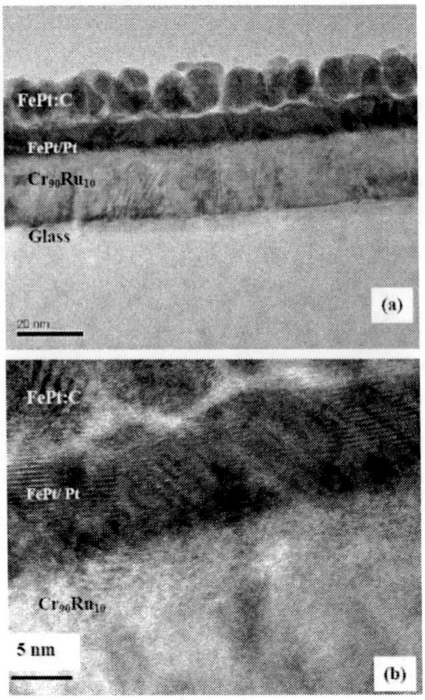

Figure 15. TEM cross-sectional views of about 15 nm thick FePt:C film doped with 10 vol.% C: (a) low and (b) high magnifications (From Ding Y.F., Chen J.S., Liu E., Lim B.C., Hu J.F., and Liu B., Thin Solid Films 517, 2638 (2009)).

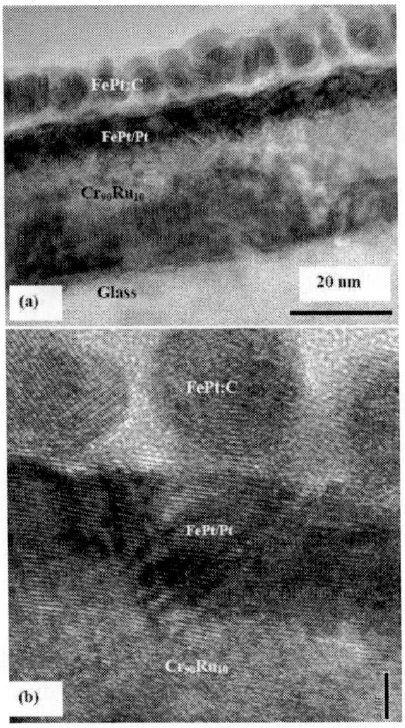

Figure16. TEM cross-sectional views of about 10 nm thick FePt:C film doped with 20 vol.% C: (a) low and (b) high magnifications (From Ding Y.F., Chen J.S., Liu E., Lim B.C., Hu J.F., and Liu B., Thin Solid Films 517, 2638 (2009)).

An interesting phenomenon is that the morphology of the FePt layers changes from column-like to ball-like with increased C content, which should be due to the surface tension and deteriorated epitaxial stress resulted from the lattice misfit at the Pt/FePt interface.

In order to confirm the suggested growth mode, the FePt:C (15 vol.% C) films with different thicknesses from 5 to 30 nm were deposited on the Glass/$Cr_{90}Ru_{10}$ (30 nm)/Pt (2 nm) under the same condition with above samples. The thickness dependence of the microstructure is investigated by taking TEM cross sectional images as shown in Figure 19. As seen from Figure 19(a), when the FePt:C film thickness is about 5 nm, the cross-sectional image does not reveal a clear FePt grain boundary, and the grains are found to have formed a continuous microstructure rather than a granular one. In addition, the contrast of the top areas of the FePt:C film [marked with arrows in Figure 19(a) and (b)] is obviously lighter than other areas, suggesting the existence of a carbon rich area on the top of the FePt:C film, which has been confirmed by X-ray photoelectron spectroscopy (XPS). This confirms the above suggestion: in the initial stage of the FePt:C film growth the carbon adatoms tend to diffuse onto the surface of the FePt layer. As the thickness of the FePt:C film increases to 15 nm, the top of the FePt:C layer shows a ball-like microstructure with a 7~8 nm thick continuous intermediate layer that consists of 2 nm Pt and about 5 nm thick continuous FePt layer. With a further increase of thickness up to about 30 nm, a columnar microstructure appears, while the thickness of the middle continuous layer with dark contrast does not change with the growth of the FePt grains only in the film normal direction. The evolution of the microstructure with increased film thickness should obey the system energy minimum principle. In this case, the thickness dependence of the microstructure should be mainly attributed to the competition between the epitaxial stress, surface tension and surface energy.

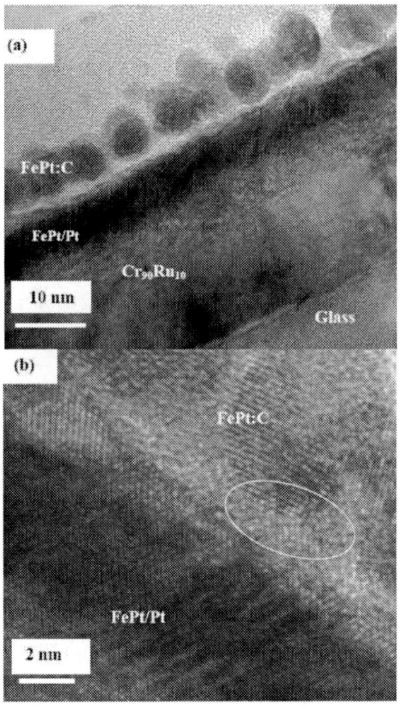

Figure 17. TEM cross-sectional views of about 10 nm thick FePt:C film doped with 30 vol.% C: (a) low and (b) high magnifications (From Ding Y.F., Chen J.S., Liu E., Lim B.C., Hu J.F., and Liu B., Thin Solid Films 517, 2638 (2009)).

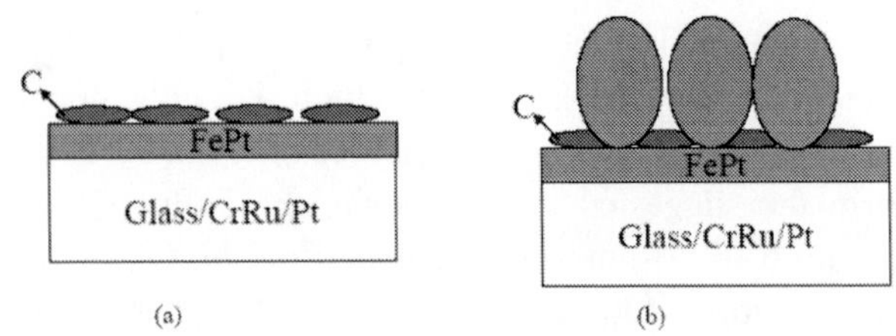

Figure 18. A two-step growth model for FePt:C films: (a) Step 1: FePt is epitaxially grown on CrRu/Pt and carbon is squeezed onto the top surface of FePt film, and (b) Step 2: with increased film thickness, carbon is accumulated to some extent and the epitaxial stress is released. At the same time, Fe and Pt atoms nucleate at some sites and then the nuclei grow larger at step 2 (From Ding Y.F., Chen J.S., Liu E., Lim B.C., Hu J.F., and Liu B., Thin Solid Films 517, 2638 (2009)).

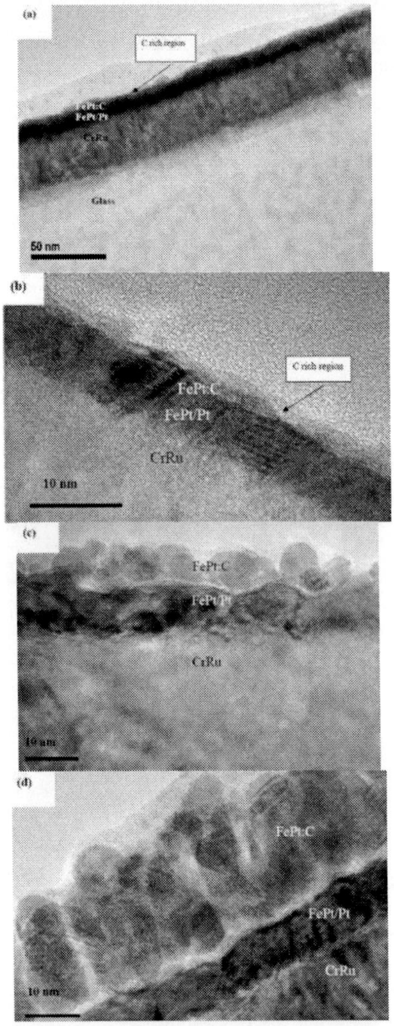

Figure 19. TEM cross-sectional views of (a) 5 nm with low magnification, (b) 5 nm with high magnification, (c) 15 nm, and (d) 30 nm thick FePt:C films (15 vol.% C) grown on a 30 nm thick $Cr_{90}Ru_{10}$ underlayer with a 2 nm thick Pt buffer layer deposited at 350 °C (From Ding Y.F., Chen J.S., Liu E., Lim B.C., Hu J.F., and Liu B., Thin Solid Films 517, 2638 (2009)).

7. Magnetic Properties of FePt:C Films

In this section, the magnetic properties such as hysteresis loops, switching field distribution and time dependent effect of the FePt:C films are discussed.

7.1. Hysteresis Loops of FePt:C Films

Figure 20 shows the out-of-plane hysteresis loops of the FePt:C films doped with various C contents as measured by polar kerr. The loops illustrate that all the FePt:C films have a perpendicular magnetic anisotropy, confirming that the easy axis is perpendicular to the film plane, which is consistent with the preceding XRD analysis.

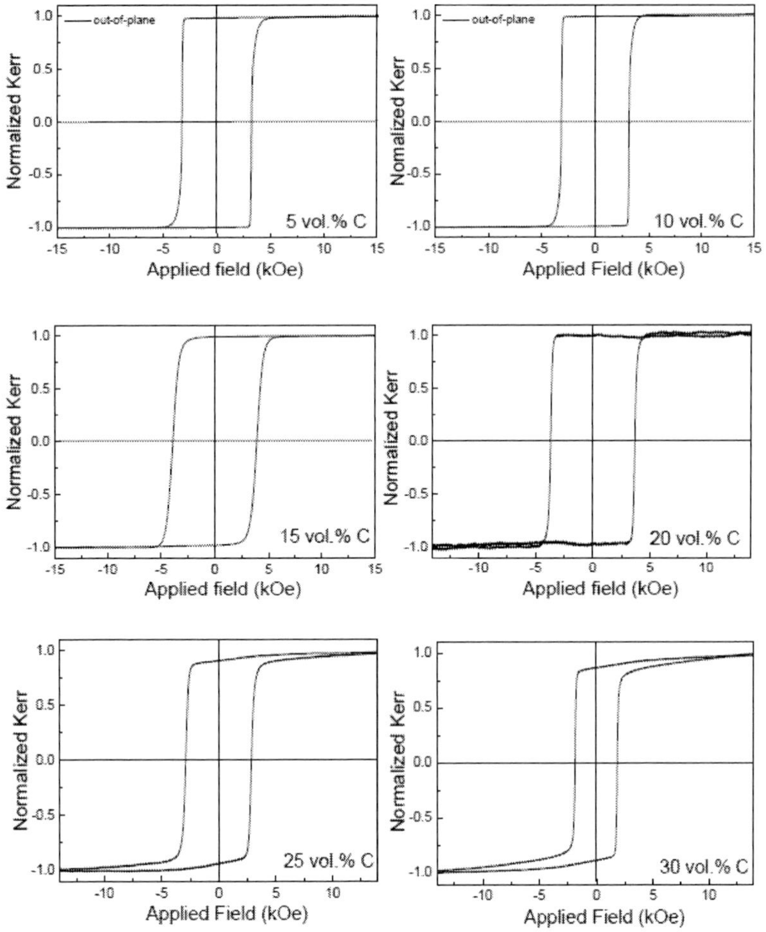

Figure 20. Out-of-plane hysteresis loops of FePt:C films with different C contents (From Ding Y.F., Chen J.S., Liu E., Lim B.C., Hu J.F., and Liu B., Thin Solid Films 517, 2638 (2009)).

7.2. Read-Write Test on FePt:C Perpendicular Media

In this section, a Guzik spin-stand (1701B) with a commercial ring head (for longitudinal media) was used to roughly characterize the recording properties of the fabricated FePt:C media. The reproduced waveforms of the individual tracks of the FePt:C perpendicular media doped with 20 vol.% C at various densities are shown in Figure 21. The square waveforms are observed at a linear density of 59 kfci, indicating that it is a perpendicular medium, though the peaks are not smooth and there exist some small fluctuations. It is well understood that the demagnetization field is quite large at a low linear density for a perpendicular medium, and a large demagnetization field will result in the reverse of magnetic domains with a small magnetic anisotropy. Thus, the fluctuations at the peaks should be mainly due to the demagnetization field and the inter-granular exchange coupling. As the linear density increases to 118 kfci, the read-back signals can still be distinguished. However, as the linear density is further increased to 221 kfci, some signals overlap in addition to the amplitude fluctuations, which should also be mainly due to the demagnetization field and exchange coupling from the media point of view, in addition to the effects from the head, head-media spacing (HMS), etc. The signal-to-noise ratio (SNR) with respect to the linear density for the pure FePt media and the FePt:C media doped with 20 vol.% C are shown in Figure 22. As seen from the graph, the SNR ratios of the FePt:C media doped with 20 vol.% C are about 10 dB better than that of the pure FePt media. It is worth noting that the SNR measurement is not absolute, but only relative. This is because the signal and noise are determined by the type of the head (e.g. sensitivity, writing capability), head-media spacing, etc. As far as the recording media are concerned, for a given track width, smaller de-coupled grains will produce a higher SNR. In this case, the decreased grain size in the FePt:C media should be one of the main reasons for the improved SNR.

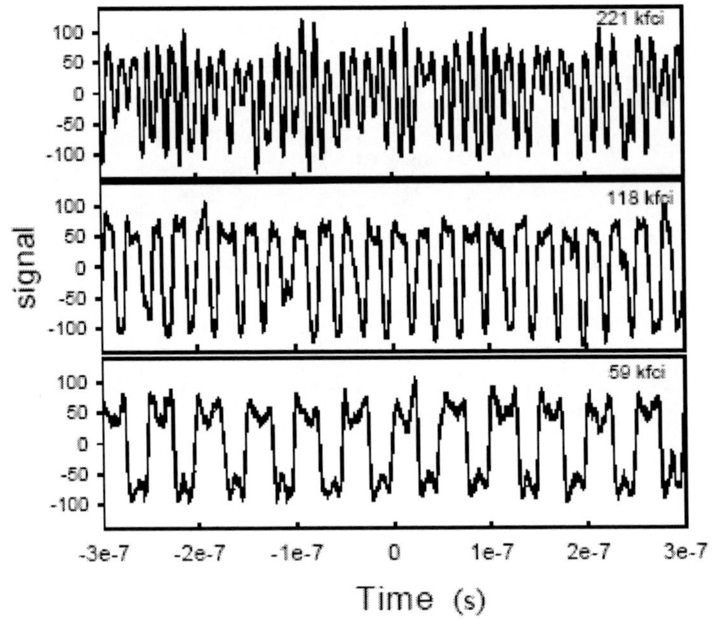

Figure 21. Reproduced waveforms of individual tracks at various densities for FePt:C media doped with 20 vol.% C (From Ding Y.F., Chen J.S., Liu E., Lim B.C., Hu J.F., and Liu B., Thin Solid Films 517, 2638 (2009)).

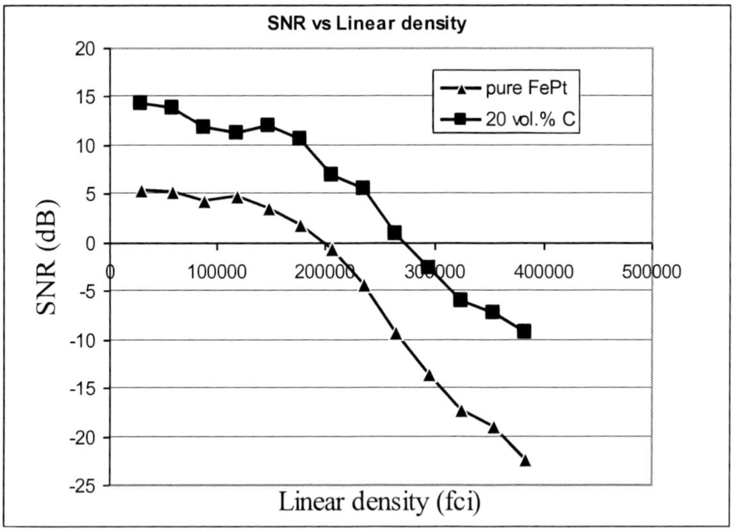

Figure 22. Signal-to-noise ratio (SNR) with respect to linear density for pure FePt media and FePt:C media doped with 20 vol.% C. (From Ding Y.F., Chen J.S., Liu E., Lim B.C., Hu J.F., and Liu B., Thin Solid Films 517, 2638 (2009))

Summary

This chapter reviewed the development of $L1_0$ (CuAu-I type structure) FePt thin films for ultra-high magnetic recording media and addressed two main issues, such as (a) how to lower the $L1_0$ FePt ordering temperature when getting the crystallographic texture controlled at the same time and (b) how to reduce the grain size and inter-granular exchange coupling of the films. Several other issues were also discussed in the chapter, such as (a) residual stress that could help expand the *a*-axis and shrink the *c*-axis of FePt crystal lattice and thus favored the chemical ordering of FePt films at relatively low temperatures; (b) texture of FePt films that strongly depended on the texture of CrX underlayers, for which the Cr (200) texture and substrate temperature (T_s) were the key parameters for the growth of the FePt (001) films; and (c) promising potential of FePt:C films for ultra-high density magnetic recording media.

References

Airson B.M., Visokay M.R., Marinero. E.E., Sinclair R, and Clemens B.M., *J. Appl. Phys.* 74, 1992 (1993).

Bertram H.N. and Williams M., *IEEE Trans. Magn.* 36, 4 (2000).

Chantrell R., Weller D., Klemmer T., Fullerton E., and Sun S., MMM Seattle, November 2001.

Chen J.S. and Wang J.P., *J. Magn. Magn. Mater.* 284, 423 (2004).

Chen J.S., Xu Y.F. and Wang J.P., *J. Appl. Phys.* 93, 1661 (2003).

Cheng J.Y., Ross C.A., Thomas E.L., Smith H.I., Lammertink R.G.H., and Vancso G.J., *IEEE Trans. Magn.* 38, 2541 (2002).

Dehlinger U. and Graf L., *Physik.* 64, 359 (1930).

Ding Y.F., Ph.D. thesis, Nanyang Technological University, Singapore, 2006.

Ding Y.F., Chen J.S., Lim B.C., Hu J.F., Liu B., and Ju G., *Appl. Phys. Lett.* 93, 032506 (2008).

Ding Y.F., Chen J.S., and Liu E., *Appl. Phys.* A 81, 1485 (2005).

Ding Y.F., Chen J.S., Liu E., Lim B.C., Hu J.F., and Liu B., Thin Solid Films, 517, 2638 (2009).

Ding Y.F., Chen J.S., Liu E., Sun C.J. , and Chow G.M., *J. Appl. Phys.* 97, 10H303 (2005).

Endo Y., Kikuchi N., Kitakami O., and Shimada Y., *J. Appl. Phys.* 89, 7065 (2001).

Fallot M., *Compt. Rend.* 199, 128 (1934).

Graf L. and Kussman A., *Physik.* 36, 544 (1935).

Hansen M, 2nd ed., McGraw-Hill, 1958.

Isaac E., Tammann G., and Anorg Z., *Chem.* 55, 63 (1907).

Ivanov O.A., Solina L.V., Demshina V.A., and Magat L.M., Fiz. Met. Metalloved. 35, 92 (1973).

Kang K., Zhang Z. G., Papusoi C., and Suzuki T., *Appl. Phys. Lett.* 82, 3284 (2003).

Kittel C., 7th ed., *John Wiley and Sons*, New York, 1996.

Ko H.S., Perumal A., and Shin S.C., *Appl. Phys. Lett.* 82, 2311 (2003).

Kussma A., Auwarter M., and Rittberg G.V., *Ann. Phys.* 4, 147 (1948).

Kussma A. and Tittberg G.V., *Metallkunde*, 42, 470 (1950).

Lairson B.M., Visokay M.R., Sinclair R., and Clemens B.M., *Appl. Phys. Lett.* 62, 639 (1993).

Lambeth D.N., Velu E.M.T., Bellesis G.H., Lee L.L., and Laughlin D.E., *J. Appl. Phys.* 79, 4496 (1996).

Lee S.R., Yang S.Y., Kim Y.K., and Na J.G., *Appl. Phys. Lett.* 78, 4001 (2001).

Lipson H., Schoenburg D., and Rittberg G.V., *J. I. Met.* 67, 333 (1941).

Luo C.P. and Sellmyer D.J., *IEEE Trans. Magn.* 31, 2764 (1995).

Luo C.P. and Sellmyer D.J., *Appl. Phys. Lett.* 75, 3162 (1999).

Naito K., Hieda H., Sakurai M, Kamata Y., and Asakawa K., *IEEE Trans. Magn.* 38, 1949 (2002).

Nemilov V.A. Izevest, Inst. Platiny. 7, 1 (1912).

Nose Y., Kushida A., Ikeda T., Nakajima H., Tanaka K., and Numakura H., *Mater. Trans.* 44, 2723 (2003).

Okamoto H., Materials Park, OH, pp 330, (1993).

Osaka K., Sakaki D., and Takama T., *Jpn. J. Appl. Phys.* 41, L155 (2002).

Pynko V.G., Komalov A.S., and Ivaeva L.V., *Phys. Stat. Sol.* 63, 127 (1981).

Ravelosona D., Chappert C., and Mathet V., *Appl. Phys. Lett.* 76, 236 (2000).

Saito S., Hoshi F., and Takahashi M., *J. Appl. Phys.* 91, 8028 (2002).

Shih J.C., Hsiao H.H., Tsai J.L., and Chin T.S., *IEEE Trans. Magn.* 37, 1280 (2001).

Stappert S., Rellinghaus B., Acet M., and Wassermann E.F., *J. Crys. Grow.* 252, 440 (2003).

Sun S., Murray C.B., Weller D., Folks L., and Moser A.,, *Science 287,* 1989 (2000).

Suzuki T., Harada K., Honda N., and Ouchi K., *J. Magn. Magn. Mater.* 193, 85 (1999).

Suzuki T. and Ouchi K., *IEEE Trans. Magn.* 37, 1283 (2001).

Takahashi Y.K., Ohnuma M., and Hono K., *Jpn. J. Appl. Phys.* 40, 1367 (2001).

Takahashi Y.K., Ohnuma M., and Hono K., *J. Magn. Magn. Mater.* 246, 259 (2002).

Watanabe M., Nakayama T., Watanabe K., Hirayama T., and Tonomura A., *Mater. Trans. JIM*. 37, 489 (1996).
Weller D., Lu B., and Kryder M., Seagate Technology, Pittsburgh, PA 15222 (2005).
Weller D., Lu B., and Kryder M., IEEE *Distinguished Talk* (2005).
Wong B.Y. and Laughlin E.E., *Appl. Phys. Lett.* 61, 2533 (1992).
Wood R., Sonobe Y., Jin Z., and Wilson B., *J. Mag. Mag. Mater*. 235, 1 (2001).

In: Magnetic Thin Films
Editor: John P. Volkerts

ISBN: 978-1-61209-302-4

Chapter 9

HIGH TEMPERATURE FERROMAGNETISM IN PRISTINE SEMICONDUCTING OXIDE THIN FILMS

Nguyen Hoa Hong*

Department of Physics and Astronomy, Seoul National University,
Seoul 151-747, South Korea

ABSTRACT

10 years ago, Dietl theoretically predicted that the ferromagnetism (FM) at high temperature could be obtained in many semiconductors such as ZnO, GaAs, GaN, etc., if we dope Mn along with a certain concentration of holes. Reports of Curie temperatures well above room temperature for wide-band gap oxides doped with a few percent of transition-metals have triggered intense interest in these materials as potential magnetic materials for spintronics.

The origin of the magnetism is debated; in some systems, the FM can be attributed to nanoparticles of a ferromagnetic secondary phase, but in others, properties are found which are incompatible with any secondary phase, and an intrinsic origin related to structural defects is implicated: Our experimental results on TiO_2, HfO_2, In_2O_3, ZnO, and SnO_2 thin films have confirmed that magnetism is certainly possible in pristine semiconducting oxides, and the observed FM is most probably due to oxygen vacancies and/or defects. The assumption for FM due to oxygen vacancies/defects in TiO_2 thin films is strongly confirmed by our X-ray magnetic circular dichroism measurements (XMCD): There is a presence of XMCD signals at both O K and Ti L2,3 edges. It shows that the FM in TiO_2 films stems from both O-2p and Ti-3d electrons. Our theoretical model also suggests that confinement effects must play a key role in shaping up magnetic properties of thin films, or more generally, of low dimension structured oxides.

A new picture of defect-based magnetism is emerging. There is a need for a type of dilute magnetic thin films— a system that is easy to prepare and reproduce. Device applications can follow by design. Once the mechanism is better understood, the next challenge will be to generate stable and controllable defect structures where we can get the benefit of this unusual high-temperature FM.

---

* E-mail: nguyenhong@snu.ac.kr

1. Introduction

After a theoretical prediction about a possibility of having ferromagnetism (FM) in transition-metal TM doped semiconductors in 2000 [1], many research groups have attempted to look for high temperature FM in TM-doped semiconducting oxides. Many studies have been done on TM-doped TiO_2, ZnO, SnO_2, and In_2O_3 and FM beyond room temperature has been obtained [2, 3]. However, in 2004, Coey's group in Dublin first announced about the observed FM in HfO_2 thin films grown on sapphire or silicon substrates [4]. This report has really called a special attention of researchers in magnetism community about a new phenomenon, so-called d^0 magnetism. Indeed, the thin film form (i.e. 2-dimension structure) might make a big difference, since defects and/or oxygen vacancies that are formed during the growth, can become a source for magnetism. The fabrication conditions actually can necessarily create oxygen vacancies, which play a similar role to an *n*-type doping. In the recent years, many experimental works have given feedbacks to the current theories with evidences showing that defects certainly can change the magnetic properties of diluted magnetic oxide thin films. For example, it was obvious that defects could intentionally introduce FM into ZnO system [5]. In some other cases, it was found that a good crystallinity can indeed destroy the ferromagnetic ordering. And it was also shown that filling up oxygen vacancies could enormously degrade the magnetic moment of those oxides [6]. One theory group has performed simulation on HfO_2 and assumed that isolated cation vacancies in HfO_2 could form high-spin defect states, and as the results, they could be ferromagnetically coupled with a rather short-range magnetic interaction resulting a ferromagnetic ground state [7]. Another experiment group could not detect FM in their HfO_2 samples, and then insisted that FM must come from some contamination of their Si substrate [8]. We ourselves obtained FM in HfO_2 thin films grown on Yttrium Stabilized Zirconia (YSZ) substrates; however the corresponding XMCD data do not show any magnetic signal on the Hf site due to big ratio of noises [9]. Later, the Dublin group claimed that the FM of their HfO_2 films is not stable but can be aging with time [10]. Another group proposed their model but it does not seem to explain well experimental results [11]. All of these controversial issues have urged us to experimentally verify the magnetic properties of several types of pristine oxides. Actually, we have evidenced that pristine oxides could be certainly room temperature ferromagnetic in nano-structured forms. On the other hand, transition - metal doping indeed plays no important role in introducing FM in those oxides. In this article, we will discuss in details about the observed FM in various pristine oxides such as TiO_2, HfO_2, In_2O_3, ZnO, and SnO_2 laser ablated thin films.

2. Transition-Metal Doped TiO_2 : It Still Seems to be a Beautiful Picture

TiO_2 is a transparent wide-gap semiconductor which has long been investigated for its rich physical and chemical properties. It has applications in several domains, ranging from photocatalysis to solar cells. Recently it has attracted a great attention from condensed matter physics community since it appears that with some appropriate doping with magnetic impurities, might exhibit FM at high temperatures [3]. This makes it one of the best candidates for spintronic devices. Many groups have reported on TM-doped TiO_2 with contradictory results: the magnetic properties for a given dopant can range from diamagnetism to ferromagnetism, going through antiferromagnetism, and the intrinsic nature of the measured magnetic properties has not been clearly established yet [2]. TM-doped TiO_2 thin films were grown by a pulsed-laser deposition (PLD) system (KrF, 248 nm) from ceramic targets on (100) $LaAlO_3$ (LAO) substrates. The targets were 99.99% pure. Iron and nickel impurities are well below 10^{-2} wt.%. The growth conditions for these films are exactly the same as those reported in Ref. [12]. Films were grown with thicknesses of 5, 10 and 220 nm. All films were colorless, shiny, and highly transparent. The structural properties of the films have been determined by X-ray diffraction one can see from Figure 1. that all films are single phase anatase (within the detection limit of the technique) and have their c-axis oriented perpendicular to the film surface. One point to notice here is that the type of dopant does not influence much on the peak positions as well as intensity. . In spite of the difference in concentration, the out-of-plane parameters of TM-doped TiO_2 films do not differ much from one to another, and it might be assumed that the dopant atoms were well substituted for Ti atoms in the TiO_2 host matrix, and a solid solution was obtained [12]. The magnetic force microscopy (MFM) data that will be shown later also rule out any assumption for a possible presence of any dopant particle or clusters. All films show FM up to and above room temperature, as shown by the macroscopic magnetization measurements, performed with a SQUID magnetometer, resumed in Figure 2. (a). Note that the diamagnetic signal from the bare LAO substrate has been subtracted from the measured magnetization in order to give the right magnitude of magnetization of films only. In some case, the magnetic moment can be very large as 4.2 μ_B per atom (see a typical example of M-H curve of V-doped TiO_2 films taken at room temperature shown in Figure 2. (b)). This finding is significant since vanadium itself is known to be non-magnetic. Thus, the observed FM was assumed to originate from the RKKY interaction but not from any kind of dopant clusters [13]. This remark is supported by the AFM-MFM measurements that were performed at room temperature on a film of V-doped TiO_2 (Figure 3).

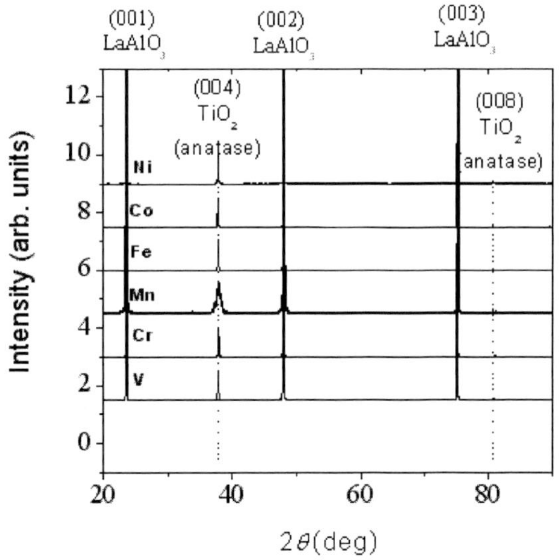

Figure 1. X-ray patterns of transition-metal doped TiO2 thin films.

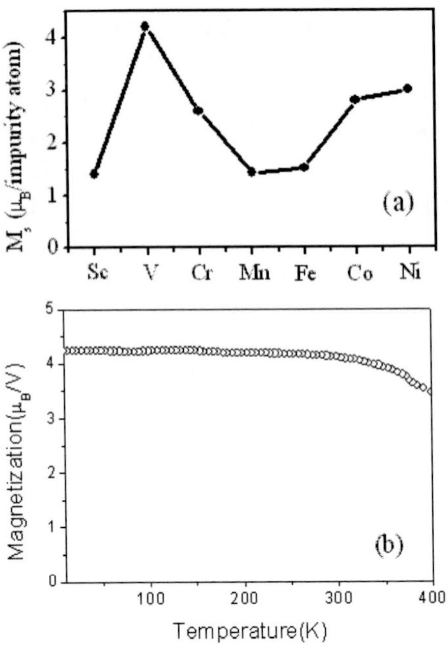

Figure 2. (a)Saturated magnetization of transition-doped TiO2 films versus element and (b) Magnetization versus temperature taken at 0.2 T for a V-doped TiO2 film.

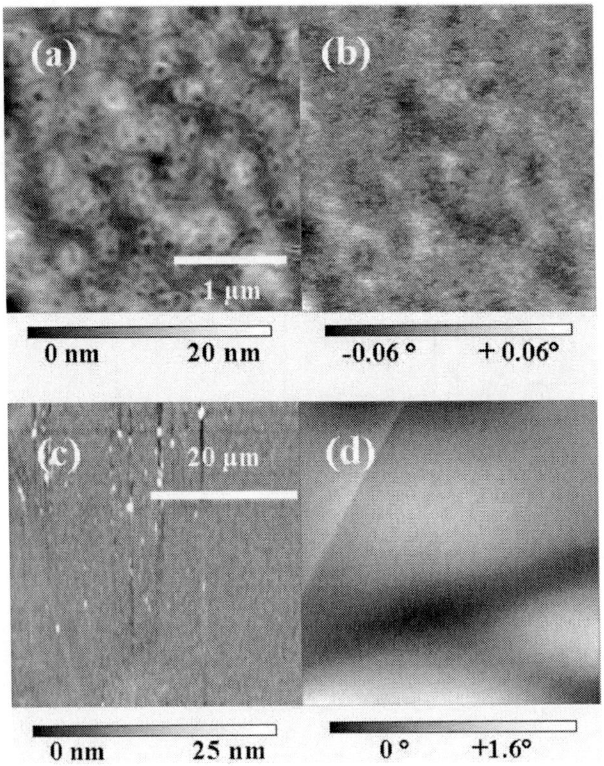

Figure 3. AFM ((a) and (c)) and corresponding MFM pictures ((b) and (d)) performed in small and large scale for a V-doped TiO2 film., respectively.

While in the small scale, one can see the magnetic homogeneity of the sample only but not the domain size, in the large scale, we obtained a direct observation of big domains that are determined to be about 5-10 μm. This has proven that the observed FM in TM-doped TiO_2 films is intrinsic and must come from the whole sample but not from any local precipitation/cluster [13]. Up to this point, people still believed that transition-metals could play a key role in introducing FM into semiconducting oxides, and then TiO_2 seems to be a good example to look at and verify what theorists have assumed.

3. Doping TM into in ZnO, First Question Comes: What should be the Real Source for Magnetism in Semiconducting Oxides?

Even though our work on TM:ZnO thin films have showed that room temperature FM could be obtained in TM-doped ZnO [14, 15], the main purpose of our study on the ZnO-based system is not to find the FM but to understand the nature of its magnetism. In comparison to the other host oxides such as TiO_2 that we have discussed earlier, doping TMs in ZnO does not result in compounds with any very big magnetic moment (in general the magnetic moment is quite modest [14], about one order smaller than that of TM:TiO_2 films). However, TM-doped ZnO appear to be very sensitive to defects and/or oxygen vacancies. Therefore, we hoped that this study can help to better understand the nature of magnetism in semiconducting oxides in general.

We first learnt from the case of V:ZnO films that by changing the substrate temperature of only 50°C, the magnetization could be changed by 1 order of magnitude [14], then it strongly suggested that growth conditions can play an important role in tuning magnetic properties of a ZnO-based system. We have then exploited its influences to try to deal with the most controversial case in the field: Mn-doped ZnO [15]. A theoretical work predicted that antiferromagnetism should be the ground state of Mn-doped ZnO [16], and Mn doping alone cannot produce FM in ZnO system; therefore, a co-doping with Cu is a must [17]. Our experimental work on Mn-doped ZnO showed that oxygen vacancies could play a more important role than that of additional carriers. We have seen that doping Mn alone could not result in room temperature FM if inappropriate conditions were applied. However, under appropriate ones, it could be absolutely possible [15]. The substrate temperature and oxygen pressure during the growth process might create defects and/or necessary oxygen vacancies that are similar to an *n*-type doping. This hypothesis is in accord with the explanations in Ref. 4 for the magnetism in HfO_2 films, which was proven to be acceptable later by a theoretical report if supposing that vacancies can be necessary ingredients to create additional bands inside the semiconducting gap that is responsible for such FM [7]. Our work on Cr-doped ZnO films have revealed that an oxygen annealing could certainly improve the crystallinity of the films [6], but simultaneously it could enormously degrade its ferromagnetic ordering (see Figure 4). It is obvious that in this system, a perfect crystallinity does not go along with FM, and filling up oxygen vacancies destroy significantly ferromagnetic ordering. Or in others words, defects and oxygen vacancies indeed must play a very important role in tuning FM [6]. A question arises here is that perhaps in diluted magnetic oxides, there is another source for magnetism rather than magnetism that comes from RKKY interaction?

4. Ferromagnetism in Undoped Semiconducting Oxides due to Defects and/or Oxygen Vacancies: A Special Feature of Low Dimension Systems

Pure TiO_2, HfO_2, ZnO, In_2O_3, SnO_2 films were deposited by a pulsed-laser deposition (PLD) system (KrF, 248 nm) from ceramic targets on (100) $LaAlO_3$ (LAO), (100) Yttrium Stabilized Zirconia (YSZ), *C*-cut Al_2O_3, (001) MgO, (100 LAO substrates, respectively. The targets were 99.99% pure. Iron and nickel impurities are well below 10^{-2} wt %. The growth conditions for those undoped oxide films are exactly the same as the optimal conditions we had found for TM-doped TiO_2 [12], Ni-doped HfO_2 [18],V-doped ZnO [14],TM-doped In_2O_3 films [19], Cr-doped SnO_2 [20]. The typical thickness of TiO_2, HfO_2, and SnO_2 films is 220 nm, while it is 600 nm for In_2O_3 films on MgO and 375 nm for ZnO films on Al_2O_3. All films of TiO_2, HfO_2, ZnO, In_2O_3 and SnO_2 are colorless, shiny and highly transparent. Magnetic moment data was basically taken when the magnetic field was applied parallel to the film plane.

All films are ferromagnetic at room temperature (see Figure 5. (a) for magnetization (*M*) versus temperature (M-T) taken at 0.2 T and Figure 5. (b) for a M versus field (M-*H*) curve taken at 300 K). While the magnetic moments for HfO_2 and TiO_2 films are rather large (M_s is almost 30 emu/cm^3 for HfO_2 and 20 emu/cm^3 for TiO_2), it is very modest for In_2O_3 films on MgO and ZnO films on Al_2O_3 (one order smaller, i.e. only about few emu/cm^3). Later, another group also obtained similar results for their laser ablated TiO_2 films [21], and the TiO_2 films made by spin coating in-air by our group also show room temperature FM (see Ref. 22, to be discussed later on). Thus, we can say confidently that the observed phenomenon is not mistaken. The large value of magnetic moment in the case of HfO_2 and TiO_2 films is hard to be attributed to any kind of impurities. What can be the source for magnetism here? For the TiO_2 case, neither Ti^{4+} nor O^{-2} is magnetic. Also for the HfO_2 case, neither Hf^{4+} nor O^{-2} is magnetic. An initial assumption is that it is due to impurities. From the viewpoint of purity of the targets, we must say that such possibility is very small, since impurities of less than 10^{-2} wt% could not create such huge magnetic moments. From the viewpoint of the structural properties of the deposited films, it is found that there is no trace of impurities that could be seen from XRD and films are single phase [23]. One can see from Figure 6. (a) that oxygen vacancies must play a key role in introducing FM: if the sample of TiO_2 is annealed in oxygen vacancies for few hours, the magnetic moment can be reduced by 1 order. When we increase the time of annealing, the TiO_2 film can turn to be diamagnetic. It means that filling up oxygen vacancies destroys FM in those films.

Returning to Figure 5. (a), one can see that films of In_2O_3 on MgO are room temperature ferromagnetic but with a rather modest magnetic moment. However, an important feature that needs to note here is that In_2O_3 films fabricated under the same conditions on Al_2O_3 substrates are diamagnetic; even if the films were well crystallized [23].There has been no report so far about In_2O_3 that could be magnetic, since In^{3+} could not be the source of magnetism.

Even though In_2O_3 tends to create oxygen vacancies [24], the fact that FM is observed on only one type of substrate but not on the other implies some sort of defects that might cause such magnetism. Similarly, in the ZnO case, neither Zn^{2+} nor O^{2-} is magnetic, thus, according to the conventional concepts; there is no source for magnetism in pure ZnO.

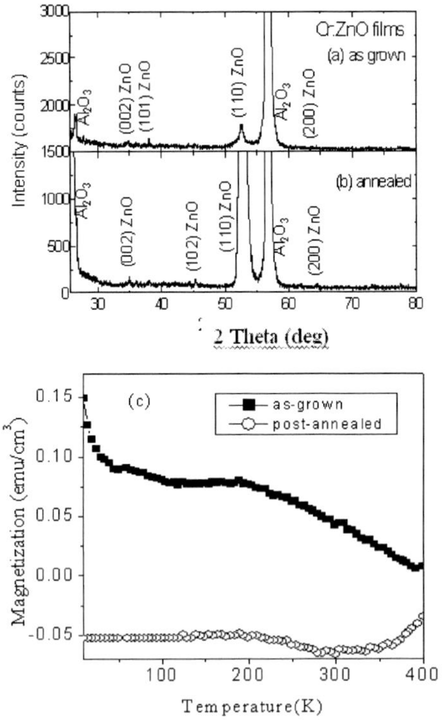

Figure 4. (a) and (b): X-ray patterns and (c) magnetization versus temperature taken at 0.2 T for as-grown and post-annealed Cr-ZnO films.

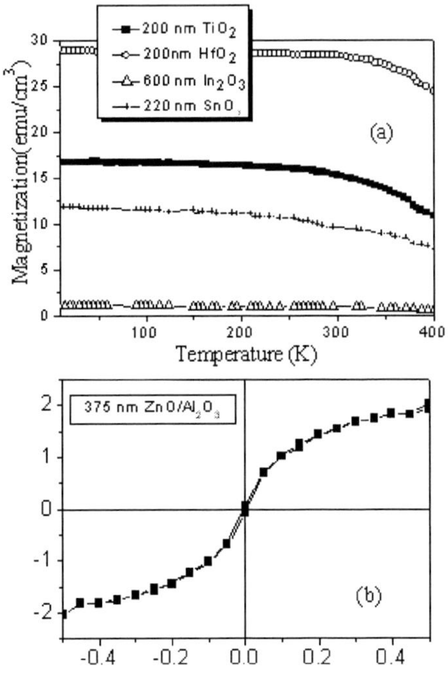

Figure 5. Magnetization (a) versus temperature taken at 0.2 T for TiO2, HfO2, In2O3 and SnO2 films and (b) versus field taken at 300K for a ZnO film.

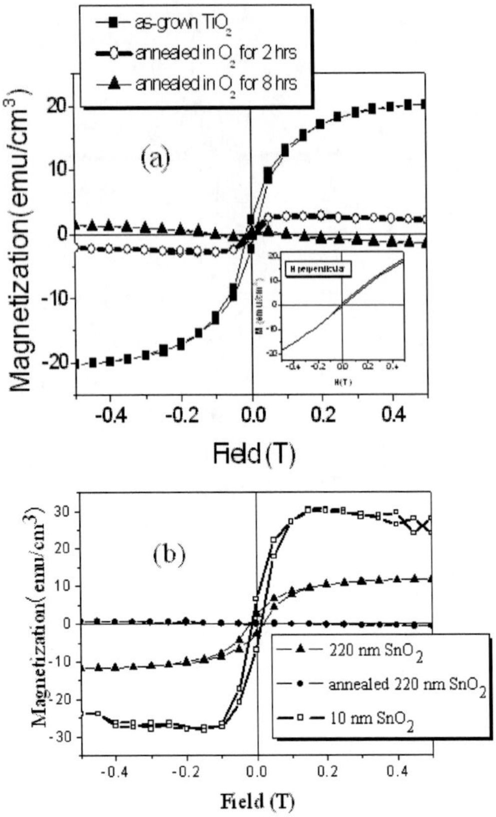

Figure 6. Magnetization versus magnetic field taken at 300 K with field parallel to the film's plane (a) for as grown and post-annealed 220nm-thick TiO2 films and (b) for as grown 220nm-thick and 10nm thick SnO2 films and the post annealed 220 nm-thick SnO2 film. The inset of Figure 6. (a) shows the M (H) curve taken in perpendicular configuration for the 220nm-thick TiO2 film.

As for pristine SnO_2 films, there is no reason to attribute the introduction of FM to any dopant, and moreover, there is no 3*d* electron involved, so that one cannot think of any interaction that may originate from that. Some groups reported that their SnO_2 films are diamagnetic, while ours is certainly ferromagnetic. Hays *et al.* reported that their nanoparticles of SnO_2 are non-ferromagnetic [25], while Bangalore group has confirmed that their SnO_2 nanoparticles were weakly ferromagnetic with some paramagnetic component [26]. To explain the observed FM in our films, we must say that most probably, oxygen vacancies formed during the growth are a key factor here, beside the presence of confinements. However, growth conditions and how to make them precisely controllable must be a standing issue. We would like to recall Ref. 26 stating that thermal treatments could drastically influence the magnetic properties of SnO_2 nanoparticles. This explained well why they got weak FM which is different from result reported in Ref. 25. It is also the reason why we could get much more pronounced FM in our SnO_2 films, while in the case of Ref. 26, the paramagnetic phase is still more dominant than the co-existed ferromagnetic one. The assumption about FM due to oxygen vacancies is supported by the data of the post-annealed SnO_2 film. One can see from Figure 6. (b) that after annealing in oxygen for 10 hrs, the SnO_2 film loses its ferromagnetic ordering, and becomes diamagnetic. It is obvious that filling up oxygen vacancies degrades magnetic ordering of those films.

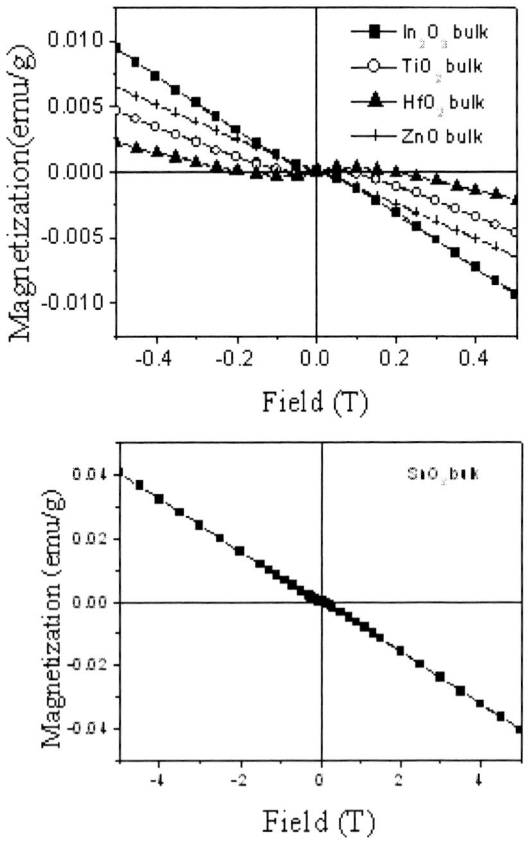

Figure 7. Magnetization versus magnetic field taken at room temperature for TiO2, HfO2, In2O3, ZnO, and SnO2 bulks.

Since the other groups reported that the FM could be found only in TM-doped SnO_2 but not in the undoped SnO_2 [25, 28, 29], we also carefully studied Mn-doped SnO_2 films, to see how a TM doping could influence the magnetic properties of the SnO_2 host. It is revealed that the Mn doping obviously reduces the magnetic moment of the SnO_2 host. No matter how much the dopant concentration is, the degradation of magnetic ordering due to TM doping is always present [27]. One should question if there is any relation between structural and magnetic properties leading to this reduction of magnetic moment. Referring to X-ray diffraction patterns of the $Mn_xSn_{1-x}O_2$ films, we found that Mn doping actually degrades the crystallinity of the SnO_2 films. When Mn content is equal or larger than 5%, the (305) peak of Mn_3O_4 clearly appears in the spectra. This formation of Mn_3O_4 clusters/particles should be the reason for the magnetic ordering of the SnO_2 host lattice to get degraded [27]. From Ref. 25 reporting on the room temperature FM of the TM-doped SnO_2, it was found that as for the Co doping, for example, it is not recommended to dope more than 1%, since it may cause an enormous distortion of the lattice that may result in a significant disordering that destroys FM. We assume that the nano-structured formation of SnO_2 should be a key factor to introduce FM, and it is even more important than the TM doping. This remark would be logical if one looks at the case of Fe-doped SnO_2 with the Fe concentration below 5% [28]. Fe doping decreases the lattice parameter of SnO_2 and increases its magnetic moment. On the other hand, if being heated at 600°C, Fe diffused to the surface and destroyed the FM of the

SnO_2. Even though only 24% of Fe content was uniformly incorporated into SnO_2 as Fe^{3+}, there was no evidence of any ion impurity phase in Fe-doped SnO_2. The authors then interpreted that the magnetic interaction in this system was most likely related to properties of the host SnO_2 and their oxygen stoichiometry. This is reinforced by the work of the other group reporting that Fe doping in fact just brings more oxygen vacancies into the SnO_2 system. And those vacancies likely play a key role in magnetic interaction (i. e. the Fe doping only acts as a catalyst) [30]. Concerning Mn-doped SnO_2, Dublin group reported that they could dope up to 28%, and the M_s obtained was really huge (for Mn content of 1 %, M_s was 20 μ_B/Mn) [29]. Theoretically, in no case M_s can be larger than 5 μ_B/Mn. Therefore, it is impossible to attribute the source for FM observed in the Mn-doped SnO_2 to only the TM doping. One should suppose that doping of 3*d* element into SnO_2 actually just helps to stabilize the low-lying magnetic excited state. Or in other words, one must say that the dopant acts as the activation factor of the defect moment [29]. This is absolutely in consistent with our results. Thus, for all of those 5 cases, we can assume that defects and/or oxygen vacancies might be the main source for the observed ferromagnetism as suggested in Refs. 4 and 23.

If defects/oxygen vacancies could cause magnetism in those films, a strong anisotropy must be observed (normally defects/oxygen vacancies are not expected to be distributed homogeneously, moreover, they cannot be symmetrically magnetic. Due to the nature of the film growth, if it is influenced by substrate effect, the effect must have a favored direction, basically in-plane).

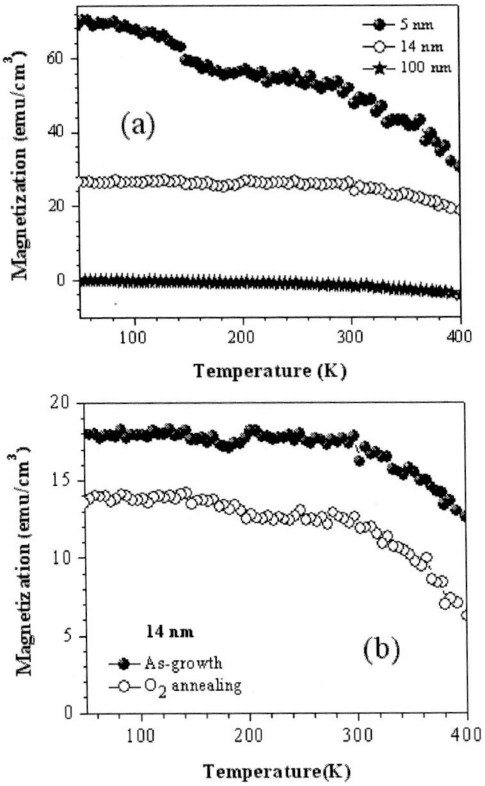

Figure 8. Magnetization versus magnetic field taken at room temperature for spin coated TiO2 films (a) as-deposited with different thickness and (b) as-deposited and post-annealed 14nm-thick.

For example, one can compare the data in perpendicular configuration for TiO_2 film in the inset of Figure 6. (a) with the one in parallel configuration shown in the main panel of Figure 6. (a), basically magnetization is in-plane. As magnetic field was applied perpendicular to the film plane, the curve shows paramagnetic state, which is completely different from the ferromagnetic signal detected in parallel configuration. The strong anisotropy enforces the assumption for magnetism due to defects/oxygen vacancies in those films. It is also found that there is a strong thickness dependence of the magnetic moment in pristine oxide thin films. As reported in Ref. 23 for TiO_2 and HfO_2 films, the 10 nm-thick films could have a magnetization of more than 20 times larger than that of the 220 nm –thick-films. Therefore, if the magnetism in these samples due to defects, then those defects must be localized mostly near the interface between the film and the substrate, or at the surface. Or in other words, oxygen vacancies/defects in the films that were fabricated by our chosen conditions must be very close to each other. A similar feature for SnO_2 film can be seen from Figure 6. (b). However in SnO_2 case, the ratio between magnetic moment of the thick- and thin ones are much less than that of TiO_2 and HfO_2 cases, indicating that in SnO_2 films, those defects and vacancies, even though are mainly located at the surface/interface, but also spread over the whole thickness of the films with smaller density [27].

In order to check if there is any contamination of the substrates that might cause the observed FM, all the substrates were measured under the same sequences as for the films. All the bare substrates showed diamagnetic behavior as expected [23, 27]. Note that all the tools we used to hand the samples are plastic, non-magnetic based, and all the pieces of straws that we used during the measurements were also checked carefully and all gave no magnetic signals. Additionally, data of bulks likely support the assumption for FM due to the thin film form only. As one can see from Figure 7, all bulk TiO_2, HfO_2, In_2O_3, SnO_2 and ZnO (i.e. pieces cut from the corresponding targets) are diamagnetic, or in other words, we must say that the room temperature FM observed in pure TiO_2, HfO_2, In_2O_3, SnO_2, and ZnO films are very unique for low dimension systems, or in other words, it is assumed that confinement effects should play some role here. The similar result obtained for very thin- films of TiO_2 made by another method-spin coating technique [22], has confirmed this speculation. One can see from Figure 8. (a) that the ultra thin film such as 5 nm-thick has a much larger magnetic moment than the 100nm-thick one. Additionally, we also see that annealing in oxygen atmosphere degrades magnetic ordering of spin coated TiO_2 films. This ensures the importance of oxygen vacancies as well as the formation of nano-sized structure in introducing FM into our undoped samples.

5. Re-Judge the Role of Transition-Metal Doping

We discuss here in details the case of Mn-doped TiO_2 films as one among typical examples of TM-doped semiconducting oxide, to see how a TM doping could influence the magnetic properties of the oxide host. The *M*(*T*) curves of $Ti_{1-x}Mn_xO_2$ films taken at 0.2 T is shown in Figure 9. (a). One can see that when the Mn concentration is small (i.e. 2-5%), the doped films could have magnetic moments, which are a bit larger than that of the undoped TiO_2 film: 30 and 23 emu/cm^3, for 2% and 5% doped Mn, respectively, in comparison to the value of 17 emu/cm^3 for the TiO_2 host). One can see clearly that even without doping, the

TiO_2 film is certainly ferromagnetic for the whole range of temperature below 400 K. It was assumed that oxygen vacancies and/or defects in the thin films must be responsible for the induced magnetism in such a system. As mention earlier, the TiO_2 bulk is diamagnetic, and the $LaAlO_3$ substrate is diamagnetic, too (recall the magnetization data for TiO_2 bulk in Figure 7)). Doping Mn into the TiO_2 in fact does not induce any FM but it can only enhance the magnetic moment, which already exists in the TiO_2 host. Looking carefully at the low *T* region below 50 K of the *M*(*T*) curves, one can see that as for Mn - doped films, there is some tendency for *M* to rise up as the temperature decreases, while that feature does not appear for the pristine TiO_2 film. It is very likely that the Mn doping disturbs the TiO_2 structure, and as consequence, adds some secondary phase such as precipitations into the system. This feature becomes more pronounced when the doping concentration increases. When the Mn content is 10%, the doping appears to influence negatively on the ferromagnetic ordering of TiO_2: magnetic moment is reduced clearly. One can see that the $Ti_{0.85}Mn_{0.15}O_2$ film is basically antiferromagnetic (with only a small amount of a weak ferromagnetic component that could be seen only at very low *T*), while for the $Ti_{0.8}Mn_{0.2}O_2$ film, the paramagnetic phase is dominant, along with another secondary phase that is present (see the jump below 50 K). In principle, the rising up of magnetic moment at low *T* indicates an existence of precipitations/clusters as discussed in Refs. 20 and 32. It is obvious that an increase of Mn content above the cut-off value just degrades, and then destroys completely the ferromagnetic ordering of the TiO_2 film, since the system loses a big volume of phase for antiferromagnetic/paramagnetic components that come from clusters. The remained part of a weak ferromagnetic phase observed in those overdoped samples most likely comes only from the ferromagnetic base of the undoped TiO_2 films, but certainly not the magnetism induced by the Mn doping (note that for 3 samples of doping 10, 15, and 20%, the magnitudes of magnetic moment at low *T* are very comparable). Note also that the magnetic moment does not decrease as expected when Mn content increases from 15% to 20%. It could be explained that for each composition, the types of clusters must not be the same. Thus, above the cut-off value, an increase of the dopant content does not simply mean that the non-ferromagnetic clusters increase in volume (if it were the case, the magnetic moment of the $Ti_{0.8}Mn_{0.2}O_2$ film would have been smaller than that of the $Ti_{0.85}Mn_{0.15}O_2$). As for the higher doping case, it is more likely that another type of precipitations (with a different type of magnetic component) could be formed. The degradation of ferromagnetic ordering of TiO_2 thin films due to the overdoping of Mn could be explained reasonably by the XRD data [33]. As the Mn content is below 5%, the TiO_2 structure could still remain, with only the anatase peaks of strong intensities appearing in the spectra. However, when the Mn content reaches 10 or 15%, the anatase peaks become very weak (with the intensity of one order smaller). When the Mn content is 20%, the TiO_2 structure is destroyed completely (note that no peak of TiO_2 appears in the spectrum). Coey *et al.* mentioned that TMs might not play any important role in introducing FM in HfO_2 [34]. As discussed in the previous paragraph, magnetism can exist in pristine semiconducting/insulating oxides under thin film forms, due to induced oxygen vacancies. TiO_2 film itself could be ferromagnetic, therefore, doping Mn can only enhance the magnetic moment of TiO_2 a bit, if the Mn content is kept as few percents. In contrast, the ferromagnetic ordering is degraded remarkably if the dopant content increases. It is certain that the TiO_2 host is the most important factor to remain the FM of the system, but not the doping. Therefore, once the TiO_2 structure can be remained, the films can still be ferromagnetic. As the Mn content increases, there is a formation of clusters in the films, as well as a destruction of their anatase structure of the host lattice.

As consequences, the FM of TiO_2 films is suppressed, or even at some critical point, completely destroyed. This remark is consistent with our preliminary X-ray magnetic circular dichroism (XMCD) measurements on the film of $Ti_{0.98}Mn_{0.02}O_2$ (Figure 9. (b)) showing that the magnetic signals from the Mn edge are well paramagnetic. One can clearly see that the dependence of the XMCD signals on magnetic field is well paramagnetic. The magnetic moment at remanence (i.e., $H = 0$) is estimated to be below 0.02 μ_B/at. for Mn by applying the magneto-optical sum rules. A very similar feature was observed also for the Co- and Cr-doped TiO_2 film [35]. These results show that the main contribution to the ferromagnetic signal is certainly that of the TiO_2 host matrix. It then explains why the FM (and T_C) in transition metal-doped TiO_2 does not depend much on the type and concentration of dopant. It appears here that new theories are required in order to understand this novel class of materials.

6. FM Due to Oxygen Vacancies / Defects in Pristine Semiconducting Oxides

To confirm above results, the chemical and orbital selectivity of X-ray absorption spectroscopy and X-ray magnetic circular dichroism were exploited, to prove that the observed ferromagnetism in laser ablated TiO_2 films is an intrinsic property of this material, directly originating from the O-2*p* and, to a lesser extent, from the Ti-3*d* electrons. This ferromagnetism stems from a surface region, of a thickness of several nanometers, which is likely to be rich in oxygen vacancies [36]. In order to find some model that fit better to explain our experimental results that are quite far from what theory of Dietl could cover, we propose a model of FM in oxide thin films of HfO_2 and TiO_2 to investigate the possibility of magnetism due to oxygen vacancies and the confinement effect. Due to HfO_2 and TiO_2 structure, each oxygen atom is surrounded by three or four Hf (or Ti) atoms and the local symmetry is D_{3h} and D_{4h} accordingly. An oxygen vacancy would result in the loosing of bonding of two *d*-electrons in the outer shell of the Hf (Ti) atom. We suppose that in the local D_{3h} or D_{4h} molecular orbital high symmetry, the two electrons do not really leave their own shell and remain as the *d*-electrons, becoming a d^2 impurity center. The exchange interaction of these two *d*-electrons with each other and with the molecular orbital field leads to splitting of energy level of the impurity band around the vacancy, so that produces a large magnetic moment. Beside the orbital local symmetry, the electrons here are considered to be confined in the two-dimensional confinement. The thin film configuration would have two effects on the structure of the impurity band: it enhances the formation of the oxygen vacancy as well as enhances the coupling between the *d*-electrons and between the *d*-electrons and the local field. The boundary conditions at the surfaces of the thin film in the strong confinement approximation make the matrix elements for the exchange interaction much larger in comparison with the bulk case. In the strong local field, the interaction of each *d*-electron with the local tetragonal orbital field is considered first to get the eigenstates and the splitting of the five-fold degenerate energy level of the single *d*-impurity center into *e*- and t_2-orbitals. Then the interaction between the two *d*-electrons is calculated according to irreducible representation of the product of E and T_2 representations, along with the spin part and the Pauli principle. By diagonalizing the Coulomb exchange matrices, we obtain the splitting of all energy levels and the states of the impurity band with a high spin ground state. We use the tight binding calculation for HfO_2, TiO_2 and In_2O_3 thin films to consider the effect of the molecular orbital field and the impurity band created by the exchange interaction of the electrons created by the oxygen vacancy and trapped around the vacancy, while taking into account the two-dimensional confinement effect of the thin films. The splitting of energy levels of this impurity band has been obtained and the two-dimensional confinement results in large exchange interaction matrix element, which is very different in comparison with the bulk's case. Our calculation shows a high spin state with a magnetic moment per vacancy of 3.18 μ_B for TiO_2, 3.05μ_B for HfO_2 and 0.16μ_B for In_2O_3.

If the vacancy concentration about 3%, we would have the magnetic moment comparable to the experimental values. According to this model, FM is quite possible for thin film configurations of pristine semiconducting oxides, especially for the configurations that favour a formation of oxygen vacancies [35]. Our suggestions for a possible FM in oxides with low dimension configurations having vacancies is recently enforced by the finding of FM in nanoparticles of various undoped oxides such as CeO_2, Al_2O_3, ZnO, In_2O_3, and SnO_2 [26].

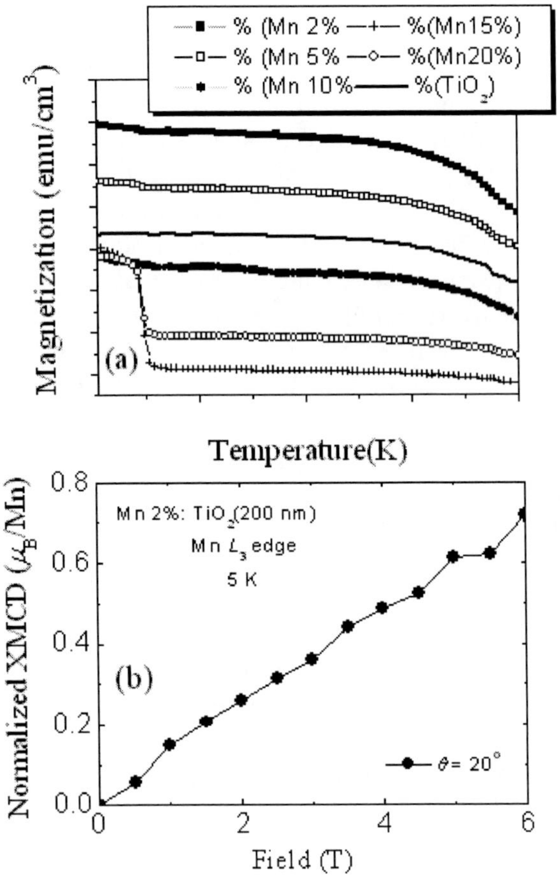

Figure 9. (a) Magnetization versus temperature taken at 0.2 T for Mn-doped TiO2 films with different Mn concentrations and (b) XMCD data for a TiO2 film doped with 2% of Mn.

CONCLUSION

Our experimental results on TiO_2, HfO_2, In_2O_3, ZnO, and SnO_2 laser ablated thin films have proven that magnetism could be obtained in pristine semiconducting oxide thin films. The observed ferromagnetism is intrinsic, and most probably due to oxygen vacancies and/or defects. The assumption for FM due to oxygen vacancies/defects in TiO_2 thin films is strongly confirmed by our X-ray magnetic circular dichroism measurements (XMCD): There is a presence of XMCD signals at both O K and Ti $L_{2,3}$ edges. It shows that the FM in TiO_2 films stems from both O-2p and Ti-3d electrons. Our theoretical model also suggests that confinement effects must play a key role in shaping up magnetic properties of thin films, or more generally, of low dimension structured systems. This finding may open a new road for searching for a novel class of materials that have many good properties for spintronic applications. By down scaling semiconducting oxides to nano-size, under appropriate conditions that create oxygen vacancies/defects, room temperature ferromagnets can be made, then spin and charge could be manipulated at the same time in the same device. How to make these things controllable should be the next challenging step in order to realistically master and exploit it fully to meet industrial needs.

ACKNOWLEDGMENTS

The author is grateful to J. Sakai, A. Barla, N. Q. Huong, N. Poirot, A. Hassini, and A. Ruyter for their co-work that has fruited into nice results that have been presented in this article.

REFERENCES

[1] T. Dietl, H.Ohno, F. Matsukura, J. Cibert, D. Ferrand, *Science* 287, 12019 (2000).
[2] R. Ranish, P. Gopal, and N. A. Spaldin, *J. Phys: Condens. Matter* 17, R657 (2005).
[3] "Magnetism in semiconducting oxides", Edited by Nguyen Hoa Hong, Transworld Research Network ISBN: 81-7895-264-5 (2007).
[4] M. Venkatesan , C. B. Fitzgerald, and J. M. D. Coey, *Nature* 430, 630 (2004).
[5] D. A. Schwartz and D. R. Gamelin, *Adv. Mater.* 16, 2115 (2004).
[6] N. H. Hong, J. Sakai, N. T. Huong, N. Poirot, and A. Ruyter, *Phys. Rev. B* 72, 45336 (2005).
[7] D. P. Pemmaraju and S. Sanvito, *Phys. Rev. Lett.* 94, 217205 (2005).
[8] D. W. Abraham, M. M. Frank, and S. Guha, *Appl. Phys. Lett.* 87, 252502 (2005).
[9] A. Barla and N. H. Hong, Unpublished data .
[10] J. M. D. Coey, M. Venkatesan, P. Stamenov, C. B. Fitzgerald, and L. S. Dorneles, *Phys. Rev. B.* 72, 24450 (2005).
[11] G. Bouzerar and T. Ziman, *Phys. Rev. Lett.* 96, 207602 (2006).
[12] N. H. Hong, J. Sakai, W. Prellier, A. Hassini, A. Ruyter, and F. Gervais, *Phys. Rev. B* 70, 195204 (2004).
[13] N. H. Hong, J. Sakai, and A. Hassini, *Appl. Phys. Lett.* 84, 2602 (2004).

[14] N. H. Hong, J. Sakai, and A. Hassini, *J. Phys.: Condens. Matter* 17, 199 (2005).
[15] N. H. Hong, V. Brizé, and J. Sakai, *Appl. Phys. Lett.* 86, 82505 (2005).
[16] Q. Wang, Q. Sun, B. K. Rao, and P. Sena, *Phys. Rev. B.* 69, 233310 (2004).
[17] N. A. Spaldin, *Phys. Rev. B.* 2004, 69, 125201.
[18] N. H. Hong, J. Sakai, N. Poirot, and A. Ruyter, *Appl. Phys. Lett.* 86, 242505 (2005).
[19] N. H. Hong, J. Sakai, N. T. Huong, A. Ruyter, and V. Brizé, *J. Phys: Condens. Matter* 18, 6897 (2006).
[20] N. H. Hong, J. Sakai, W. Prellier, and A. Hassini, *J. Phys.: Condens. Matter* 17, 1697 (2005).
[21] S. D. Yoon, Y. Chen, A. Yang, T. L. Goodrich, X. Zuo, D. A. Arena, K. Ziemer, C. Vittoria, and Vincent G Harris*, J. Phys.: Condens. Matter* 18, L355–L361 (2006).
[22] A. Hassini, J. Sakai, J. S. Lopez, and N. H. Hong, *Phys. Lett. A* 372, 3299 (2008).
[23] N. H. Hong, J. Sakai, N. Poirot, and V. Brizé, *Phys. Rev. B* 73, 132404 (2006).
[24] "Semiconducting transparent thin films", H. L. Hartnagel, A. L. Dawar, A. K. Jain, C. Jagadish, IOP Publishing 1995, Bristol and Philadelphia.
[25] J. Hays, A. Punnoose, R. Baldner, M. H. Engelhard, J. Peloquin, and K. M. Reddy, *Phys. Rev. B* 72, 075203 (2005).
[26] A. Sundaresan, R. Bhagravi, N. Rangarajan, U. Siddesh, and C. N. R. Rao, *Phys. Rev. B.*74, 161306 (R) (2006).
[27] N. H. Hong, N. Poirot, and J. Sakai, *Phys. Rev. B.* 77, 33205 (2008).
[28] A. Punnoose, J. Hays, A. Thurber, M. H. Engelhard, R. K. Kukkadapu, C. Wang, V. Shutthanandan, and S. Thevuthasan, *Phys. Rev. B.* 72, 054402 (2005).
[29] C. B. Fitgerald, M. Venkatesan, L. S. Dorneles, R. Gunning, P. Stamenov, and J. M. D. Coey, *Phys. Rev. B.* 74, 115307 (2006).
[30] H. J. Meng, D. L. Hou, L. Y. Jia, X. J. Ye, H. J. Zhou, and X. L. Li, *J. Appl. Phys*. 102, 073905 (2007).
[31] Y. Xie and J. Blackman, *J. Phys.: Condens. Matter* 16, 4373 (2004).
[32] N. H. Hong, A. Ruyter, W. Prellier, Joe Sakai, and N. T. Huong, *J. Phys.: Condens. Matter,* 17, 6533 (2005).
[33] N. H. Hong, J. Sakai, A. Ruyter, and V. Brizé, *Appl. Phys. Lett.* 89, 252504 (2006).
[34] J. M. D. Coey, M. Venkatesan, P.Stamenov, C. B. Fitzgerald, and L. S. Dorneles, *Phys. Rev. B.* 72, 024450 (2005).
[35] N. H. Hong, A. Barla , J. Sakai, and N. Q. Huong, *Phys. Stat. Solidi* (c) 4, No. 12, 4461 (2007).
[36] A. Barla, N. H. Hong, E. Beaurepaire, P. Imperia, J. P. Kappler, J. C. Cezar, N. B. Brookes, and J. Sakai, *Unpublished data.*

In: Magnetic Thin Films
Editor: John P. Volkerts
ISBN: 978-1-61209-302-4

Chapter 10

FePt Thin Films: Magnetic Properties and Domain Wall Propagation

J. P. Attané
CEA, Inac, SP2M, 38054 Grenoble, France
Université Joseph Fourier, 38041 Grenoble, France

Abstract

In this chapter, we present the magnetic properties of thin epitaxial FePt layers (5-40 nm) deposited by molecular beam epitaxy on MgO(001) and Pt(001) samples. The tuning of growth conditions, as well as the use of ion irradiation, can be used to control the ordering of the FePt alloy within the $L1_0$ phase, and thus the uniaxial magnetic anisotropy of these samples.

We then focus on samples with high anisotropy, which possess ultra-thin domain walls. We present the magnetic properties of these samples, as well as some of their transport properties. Using microstructural characterizations, domain observations and micromagnetic calculations, we show that hysteresis properties are linked to the interaction of domain walls with structural defects. In these layers, the magnetization reversal can be analyzed as an invasion percolation process without trapping, leading to fractal geometries of the reversed domain. This thermo-activated growth process involves domain wall motion by avalanches, where one single depinning event controls the propagation dynamics over large distances.

We show that this property can be used in nanostructures to observe the interaction of a domain wall with a single defect. This allows realizing current-induced depinning in nanostructures, and measuring the spin-transfer efficiency in FePt samples.

Introduction

The interest of the nanomagnetism community towards perpendicularly magnetized magnetic materials increased during the last decade. The first reason of such an increase was linked to the choice of the hard disk drive industry to change the magnetic storage process from planar to perpendicular magnetic recording after 2004 [TOS04].

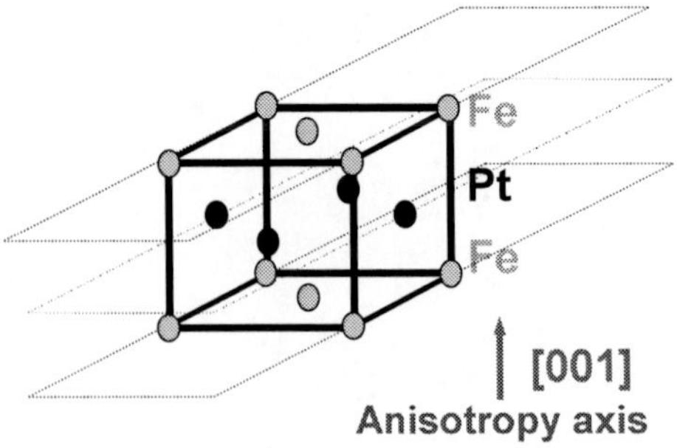

Figure 1. Elementary cell (face-centered and slightly tetragonal) of the FePt alloy ordered within the $L1_0$ phase.

The interest towards perpendicularly magnetized materials is still important for the magnetic recording industry, notably because the use of magnetic nanoparticles with high anisotropy could allow minimization of the information bit at the nanoscale, by repelling the supermagnetic limit [FRE09].

The second reason is linked to current-induced magnetization reversal. Torques appear between charge carrier spins and local moments in regions of ferromagnetic media where spatial magnetization gradients occur, such as a Domain Walls (DW). These torques are able to induce magnetization reversal, and thus could lead to the creation of new classes of spintronic devices, as racetrack memory-like storage devices [PAR08], domain-wall MRAM [NEC09] or logical computing devices [ALL05].

In this context, the use of materials with strong anisotropy became recently very popular: as they possess thin DWs, the magnetization gradients are great, and the critical currents expected to displace the DWs should be lower than those obtained in permalloy-like materials, which are up to now too high for applications. Among perpendicularly magnetized materials, like Co/Pt or Co/Ni multilayers, the chemically ordered FePt alloy possesses a huge magnetocristalline anisotropy, with an anisotropy field approximately equal to 10 T. This magnetic anisotropy appears only if the alloy is ordered within the $L1_0$ phase along the [001] direction. The perfectly ordered $L1_0$ phase can be described as an alternance of pure Fe and pure Pt atomic planes (cf. figure 1). The in- and out-of-plane lattice parameters of this tetragonal structure are, respectively, 0.386 and 0.379 nm.

In this chapter, we sum up results obtained in the study of monocrystalline FePt thin layers with high perpendicular magnetocrystalline anisotropy, deposited by molecular beam epitaxy either on Pt(001) or MgO(001) substrates. The deposit method is presented, as well as the basic hysteresis and transport properties of our layers. The origin of the coercivity, linked to DW pinning, is explained detailedly, and the magnetization reversal of these layers is analysed using tools derived from percolation theory. Finally, we show that nanowires can be lithographed in these layers, which allow realizing field and current-induced DW depinning from a single defect.

1. Deposit Technique and Chemical Order

The samples are prepared by molecular-beam epitaxy under ultrahigh vacuum (10^{-7} Pa), using commercial MgO(001) successively washed in trichloroethane, acetone and ethanol.

Different state of chemical order of the FePt alloy can be obtained, depending on the growth conditions (temperature deposit [HAL04b,BAR05]) and on post-deposit treatments like high temperature annealing [HAL04a] or ion irradiation [BER03,RAV00]. Usually the highest chemical order is looked for, in order to obtain a high magnetic anisotropy [BAR05]. The long-range order S of our samples can be measured by x-ray diffraction. If $S=n_{Fe}-n_{Pt}$, where $n_{Fe/Pt}$ is the site occupancy by one of the elements on the Fe or Pt sublattice, S ranges from zero (disordered film) to unity (perfectly ordered film).

The intensities of the (001) and (003) superstructure peaks identify the variant of the $L1_0$ ordered structure with the quadratic c axis perpendicular to the film surface. Integrating these peaks and the fundamental (002) and (004) peaks, we calculated [GEH98,ATT03] the long-range order parameter S of our FePt films. Note that low values of S do not preclude a high degree of short-range order, associated to a significant in-plane homocoordination and a high anisotropy. In [BER03], we also measured by Mössbauer spectroscopy the directional short range order of our samples.

The simplest method to deposit a FePt layer is to realize the co-deposition of Fe and Pt at room temperature. The incident fluxes of Fe and Pd are fixed using quartz crystal sensors, which are calibrated prior to the deposit using RHEED oscillations [GEH98].

However, in this case the temperature is too low to induce the chemical order, and the corresponding samples posses an in-plane magnetization and an anisotropy constant close to zero. To obtain a very high degree of uniaxial chemical order, deposits are realized at 770K. The long-range order parameter S then becomes very high (S=0.9). Fits of Mössbauer spectra shows that in those samples the percentage of Fe atoms whose close neighbour corresponds to the $L1_0$ phase is of 94% [BER03] (cf. Figure 2). The corresponding anisotropy constant, measured by classical magnetometry, is $Ku=5\times10^6$ $J.m^{-3}$ [ATT01].

Another method to induce chemical order of the Fept layer is to use the “layer-by-layer” deposition method. The layer-by-layer FePt films are obtained at room temperature by masking the Fe or Pt atomic fluxes in a molecular beam epitaxy system, growing a biatomic period multilayer of alternating pure Fe and pure Pt atomic planes, thus mimicking the structure of the $L1_0$ phase.

In such case, the magnetization is out-of-plane, but the anisotropy is about one half of those of codeposits at 770K. Indeed, for layer-by-layer samples the percentage of Fe atoms whose close neighbour corresponds to the $L1_0$ phase is only of 74%.

One way to obtain chemical order is to use He ion irradiation [BER03] at moderate temperatures on layer-by-layer samples. Indeed, the presence of a directional short range order in the pre-irradiated sample provides the basis for a mechanism by which practically complete ordering perpendicular to the film plane can be obtained (cf. figure 2).

In the following, we will focus on two types of samples: FePt/MgO and FePt/Pt. Both are obtained by co-deposition of Fe and Pt. FePt/MgO samples were obtained by direct deposition of FePt on a MgO(001) substrate, whereas FePt/Pt samples were grown according to the following process: a thin seed layer of Cr (~3nm) is deposited on a MgO(001) substrate, and then covered by a Pt buffer layer, thick enough (40 nm) to let the epitaxial strain relax and to get a smooth surface after annealing at 820 K. In both cases, the FePt layer is then grown at 770 K, and is eventually protected from oxidation by a 3 nm thick Pt capping layer.

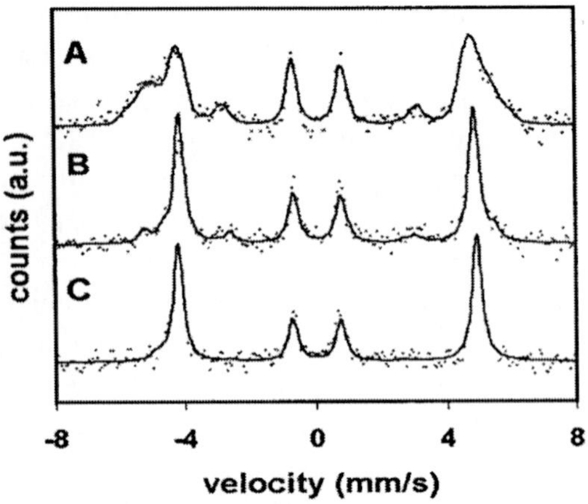

Figure 2. Mössbauer spectra of "layer-by-layer" FePt samples, taken before (A) and after (B) irradiation at 623 K by 130 keV He ions (fluence $2x10^{16}$ ions/cm^2). The latter is almost identical to the spectrum of a high temperature (770 K)-codeposited sample (C), whose perpendicular anisotropy corresponds to nearly complete chemical ordering. Lines are fits to the hyperfine field distributions, which give a percentage of Fe atoms corresponding to the $L1_0$ phase of 92% for the irradiated sample (B), close to the 94% observed for the codeposited sample (C). Figure extracted from ref. [BER03].

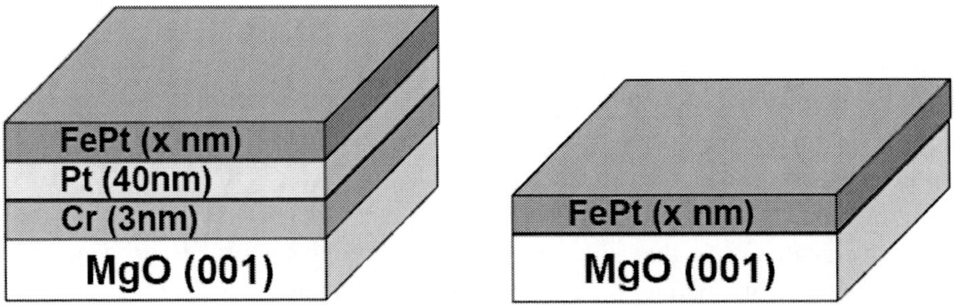

Figure 3. Structures of the FePt/Pt and FePt/MgO samples.

2. HYSTERESIS AND TRANSPORT PROPERTIES

Figure 4.a. presents the in-plane and out-of-plane hysteresis loop of a FePt(40 nm)/MgO, and shows that the easy axis of magnetization is clearly out-of-plane. The measured saturation magnetization is $Ms=1.03 \times 10^6$ $A.m^{-1}$, and the exchange constant is equal to $A=6.9 \times 10^{-12}$ $J.m^{-1}$. The value of the anisotropy constant extracted from these measurements is $Ku=5 \times 10^6$ $J.m^{-3}$, corresponding to an anisotropy field $B_A=2Ku/\mu_0 M_s \sim 10T$, and to a DW width in the nanometer range ($\Delta=\sqrt{A/K} \sim 1.2$ nm). This quite high value of the anisotropy constant underlines the good quality of the chemical order. The hysteresis loop in perpendicular field exhibits full magnetic remanence at zero field, and a large coercive field. Also, this strong coercivity allows stabilizing the sample at any stage of magnetization reversal, simply by returning to zero field (cf. figure 4.b).

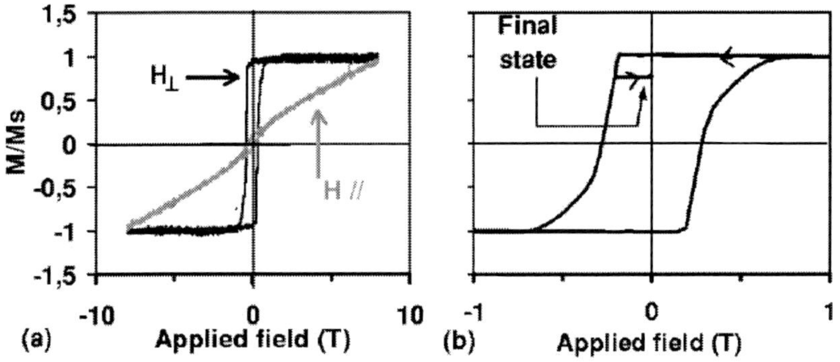

Figure 4. (a) In-plane and out-of-plane hysteresis loops of a FePt(40 nm)/Pt/MgO sample, obtained using a vibrating sample magnetometer at room temperature. The data have been corrected from the diamagnetic contribution of the substrate. (b) Extraordinary Hall effect measurement demonstrating the ability to stabilize the sample in a partially magnetized state (here the minor loop leads to M/Ms~0.75). Figure published in ref. [ATT01].

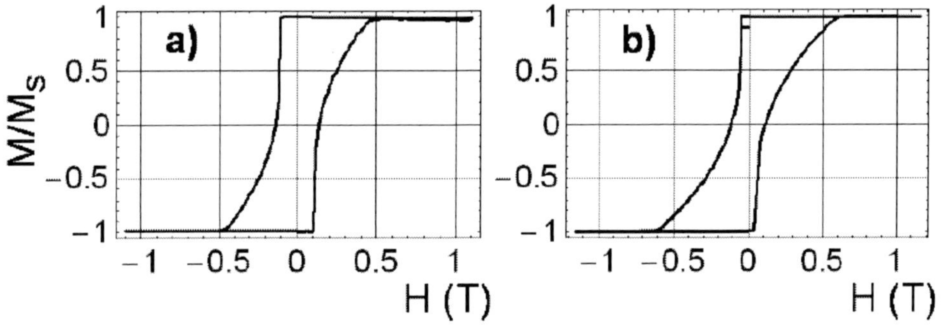

Figure 5. Hysteresis loops measured in perpendicular field by extraordinary hall effect, in a (a) FePt(32nm)/Pt(40nm)/MgO(001) sample and a (b) FePt(32nm)/MgO(001) sample. In the (b) case, the major loop is followed by a partial reversal of magnetization and a decrease of the magnetic field to zero. Figure published in ref. [ATT10].

Figure 5. compares typical hysteresis loops of FePt/MgO and FePt/Pt samples of equal thicknesses, obtained by extraordinary Hall effect (EHE) measurements. These hysteresis loops samples are qualitatively very similar, and possess the same remanence properties.

Concerning the basic transport properties of our samples, the mean resistivity of the FePt/MgO samples, measured using the Van der Paw method, is about 400×10^{-9} Ω.m. The EHE angle, measured within Hall crosses, is equal to 4.1×10^{-2}. In the following, we study the magnetoresistance of these samples.

3. Magnons Magnetoresistance

The development of spintronics is based upon a few electronic spin-dependent transport properties, mainly anisotropy magnetoresistance (AMR), giant magnetoresistance (GMR) and tunneling magnetoresistance (TMR), and at a lower level Extraordinary Hall Effect (EHE) and intrinsic domain wall resistivity (DWR, see Refs. [BAR02],[SCH05],[GER02] for review articles). By measuring the dependence of the electrical resistivity on the perpendicular magnetic field in FePt(10–32 nm)/MgO layers, we showed that the magnetoresistance of FePt samples is linked to electron-magnon diffusion, and that this "Magnon Magneto-Resistance" (MMR) allows the detection of magnetization reversal [MIH08]. Room temperature EHE and resistivity measurements were performed on FePt/MgO samples, using a very sensitive electrical transport setup using lock-in techniques, with an external magnetic field applied perpendicularly to the layer. When the sample is magnetically saturated, one can see a linear variation of the resistivity with the applied field (cf. figure 6). This resistivity decrease has been assigned by Raquet *et al.* [RAQ02] to spin-wave damping in high fields, which corresponds to a decrease of the intrinsic spin disorder. The field increase thus leads to a diminishing of the electron-magnon scattering. Figure 7. presents both the EHE and the corresponding resistivity measurements of a FePt(10nm)/MgO(001) sample. In these measurements, we also observe a linear dependence of the MR on the applied magnetic field for low positive fields, and even for negative applied fields (see Figure 1), *i.e.* in the state where magnetization and applied field are antiparallel. A sharp decrease in resistivity, of about 0.17%, is noticed at the coercive field, in both positive and negative half loops, this decrease being clearly related to the magnetization reversal. This unexpected behaviour can be explained qualitatively. When the applied field and magnetization directions are antiparallel, the high anisotropy field tends to maintain the magnetization direction, while the applied field tends to destabilize it, thus increasing the magnon population. As a result, when the magnetization and applied field are antiparallel, a linear increase of the resistance with the field absolute value can be observed. When the coercive field is attained, the magnetization switches from a saturated state to the opposite one, and the relative orientation of the applied field and the magnetization switches from antiparallel to parallel. This leads to a sudden decrease of the magnon population and thus to the observed decrease of resistivity.

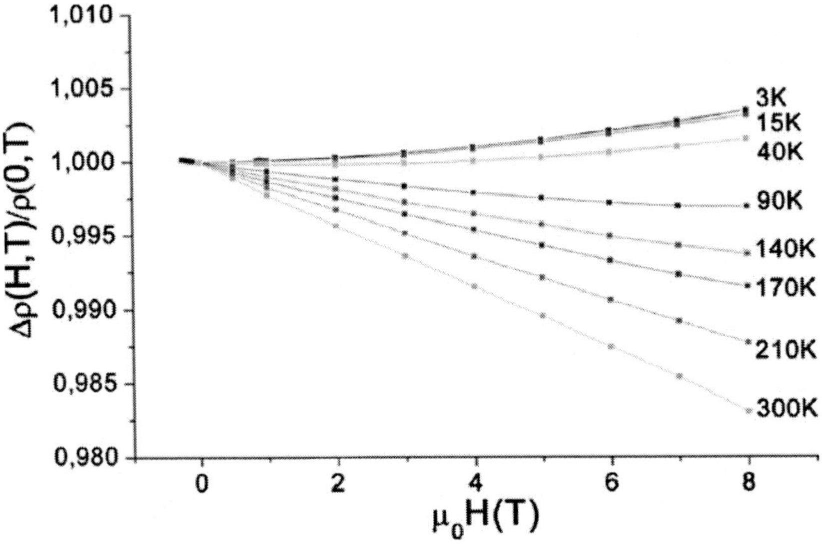

Figure 6. Low to room-temperature magnetoresistance measurements performed on a FePt(32 nm) /MgO(001) sample. The resistivity values are normalized by their values at zero field. This figure is extracted from ref. [MIH08].

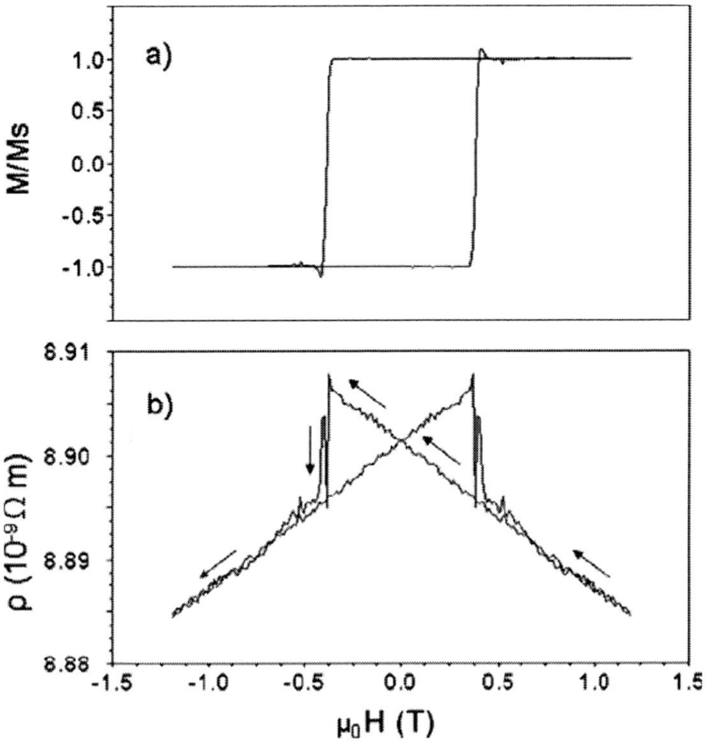

Figure 7. (a) Hysteresis loop measured in perpendicular field by EHE, and (b) corresponding resistivity measurement, performed on a FePt(10 nm)/MgO(001) sample. The data have been symmetrized and antisymmetrized, respectively, to separate properly the magnetoresistance and EHE contributions. This figure has been published in ref. [MIH08].

Now, what happens in a partially reversed (*i.e.* multidomain) state of magnetization? For a positive applied field, up domains possess a high resistivity, whereas down domains possess a lower resistivity. The relative proportions of the area covered by the reversed and unreversed domains are given, respectively, by $M/M_S+1\ /2$ and $1-M/M_S/2$. Let us denote by ρ_{MMR} the contribution of the MMR to the resistivity, and $\alpha(T)$ the absolute value of the $\rho(B)$ slope taken for a single-domain state, which can be easily obtained from measurements above the magnetization saturation. Using a simple analysis, we get in a partially reversed state

$$\rho_{MMR} = -M/M_S\ \alpha(T)\ B.$$

As ρ_{MMR} depends linearly on M_S, it is possible to use MMR as a quantitative tool to detect magnetization reversal. One can indeed see directly in Figure 7. that the MMR measurement reveals basic features of the magnetization reversal (remanence, coercive field). Note that in this analysis, we neglect the DW resistance (intrinsic and AMR contributions), whereas, it is often necessary to take DWR into account [DAN02]. MMR gets broadly the same kind of information (signal proportional to M/M_S) as TMR and GMR, with a lower signal, but with a simpler layer structure, which can be useful in experiments where the use of multilayers would complicate the results analysis. In comparison with the MMR, the EHE is a very accurate detection technique, but requires a more complicated geometry. Finally, the DWR offers only quantized information (the presence or not of DWs between contacts).

Thus, the MMR might be in some cases the most interesting tool to detect magnetization reversal.

4. ORIGIN OF THE COERCIVITY

If intrinsic magnetic properties (saturation magnetization, anisotropy, etc.) can be analyzed using *ab initio* calculations, the coercivity of magnetic materials is more difficult to study, as it is often linked to extrinsic properties. This coercivity can originate from nucleation, being linked to the field needed to nucleate a reversed domain, the nucleation event being followed by an easy propagation of magnetic DWs. It can also be directly linked to DW propagation. The nucleation then occurs at fields lower than reversal fields, but the DW propagation is limited by the pinning on structural defects. It is often impossible to determine experimentally the nature of the pinning sites, especially in bulk materials where domain observation is difficult, and in polycrystalline thin layers, where the structural disorder is too strong. Moreover, once the defect is identified, one has to know the specific magnetic properties associated to the defect, in order to calculate its pinning strength. Thus, if several attempts have succeeded in the theoretical determination of depinning fields using well-known artificial defects as constrictions in nanowires [KLA05], [ALL04],[HIM05], few studies focus on the link between microstructural defects and coercivity. Most of them are related to pinning at grain boundaries in permanent magnets such as $Sm(Co,Fe,Cu,Zr)_z$ ([KRO01],[SCH02]). In our FePt/MgO and FePt/Pt samples, the DW pinning mechanisms have been studied in details.

FEPT/MGO

Using the remanence of our samples, it is possible to observe the magnetic domain configuration during the reversal by Magnetic Force Microscopy (MFM), simply by reversing part of the magnetization and returning to zero field. Figure 8. presents the MFM image of a FePt/MgO sample. The reversed domain is connected, *i.e.* it is a single domain, the reversal occuring through DW propagation from scarce nucleation centers. The magnetization is stabilized in an out-of equilibrium state by the pinning of DWs. In FePt/MgO samples, two kinds of defects are possible candidates as pinning centers: interface dislocations and antiphase boundaries. On the first hand, structural studies [HAL01] and [HAL02] showed that dislocations are introduced at the beginning of the growth, in order to relax the epitaxial strain between the FePt layer and the MgO substrate. According to Halley [HAL01], these perfect dislocations possess a ½[101] Burgers vector, and form a lattice with a parameter of approximately 2 nm (1.7 nm in FePd/MgO layers, whose structural and magnetic properties are very similar to those of FePt/MgO layers). These dislocations generate a long-distance stress field which can, via magneto-elastic coupling, interact with the DW. On the other hand, TEM observations showed also that antiphase boundaries are present in our $L1_0$ samples. We realized numerical simulations [JOU07, JOU09] which showed that antiphase boundaries can pin efficiently the DW in FePt layers (cf. figure 9).

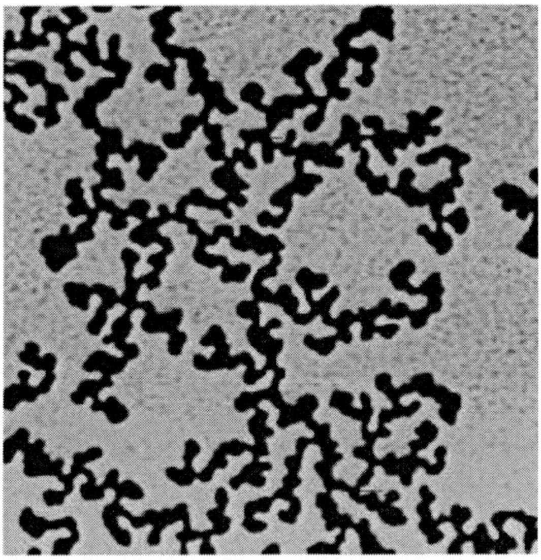

Figure 8. 8μm × 8μm image of the magnetic domains at the beginning of the magnetization reversal, observed by MFM on a FePt/MgO sample. Image published in ref. [ATT10].

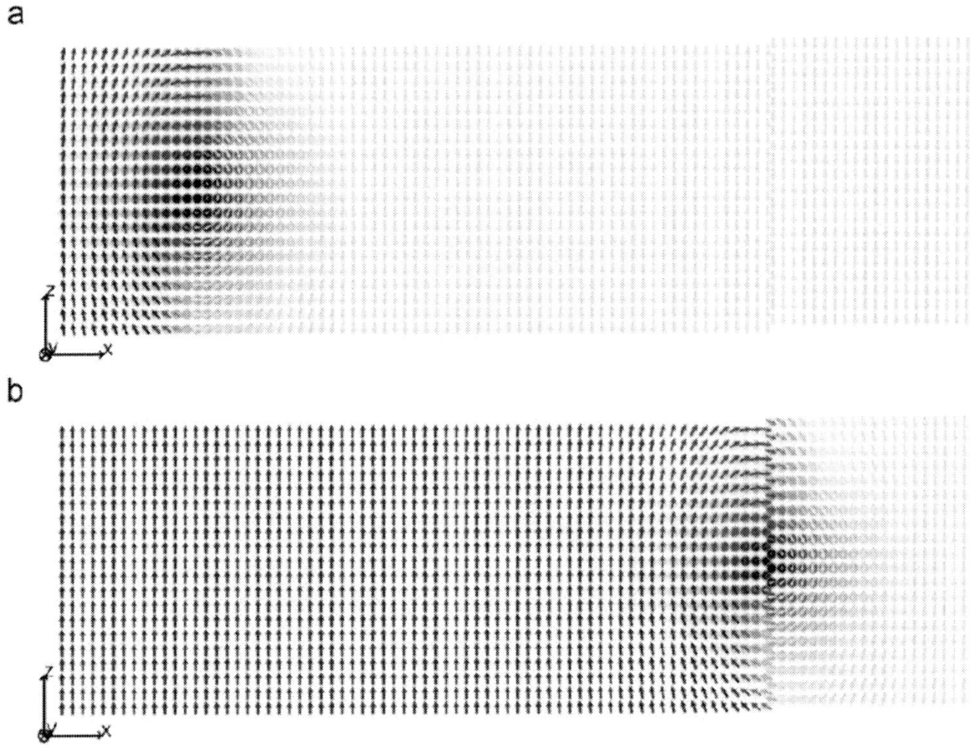

Figure 9. Cross sections of magnetic configurations obtained by numerical simulations: Bloch wall possessing Néel caps in a FePt layer, far from the antiphase boundary (a) and pinned on the antiphase boundary (b). Arrows indicate the local direction of magnetization. Extracted from [JOU09].

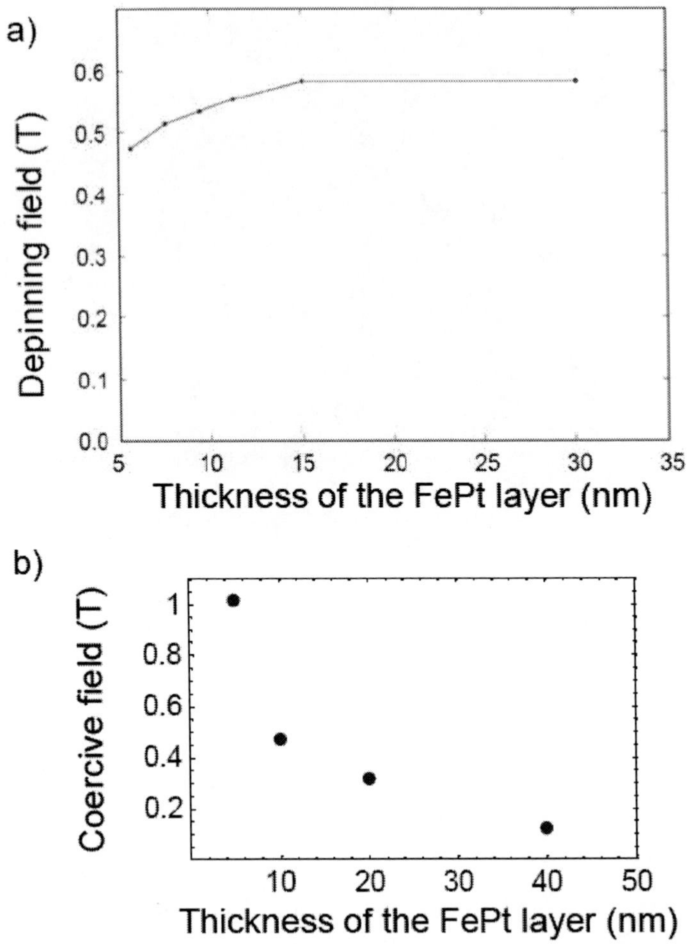

Figure 10. a) Depinning field of a domain wall from an antiphase boundary, as a function of the thickness of the layer, obtained by numerical simulations. The uncertainty on the field values is 5 mT. b) Experimental variations of the coercive field H_c vs. the thickness of the FePt layer in FePt/MgO. Data analysed in ref. [JOU09].

Experimentally, the coercivity decreases with the thickness of the FePt layer (cf. Figure 10.b). This is incompatible with simulations based on the hypothesis of a pinning on antiphase boundaries, which predict a weak dependence of the coercive field with the film thickness (cf. figure 10.a). In comparison, the hypothesis stating that the coercivity is linked to interface dislocations is qualitatively much more consistent with the observed evolutions: as they are interface defects, their effect on DWs and especially on the coercivity is supposed to decrease with an increase in the layer thickness.

FePt/Pt

The origin of the coercivity in FePt/Pt samples is more clearly determined: using scanning probe microscopies, we are able to correlate straightforwardly the magnetic domains configuration to the microstructure of the sample. FePt/Pt samples have the property to be nanostructurated by a quasi-square lattice of structural defects: the microtwins [HAL02],[ATT01]. These defects relax the tensile misfit (~1%) between the chemically ordered alloy and the Pt(001) substrate. They result from the gliding of 1/6<112> partial dislocations on adjacent (111) atomic planes. They are almost bidimensional defects, up to a few nm wide (up to 20–25 piled dislocations), a few 100 nm long, and extend from the bottom FePt-Pt interface to the upper Pt capping. Inside the microtwins, TEM observations showed that $L1_0$ order is also present [HAL02], but with a magnetocrystalline anisotropy axis oriented at 70.52° from the perpendicular to the sample surface.

There are surface steps left at the emergence of the microtwins (along one of the <110> direction), that can be easily localized by atomic-force microscopy (AFM) [ATT01]. Indeed, images reveal a nanostructuration of the FePt layer within a pseudorectangular lattice, where the average distance between neighboring defects is 70 nm (cf. figure 11.a). The height of the surface step varies from one microtwin to another, as it depends on the number of constitutive dislocations. The MFM images thus shows that the number of dislocations per microtwin follows a relatively widely spread statistical distribution.

Using the procedure described previously, the domain configuration of partially reversed states of magnetization has been observed by MFM (cf. figure 11.b).

Similarly to what was observed in FePt/MgO samples, all reversed domains are connected, suggesting that the reversal process involves very few nucleation events. The peculiar square geometry of the reversed domain is due to the pinning of the DWs on the microtwins, along the <001> directions. One can note that the reversed domain mean size is much bigger than the mean distance between microtwins, which suggest that the DWs gets pinned only on the microtwins possessing the strongest pinning strengths. The pinning of a DW on a microtwin has been studied by micromagnetic [ATT04] and numerical [JOU09] simulations, based on the hypothesis that the microtwin is a volume of tilted anisotropy. These simulations shows that the pinned DW adopts a specific micromagnetic configuration, lying in the (111) plane of the microtwin (cf. figure 12).

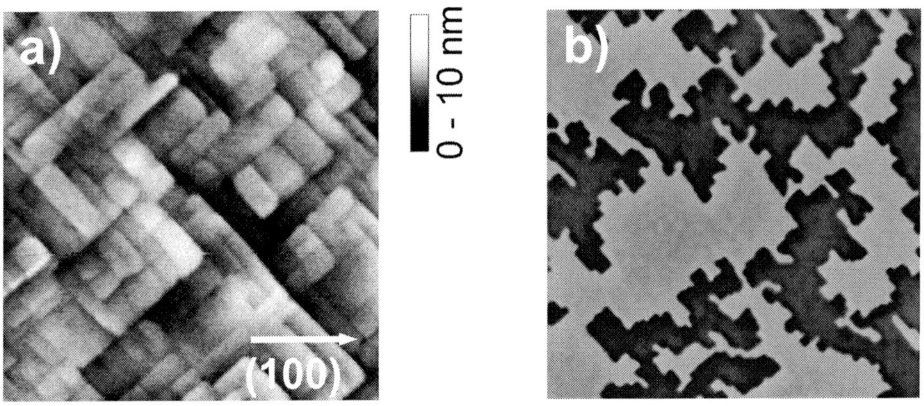

Figure 11. a) 1μm × 1μm AFM image of the surface of a FePt/Pt samples. Height scale is given on top right. Height discontinuities correspond to the emerging microtwins. b) 8μm × 8μm image of the magnetic domains at the beginning of the magnetization reversal, observed by MFM on a FePt/Pt sample. The reversed domain appears in black. The DWs are pinned along the <110> directions. Images extracted from refs. [ATT01] and [ATT10].

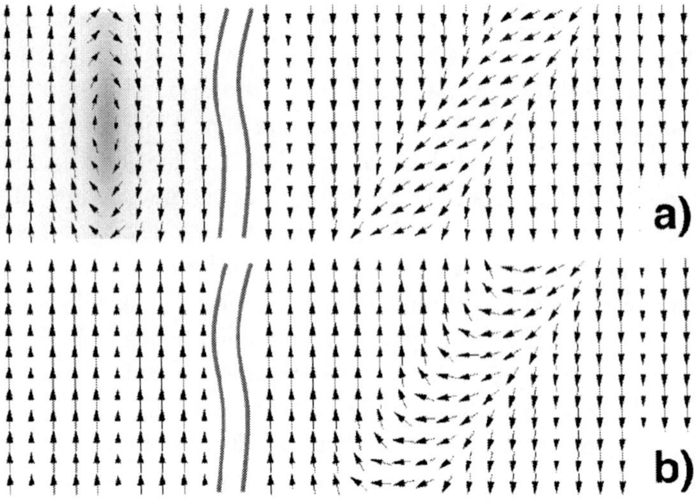

Figure 12. Cross sections of magnetic configurations, obtained by micromagnetic simulations, of the pinning of a DW on a microtwin. The FePt layer is 10 nm thick. Only the left and right parts of the calculus box (10 nm × 80nm) are displayed. The dark gray area accounts for magnetization pointing out of the simulation plane. (a) Relaxed configuration for a Bloch wall (left) far away from the microtwin (right). (b) Configuration where the wall is pinned on the microtwin. Figure extracted from ref. [ATT04].

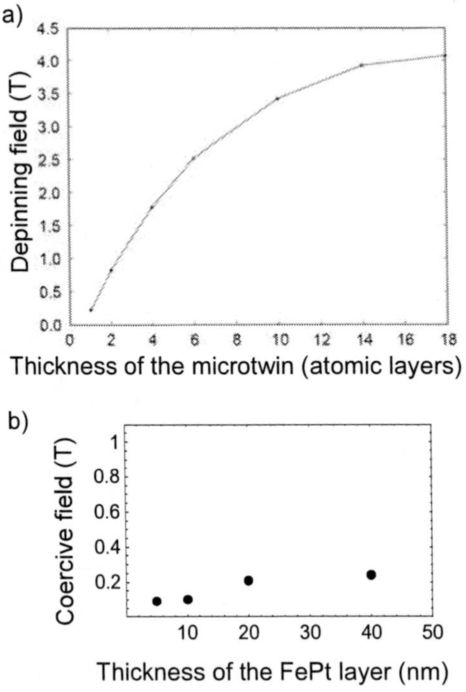

Figure 13. a) Variation of the depinning field as a function of the width of the microtwin (in atomic layers). The layer is 15 nm thick. The uncertainty on the values is 5mT. b) Experimental variations of the coercive field H_c vs. the thickness of the FePt layer in FePt/Pt samples. Data analysed in ref. [JOU09].

They also show that the depinning field increases strongly with the number of dislocations of the microtwin. Experimentally, it has been shown [HAL02] that the average number of atomic layers in the microtwin increases quasi-linearly with the alloy thickness, up to approximately 25 atomic planes in 40 nm thick FePt layers [ATT01].

Considering our simulations, one therefore expects that the depinning coercivity increase with the FePt thickness. The variation of the depinning field as a function of the width of the microtwin is shown in figure 13, where a strong increase in the depinning field is observed. The global rise in coercivity in FePt/Pt layers is thus likely linked to the widening of microtwins as the thickness of the layer increases.

5. Percolation and Magnetization Reversal

The magnetization reversal process of both thin epitaxied FePt/Pt(001) and FePt/MgO(001) layers occurs through rare nucleation events followed by domain-wall propagation.

Here, we will focus on the beginning of the magnetization reversal, which in both cases consists in a very sharp drop of the magnetization (cf. figure 5). The reversal field is applied very slowly (~10^{-3} $T.s^{-1}$), which means that the domain structure results from slow thermo-activated propagation of the domain wall rather than viscous precession-related motion.

Low-scale observations show that these systems possess very different domain-wall geometries, with isotropic patterns of smoothly curved DWs in FePt/MgO samples, and square patterns of DWs pinned on microtwins in FePt/Pt samples (figures 8 and 11.b).

In both kind of layers and regardless of the thickness of the sample, the structure of the reversed domain is fractal, *i.e.* there are unreversed (white) part of the sample of all sizes: one can easily find zones larger than (120 × 120) μm^2 (maximal scanning size of our MFM) where there is no reversed domain. In [ATT04] and [ATT10], it was shown that the geometry of the reversed domain was analogous to those of a percolating cluster near the percolation threshold. The fractal dimension of the domain, measured using a box-counting method [FOR99], is for both kind of sample D_f~1.88, very close to the theoretical value of 91/48=1.896 obtained for a percolating cluster in standard percolation theory [BUN91].

It has been shown that the similarity of the large-scale geometries of the reversed domain in FePt/Pt and FePt/MgO layers can be understood on the basis of standard domain-growth models: Invasion Percolation Without Trapping [WIL83], Random Field Ising Model at zero temperature [JI91],[JI92], cell models [LYB00A]. Basically, the similarity of domain geometries at large-scale, which contrast with the observed differences at low scale, is a manifestation of universality associated with two-dimensional percolation [ATT10].

Indeed, in a percolation problem the fractal dimension of the percolation cluster depends only on critical exponents, which are insensitive to the details of the lattice. In FePt samples, the geometry of the reversed domain thus does not depend on the nature of the pinning centers. This provides an original illustration of universality, as well as an application of percolation theory concepts to magnetic thin films.

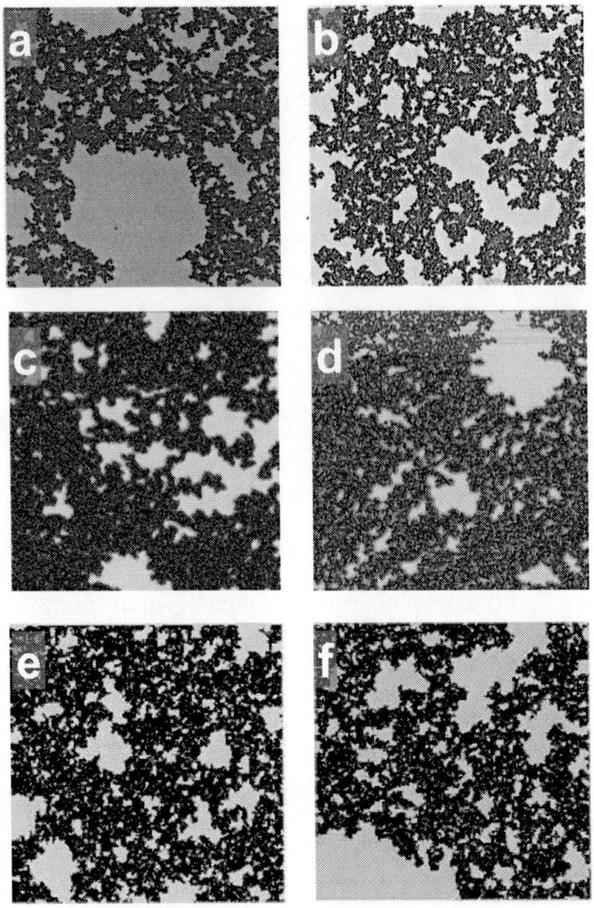

Figure 14. All these images correspond to the beginning of the magnetization reversal, the reversed domain appearing in black. (a) and (b): 64 × 64 μm2 MFM images of a FePt/Pt sample. (c) and (d): 64 × 64 μm2 MFM images of a FePt/MgO sample. The domains are fractals, with white parts (unreversed parts) of all sizes. (e) and (f) Simulated geometries of percolating clusters. Although in low scale images, domain geometries differ from one kind of sample to another, large-scale images (a), (b), (c), and (d) exhibit for both FePt/Pt and FePt/MgO samples the classical geometry of a percolating cluster shown in (e) and (f). These images have already been published in ref. [ATT10].

6. Pinning on a Single Defect

All these models that can be used to describe the magnetization reversal of our layers share the property to induce domain growth by avalanches. In such a growth process, the DW is pinned on several defects. The field increases until it overcomes the pinning field of one of these defects. The depinning of the DW from this single defect can be followed by an avalanche over large distances (cf. figure 15). Interestingly, avalanches also occur in nanowires processed in FePt/MgO and FePt/Pt samples. If the width of a nanowire is smaller than around 200 nm, the DW can not deform itself in order to bypass the defects. It thus pins successively on every defect, and the overall coercivity corresponds to the pinning on the most important defect. This main defect therefore governs the kinetic of the DW propagation over the whole nanowire, since it gives rise to a much larger depinning time.

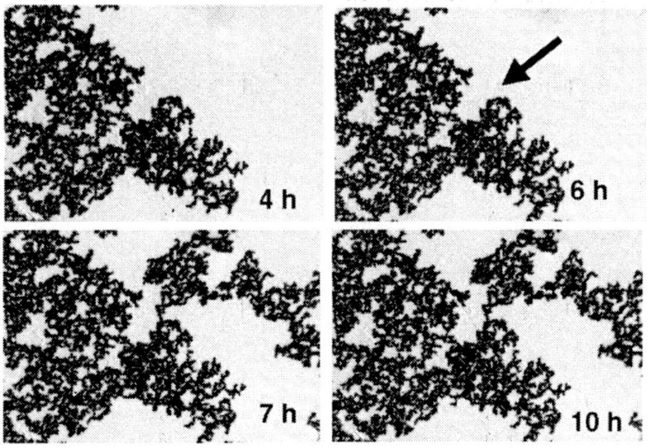

Figure 15. Magneto-optical polar Kerr images (92 × 61 μm) of a FePt(40 nm)/Pt layer. After saturating the sample in 1.2 T, a long time dynamics was followed: a -0.065 T magnetic field was applied and cancelled each 10 min to perform imaging. The magnetic configuration remained unchanged for hours, until the DW depins from a single microtwin, leading to an avalanche extending over several tenth of micrometers. The arrow (at 6 h) points to the likely origin of the avalanche. This figure has been published in [ATT04]

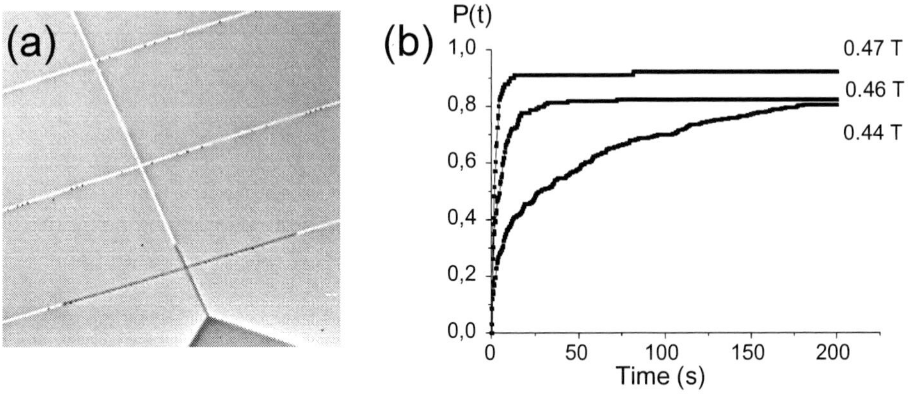

Figure 16. (a) 16 μm × 16 μm MFM image of a typical 50 nm wide nanowire, lithographed in a FePt/MgO sample. Such a nanostructure is used to study DW depinning from a single defect. The reversed domain (in dark) nucleates in the nucleation pad (bottom). Here, a DW is pinned in the propagation (nearly vertical) wire. The lateral contacts, where one can also see DWs, are only used to detect electrically the DW propagation. (b) Probability to be depinned after time t, for three different values of applied field. The behaviour is clearly stochastic, and the mean pinning time decreases when the applied fields increases.

We used this property to measure the pinning time of a DW on a single defect, on both FePt/Pt and FePt/MgO samples. Figure 16.a. presents the geometry of the lithographed nanowires, containing a nucleation pad to inject the DW in the nanowire, and electrical contacts to detect the DW propagation using EHE [ATT06]. Similar devices have also been used on FePt-based spin-valves, using Giant MagnetoResistance to detect the DW propagation [BUR09]. The key result is that, when repeating the experiment in unchanged experimental conditions (T, Happl), we observed a very wide distribution of the propagation time (ranging from less than 0.1 s to a few hours). Thus, to get a statistically meaningful distribution, the experiment has to be repeated over 400 times for each set of parameters (T, Happl), recording experimental propagation times ranging up to 300 s (above, the experimental sequence was deliberately stopped).

This stochasticity is consistant with theoretical expectations. Indeed, thermal activation over a single energy barrier is theoretically described by an Arrhenius law, whereby the crossing of the barrier E_b follows an exponential probability law, and is characterized by a time constant τ [MAR07, ATT06]. The probability to be depinned after an elapsed time t is given by :

$$P(t) = 1 - \exp\left[-\frac{t}{\tau}\right]$$

The time constant τ corresponds to the mean value of the pinning time, and theoretically follows an Arrhenius law :

$$\tau = \tau_0 \exp\left[\frac{E}{k_B T}\right]$$

where τ_0^{-1} is the attempt frequency, E the energy barrier, k_B the Boltzmann constant and T the temperature. The energy barrier E is a function of the applied field, and eventually of an applied current for spin transfer experiments [BUR09].

Exemples of the measured probability to be depinned after time t are given in figure 16.b, and are qualitively consistant with the previous equations, even if the function P(t) does not asymptotically approach 1 as t increases. This latter phenomenon also appeared in similar experiment in other materials [BRI08,BUR08]. It seems to be due to the fact that the DW can be pinned on a single defect along different micromagnetic configurations. This point is thoroughly discussed in ref. [ATT06].

The evolutions of E with applied fields and currents are discussed in details in refs. [ATT06], [GAR09] and [BUR09]. They allow extracting precise informations about current-induced depinning in samples with narrow DWs, especially the efficiency of spin transfer in FePt samples (i.e. the equivalence for DW depinning between an applied current density and an applied field), which appears to be quite high ($\sim 10^{-13}$ T/A.m^{-2}).

CONCLUSION

The interest of FePt samples, from both an applied and a fundamental point of view, is mainly due to its anisotropy. The narrowness of the DWs, in comparison with the classical lengths of magnetotransport (mean free path, spin diffusion length), is such that FePt layers are a model system to study current-induced magnetization reversal in the limit of large magnetization gradients. Also, the use of percolation theory to analyse domains geometries, and the understanding of magnetoresistance curves in terms of electron-magnon interactions are general enough to be applied to other magnetic thin layers.

Because of the narrowness of its DWs, FePt is also a good candidate for applications based on DW manipulation (racetrack memory-like storage devices, DW-MRAM). However, if the high anisotropy allows maintaining the thermal stability at small sizes, the coercivity of usual FePt samples constitute a major problem, leading up to now to high critical currents at zero field. Also, in comparison with permalloy-like materials, the strength of the pinning sites in FePt is such that it is more difficult to control pinning positions: using artificial defects (constrictions…) might lead to pinning fields lower than the intrinsic coercivity. Thus, today's most interesting prospect on this topic is a very challenging one, as it consists in developing a way to lower the coercivity of FePt samples while maintaining their anisotropy.

REFERENCES

[ALL04] D.A. Allwood, G. Xiong and R.P. Cowburn, *Appl. Phys. Lett.* 85, 2848 (2004).

[ALL05] D.A. Allwood, G. Xiong, C. Faulkner, D. Atkinson, D. Petit, R. P. Cowburn, Science 309, 1688 (2005).

[ATT01] J. P. Attané, Y. Samson, A. Marty, D. Halley, C. Beigné, *Appl. Phys. Lett.* 79, 794 (2001).

[ATT04] J. P. Attané, Y. Samson, A. Marty, J. C. Toussaint, G. Dubois, A. Mougin, J. P. Jamet, *Phys. Rev. Lett.* 93, 257203 (2004).

[ATT03] J. P. Attané, Ph.D. thesis, Université de Grenoble (France).

[ATT06] J. P. Attané, D. Ravelosona, A. Marty, Y. Samson, C. Chappert, *Phys. Rev. Lett.* 96, 147204 (2006).

[ATT10] J. P. Attané, M. Tissier, A. Marty, L. Vila, *Phys. Rev. B* 82, 024408 (2010).

[BAR02] A. Barthélémy, A. Fert, J-P. Contour, M. Bowen, V. Cros, J. M. De Teresa, A. Hamzic, J. C. Faini, J. M. George, J. Grollier, F. Montaigne, F. Pailloux, F. Petroff, C. Vouille, *J. Magn. Magn. Mater.* 242, 68 (2002).

[BAR05] K. Barmak, J. Kim, L. H. Lewis, K. R. Coffey, M. F. Toney, A. J. Kellock, J.-U. Thiele, *J. Appl. Phys.* 98, 033904 (2005).

[BER03] H. Bernas, J. P. Attané, K.-H. Heinig, D. Halley, D. Ravelosona, A. Marty, P. Auric, C. Chappert, Y. Samson, *Phys. Rev. Lett.* 91, 077203 (2003).

[BRI08] J. Briones, F. Montaigne, D. Lacour, M. Hehn, M. J. Carey, J. R. Childress, *Appl. Phys. Lett.* 92, 032508 (2008).

[BUN91] A. Bunde and S. Havlin, Fractal and Disordered Systems (Springer-Verlag, Berlin, 1991), Chap. 2.

[BUR08] C. Burrowes, D. Ravelosona, C. Chappert, S. Mangin, Eric E. Fullerton, J. A. Katine, B. D. Terris, *Appl. Phys. Lett.* 93, 172513 (2008).

[BUR09] C. Burrowes, A. P. Mihai, D. Ravelosona, J.-V. Kim, C. Chappert, L. Vila, A. Marty, Y.Samson, F. Garcia-Sanchez, L. D. Buda-Prejbeanu, I. Tudosa, E. E. Fullerton, J. P. Attané, *Nature Physics* 6, 17 (2010).

[DAN02] R. Danneau, P. Warin, J. P. Attané, I. Petej, C. Beigne, C. Fermon,O. Klein, A. Marty, F. Ott, Y. Samson, M. Viret, *Phys. Rev. Lett.* 88, 157201 (2002).

[FRE09] N. A. Frey, S. Peng, K. Cheng, S. Sun, *Chem. Soc. Rev.* 2532 (2009).

[GAR09] F. Garcia-Sanchez, H. Szambolics, A. P. Mihai, L. Vila, A. Marty, J.-P. Attané, J.-Ch. Toussaint, L. D. Buda-Prejbeanu, *Phys. Rev. B* 81, 134408 (2010).

[GEH98] V. Gehanno, C. Revenant-Brizard, A. Marty, B. Gilles, *J. Appl. Phys.* 84, 2316 (1998).

[GER02] A. Gerber, A. Milner, M. Karpovsky, B. Lemke, H.-U. Habermeier, J. Tuaillon-Combes, M. Negrier, O. Boisron, P. Melinon, and A. Perez, *J. Magn. Magn. Mater.* 242/245, 90 (2002).

[HAL01] D. Halley, PhD, Université de Grenoble (France) (2001).

[HAL02] D. Halley, A. Marty, P. Bayle-Guillemaud, B. Gilles, J. P. Attane, Y. Samson, *Phys. Rev. B* 65, 205408 (2002).

[HAL04a] D. Halley, B. Gilles, P. Bayle-Guillemaud, R. Arenal, A. Marty, G. Patrat, Y. Samson, *Phys. Rev. B* 70, 174437 (2004).

[HAL04b] D. Halley, A. Marty, P. Bayle-Guillemaud, B. Gilles, J. P. Attane, Y. Samson, *Phys. Rev. B* 70, 174438 (2004).

[HIM06] A. Himeno, T. Okuno, S. Kasai, T. Ono, S. Nasu, K. Mibu, *J. Appl. Phys.* 97, 066101 (2005).

[JI91] H. Ji and M. O. Robbins, *Phys. Rev. A* 44, 2538 (1991).

[JI92] H. Ji and M. O. Robbins, *Phys. Rev. B* 46, 14519–14527 (1992).

[JOU07] T. Jourdan, F. Lançon, A. Marty, *Phys. Rev. B* 75, 094422 (2007).

[JOU09] T. Jourdan, J. P. Attané, F. Lançon, C. Beigné, L. Vila, A. Marty, *J. Magn. Magn. Mater.* 321, 2187 (2009).

[KLA05] M. Klaui, H. Ehrke, U. Rudiger, T. Kasama, R. E. Dunin-Borkowski, D. Backes, L. J. Heyderman, C. A. F. Vaz, J. A. C. Bland, G. Faini, E. Cambril, W. Wernsdorfer, *Appl. Phys. Lett.* 87, 102509 (2005).

[KRO01] H. Kronmüller and M. Bachmann, *Physica B* 306, 96 (2001).

[LYB00A] A. Lyberatos, *J. Phys. D : Appl. Phys.* 33, R117 (2000).

[MAR07] E. Martinez, L. Lopez-Diaz, O. Alejos, L. Torres, C. Tristan, *Phys. Rev. Lett.* 98, 267202 (2007).

[MIH08] A. P. Mihai, J. P. Attané, A. Marty, P. Warin, Y. Samson, *Phys. Rev. B* 77, 060401 (2008).

[MIH09] A. P. Mihai, J. P. Attane, L. Vila, C. Beigne, J. C. Pillet, A. Marty, *Appl. Phys. Lett.* 94, 122509 (2009).

[NEC09] NEC press release 2009, "NEC Develops Scalable High-Speed MRAM Technology Suitable for System LSI Embedding", http://www.nec.co.jp/press/en/0906/1705.html

[PAR08] S. S. P. Parkin, M. Hayashi, L. Thomas, *Science* 320, 190 (2008).

[RAV00] D. Ravelosona, C. Chappert, V. Mathet, H. Bernas, Appl. Phys. Lett. 76, 236 (2000)

[RAQ02] B. Raquet, M. Viret, E. Sondergard, O. Cespedes, R. Mamy, *Phys. Rev. B* 66, 024443 (2002).

[SCH02] W. Scholz, J. Fidler, T. Schrefl, D. Suess, T. Matthias, *J. Appl. Phys.* 91, 8492 (2002).

[SCH05]A. Schuhl, D. Lacour, C. R. *Physique* 6, 945 (2005).

[TOS04] Toshiba press release 2004, "Toshiba Leads Industry in Bringing Perpendicular Data Recording to HDD--Sets New Record for Storage Capacity With Two New HDDs", http://www.toshiba.co.jp/about/press/2004_12/pr1401.htm

[WIL83] D. Wilkinson and J. Willemsen, *J. Phys. A: Math. Gen.* 16, 3365 (1983).

In: Magnetic Thin Films
Editor: John P. Volkerts
ISBN: 978-1-61209-302-4

Chapter 11

IMPROVEMENT OF MAGNETOIMPEDANCE EFFECT IN MAGNETIC THIN FILMS BY LAMINATING PERMALLOY LAYER

Zhiyong Zhong*[1]*, Huaiwu Zhang*[1]*, Xiaoli Tang*[1]*, Hu Su*[1]*,
***Feiming Bai*[1] *, Lijun Jia*[1] *and Shuang Liu*[2]**

[1]State Key Laboratory of Electronic Thin Films and Integrated Devices, University of Electronic Science and Technology of China, Chengdu, 610054, China

[2]School of Optoelectronic Information, University of Electronic Science and Technology of China, Chengdu, 610054, China

ABSTRACT

The magnetoimpedance (MI) effect is defined as the change of the impedance experienced by an ac current flowing through magnetic materials when an external dc magnetic field is applied. This effect is promising in the application of micromagnetic field sensors with high sensitivity and quick response.

If the film is magnetically soft and has a well-defined anisotropy axis and large saturation magnetization, it will help enhance the MI effect due to the increased interaction with the external magnetic field. It is well known that FeCo based alloy thin films exhibit very large saturation magnetization.

However, FeCo based thin films also have very large saturation magnetostriction. Such large saturation magnetostriction causes the degradation of soft magnetic properties and no-detectable magnetoimpedance effect. This paper used a soft magnetic layer or nonmagnetic layer as interlayer to improve the soft magnetic properties of FeCo based films effectively, and investigated the effect of laminating layer on the microstructure and magnetoimpedance effects of FeCo based single or sandwiched thin films. Results show that the MI ration of single or sandwiched thin films with laminating Permalloy layer is enhanced obviously. The improvement of MI ratio of single or sandwiched films with Permalloy laminating layer can be explained by exchange induced ripple reduction mechanism.

1. INTRODUCTION

The magnetoimpedance (MI) effect is the change of the impedance experienced by an ac current flowing through magnetic materials when an external dc magnetic field is applied. This effect is quite promising for small field sensors with high sensitivity and quick response [1, 2]. Most of the studies performed on MI have been carried out on amorphous or nanocrystalline soft magnetic wires or ribbons. The use of thin film technology for MI is preferable in many applications because of its compatibility with integrated circuit technology. The investigation of MI in thin films systems includes a number of sandwiched systems with different composition, magnetic structure, size [3-7]. But compared to wires or ribbons, the single magnetic thin films typically exhibit a lower MI sensitivity because of a higher anisotropy field induced the fabrication process and annealing [8]. A high sensitivity MI has been reported to occur in ferromagnetic layer(F)/non-ferromagnetic metal layer(M)/ferromagnetic layer(F) sandwiched films, in which the impedance change ratio is several times larger than that in a similar ferromagnetic single-layer films[4-7,9]. For example, in CoSiB/Cu/CoSiB sandwiched films of 7μm thick , the MI ratio is 340% for a frequency at 10MHz and DC magnetic field of 9Oe[8]. In the sandwiched structure F/M/F, a very large change in impedance which is related to the outer magnetic layers becomes larger than the resistance determined mainly by the inner conductor [10]. Due to this advantage, MI in sandwiched films has a potential to be used in developing sensitive micro magnetic sensors and magnetic heads for high-density magnetic recording [11].

The MI effects of sandwiched films made of a soft ferromagnetic alloy of composition such as Co-riched CoSiB, CoFeSiB, CoNbZr,and Permalloy, etc. have been extensively studied, but all these films with above mentioned composition have relatively low saturation magnetization. If the ferromagnetic films are magnetically soft and have a well-defined anisotropy axis, these films with a large saturation magnetization will be beneficial to enhance the MI effects due to increasing the interaction with the external magnetic field [12]. It is well known that FeCo alloy thin films exhibit very large saturation magnetization. However, the FeCo thin films have also a very large saturation magnetostriction [13]. Such a large saturation magnetostriction causes a degradation of soft magnetic properties. It has been proved that adding the third element such as B, N to form FeCo-based amorphous or nanocrystalline materials can exhibit excellent soft magnetic properties. $Fe_{67}Co_{18}B_{15}$ (at%) amorphous thin film is one of these materials has been reported to sputter on glass substrate [14], but the dispersion of the induced magnetic anisotropy was too high to obtain high sensitivity for MI elements [12]. Wang et al [15] has demonstrated that single layer thin film of Fe-Co-X (X=N, B, etc) can be made magnetically soft by sandwiching the Fe-Co-X layer between two Permalloy layers. It was found that the Permalloy underlayer, rather than the cap layer, is more important in achieving the low coercivity field and relatively low dispersion of magnetic anisotropy in Fe-Co-X films [16].

Recently, lamination using a thin soft magnetic layer or non-magnetic layer as interlayer effectively improves the soft magnetic properties, and avoids stripe domains of thicker N-doped Fe-based thin films with high B_S [9-11]. In this paper, We use a soft magnetic layer or nonmagnetic layer as interlayer effectively improves the soft magnetic properties of FeCo based films, and investigated the effect of laminating layer on the microstructure and magnetoimpedance effects of FeCo based single and sandwiched thin films.

2. Experimental Methods

The thin films were deposited by RF/DC magnetron sputtering method. The detailed fabrication information can be found in references [17-19]. The topography and magnetic structure was obtained by atomic force microscopy (AFM), magnetic force microscopy (MFM) and x-ray diffraction (XRD). A vibrating sample magnetometer (VSM) was applied to measure magnetic properties of the film. The resistivity of the film was measured by a four-point probe method. The MI effect was investigated with a strip-type sample using a HP4294A or HP4291B impedance analyzer. The amplitude of AC current was fixed at 10mA during measurements. The external magnetic field was generated by Helmholtz coils, and its direction was along the length direction of the measured samples. In order to avoid the effect of earth's magnetic field, the magnetic field of the solenoid was perpendicular to the earth field direction. In order to avoid the negative value, the MI ratio is defined as [Z(H)-Z(0)]/Z(0), where Z(H) and Z(0) are the magnetoimpedance with and without the applied DC magnetic field, respectively.

3. Results and Discussion

A. Effects of Nonferromagnetic and Ferromagnetic Laminating Layer on Magnetoimpedance of Fealn Films

The amount of N doping is closely related to soft magnetic properties and resistivity of FeAlN films. Firstly, we alter N_2 partial pressure from 0 to 15% to optimize magnetic properties of the 50nm-thickness FeAlN films. The optimum N_2 partial pressure is 8%. Under the optimized deposition condition, a FeAlN film can be achieved on a natural oxidized Si (100) wafer, with the resistivity of 40μΩ.cm, the coercivity of easy axis 3.1 Oe, and hard axis 1.2Oe.

After optimizing the properties of FeAlN single film with 50nm thickness, we fabricated the single FeAlN film with 1μm thickness, and the laminating [FeAlN(50nm)/interlayer (10nm)]$_{20}$ films, where the interlayer is Cu or NiFe. Then the film was cut into 10mm×2mm strips to measure its soft magnetic properties and MI ratio. The shorter side of the rectangular sample was along the easy axis, which was formed by sputtering in magnetic field. Figure 1. shows hysteresis loops of the films along the easy axis and the hard axis, respectively. The applied field is in plane while measuring. Among them, Figure 1.a. is a typical hysteresis loop with a weak perpendicular anisotropy, which will cause formation of stripe domain structures [20]. The easy axis loop is similar to the hard axis one due to large coercivity of 20Oe. While laminating the interlayer, seen from Figure 1.b. and c, the softness of FeAlN films is improved significantly (The easy axis coercivity of both the laminating films is 1.8Oe), and the samples also exhibit a clearly induced uniaxial anisotropy (6Oe), compared to the single layer film without interlayer. Furthermore, the sample with the NiFe spacer shows somewhat reduced hard axis coercivity H_{ch} compared with the Cu spacer, and demonstrates more rectangular easy axis loop. To explain the effect of the spacer layer on the magnetic properties of FeAlN, XRD patterns of the film were measured, as shown in Figure 2. The laminating

FeAlN thin films with spacer (NiFe or Cu) have a slightly stronger (110) diffraction peak, compared to the FeAlN single thin film.

It suggests that the spacer layer induced a (110) fiber texture in laminating FeAlN thin films. It is reported that some FCC metallic layer such as Cu and NiFe with (111) fiber texture significantly improves soft magnetic properties of FeXN thin films [21-22]. The possible explanation is that nanosized FeXN (110) grains with 120^0 to each other epitaxially grow on a larger NiFe (111) or Cu (111) seed layer grain. The growth results in partial cancellation in both crystalline and magnetoelastic anisotropy [23]. The other reason may be that the interlayer disrupts the columnar growth of FeAlN nanocrystalline grains, thus smaller grain size is obtained.

Of course, the layer thickness of the FeAlN in the laminated samples is strongly reduced in comparison to the single layer. This will have a strong impact on the magnetic properties, as for thin films of 50nm, which is the case in the laminated layer, the shape anisotropy forces the magnetic moments into the plane. The effect will contribute to improve the magnetic properties of the laminating sample.

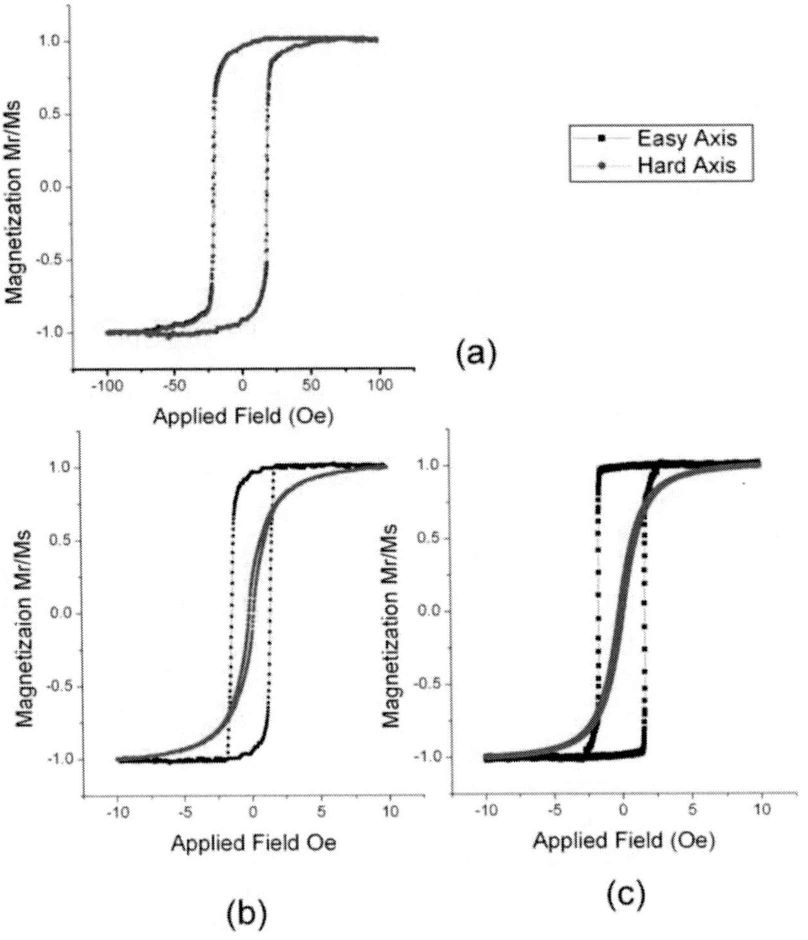

Figure 1. Hysteresis loops of films along the easy axis and hard axis, (a) single FeAlN film, (b) laminating FeAlN film with Cu interlayer, (c) laminating FeAlN film with NiFe interlayer.

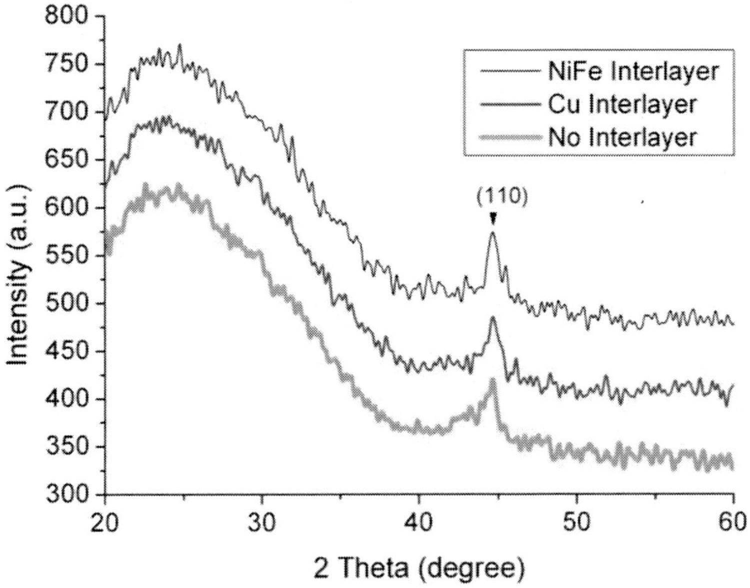

Figure 2. XRD pattern of films.

Figure 3. shows the dependence of the MI ratio on the applied magnetic field for those two laminating samples at 110 MHz. There is no detectable MI effect in single FeAlN films. It could be explained by the relatively bad soft magnetic properties and weak perpendicular anisotropy of the film, shown in the domain structure in Figure 4.a. Figure 4. presents magnetic force microscopy (MFM) images of those films.

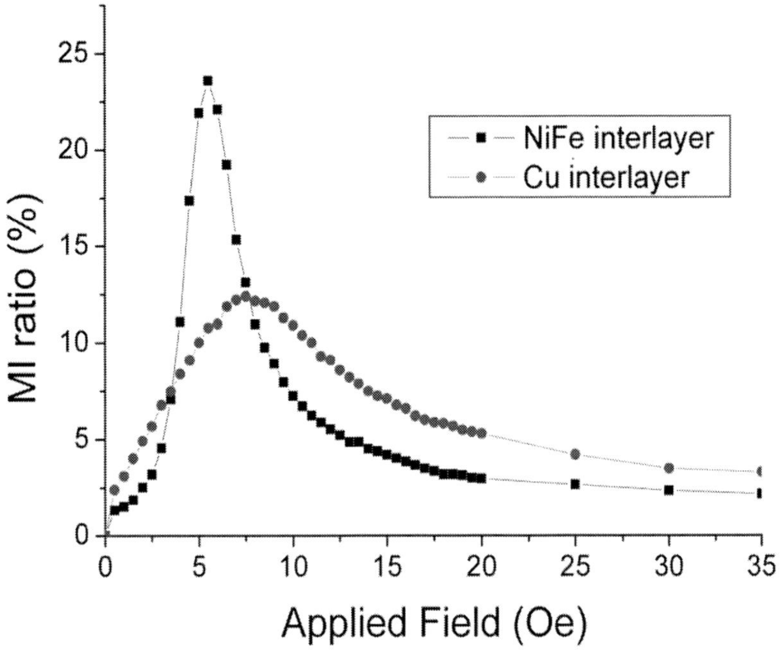

Figure 3. Dependence of MI ratio on the applied field at 110MHz in laminating FeAlN films.

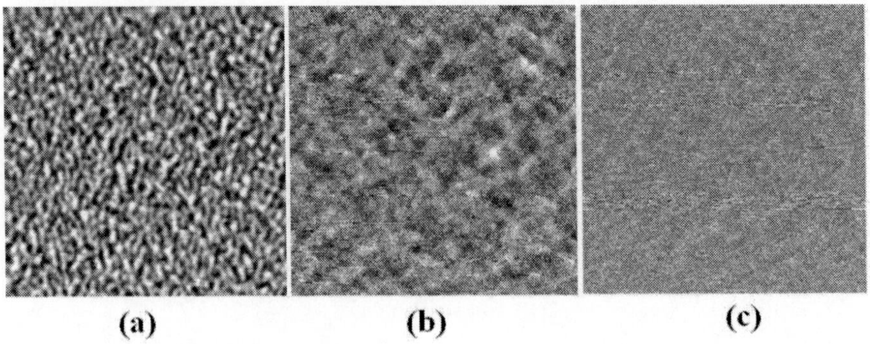

Figure 4. MFM images, (a) single FeAlN film, (b) laminating FeAlN film with Cu interlayer, (c) laminating FeAlN film with NiFe interlayer.

Although the laminating FeAlN thin films with NiFe or Cu have excellent soft magnetic properties and induced similar transverse uniaxial anisotropy, their field dependence of the MI ratio is different. The variation of MI ratio of the laminating FeAlN thin film with NiFe is sharper than that with Cu. The maximum MI ratios are 12.4% and 23.6% for the Cu and NiFe spacer layer, respectively. This difference can be attributed to the dispersion of the induced transverse anisotropy, and Sun et al proposed an exchange induced ripple reduction mechanism to explain it [24]. Because the NiFe interlayer has an excellent softness with a well-defined anisotropy, the FeAlN layer will be exchange-coupled with NiFe layer during growth. It would be beneficial from the same local moment alignment and low dispersion angle manifested in the NiFe layer. This point is also demonstrated by the MFM images of those two laminating films at the hard axis remanence state in Figure 4.b. and c, respectively. The ripple structure appears in both the laminating FeAlN films. Compared with the laminating FeAlN film with Cu, the magnetic ripple in the laminating FeAlN film with NiFe was suppressed, which means low ripple angle, therefore low dispersion of induced anisotropy based on ripple theory [25]. The dispersion of the induced anisotropy leads to the smearing of the peak and the decrease of the MI ratio [26].

B. Magnetic Microstructure and Magnetoimpedance Effect in Nife/Fealn Multilayer Films

In order to investigate the effect of the thickness of FeAlN sublayer on the microstructure and magnetic domain structures, double layer NiFe/ FeAlN (x) films, where x=10 nm, 50 nm, 100 nm, 200 nm, respectively, were fabricated firstly. Figure 5. shows AFM topography of these double layer films. The grain sizes of the double layer films with FeAlN thickness of 10 nm, 50 nm, and 100 nm are grown gradually, but the largest grain size is not greater than 25 nm, which estimates from AFM images and less than the ferromagnetic exchange of α-Fe grains [27]. In addition, it is interesting to note that the grain boundaries in films with FeAlN thickness of 50 nm and 100 nm seem more well defined than that of film with thinnest (10 nm) FeAlN sublayer. It means that the growth of FeAlN grain is suppressed if the FeAlN sublayer thickness is thin. For the film with FeAlN thickness of 200 nm, the grain size becomes larger than 35nm. It is reasonable to assume that the long sputtering duration time improves the crystallization conditions.

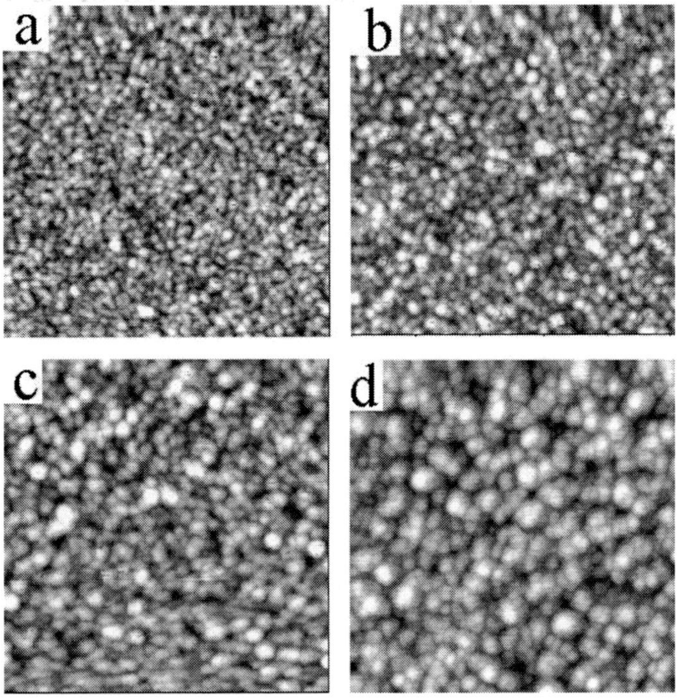

Figure 5. AFM images of samples with different thickness of FeAlN sublayer, (a) 10 nm, (b) 50 nm, (c) 100 nm, (d) 200 nm. The each image size is 1μm×1μm.

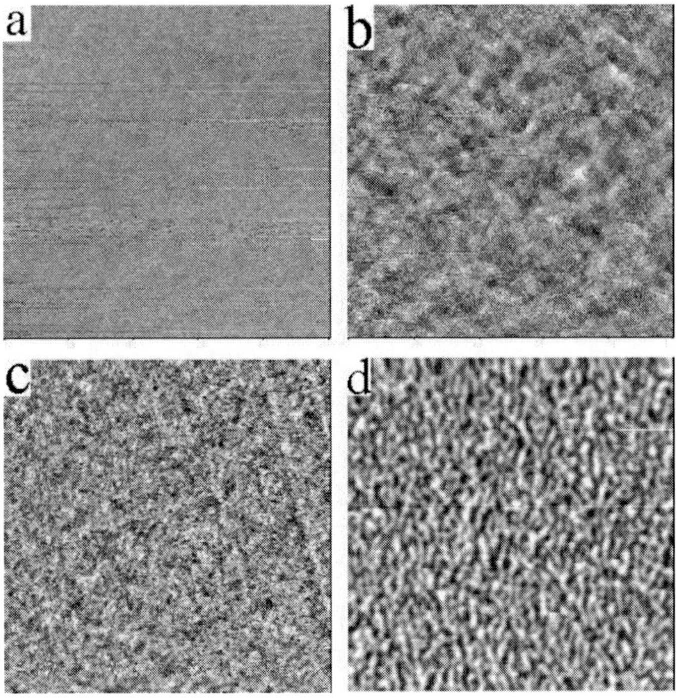

Figure 6. MFM images of samples with different thickness of FeAlN sublayer. (a) 10 nm, (b) 50 nm, (c) 100 nm, (d) 200 nm. The each image size is 5μm×5μm.

Table 1 Magnetic property of the fabricated multilayer films

Multilayer films with different thickness of FeAlN sublayer	Coercivity (Oe)		Induced anisotropy H_k (Oe)
	H_{ce}	H_{ch}	
10 nm	1.2	0.6	6.5
50 nm	1.8	0.8	6
100 nm	2.5	1.4	7
200 nm	10.8	10.2	-

- H_{ce} and H_{ch} represent the coercivity along easy axis and hard axis loops, respectively.
- The value of H_k is extracted from the two direction loops.

Figure 6. shows the corresponding remanent magnetic microstructure of those double films. Because the MFM tip is perpendicularly magnetized, so the tip is sensitive to stray field gradients from z-orientated domains or the z-components of domains oblique to the sample surface. The weak stripe domain, shown in Figure 6.(d), where bright to dark contrast arises from the magnetization canted up or down out of the films plane, confirms the presence of perpendicular anisotropy. The low contrast MFM images, shown in Figure 6. (a), (b) and (c), are characteristics of magnetic films which magnetization mainly lies in the film plane [28]. The magnetic ripple structure, which is explained by local variations of the induced anisotropy due to a superposition of the uniaxial anisotropy of the film and the crystalline anisotropy of single crystallites [29], appears in these samples. In comparison with samples with thicker FeAlN sublayer, the magnetic ripple in film of samples with thinner FeAlN sublayer is suppressed. This means that the dispersion of the induced magnetic anisotropy in samples with thinner FeAlN sublayer is smaller. The hysteresis loops of the multilayer films with different thickness of FeAlN sublayer along the easy axis and the hard axis in the plane of the sample, respectively, were measured by VSM. The magnetic properties were extracted from these hysteresis loops, and shown in Table 1. From the Table 1, we can see that the soft magnetic properties of the multilayer films with thinner FeAlN sublayer are improved substantially, compared to that with thicker FeAlN sublayer. XRD pattern of the multilayer film with 50 nm thickness of FeAlN sublayer shows that the multilayer has a slightly strong (110) diffraction peak, compared to the FeAlN single thin film with the same thickness [17]. The XRD patterns of the multilayer films with other thickness FeAlN sublayer have same characteristics, except for stronger (110) diffraction peak with increase of the thickness of FeAlN sublayer. This means that the grain grows with thickness, which is in according with the AFM images of Figure 5. The stronger (110) diffraction peak suggests that the NiFe spacer layer induced a (110) fiber texture in the FeAlN multilayer films. It is reported that some FCC metallic layer such as Cu and NiFe with (111) fiber texture significantly improves soft magnetic properties of FeXN thin films [23]. The possible explanation is that nanosized FeXN (110) grains with 120^0 to each other epitaxially grow on a larger NiFe (111) or Cu (111) seed layer grain. The growth results in partial cancellation in both crystalline and magnetoelastic anisotropy [23]. The other reason may be that the NiFe spacer layer disrupts the columnar growth of FeAlN nanocrystalline grains, thus smaller grain size is obtained. Of course, the layer thickness of the FeAlN in the multilayer with thinner FeAlN sublayer is strongly reduced, and has a strong impact on the magnetic properties. The shape anisotropy forces the magnetic moments into the plane. The effect will contribute to improve the magnetic properties of the multilayer film sample.

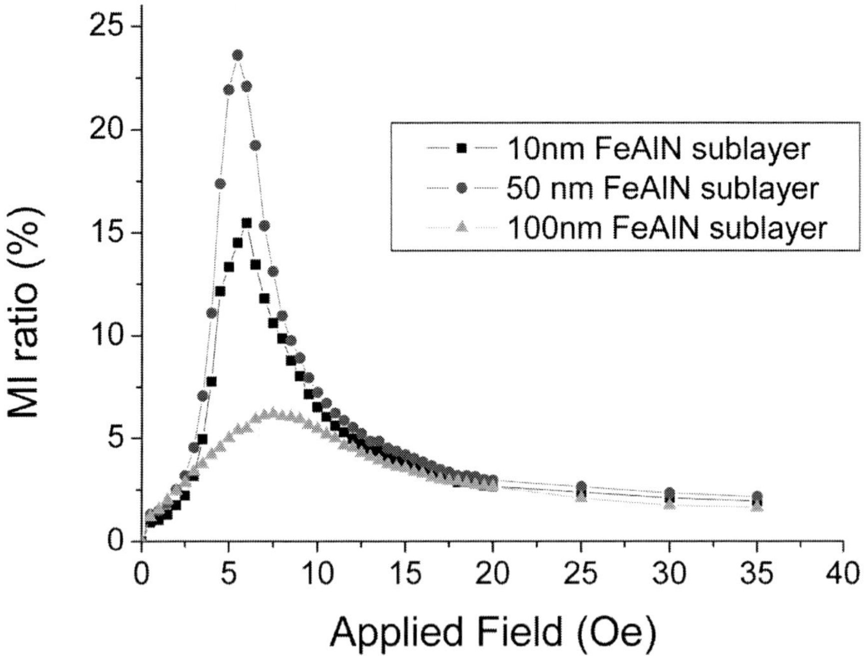

Figure 7. Dependence of MI ratio on the applied field at 110MHz with different thickness of Permalloy layer.

The multilayer films were also cut into 10 mm×2 mm strips to measure its soft magnetic properties and MI ratio. The shorter side of the rectangular sample was along the direction of easy magnetization, which was formed by sputtering in magnetic field. When measuring the MI ratio, the applied magnetic field was along the direction of hard magnetization that was along the longer side of measured units. Figure 7. shows the dependence of the MI ratio on the applied magnetic field for those multilayer samples at 110 MHz. There is no detectable MI effect in the sample with 200 nm thickness of FeAlN sublayer. The maximum of MI ratio of the samples with 10 nm and 50 nm thickness of FeAlN sublayer is 15.4% and 23.6%, respectively. The peak of MI ratio of the sample with 100nm thickness of FeAlN sublayer is broader than that of that of the samples with 10nm and 50nm thickness of FeAlN sublayer, and the MI ratio is decreased to 6.2%.

The rigorous and quantitative analysis of MI effects in thin films should be dealt with by combining Landau-Lifshitz dynamics with Maxwell equations [30]. Here we use a quasistatic model to discuss the above results qualitatively. The MI effect in thin films is proportional to $(\omega\mu_t)^{1/2}$, where μ_t is rotational transverse permeability at high frequency. When there is only in-plane anisotropy in magnetic film, the transverse permeability is [31]

$$\mu_t = \frac{M_s \sin^2(\theta+\theta_k)}{H_{ext}\sin^2(\theta+\theta_k)+H_K\cos 2\theta}+1 \tag{1}$$

where H_{ext} is applied external magnetic field, θ_k, which represents the dispersion of the induced anisotropy, is the angle between transverse direction of measured unit and the direction of induced anisotropy. The equilibrium angle θ of the magnetization with respect to the transverse direction can be obtained by minimizing the free energy in the film.. From the equ.1, we can see that the MI ratio curve exhibits peak near H_k for small dispersion of the induced anisotropy. As the dispersion increases, the peak decreases and the curve becomes much broader [31]. Because the NiFe spacer layer has an excellent softness with a well-defined anisotropy, the FeAlN layer will be exchange-coupled with NiFe layer during growth. It would be beneficial from the same local moment alignment and low dispersion angle manifested in the NiFe spacer [32]. But the exchange coupling decreases while the thickness of FeAlN sublayer increases, as demonstrated in Figure 6. So, the maximum MI ratio of sample with 100nm thickness of FeAlN sublayer is smaller than that of samples with thinner FeAlN sublayer. The reason that the MI ratio of sample with 10nm thickness of FeAlN sublayer is smaller than that with 50nm thickness of FeAlN sublayer is that the saturation magnetization drops substantially for thin FeAlN sublayer. When a magnetic film presents perpendicular and in-plane anisotropy simultaneously, at first-order approximation, the transverse permeability is [33]

$$\mu_t = \begin{cases} \dfrac{M_s}{H_{kp} - H_{ks}} + 1 & for \quad H_{ext} < H_{kp} \\ \dfrac{M_s}{H_{ext} - H_{ks}} + 1 & for \quad H_{ext} > H_{kp} \end{cases} \tag{2}$$

where H_{kp} is effective perpendicular anisotropy field and H_{ks} is effective in-plane anisotropy field. In order to maintain the magnetization lies in out-of-plane of the film, the H_{kp} must be several hundreds Oe to overcome the demagnetizing field, so we cannot observe MI variation in the sample with 200nm thickness of FeAlN sublayer in present measuring field range.

C. Magnetoimpedance Effect in Fecob/Cu/Fecob Sandwiched Films with Permalloy Underlayer

The as-deposited and post-annealed FeCoB(100 nm)/NiFe (t) with t's range form 0 to 10 nm films investigated here were shown to have an amorphous structure, which was confirmed by XRD. The dc magnetization loops of the as-deposited and post-annealed FeCoB(100 nm)/NiFe (t) films were carried out along different directions in the film plane. It is found that there exists a strong uniaxial anisotropy in the film plane, in which an easy axis is shown to be parallel to the direction of applied field during deposition, and a hard axis perpendicular to the easy axis. All the films exhibit excellent soft magnetic properties. However, the as-deposited single FeCoB film is found to have a easy axis coercivity of ~3.5 Oe and shows a uniaxial anisotropy H_k of ~40 Oe in the film plane, but upon annealing at 300°C for 0.5 hour, the magnetic properties of the films undergoes significantly changes. The easy axis coercivity of films decreases to a much smaller value of ~1.8 Oe, the anisotropy field decreases to 23 Oe. Fixing the thickness of FeCoB films at 100nm, and varying the thickness of Permalloy underlayer, it is clear that the easy axis coercivity drops from 1.8 Oe to about 0.6 Oe when a permaloy underlayer as thin as 1nm is put beneath the FeCoB film, and then it stays between 0.5-0.8 Oe when the Permalloy thickness is less than 4nm.

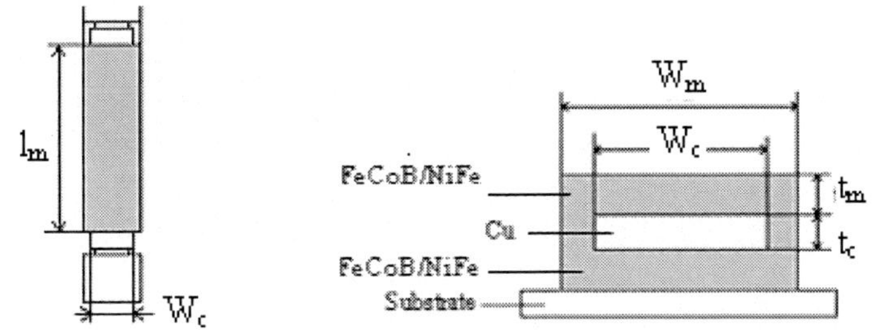

Figure 8. Schematic drawing of sandwiched GMI element.(a).top view, and (b) cross-section view.

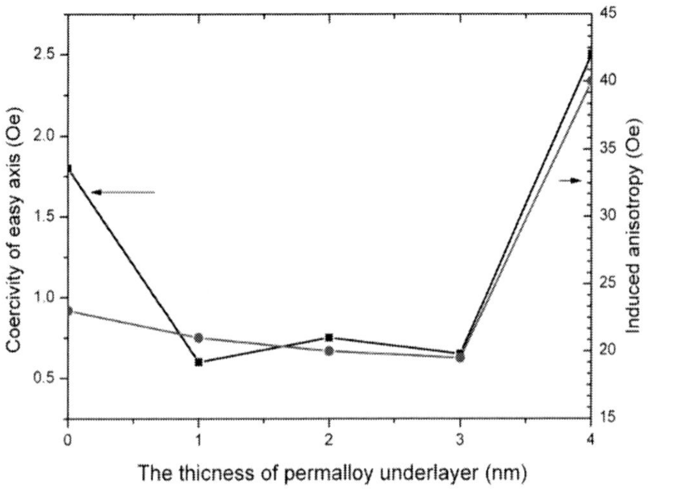

Figure 9. The dependence of easy axis coercivity and induced magnetic anisotropy field of the after stress-eliminated annealing films on Permalloy underlay.

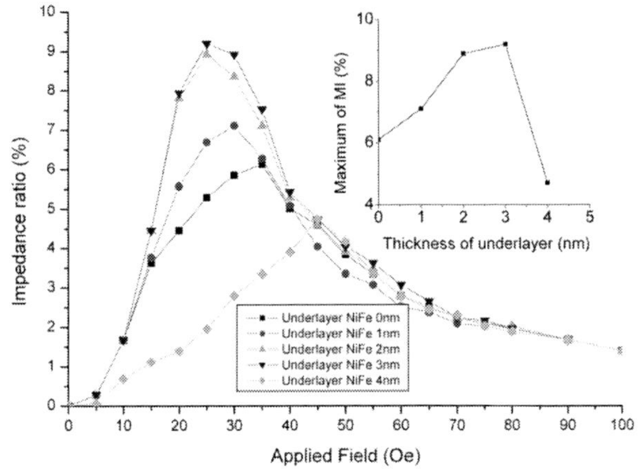

Figure 10. The dependence on magnetoimpedance effects of post-annealed FeCoB with different thickness of Permalloy underlayer.

The difference of the induced anisotropy between single FeCoB film and FeCoB films with Permalloy underlayer is small when the thickness of Permalloy underlayer is thinner than 4nm. However, when the thickness of the Permalloy underlayer exceeds 4nm, the induced anisotropy increases dramatically to 40Oe (as shown in Figure 9). The improvement of soft magnetic materials of the post-annealed films can attribute to elimination stress formed by sputtering process. The variation of the induced anisotropy of the FeCoB films with underlayer could be related with change in domain structure in films, and will discuss it in the following section.

We have measured the evolution of impedance change as a function of frequency for F/M/F sandwich films without Permalloy underlayer, and found that the maximum change of impedance appeared at 450 MHz. Therefore, the MI ratio of all the samples was measured fixed at 450 MHz. Figure 10. shows the relationship between the impedance change and the applied field at 450 MHz for the post-annealed sputtered FeCoB with different thickness of Permalloy underlayer. As observed from Figure 10, one can see that the change in the MI ratio first increases with the applied field, reaching a maximum value at a certain applied field and then gradually drops with further increase of the applied field. This is a typical profile of the impedance characteristics of films with transverse anisotropy. For the case of the well-defined transverse anisotropy, the maximum MI ratio should appear at near H_k [34]. As pointed from Figure 9, the H_k of the FeCoB single layer and the FeCoB/Permalloy underlayer (when the thickness<4nm) measured by VSM are very close, about 20 Oe. But in Figure 10, we can see that the peak fields, which corresponding to the maximum MI ratio are quite difference. This phenomenon can be explained by the dispersion of transverse anisotropy or transverse permeability.

It has been pointed in [35]

$$Z = R_m(1 - 2j\mu_t \frac{d_2 d_1}{\delta_1^2}) \tag{3}$$

where Rm is the resistance of the conductive lead, μt the transverse permeability, 2d1,d2 the thickness of the conductive and magnetic layers, respectively, δ1 is the skin-depth in the inner conductor. Meanwhile, based on ripple theory, the transverse permeabilityμt can be described by [37]

$$\mu_t = \frac{4\pi M_s}{H_K \times (h_0 + B \times h_0^{-1/4})} \tag{4}$$

where, $h_0=(H/H_k)-1$, B is the ripple stray field parameter, which is related to the dispersion angleαof the local magnetic moments, and H is the applied magnetic field. The dispersion angle α is defined as the angle between the hard axis direction and the applied field at which the remanence is 90% during magnetic reversal, and this parameter can be obtained by B-H loops . Figure 11. shows the dispersion angleαof the films. The dispersion angle is about 2.0^0 for the single layer FeCoB, while for the peramlloy underlayer thickness in the range of 2-3nm, the dispersion angle of film shows a minimum value of about 0.8^0 for the FeCoB/Permalloy film.

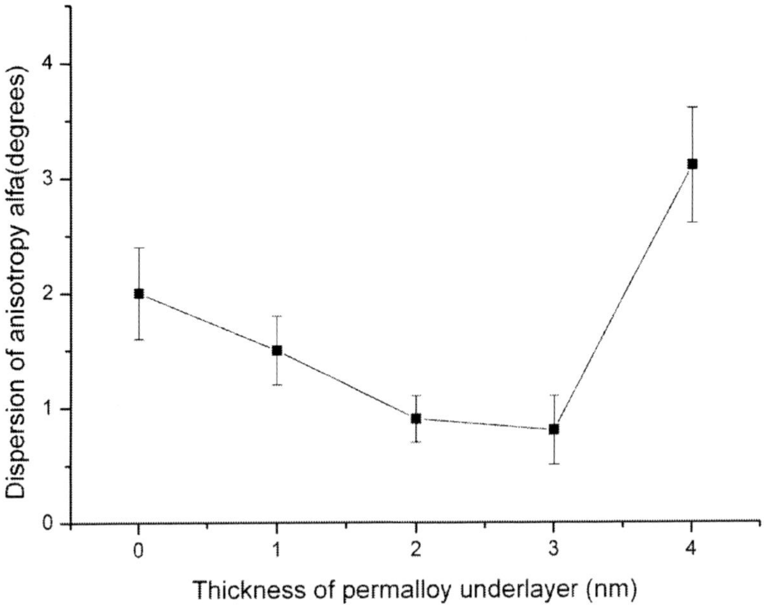

Figure 11. Dispersion angle of the FeCoB film on Permalloy underlayer, with error bar showing the range of data scattering.

Because the ripple stray field parameter B is related to the dispersion angle, the smaller dispersion angle corresponds to a smaller B, according to Eq.(4), the film with a Permalloy underlayer thickness of 2-3nm will have the highest transverse permeability, therefore will have biggest MI ratio according to Eq.(3). While increasing the dispersion angle, the ripple stay field parameter B will increase, then lead to decrease the transverse permeability and increase the blocking field, which corresponding to the maximum transverse permeability [36]. From the inset of Figure 10, one can observe that the maximum magnetoimpedance change for the sandwiched film in our experiments appear at 2-3 nm thickness of Permalloy underlayer, and the maximum of MI ratio is 9.2%. The reason for the Permalloy underlayer improving the local moment alignment in the FeCoB films can ascribe to exchange coupling effect between the FeCoB layer and the Permalloy underlayer, which is intrinsically soft and well aligned with a low dispersion angle [37].

Regard to the 4 nm thick Permalloy underlayer, the magnetoimpedance drops to 4.7%. For thicker Permalloy underlayer, in-plane and out-of-plane XRD studies [38] clarified that the lattice spacing of planes along the easy axis direction was expanded than that along the hard axis direction. The difference would result in that a compressive stress along the hard axis direction of the film, which plays an important role to induce a high magnetic anisotropy field in the double layer.

Conclusion

Laminating FeCo based films with Permalloy layer can enhance its magnetoimpedance obviously. The single FeAlN film (1μm) exhibits no detectable MI effect, while it is distinct for the laminating films [FeAlN/NiFe]. Through optimizing the thickness of Permalloy underlayer, it can be obtained an enhanced MI ratio in FeCoB/NiFe double layers, which is favorable for improvement the magnetoimpedance effect of F/M/F sandwiched films. The MI enhancement MI of laminating FeCo based films with Permalloy interlayer can be explained by the exchange induced ripple reduction mechanism. Because the soft NiFe interlayer is exchange-coupled with FeCo based layer, it decreases the dispersion angle of the induced anisotropy, and thus improves the MI ratio of the FeCo base single or sandwiched films.

ACKNOWLEDGMENTS

This work is partly supported by the NSFC under Grant Nos. 61021061 and 61071028, the Specialized Research Fund for the Doctoral Program of Higher Education of China under Grant No.20100185110024, and the Program for New Century Excellent Talents in University under Grant No. NCET-08-0089.

REFERENCES

[1] R.S. Beach, and A.E. Berkowitz, *Appl. Phys. Lett.* 64(1994)3652.
[2] M. Knobel, and K.R. Pirota, *J. Magn. Magn. Mater.* 242-245(2002)33.
[3] M. Senda, O. Ishii, Y. Koshimoto, and T. Toshima, *IEEE Trans. Magn.* 30(1994)3400.
[4] T. Morikawa, Y. Nishibe, H. Yamadera et al, *IEEE Trans. Magn.* 32(1996)4965.
[5] Y. Nishibe, H. Yamadera, N. Ohta et al, *IEEE Trans. Magn.*39(2003)571.
[6] I. Giouroudi, H. Hauser, L. Musiejovsky et al, *J. Appl. Phys.* 97 (2005)10M109.
[7] D. de Cos, L.V. Panian, N. Fry et al, *IEEE Trans. Magn.* 41 (2005)3697.
[8] T. Uhiyama, K. Mohri, L.V. Panina, and K. Furuno, *IEEE Trans*, Japan 19(115-A) (1995)949.
[9] T. Morikawa, Y. Nishibe, and H. Yamadera, *IEEE Trans. Magn.* 33(1995)4367.
[10] K. Hika, L.V. Pania, and K. Mohri, *IEEE Trans. Magn.*33(1995)4367.
[11] L.V. Panina, and K. Mohri, *Sensors and Actuators* 81(2000)71.
[12] C. Tannous, and J. Gigeraltowski, *Journal of Materials Science* 15 (2004)125.
[13] R.M. Bozorth, Ferromagnetism, New York: *IEEE press*, 1993, p.194.
[14] L.H. Chen, T.J. Kiemmer, K.A. Ellis, R.B. van Dover, and S. Jin, *J. Appl. Phys.* 87(2000) 5858.
[15] S.X. Wang, N.X. Sun, M. Yamaguchi, and S. Yabukami, *Nature*, 407(2000)6801.
[16] N.X. Sun, S.X. Wang, Chin-ya Hung, Chester X. Chien, and Hua Ching Tong, *Mater. Res. Soc. Symp. Proc.* 614(2000)F9.2.
[17] Z.Y.Zhong, H.W.Zhang, Y.L.Jin, X.L.Tang, S.Liu, *IEEE Trans. Magn.*,43(2007)2989.
[18] Z.Y.Zhong, H.W.Zhang, Y.L.Jin, X.L.Tang, S.Liu, *Sensors and Actuators* A, 141(2008)29.
[19] Z.Y.Zhong, H.W.Zhang, Y.L.Jin, X.L.Tang, L.Zhang, S.Liu, *Vacuum*, 82(2008)491.
[20] J.Ben Youssef , N.Vukadinovic, D.Billet, M.Labrun, *Phys. Rev.*B, 69(2002)174402.
[21] V.R.Inturi, J.A.Barnard, *J. Appl. Phys.*, 79(1996)5904.
[22] H.Jiang, K. Sin, Y.J. Chen, *IEEE Trans. Magn.* 41(2005)2896.
[23] H.Jiang,Y.J.Chen, L.F. Chen, Y.M. Huai, *J. Appl. Phys.*, 91(2002)6821.
[24] N.X.Sun, S.X. Wang, *J. Appl. Phys.* , 92(2002)1477.
[25] H. Hoffmann, *Thin Solid Films*,, 58(1979)223.

[26] L.Kraus, *J. Magn. Magn. Mater.* , 195(1999)764.
[27] K.H.Buschow, Handbook of Magnetic Materials, Vol.10, Amsterdam, North Holland, 1997, p433.
[28] J.Shih, T.Chin, Z.Sun, H.Zhang, B.Shen, *IEEE Trans. Magn.*, 37 (2001) 2681.
[29] M.Löhndorf, A.Wadas, R.Wiesendanger, *Appl. Phys.A Mater. Sci.and Proc.* 65, (1997) 511.
[30] L. Kraus, *Sensors and Actuators* A , 106 (2003) 187.
[31] S.S.Yoon, S.C.Yu, G.H.Ryu, C.G.Kim, *J.Appl.Phys.,* 85 (1999) 5432.
[32] N.X.Sun, S.X. Wang, *J. Appl. Phys.*, 92 (2002)1477.
[33] M.Rivas, J.F.Calleja, M.C.Contreras, *J.Magn.Magn.Mater.*, 166 (1997) 53.
[34] R.C. O'Hnadley, Modern Magnetic Materials—Principles and Applications, *John Wiely and Sons*, Inc, 2000, p505.
[35] D.P. Makhnovsky, and L.V. Panina, Sens. *Actuators A: Phys.*, 81(2000)91.
[36] H. Hoffman, *Thin Solid Films* , 373,(2000)107.
[37] R. Nakatani, T. Kobayashi, S. Ootomo, and N. Kumasaka, *Japn. J. Appl. Phys.*, 27(1988)937.
[38] S.H. Kong, T. Okamoto, and S. Nakagawa, *J. Magn. Magn. Mater.*, 272-276(2004)2184.

Reviewed by G.W.Qin, Professor of Northeastern University, China. It is the author's responsibility to contact the reviewer and take into account any suggestions or comments.

In: Magnetic Thin Films
Editor: John P. Volkerts

ISBN: 978-1-61209-302-4

Chapter 12

PREPARATION OF THIN FERRITE FILMS ON SILICON SUBSTRATES

M. R. Kobliscka,[1] A. Kobliscka-Veneva,[2] V. Skumryev[3] and U. Hartmann[1]

[1]Institute of Experimental Physics, Saarland University, D-66041 Saarbrücken, Germany
[2]Institute of Functional Materials, Saarland University, D-66041 Saarbrücken, Germany
[3]Institut Català de Recerca i Estudis Avançats (ICREA), Barcelona, Spain

ABSTRACT

Thin-films of (Ni,Zn)Fe_2O_4 were grown by means of RF sputtering on Si (100) and (111) substrates. Films with a thickness up to 100 nm were prepared for analysis purposes, enabling the optimization of the sputter process. The purpose of these ferrite thin films is the preparation of MFM cantilever-coatings for use with a high-frequency magnetic force microscope (HF-MFM). As a basis for these probes, we employ commercial, micromachined silicon cantilevers which exhibit (100)-oriented Si surfaces on the shank, and (111)-oriented surfaces on the pyramid-like tip end. The substrates were not additionally heated during the evaporation.

A slow cooling enabled the grain growth, so that grains of about 30 – 50 nm diameter were produced. Hysteresis loops of the (Ni,Zn)Fe_2O_4 films were obtained using a VSM magnetometer at room temperature. The films were found to exhibit the behaviour of a soft magnetic material, which is well suited for the HF-MFM imaging of harddisk recording heads up to a carrier frequency of 2 GHz. The material properties of the (Ni,Zn)Fe_2O_4 thin films are characterized by means of transmission electron microscopy (TEM). An electron backscatter diffraction (EBSD) analysis of the individual grain orientations is carried out on this type of ferrite films. This enables a determination of the magnetization direction of individual grains, which contributes to the modelling of the tip properties. The current goal for further research is the preparation of even thinner ferrite coatings in order to maintain a small tip radius at the tip end of the cantilever, which is an important factor for the achievable spatial resolution of a MFM image.

Keywords: ferrite thin films, silicon, MFM, cantilevers.

1. INTRODUCTION

Ferrite thin films are, in contrast to metal alloys, of great interest for high-frequency applications due to the small conductivity and the resulting small ac-losses. Important materials of this class are the (Ni,Zn)-ferrites with spinel structure ($NiZnFe_2O_4$) and the Co_2 Z-type hexaferrites ($Ba_3Co_2Fe_{24}O_{41}$, BCFO) [1-3]. Especially BCFO is employed in the MMIC (microwave monolithic integrated circuits) technology, as the cut-off frequency for bulk samples is above 2 GHz [4]. In contrast to this, the cut-off frequency for bulk (Ni,Zn)-ferrite is only about 300 MHz, but in form of thin films the cut-off frequency was found to increase up to 1.2 GHz [5].

The magnetic force microscope (MFM) technique is a method enabling a high spatial resolution even in ambient conditions [6-8]. Using low-moment magnetic tips, a resolution down to the 20 nm regime in ambient conditions could be reached [9-12]. However, the high frequency (microwave) response of magnetic materials and fast time-domain phenomena are fundamental properties that are also closely related to applications such as magnetic data storage. Therefore, the ability to evaluate in practice the high-frequency performance of magnetic recording heads is crucial to magnetic disc drive designs for high data-rate application. To achieve this goal, the MFM technique was further developed into the high-frequency MFM (HF-MFM) technique enabling the characterization of the harddisk writer pole field at high frequencies [13-19]. The "standard" MFM cantilever coatings employed in the HF-MFM measurements consist mainly of CoCr, partly also CoPtCr [6-8]. In all former experiments carried out using the HF-MFM technique, cantilevers coated with layers of CoCr were employed like for conventional MFM [13-19]. As the essential reason for the signal detection in HF-MFM is the non-linearity of the magnetic material provided by the hysteresis [20], it is important to employ a magnetic coating which still exhibits a hysteresis even at high frequencies. Therefore, in order to find an optimally suited magnetic cantilever coating for the HF-MFM experiments, we already employed a Z-type hexaferrite (BCFO) for imaging [21]. This material is commercially used at frequencies in the range up to 2 GHz [5], which implies that this range covers the current operating frequencies in commercial harddisks. Additionally, we prepared thin films and cantilever coatings of (Ni,Zn)-ferrite, which offers an easier preparation due to the less complicated structure of the unit cell [2].

2. EXPERIMENTAL PROCEDURES

Sample Preparation

Ferrite thin films are generally prepared employing various techniques including RF-sputtering, which is described for (Ni,Zn)-ferrites [22,23] as well as for BCFO [24,25]. In both cases, a subsequent annealing in air is required in order to form the desired crystal structure.

By means of RF-sputtering, we prepared ferrite thin-films (thickness between 50 and 100 nm) on Si substrates analoguous to the cantilevers, i.e., on (1 0 0) and (1 1 1) surfaces. The Si substrates and cantilevers were cleaned in acetone and isopropanol, followed by an etching step, similarly to the cleaning process of Si wafers. At a working pressure of 3×10^{-3} mbar in Ar and a sputter power of 50 W, the time for sputtering was about 55 min for a thickness of 50 nm. The substrates were not additionally heated during the evaporation. The ferrite coatings were subsequently annealed in air at 1050 °C (Ni,Zn-ferrite) and 800 °C (BCFO) [26]. These ferrite films on Si substrates were employed for magnetic measurements, in order to improve the magnetic properties of the ferrite films and to find the optimum preparation conditions. After determining these conditions, we prepared a nominally 50 nm thick coating on micromachined Si cantilevers (Nanoworld Services GmbH, 2 – 3 Nm^{-1}, $F_{res} \approx 70-80$ kHz) [27]. The required high annealing temperature of 1050 °C did not influence the mechanical properties of the cantilevers. The measured resonance frequencies of the coated and uncoated cantilevers varied only by about 0.14%.

The material properties of the (Ni,Zn)-ferrite thin films are characterized by means of transmission electron microscopy (TEM), performed on slices cut by means of focussed ion-beam (FIB) milling directly from a ferrite-coated cantilever, enabling the analysis of the growth mode on both types of surfaces, (100) and (111). Details of this procedure can be found in Ref. [28]. The slice is then fixed on a Cu grid, and received a Pt-cap layer for protection. The TEM analysis is performed at 200 kV, using a LaB_6-cathode.

Furthermore, electron backscatter diffraction (EBSD) analysis of the individual grain orientations is carried out for the first time on this type of ferrite films.

For the transmission electron microscopy (TEM) analysis, we prepared TEM slices by means of focused-ion beam (FIB) milling.

MFM Imaging

For MFM and HF-MFM imaging, a commercial AFM system (Veeco/DI Nanoscope models IIIa and IV) was employed. All measurements shown here were performed in ambient conditions. Details of the HF-MFM imaging technique can be found in Refs. [21,29].

As basis for the MFM probes, we employ commercial, micromachined silicon cantilevers (Nanoworld Services, type PPP, 2 – 3 Nm^{-1}, $F_{res} \approx 70-80$ kHz [27]) which exhibit (1 0 0)-oriented Si surfaces on the shank, and (1 1 1)-oriented surfaces on the pyramid-like tip end. The Si substrates and cantilevers were cleaned in acetone and isopropanol, followed by an etching step, similarly to the cleaning process of Si wafers.

Electron Backscatter Diffraction (EBSD)

The EBSD system employed consists of a FEI dual beam workstation (Strata DB 235) equipped with a TSL OIM analysis unit [30]. The Kikuchi patterns are generated at an acceleration voltage of 20 kV, and are recorded by means of a DigiView camera system, allowing a maximum recording speed of the order of about 14 pattern/s. The time employed in the case of a multi-phase scan is somwhat longer, of the order of 10 pattern/s, as a higher image quality/confidence index is required. To produce a crystallographic orientation map, the electron beam is scanned over a selected surface area and the resulting Kikuchi patterns are indexed and analyzed automatically (i.e., the Kikuchi bands are detected by means of the software). An image quality (IQ) parameter and a confidence index (CI) is recorded for each such Kikuchi pattern.

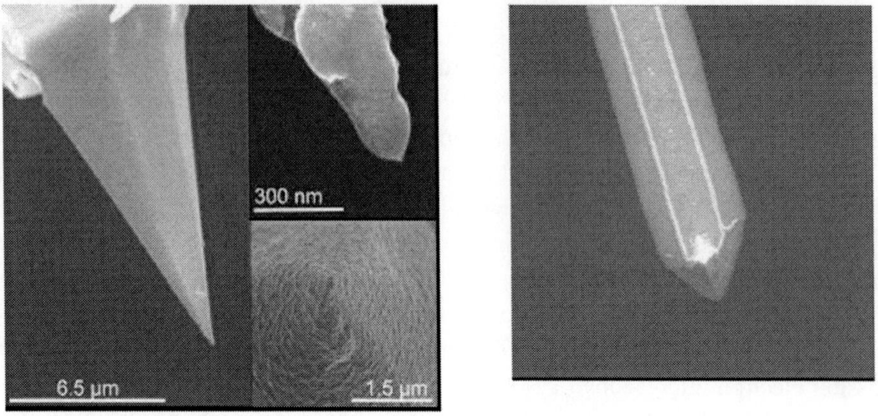

Figure 1. (left) SEM images of a ferrite-coated cantilever tip. Left: side view, and details of the cantilever pyramid. Right: View of the entire cantilever. This cantilever was coated with a (Ni,Zn)-ferrite film.

The dimensionless IQ parameter is the sum of the detected peaks in the Hough transform employed in the image recording; the CI value yields information about how exact the indexation was carried out. The CI value ranges between 0 and 1 [31]. Based on the analysis of the recorded CI value, a multi-phase analysis is realized. A detailed description of the measurement procedure can be found in Refs. [32,33]. The results of the EBSD measurement are presented in form of maps, the most important thereof are the so-called inverse pole figure (IPF) maps, indicating the crystallographic orientation of each individual point. Automated EBSD scans were performed with a minimum step size of 30 nm; the working distance was set to 10 mm.

A detailed knowledge of the grain orientation within a ferrite film on the cantilever enables a determination of the magnetization direction of individual grains, which contributes to the modelling of the tip properties.

3. Results and Discussion

3.1. Microstructural Characterization by SEM and TEM

In Figure 1, we present SEM images of a Si cantilever coated with (Ni,Zn)-ferrite, together with some detail views of the cantilever tip.

The ferrite film thickness is 50 nm; the typical thickness of the "standard" CoCr-films employed for MFM imaging is around 30 nm.

The tip radius is still quite sharp. The ferrite forms a uniform film, which can also be seen in the TEM cross section view as presented in Figure 2. The inset to Figure 2. gives the diffraction pattern, indicating a polycrystalline grain growth, with no amorphous layer being present. The slow-cooling after reaching the maximum annealing temperature is initializing the grain growth, so no amorphous phase is present as in earlier experiments. The grain sizes determined from the TEM cross section view are in the range of 50 – 100 nm [34].

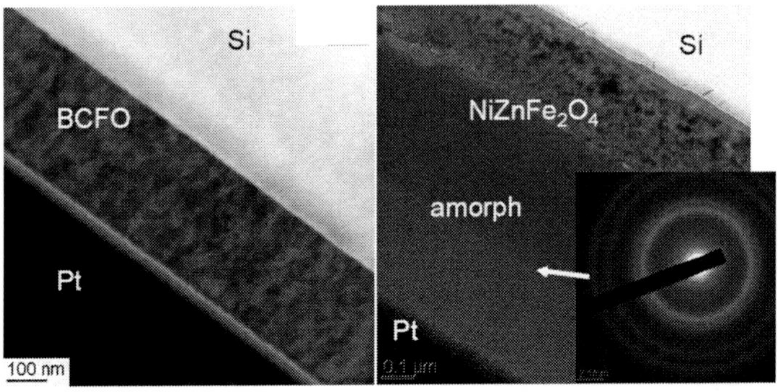

Figure 2.TEM analysis of the two types of ferrite films prepared on the Si cantilevers. (left) Bright view of the BCFO ferrite film. BCFO forms a fully crystalline layer right from the Si surface. In contrast to that, (right) shows TEM image of the (Ni,Zn)-ferrite, which exhibits the presence of a large amorphous layer on top of the first layer, which is polycrystalline. This is also indicated by the diffraction pattern obtained from this layer (inset).

3.2. MFM and AFM Analysis

Figure 3. presents the results of MFM measurements on a (Ni,Zn)-ferrite film prepared on a (100)-oriented Si substrate. The overview measurement (a) shows a maze-like domain pattern. In images (b) and (c), some domains can be observed which span many small grains. From this observation, we can conclude that a grain coupling is present in order to allow the formation of larger magnetic domains. The inset to (a) gives a magnetization loop of such a film; indicating the soft magnetic character. The soft character of the ferrite film is well suited for the demands of the HF-MFM imaging technique, where a hysteresis at high frequencies *and* a fast switching is required for optimum performance of such a ferrite-coated HF-MFM cantilever [20].

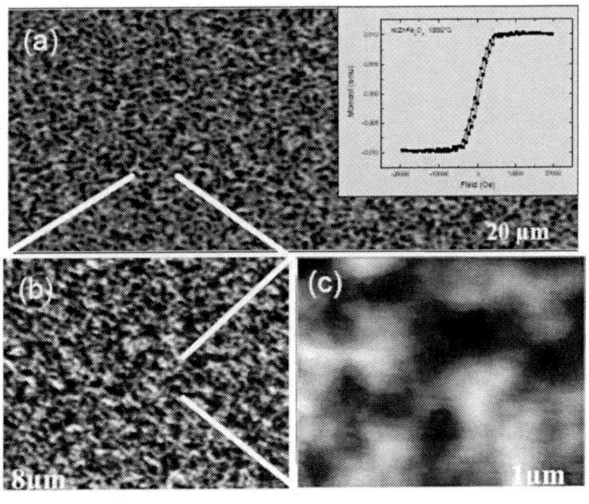

Figure 3. MFM images of the domain structure of a (Ni,Zn)- ferrite film on (111)-oriented Si substrate, remnant state after saturation. The edge length of the images is 20 μm (a), 8 μm (b) and 1 μm (c). The inset to (a) shows a magnetization loop as measured by VSM.

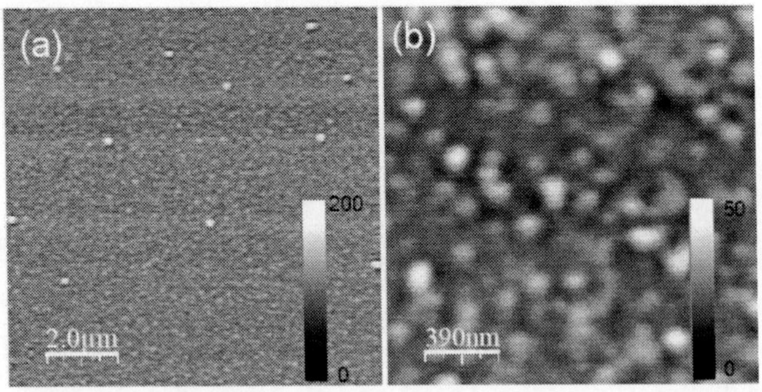

Figure 4.AFM topography images of the surface of a (Ni,Zn)-ferrite film on a (100) Si substrate. The ferrite grains are about 200 nm in diameter.

Figure 4. gives AFM topography scans on the same (Ni,Zn)-ferrite film as shown before, revealing the ferrite grains to be in the 200 nm-range.

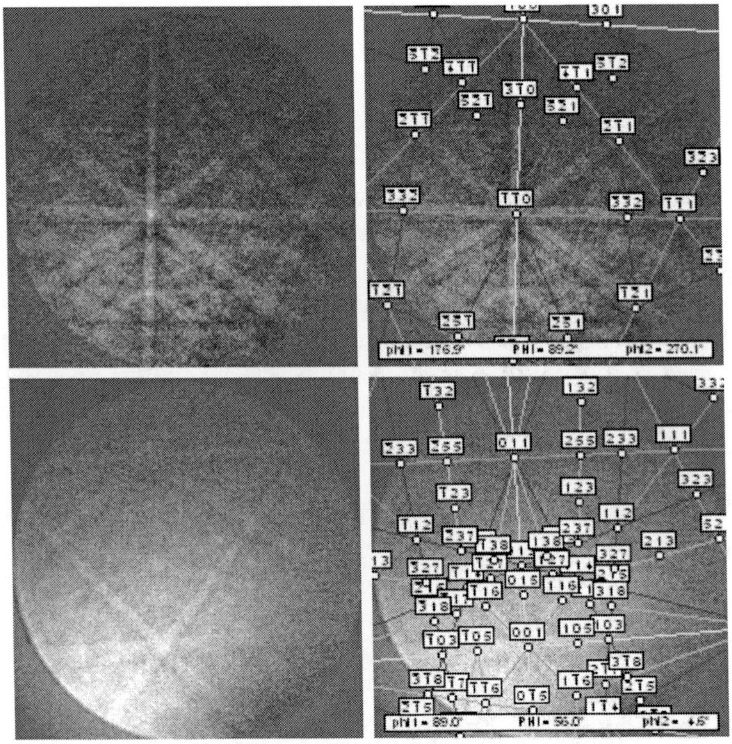

Figure 5. Kikuchi patterns and indexation of (Ni,Zn)-ferrite; showing two different orientations (a) φ_1 = 176.9°, Φ = 89.2°, φ_2 = 270.1° and (b) φ_1 = 89.0°, Φ = 56.0°, φ_2 = 4.6°.

3.3. EBSD Analysis of the Ferrite Films

Now, we will turn to the EBSD analysis of the present sample. First of all, it is necessary to properly identify the Kikuchi patterns of (Ni,Zn)-ferrite, as these patterns are recorded for each individual point of an EBSD map.

Figure 5. presents two measured Kikuchi patterns and their corresponding indexation (Eulerian angles φ_1, Φ, φ_2) performed by the analysis computer. Two different orientations (a) $\varphi_1 = 176.9°$, $\Phi = 89.2°$, $\varphi_2 = 270.1°$ and (b) $\varphi_1 = 89.0°$, $\Phi = 56.0°$, $\varphi_2 = 4.6°$ are observed. The corresponding confidence index, CI, is close to the maximum value of 1; indicating a perfect indexation. This high quality of the image enables either fast EBSD scans or even a multiphase analysis.

Finally, in Figure 6. we present EBSD maps of the ferrite film. The first map (a) is a grain colour map, where each detected grain is coloured randomly with respect to its neighbours. The second map is an inverse pole figure (IPF) map in (001) direction (i.e., perpendicular to the sample surface), giving the crystallographic orientation of the detected grains. The colour code for this map is given in the stereographic triangle below. No texture of the ferrite grains is observed; the grain orientation is random. The grain sizes as determined by EBSD range between 100 and 500 nm, which is illustrated in detail in Figure 7.

The slow cooling applied during the preparation of the ferrite films enables the grain growth, so no amorphous layer is obtained as in previous experiments [26]. Therefore, a more complete analysis of the microstructure can now be performed as EBSD analysis is possible using the present ferrite films. This will enable a detailed interpretation of MFM images combined with the EBSD grain orientation information.

3.4. Magnetic Characterization and Performance as (HF-)MFM Tips

MFM cantilevers coated with such ferrite films were found to perform well for MFM and HF-MFM measurements as demonstrated in Ref. [35], where HF-MFM measurements could be performed up to 2 GHz carrier frequency. Figure 8. gives the magnetization loops obtained by means of VSM on films on Si substrates. The measurements reveal a very soft behavior for the (Ni,Zn)-ferrite film, whereas a clear hysteresis is observed for the BCFO ferrite. During MFM imaging, the (Ni,Zn)-ferrite coated tip may be remagnetized by a hard magnetic sample, whereas the BCFO-coated tip is well suited for measuring such samples. For HF-MFM especially the high-frequency properties (susceptibility) of the ferrite material is essential. Firstly, we compare a (Ni,Zn)-ferrite coated tip to a "standard" CoCr-coated tip which is often used in MFM imaging. Figure 9. shows two HF-MFM images of a SEAGATE harddisk head (longitudinal type) taken by two different cantilevers. The left image is showing the stray field emanating from the writer pole at 1 GHz as taken by the (Ni,Zn)-ferrite coated tip, while the right image gives the same head, but recorded with the CoCr-tip. The corresponding profiles taken across the head structure clearly indicate that the ferrite tip presents a true image of the emanating stray field, while the CoCr-tip practically measures only at the noise level.

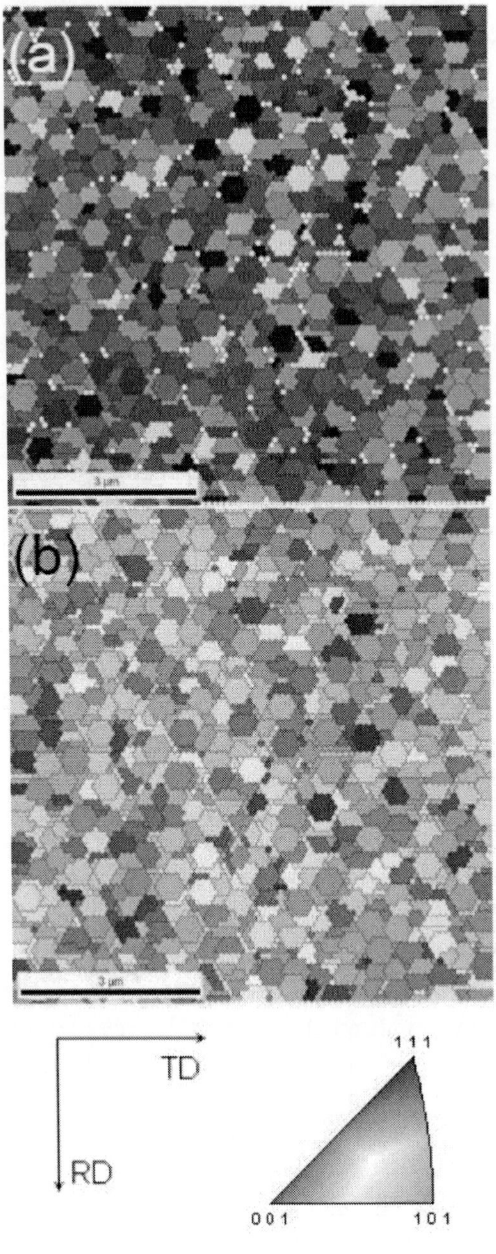

Figure 6.(a) EBSD unique grain color map which illustrates the polycrystalline growth of the ferrite grains. (b) Inverse pole figure (IPF) map in (001) direction, giving the local grain orientation. The colour code for this map is given in the stereographic triangle. The scale bar of the EBSD map is 3 μm long.

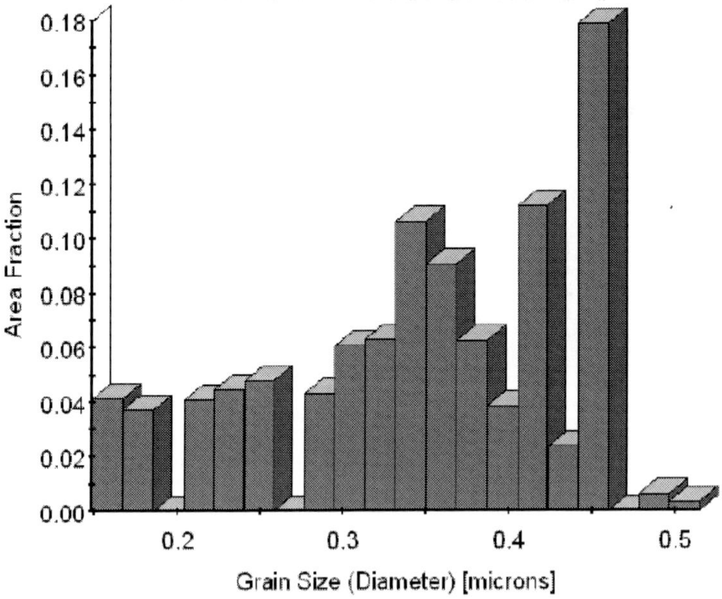

Figure 7. Area fraction of the grains as a function of the grain size (diameter) as determined by EBSD.

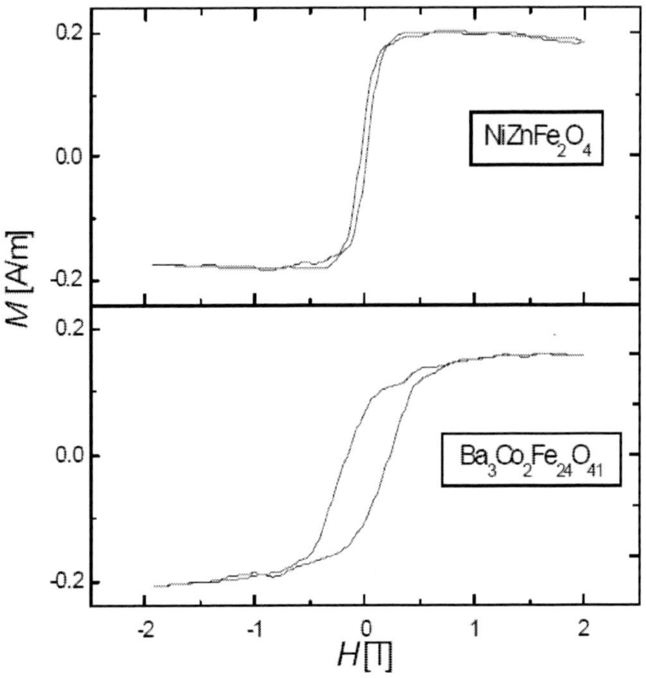

Figure 8. Magnetic hysteresis loops of ferrite thin films on (1 1 1) Si substrates recorded by means of VSM at room temperature. Upper curve (Ni,Zn)-ferrite, and lower curve BCFO.

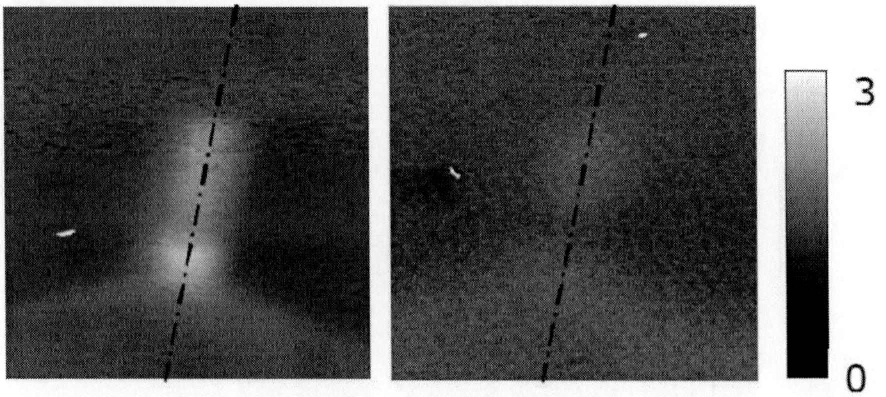

Figure 9. HF-MFM images of a Seagate write head at 1 GHz. Left and right images are made by (Ni,Zn)-ferrite and CoCr-coated MFM-tips, respectively. The corresponding down track profiles of stray fields from write head are presented under the images. The scale bar is shown on the right and the size of images is 4×4 μm2.

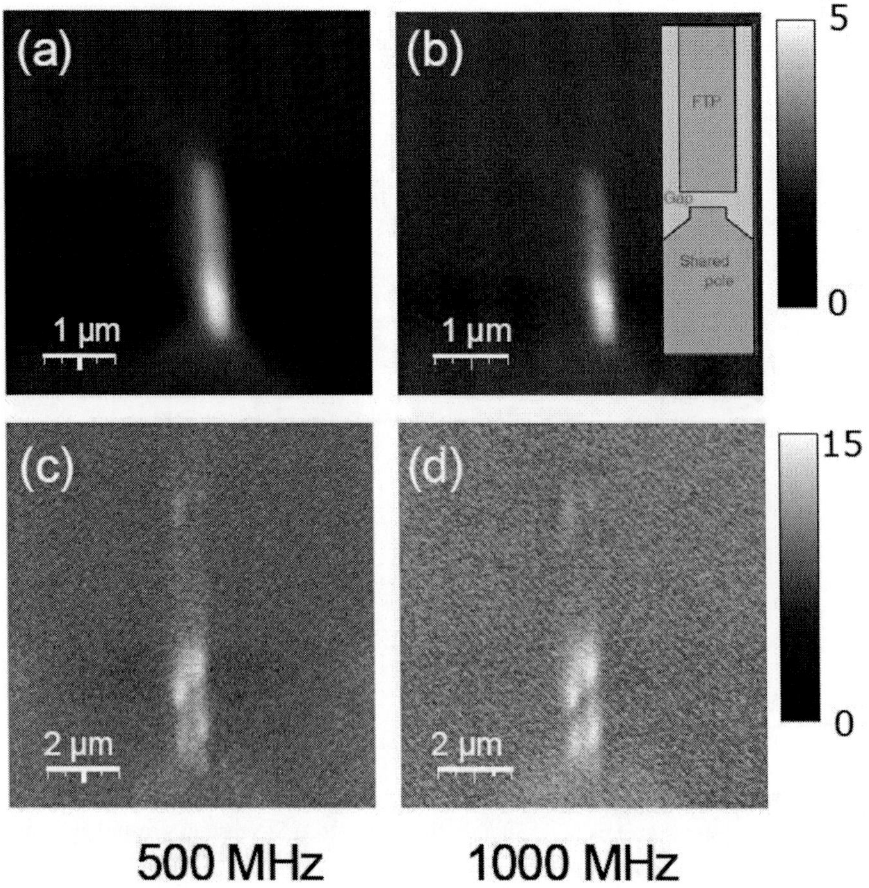

Figure 10. HF-MFM images of harddisk writer poles obtained employing (Ni,Zn)-ferrite coatings (a,b), and BCFO-coated cantilevers (c,d) at two different carrier frequencies, 500 MHz and 1000 MHz. The inset in (b) shows the structure of a harddisk writer pole for comparison.

When comparing the two types of ferrites employed here directly to each other, the differences in the cut-off frequency will appear. In figure 10. (a—d), we present such a comparison of the two types of ferrite films. As sample, we employ again a longitudinal writer pole from SEAGATE. A schematic sketch of the head arrangement is given as an inset to (b). Note that the images are elongated in the y-direction for clarity. The HF-MFM measurements are performed at two different carrier frequencies, 500 MHz and 1 GHz, and the modulation frequency is in all cases 1 kHz. Figures (a) and (b) are taken with a (Ni,Zn)-ferrite coated tip, while images (c) and (d) are recorded with a BCFO-coated tip. The MFM phase signals achieved by the (Ni,Zn)-ferrite coated tip are much smaller as compared to that of the BCFO-coated tip. Furthermore, the resulting image lools much more blurred and not so many details of the stray field structure are resolved [36,37].

The images of Figure 10. clearly reveal the improved HF-properties of the BCFO-coated tip, which yields a clear structured MFM image. The BCFO-coated tips enable further to extend the range of carrier frequencies up to 2 GHz [38]. All our observations concerning the frequency behavior of the tips are summarized in figure 11.

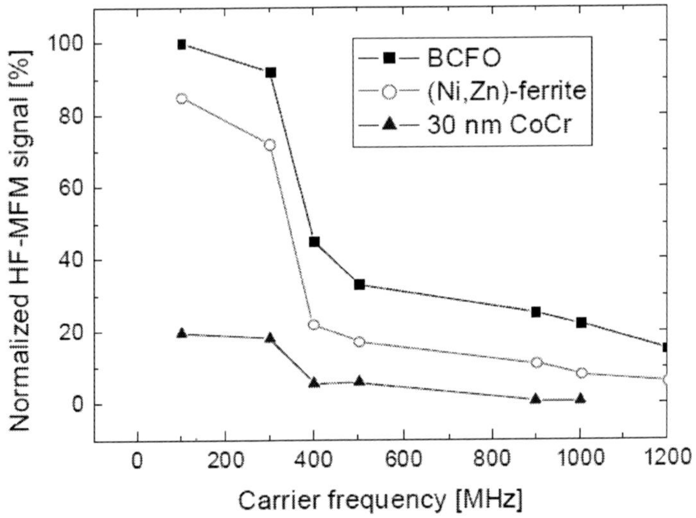

Figure 11. HF-MFM signal strength as a function of the carrier frequency for BCFO, (Ni,Zn)-ferrite and CoCr with 26 mA feeding current. The lines are guidelines for the eyes. The "standard" CoCr-coated cantilevers work only up to carrier frequencies of about 500 MHz. The jump at 400 MHz is related to the design of the writer pole.

Figure 11. clearly reveals that both types of ferrite cantilevers are suited for HF-MFM imaging. Due to the easier fabrication process and the better homogeneity of the (Ni,Zn)-ferrite films, these cantilevers may be used for most HF-MFM experiments. Of course, as soon as the carrier frequency is higher than 1 GHz, the BCFO-coated tips are required.

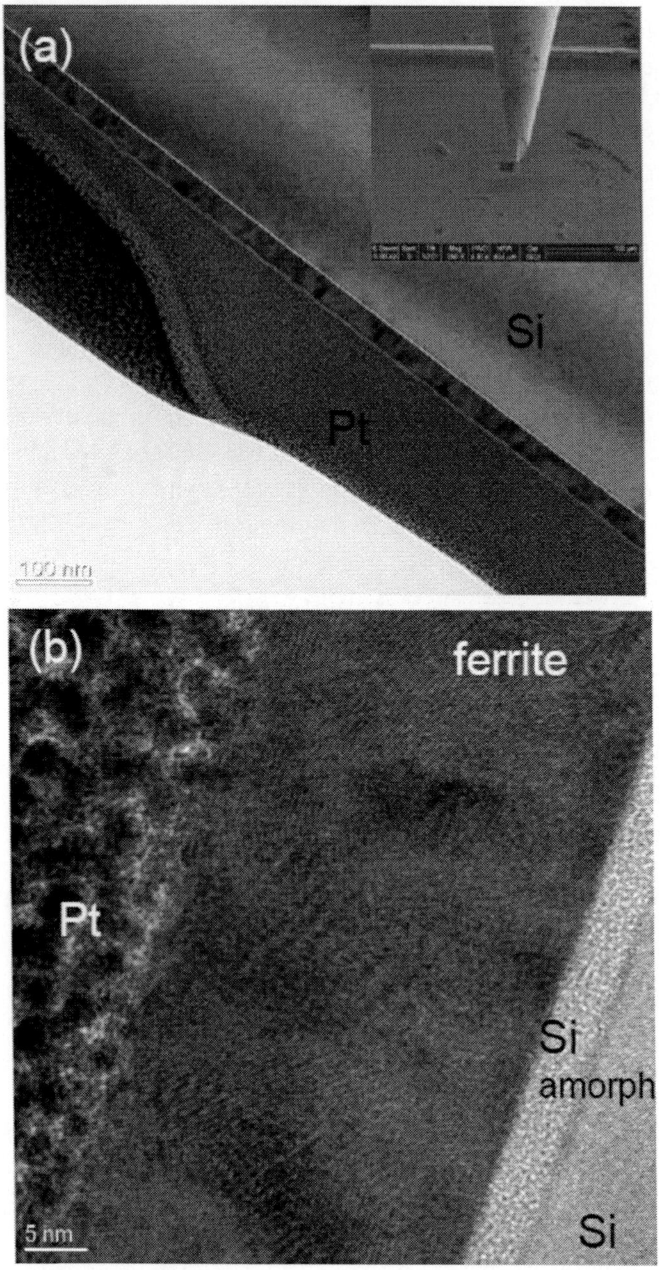

Figure 12. TEM image of the cross section of the ferrite film on the Si cantilever. (a) gives an overview, while (b) presents a high-resolution TEM image. Between the Si of the cantilever and the ferrite is a thin amorphous layer of Si/SiO_2. The polycrystalline character of the ferrite film is clearly visible. The inset to (a) shows he cut TEM slice from the cantilever shank before being mounted on the Cu grid.

3.5. Microstructural Investigation of a Thin Co-ferrite Film

Thin Co-ferrite films on Si substrates were prepared in order to reduce the thickness of the magnetic layer on the cantilever. This should lead to an improved magnetic resolution of such a MFM tip as the resulting tip radius is much smaller. Such thin ferrite films with thicknesses ranging between 2 and 30 nm are nowadays routinely prepared for e.g., magnetic tunnel junctions [39-41]. As a starting point, we have chosen a 30 nm-thick Co-ferrite film prepared on Si (100) and on the Si cantilevers. Figure 12. (a) gives a TEM image of the film cross section. The upper part of the image is the Si of the cantilever, followed by a thin, white line. Then comes the ferrite film, the thickness of which was determined from the TEM image to be ~31 nm. The inset to (a) shows the FIB milling and manipultion of the TEM slice. The white line is an amorphous layer of Si/SiO_2, which has survived the cleaning procedure. The presence of this layer has, however, no influence on the growth of the ferrite grains as shown in detail in (b). The resulting ferrite film is polycrystalline, with a nearly random orientation of the individual ferrite grains. The TEM-determined grain sizes range between 5 and 20 nm, which is too small for an electron backscatter diffraction analysis as performed in Ref. [34] on a 100 nm-thick $NiZnFe_2O_4$ film. Nevertheless, the random orientation of the grains is similar to that of the much thicker $NiZnFe_2O_4$ film. The amorphous character of the Si/SiO_2-layer is clearly resolved in the high resolution image. In order to achieve an even more oriented grain growth, this layer should be completely removed.

These cross section views clearly demonstrate that it is possible to grow polycrystalline ferrite films directly on Si without a buffer layer.

Figure 13. gives a SEM view of a cantilever pyramid covered with a 30 nm thick ferrite film. The resulting coating is quite uniform also on the (1 1 1)-oriented surfaces, indicating that even thinner coatings on cantilevers could be fabricated in this way.

Figure 13. SEM image of the pyramid ((1 1 1)-oriented Si surfaces) of a PPP-type Si cantilever covered with a 30 nm Co-ferrite coating. This image indicates that a quite homogeneous coating is achieved.

The magnetic properties of such thin ferrite films are very important for the imaging capabilities of the coated cantilevers. For very thin magnetic coatings an unwanted spatial frequency doubling effect may occur when imaging bits on a hard disk recording medium [42]. Furthermore, for HF-MFM, the initial susceptibility of the cantilever coating plays an important role, as this quantity is entering the equation for the force acting on the cantilever as a linear term [20]. Figure 14. presents the magnetization data recorded by a SQUID magnetometer. The Co-ferrite film on the Si substrate shows a soft magnetic behavior as intended for a fast moment switching, but a relatively high initial susceptibility χ, which is the highest one for all ferrite films studied by us; the one of the $NiZnFe_2O_4$ is comparable in size, and χ for BCFO is about half of it [26]. For high frequency applicatons ferrite thin films are, in contrast to metal alloys, of great interest due to the small conductivity and the resulting small ac-losses [1-3].

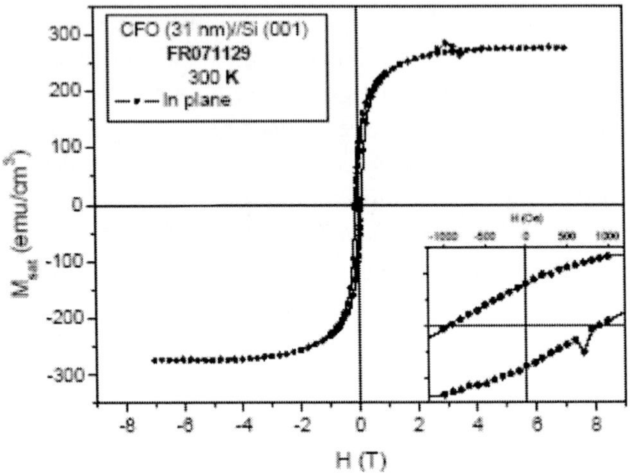

Figure 14. Magnetization loop recorded at 300 K for a 30 nm thick Co-ferrite coating on a 10 × 10 mm^2 Si (0 0 1) substrate. The inset gives a magnification of the region around zero applied field. A small hysteresis is obtained, but a high initial susceptibility results [43].

Figure 15. MFM-image of bits in a recording medium with 300 kilo flux changes per inch (kfci), corresponding to 85 nm/flux change. This image was obtained using a Co-ferrite coated cantilever (30 nm ferrite thickness).

Finally, in Figure 15. we present a MFM image of a bit structure in a rcording medium, taken by a ferrite-coated cantilever. The soft magnetic character of such a tip is visible from the image – the tip is remagnetized by the stray fields from the bits, yielding an uniform bit structure. Therefore, the thin Co-ferrite films do have the required properties for HF-MFM imaging (soft material, high initial susceptibility). The imaging performance at high frequencies will be tested employing hard disk recording writer poles. A further decrease of the ferrite thickness to about 10 nm is planned for the coating of commercial high-aspect ratio tips. In summary, we have performed a thorough microstructural analysis of various ferrite films on Si. (Ni,Zn)-ferrite coatings were prepared directly on Si substrates and cantilevers. An additional slow-cooling after reaching the maximum annealing temperature enabled the growth of grains in the size of 50 – 120 nm. This enables a more thorough investigation of the microstructure, as now a detailed EBSD analysis became possible. Even though the cut-off frequency of bulk (Ni,Zn)-ferrite samples is reported to be much lower, ferrite-coated tips perform well up to 1 GHz. In order to reach even higher frequencies, which is important for the basic understanding of e.g., magnetic recording materials, BCFO ferrite thin films were also prepared on Si substrates. With these BCFO-ferrite coated cantilevers, HF-MFM investigations up to carrier freuquencies of 2 GHz are possible. Finally, 30 nm-thick Co-ferrite coatings on commercial, micromachined Si cantilevers. The ferrite growth occurs in a polycrystalline fashion, with ferrite grain sizes ranging between 5 nm and 20 nm. The coatings exhibit soft magnetic behaviour with a high initial susceptibility, which is ideal for HF-MFM imaging.

ACKNOWLEDGMENTS

We would like to thank F. Rigato and J. Fontcuberta (Barcelona) for their collaboration concerning the Co-ferrite thin films, and J. Schmauch (Saarland University) for the TEM imaging. This work was financially supported by the EU-project "ASPRINT" (contract no. NMP4-CT-2003-1601), which is gratefully acknowledged.

REFERENCES

[1] S. Krupicka, *Physik der Ferrite und der verwandten magnetischen Oxide* (Braunschweig, Vieweg and Sohn, 1973).
[2] J. Smit and H.P.J. Wijn, *Ferrite* (Philips Technical Library, Eindhoven, 1960).
[3] A. Goldman, *Modern ferrite technology* (Van Nostrand, New York, 1990).
[4] N. Matsushita, C. P. Chong, T. Mizutani, and M. Abe, *J. Appl. Phys.* 91, 7376 (2002).
[5] M. Abe, and Y. Tanaka, *IEEE Trans. Magn.* 40, 1708 (2004).
[6] H. K. Wickramasinghe, *Acta Mater.* 48, 347 (2000).
[7] U. Hartmann, *Annu. Rev. Mater. Sci.* 29, 53 (1999).
[8] G. D. Skidmore, and E. Dan Dahlberg, *Appl. Phys. Lett.* 71, 3293 (1997).
[9] M. R. Koblischka, and U. Hartmann, *Ultramicroscopy* 97, 103 (2003).
[10] M. R. Koblischka, U. Hartmann, and T. Sulzbach, *J. Magn. Magn. Mat.* 272-276, 2138 (2004).
[11] Choi, Sang-Jun, Kim, Ki-Hong, Cho, Young-Jin, Lee, Hu-san, Cho, Soo-haeng, Kwon, Soon-Ju, Moon, Jung-hwan, Lee, Kyung-Jin, *J. Magn. Magn. Mat.* 322, 332 (2010).

[12] F. Wolny, T. Mühl, U. Weissker, A. Leonhardt, U. Wolff, D. Givord, B. Büchner, *J. Appl. Phys.* 108, 013908 (2010).
[13] R. Proksch, P. Neilson, S. Austvold, and J. J. Schmidt, *Appl. Phys. Lett.* 74, 1308 (1999).
[14] M. Abe, and Y. Tanaka, *J. Appl. Phys.* 89, 6766 (2001).
[15] S. Li, S. Stokes, Y. Liu, S. Foss-Schrader, W. Zhu, and D. Palmer, *J. Appl. Phys.* 91, 7346 (2002).
[16] M. Abe, and Y. Tanaka, *IEEE Trans. Magn.* 38, 45 (2002).
[17] M. Abe, and Y. Tanaka, *IEEE Trans. Magn.* 40, 1708 (2004).
[18] A.V. Nazarov, M. L. Plumer, and B. B. Pant, *J. Appl. Phys.* 97, 10N902 (2005).
[19] S. Li, in: *Science,* Technology and Education of Microscopy: an Overview, (Formatex, Badajoz, 2003).
[20] M. R. Kolischka and U. Hartmann, *IEEE Trans. Magn.* 45, 3228 (2009).
[21] M. R. Koblischka, J. D. Wei, M. Kirsch, and U. Hartmann, Jpn. J. Appl. Phys. 45, 2238 (2006).
[22] J. Gao, Y. Cui, and Z. Yang, *Mat. Sci. Eng.* B 110, 111 (2004).
[23] D. Guo, Zh. Zhang, M. Lin, X. Fan, G. Chai, Y. Xu and D. Xue, *J. Phys. D: Appl. Phys.* 42, 125006 (2009).
[24] A.S. Kamzin, F. Wei, Z. Yang, and X. Liu, *Phys. Solid State* 44, 1635 (2002).
[25] E. S. Ramakrishnan, K. D. Cornett, and G. Srinivasan, *J. Appl. Phys.* 81, 5162 (1997).
[26] M. Kirsch, M. R. Koblischka, J. D. Wei, and U. Hartmann, *J. Vac. Sci.* B 25, 1679 (2007).
[27] Datasheet Pointprobe cantilevers, Nanoworld Services GmbH, Erlangen, Germany.
[28] S. Bals, W. Tirry, R. Guerts, Z. Q. Yang, and D. Schryvers, *Microscopy and Microanalysis* 13, 80 (2007).
[29] M. R. Koblischka, J. D. Wei, T. Sulzbach, A. D. Johnston and U. Hartmann, *IEEE Trans. Magn.* 43, 2205 (2007).
[30] *Orientation Imaging Microscopy (OIM) software version V4.0*, user manual, TexSEM Laboratories (TSL), Draper, UT, 2004.
[31] F. J. Humphreys, *Scripta Materialia* 51 (2004) 771.
[32] A.Koblischka-Veneva, M. R. Koblischka, J. D. Wei, Y. Zhou, S. Murphy, F. Mücklich, U. Hartmann, I. V. Shvets, *J. Appl. Phys.* 101, 09M507 (2007).
[33] Koblischka-Veneva, F. Mücklich, M. R. Koblischka, N. Hari Babu, D. A. Cardwell, *J. Am. Ceram. Soc.* 90, 2582 (2007).
[34] M. R. Koblischka, M. Kirsch, M. Brust, A. Koblischka-Veneva, and U. Hartmann, *Phys. Stat. Solidi* (a) 205, 1783 (2008).
[35] M. R. Koblischka, J. D. Wei, and U. Hartmann, *J. Magn. Magn. Mat.* 322, 1694 (2010).
[36] M. R. Koblischka, M. Kirsch, J. Wei, and U. Hartmann, *J. Magn. Magn. Mat.* 316, e666 (2007).
[37] M. R. Koblischka, J. D. Wei, M. Kirsch, M. Lessel, R. Pfeifer, M. Brust, U. Hartmann, C. Richter, and T. Sulzbach, *J. Phys: Conf. Ser.* 61, 596 (2007).
[38] J. Wei, M. Kirsch, M. R. Koblischka, and U. Hartmann, *J. Magn. Magn. Mat.* 316, 206 (2007).
[39] Ramos, A. V., Guittet, M.-J., Moussy, J.-B., Mattana, R., Deranlot, C., Petroff, F., Gatel, C., *Appl. Phys. Lett.* 91, 122107 (2007).
[40] M. G. Chapline and S. X. Wang, Phys. Rev. B 74, 014418 (2006).

[41] F. Rigato, J. Geshev, V. Skumryev, and J. Fontcuberta, *J. Appl. Phys*. 106, 113294 (2009).

[42] J. R. Kirtley, Z.Deng, L. Luan, E. Yenilmez, H. Dai, and K. A. Moler, *Nanotechnology* 18, 465506 (2007).

[43] M. R. Koblischka, M. Kirsch, R. Pfeifer, S. Getlawi, F. Rigato, J. Fontcuberta, T. Sulzbach, and U. Hartmann, *J. Magn. Magn. Mat*. 322, 1697 (2010).

In: Magnetic Thin Films
Editor: John P. Volkerts

ISBN: 978-1-61209-302-4

Chapter 13

DEPENDENCE OF TEXTURE OF MAGNETITE THIN FILMS ON DIFFERENT SUBSTRATES AND ORIENTATIONS

A. Kobischka-Veneva[1] and M.R. Koblischka[2]
[1]Institute of Functional Materials, Saarland University,
D-66041 Saarbrücken, Germany
[2]Institute of Experimental Physics, Saarland University,
D-66041 Saarbrücken, Germany

ABSTRACT

The knowledge about microstructure and grain morphology of magnetite materials is very important in order to understand the partially puzzling magnetic properties. We have, therefore, investigated a variety of magnetite samples concerning details of their respective microstructures. (001)- and (111)-oriented magnetite thin films were grown on MgO substrates (film thickness 200 nm) by means of oxygen-plasma-assisted molecular beam epitaxy and by laser-ablation.

Further samples investigated in this study are electroplated magnetite thin films on Si/copper substrates and films from biogenic magnetite. The achieved grain orientations are analyzed by means of the electron backscatter diffraction (EBSD) technique. The EBSD technique enables the crystallographic orientation of individual grains to be determined with a high spatial resolution of up to 40 nm on such ceramic materials.

A high image quality of the recorded Kikuchi patterns was achieved enabling multi-phase scans (e.g., Fe_3O_4, MgO, γ-Fe_2O_3) to be performed. The facets of individual grains, which exhibit sizes of several tens of nanometers in the case of electroplated and biogenic magnetite, are analyzed in detail.

Furthermore, the (001)-surface of the films grown by molecular beam epitaxy after a short annealing in air (1 min, 250 °C) is characterised by the presence of tiny (diameter 100-200 nm) misoriented islands, which may have an influence on the anti-ferromagnetic coupling within the film.

The properties of these interfaces are studied in detail, including magnetic force microscopy (MFM) measurements in applied magnetic fields. In the (111)-oriented films and the laser-ablated films, such defects are found to be absent, and the films show a very homogeneously oriented surface.

Keywords: magnetite, thin films, EBSD analysis.

1. INTRODUCTION

Magnetite thin films have recently shown a peculiar magnetic behaviour [1-4], which is not yet understood in detail. Magnetite (Fe_3O_4) as a half-metallic ferromagnet is an interesting system for various applications because of its unusual magnetic properties for electron spin-manipulating devices [1,2]. The high Curie temperature (T_c = 860 K) enables such devices to be operated at room temperature.

However, the resulting magnetic properties of magnetite thin films grown on MgO substrates were found to be pretty peculiar [3-8], so a detailed investigation of the microstructure of these films is required. Several investigations have revealed that the microstructure of such films is not trivial [5-8], and also may influence the magnetic behaviour [9]. An important tool for the detailed investigation of the microstructure of magnetite films is the electron backscatter diffraction (EBSD) technique, which enables a spatially highly resolved (resolution of about 20 nm) phase and texture analysis [10-15]. Using the electron-backscatter diffraction (EBSD) technique, we could already analyze the crystallographic orientation of (001)-oriented magnetite films and identify the typical defects within the sample surface [7,8]. These defects can play an important role as pinning sites as observed in magnetic force microscopy (MFM) experiments [9]. In the present paper, we investigate magnetite films grown by different techniques (oxygen-plasma-assisted molecular beam epitaxy and pulsed laser deposition in UHV) and in different crystallographic orientations [(001)- and (111)-orientation on MgO substrates] by means of EBSD. Especially the occurrence of typical defects in the (001)-oriented films grown by the molecular beam epitaxy, which were also found in MFM domain measurements, triggered this study.

2. EXPERIMENTAL PROCEDURES

Sample Types and Preparation Methods

The magnetite samples studied are prepared employing different methods, yielding four types of samples:

Type I: Epitaxial magnetite thin films with a thickness of 100 nm were grown on MgO single-crystal substrates cut along the (100) direction within ±0.1° by oxygen-plasma-assisted molecular beam epitaxy [16,17]. The magnetite layer was deposited by means of *e*-gun evaporation from Fe pellets with a purity of 99.995% in a plasma oxygen environment of 1×10^{-5} Torr with a substrate temperature of 250 °C. The samples were subsequently annealed in air at 250 °C for 1 min.

Type II: Magnetite films on MgO substrates produced by pulsed-laser deposition in a UHV chamber (background pressure 5×10^{-9} Torr), with a thickness ranging between 4 and 80 nm. The films show an excellent crystallinity as determined by X-ray analysis [18].

Type III: Magnetite electro-plated on Cu substrates as described in Ref. [19]. The resulting magnetite film thickness is ~ 150 nm. The magnetite thin film samples were prepared on polycrystalline copper substrates by means of electrodeposition. Copper as a conductive underlayer is employed because of its good electrical and diamagnetic properties. Polycrystalline copper substrates with a purity of 99.995% and 0.125 mm in thickness were purchased from Advent Research Materials, Ltd. (Oxford, U.K). The copper substrates were mechanically polished and all substrates were ultrasonically cleaned in acetone and subsequently rinsed with deionized water. The Fe_3O_4 preparation is based on the ferrite-plating method as reported by Abe and Tanamura [20] and uses a conventional three-electrode cell apparatus under galvanostatic operation. The electrodeposition bath was kept at a constant temperature and purged with argon under vigorous stirring. A constant anodic current was applied using a 263A potentiostat/galvanostat from EGandG Princeton Applied Research. The bath temperature for the present sample series was 80 °C, using a 0.01 mol l^{-1} $(NH_4)_2Fe_2(SO4)_2 \cdot 6H_2O$ and 0.05 mol l^{-1} CH_3COOK electrolyte. The reference electrode was Au, and the counterelectrode Pt. More details of this deposition process and an analysis of the samples by XRD, SEM and TEM is given in Ref. [21].

Type IV: Thin films of biogenic magnetite. The extraction of the magnetosomes is described in detail in Ref. [22]. The purified magnetosomes were adsorbed on the substrates (typically Mica) in 10mMTris·HCl (pH 8.0). After 3 min, the sample was rinsed with 10mMTris·HCl (pH 8.0). For chemical fixation, the sample was incubated with 1% glutaraldehyde for 3 min following the description in Ref. [23].

Electron Backscatter Diffraction (EBSD)

The EBSD system employed consists of a FEI dual beam workstation (Strata DB 235) equipped with a TSL OIM analysis unit [24]. The Kikuchi patterns are generated at an acceleration voltage of 20 kV, and are recorded by means of a DigiView camera system, allowing a maximum recording speed of the order of about 14 pattern/s. The time employed in the case of a multi-phase scan is somwhat longer, of the order of 10 pattern/s, as a higher image quality/confidence index is required. To produce a crystallographic orientation map, the electron beam is scanned over a selected surface area and the resulting Kikuchi patterns are indexed and analyzed automatically (i.e., the Kikuchi bands are detected by means of the software). An image quality (IQ) parameter and a confidence index (CI) is recorded for each such Kikuchi pattern. The dimensionless IQ parameter is the sum of the detected peaks in the Hough transform employed in the image recording; the CI value yields information about how exact the indexation was carried out. The CI value ranges between 0 and 1 [24]. Based on the analysis of the recorded CI value, a multi-phase analysis is realized. A detailed description of the measurement procedure can be found in Refs. [25-28]. The results of the EBSD measurement are presented in form of maps, the most important thereof are the so-called inverse pole figure (IPF) maps, indicating the crystallographic orientation of each individual point. Automated EBSD scans were performed with a minimum step size of 30 nm; the working distance was set to 10 mm.

Results and Discussion

In Figure 1, EBSD-mappings are presented for a *type I* sample in (001)-orientation. The colour code for the orientation is given in the stereographic triangle. The inverse pole figure (IPF) map in (001) direction reveals the presence of several misoriented (~30°) islands (indicated as light blue, yellow in the IPF map); and the orientation of these unit cells is indicated by the wire frames. The size and distribution of these islands corresponds to that seen in magnetic measurements [9]. A multi-phase EBSD analysis indicates that these misoriented islands are due to the presence of maghemite particles. This type of defect was seen in a variety of samples of this type. Furthermore, there is an overall 3° - 5° misorientation of the entire film (shade of red), which is also not expected from the X-ray analysis.

Figure 2. gives an IQ- (a) and an IPF map (b) of a (111)-oriented magnetite film (*type I*) grown on MgO. The IPF map reveals several misoriented islands, which are induced by the scratches within the substrate. Otherwise, the sample is free of the defects found in the (001)-oriented films. The shade of blue in the (001)-map also indicates the presence of a 0° - 3° misorientation, similar to the (001)-oriented sample. This type of misorientation is obviously typical for this preparation technique.

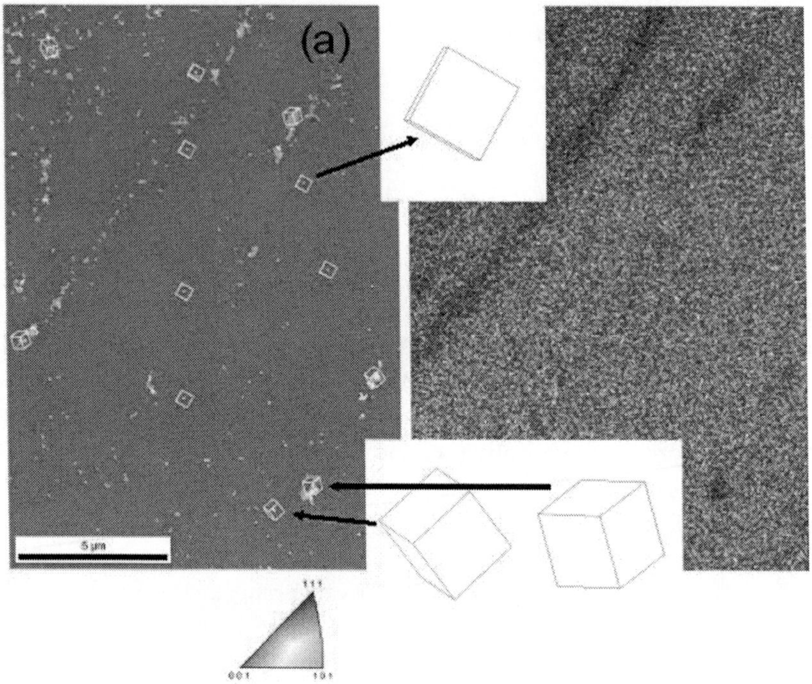

Figure 1. EBSD analysis of a type I sample in (001) direction. (a): IPF map with some specific orientations indicated by the unit cells; the color code for the orientation is given in the stereographic triangle. (b): IQ-map, which resembles a backscattered electron image under EBSD conditions.

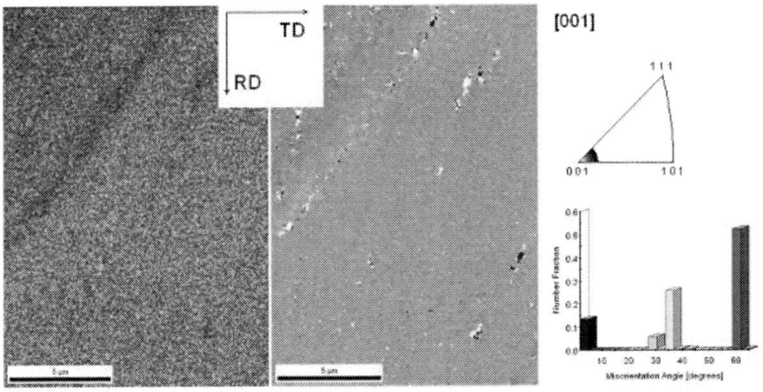

Figure 2. IQ- and crystal direction map of (001)-oriented magnetite ***type I.*** In the IQ-map, misorientation angles are marked according to the colour code presented in the histogram.

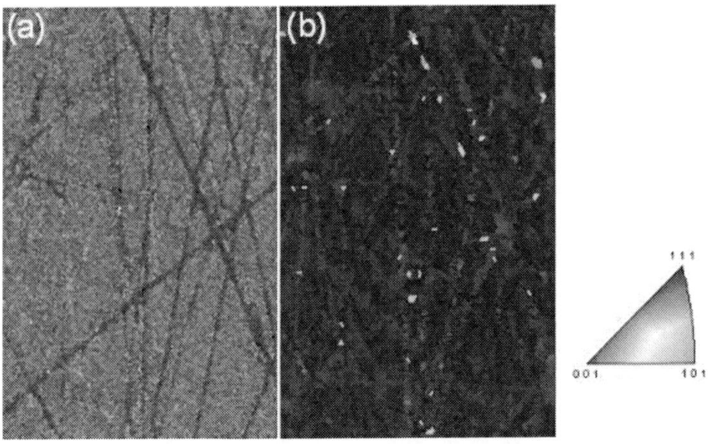

Figure 3. EBSD IQ-map (a) and IPF map (b) of a magnetite film grown on a (111)-oriented substrate (***type I***) in (001)-direction. The color code for (b) is given in the stereographic triangle.

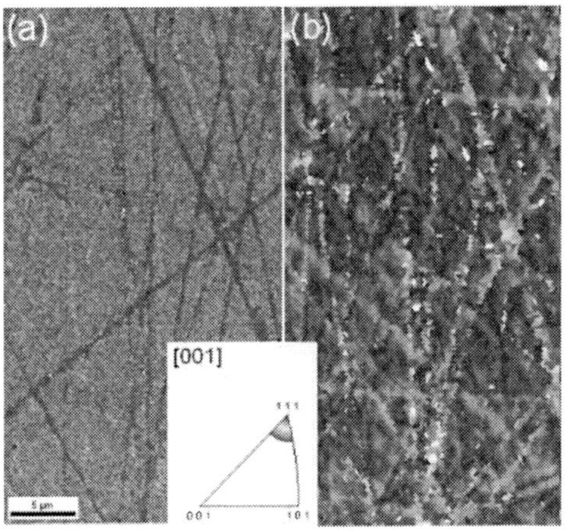

Figure 4.IQ- and crystal direction (CD) map of (111)-oriented magnetite ***type I***. The color code for the CD map is given in the inset.

Figure 1. gives an IQ- and a crystal direction (CD) map for a (001)-oriented magnetite film on MgO. The sample is fully oriented in (001)-direction, but there are some significant white spots; the orientation of which does not fall in the 10° range around the (001) direction of the CD map. Note that the overall colour of the map is orange, not yellow, which indicates a typical 2° – 3° misorientation of the magnetite grains. A multiphase EBSD analysis revealed that these misorientations are induced by the presence of maghemite grains. In contrast to that, Figure 2. presents the same analysis on a (111)-oriented film of type I. The dominant colour of the map is blue, indicating a perfect (111)-orientation, which is disturbed by the presences of scratches within the substrate leading also to a 3° – 5° misorientation. Figure 3. shows that the magnetite of *type II* is perfectly oriented in (001) direction; only the average kernel misorientation (c) reveals some structure of around 0.25°.

In Figure 3, an IQ-map (a), an IPF-map (b) and a crystal direction map (c) is presented for a sample of type II. The maps (a) and (b) are quite homogeneous and do not reveal any kind of misoriented islands being present. This indicates that the 30°-misoriented islands of the type I samples are typical for their preparation method, not for the growth of magnetite in general. To elucidate this further, (c) presents a crystal direction map, where the colours are representing a 10° range around the (001) pole. In this way, also here a structure gets visible; however, there are no such misoriented islands in this type of sample. The CD map indicates the presence of misorientations of the order of 2°, otherwise the sample is perfectly oriented in (001) direction.

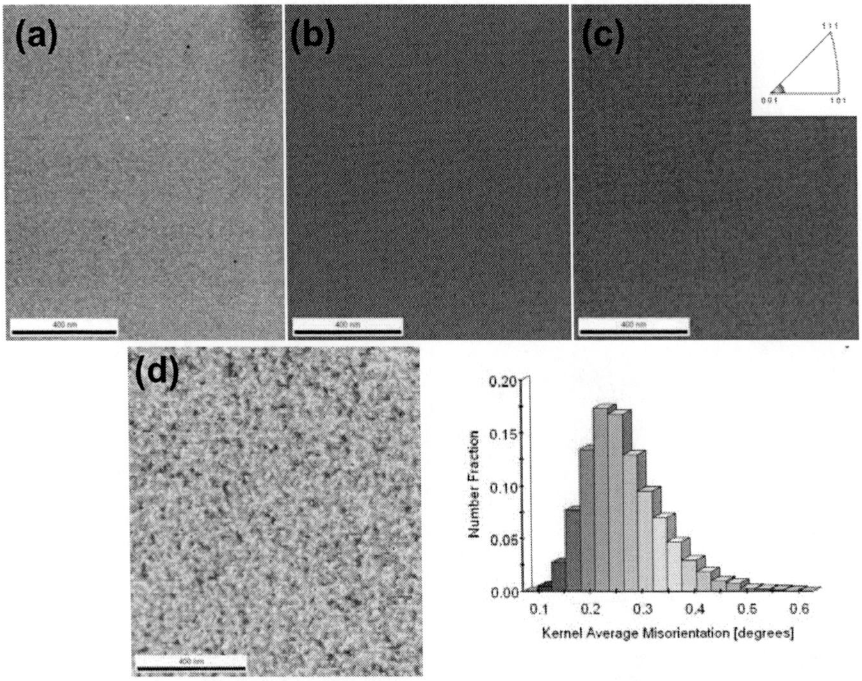

Figure 5. (a). IQ-map, (b) IPF map in (001)-direction and (c) crystal direction map of a magnetite sample of *type II*. The color code for (c) is given in the triangle (inset); i.e. around the (100) direction. Dark blue indicates the perfect (001)-orientation. Finally, (d) gives the average kernel misorientation map – the color code for this map corresponds to the colors in the histogram. The scale bar in all the maps is 400 nm long.

The spatially highly resolved EBSD analysis provides the possibility to compare the microstructure of various magnetite films; the similar resolution even allows a direct comparison with MFM images. Distinct features of the different preparation techniques are found. The laser-ablated magnetite films are very homogeneous and well oriented without exhibiting defects.

The magnetite thin film samples of *type III* were prepared on polycrystalline copper substrates by means of electrodeposition; the properties of the samples studied here are summarized in Table 1. More details of this deposition process and an analysis of the samples by XRD, SEM and TEM is given in Ref. [22]; a preliminary EBSD analysis in Ref. [29]. The resulting magnetite film thickness is ~ 150 nm. The resulting magnetite films consists of independent magnetite grains.

Table 1. Samples selected for study. The deposition voltage is varied between 15 mV and 400 mV

Sample	X001	X004	X005	X006
Voltage [mV]	400	50	25	15
Grain size (max.) [μm]	0.7	0.5	0.3	0.2

Figure 6. (a) SEM image and Kikuchi pattern obtained on sample X001; only a faint pattern is achieved (b): SEM image and Kikuchi pattern of sample X005. The samples X004 – X006 exhibit smooth surfaces; while the Kikuchi pattern of X001 indicates that the material is *not* magnetite, but a complex α-FeOOH+Fe_3O_4.

The sample X001 (see Table 1) consist of relatively large grains, thus leading to a relatively rough sample surface. Nevetheless, an EBSD analysis can be performed on the as-grown grain edges. The samples X004 – X006, prepared at much smaller voltages, exhibit a much smoother surface, even though also in this case, the individual grains can be clearly distinguished.

In Figure 1, a SEM image and a Kikuchi pattern is shown for sample X001 (a) and X005 (b). The surfaces of the samples X004 – X006 are smooth, and some rectangular crystallites are on the surface. The sample X001 consists of individual grains clustered on the substrate, and the Kikuchi pattern obtained reveals that the material is not magnetite, but a α-FeOOH+Fe_3O_4-complex, which can not be indexed at present time.

Figure 4. presents an IQ-map of magnetite type III, grown by electroplating on a Cu-substrate. Here, we have a polycrystalline sample, where individual grains with their facettes are visible in the IQ-map. The orientation of the grains is fully random. The average grain size is about 80 nm, so the achieved EBSD resolution enables to measure the orientation of individual grain facettes. A detailed EBSD analysis of this type of sample was presented in Ref. [29].

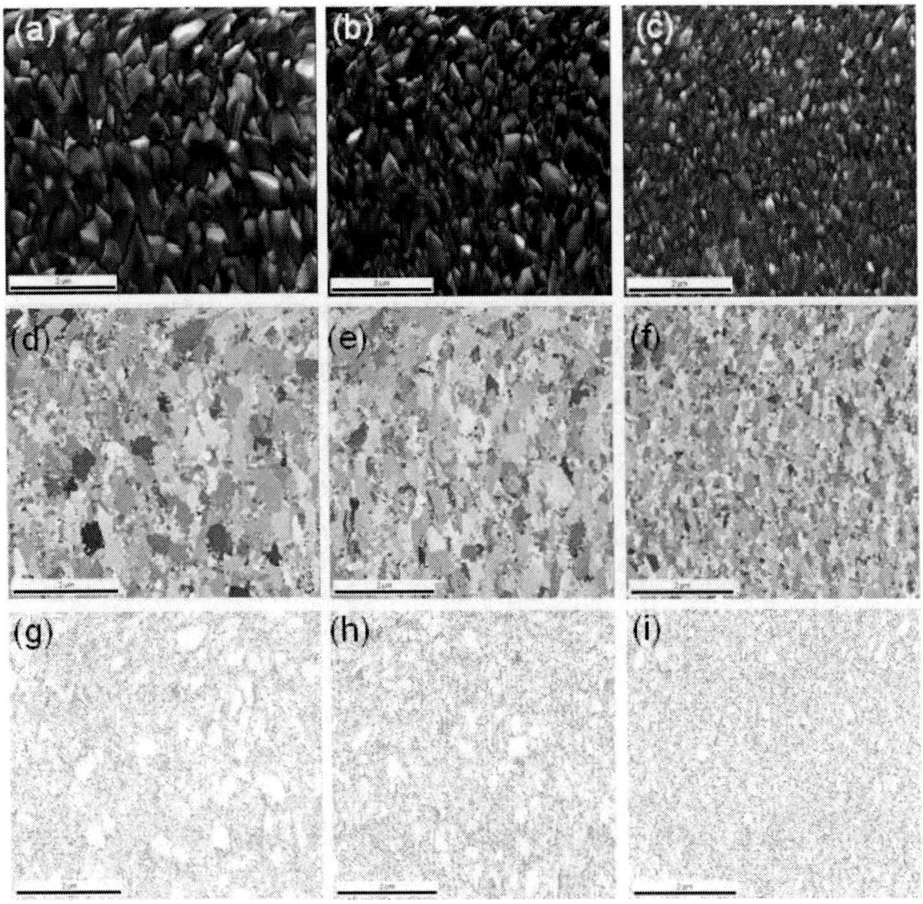

Figure 7. EBSD IQ-maps (a-c), IPF-maps in (001)-direction (d-f), and misorientation mappings (g-i). The marker in all maps is 2 μm long.

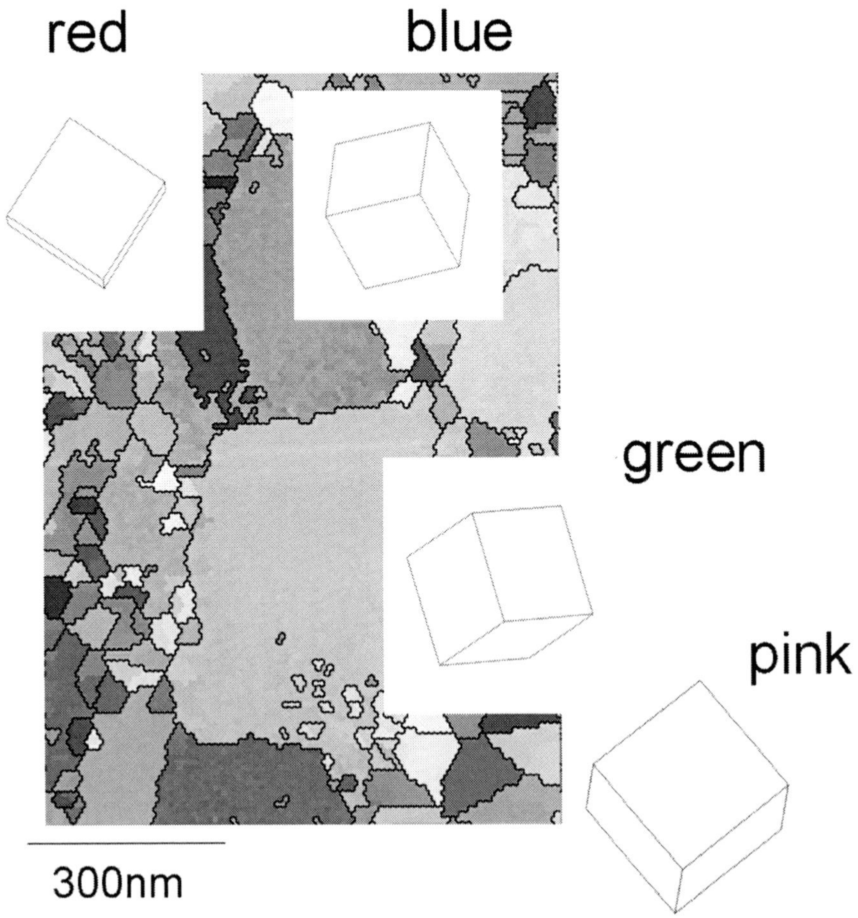

Figure 8. IPF map in (001)-direction of magnetite type IV (biogenic magnetite). Additionally indicated are the different magnetite unit cells.

Finally, Figure 5. presents an EBSD analysis of individual magnetite grains from a biogenic magnetite thin film of *type IV*. The magnetite particles employed here stem from magnetosome bacteria as described in detail in Ref. [22]. Here, on the films prepared from a dried magnetite ferrofluid, only individual grains are observed by means of EBSD. The observed situation is similar to the electroplated magnetite samples (sample X001-3). The spatial resolution of EBSD is sufficient to observe crystallographic details on the individual grains. Overall, these grains behave similar to the electroplated magnetite films prepared at higher deposition voltages; that is, similar crystallographic orientations are obtained in the EBSD measurements.

Conclusions

Using the EBSD analysis technique, we can show microstructural differences in a variety of magnetite film samples, with a relatively high spatial resolution of ~20 nm. This high spatial resolution of the EBSD images enables further a direct comparison with magnetic force microscopy images. (100)-oriented magnetite thin films on MgO substrates (*type I*) exhibit a very characteristic defect structure, which also plays an important role in the resulting magnetic properties. In (111) oriented samples, these defects are not observed, but the same orientation spread as in the (100)-oriented samples is observed in the EBSD measurements. Magnetite samples prepared by means of laser ablation (*type II*) are found to be very homogeneous and free of crystallographic defects. Magnetite thin films on copper substrates prepared by electrodeposition (*type III*) exhibit a completely different type of microstructure. Depending on the deposition voltage, the grain size varies and hence, also the resulting surface roughness. In all cases, the individual grains can be clearly resolved. Finally, thin films of biogenic magnetite (*type IV*) are found to be similar to the elctrodeposited ones; also here, individual grains can clearly be resolved and similar crystallographic orientations are observed.

ACKNOWLEDGMENTS

We would like to thank J. Zhou and S. K. Arora (Dublin), S. Morellon (Zaragoza) and J. Wei (Braunschweig) for providing us with the various magnetite thin film samples. This work is part of DFG project Mu959/19, which is gratefully acknowledged.

REFERENCES

[1] D. T. Margulies, F. T. Parker, M. L. Rudee, F. E. Spada, J. N. Chapman, P. R. Aitchison, A. E. Berkowitz, *Phys. Rev. Lett.* 79, 5162 (1997).

[2] D. T. Margulies, F. T. Parker, F. E. Spada, R. S. Goldman, J. Li, R. Sinclair, A. E. Berkowitz, *Phys. Rev. B* 53, 9175 (1996).

[3] W. L. Zhou, K.-Y. Wang, C. J. O'Conner, *J. Tang, J. Appl. Phys.* 89, 7398 (2001).

[4] W. Eerenstein, T. T. M. Palstra, T. Hibma, S. Celotto, *Phys. Rev. B* 68, 014428 (2003).

[5] Koblischka-Veneva, M. R. Koblischka, F. Mücklich, S. Murphy, Y. Zhou, and I. V. Shvets, *IEEE Trans. Magn.* 42, 2873 (2006).

[6] Koblischka-Veneva, M. R. Koblischka, Y. Zhou, S. Murphy, F. Mücklich, U. Hartmann, and I. V. Shvets, *J. Magn. Magn. Mat.* 316, e663 (2007).

[7] Koblischka-Veneva, M. R. Koblischka, J. D. Wei, Y. Zhou, S. Murphy, F. Mücklich, U. Hartmann, and I. V. Shvets, *J. Appl. Phys.* 101, 09M507 (2007).

[8] Koblischka-Veneva, M. R. Koblischka, S. Murphy, S. K. Arora, F. Mücklich, U. Hartmann and I. V. Shvets, *Microsc. Microanal.* 13 (Suppl. 3), 362 (2007).

[9] J. D. Wei, I. Knittel, Y. Zhou, S. Murphy, F. T. Parker, I. V. Shvets, U. Hartmann, Appl. Phys. Lett. 89, 122517 (2006); I. Knittel, J. D. Wei, Y. Zhou, S. K. Arora, I. V. Shvets, M. Luysberg, and U. Hartmann, *Phys. Rev. B* 74, 132406 (2006).

[10] T. Tepper, C. A. Ross, and G. F. Dionne, *IEEE Trans. Magn.* 40, 1682 (2004).

[11] K. Z. Baba-Kishi, *J. Mater. Sci.* 37, 1715 (2002).

[12] F. J. Humphreys, *Scripta Materialia* 51 (2004) 771.

[13] Koblischka-Veneva, M. R. Koblischka, J. D. Wei, Y. Zhou, S. Murphy, F. Mücklich, U. Hartmann, I. V. Shvets, *J. Appl. Phys.* 101, 09M507 (2007).

[14] Koblischka-Veneva, M. R. Koblischka, F Mücklich, K. Ogasawara, and M. Murakami, Supercond. *Sci. Technol.* 18, S158 (2005).

[15] Koblischka-Veneva, F. Mücklich, M. R. Koblischka, N. Hari Babu, D. A. Cardwell, *J. Am. Ceram. Soc.* 90, 2582 (2007).

[16] Y. Zhou, X. Jin, and I. V. Shvets, *J. Magn. Magn. Mat.* 286, 346 (2005).

[17] V. O. Golub, V. V. Dzyublyuk, A. I. Tovstolytkin, S. K. Arora, R. Ramos, R. G. S. Sofin, and I. V. Shvets, *J. Appl. Phys.* 107, 09B108 (2010).

[18] Fernandez-Pacheco, J.M. de Teresa, J. Orna, L. Morellon, P. A. Algarabel, J. A. Pardo, M. A. Ibarra, C. Magen, and E. Snoeck, *Phys. Rev. B* 78 (2008) 212402.

[19] L. Teng, M. P. Ryan, *Electrochem. Solid State* 10, D108 (2007).

[20] M. Abe and Y. Tanamura, *Jpn. J. Appl. Phys.* 22, 511 (1983).

[21] D. Carlier, C. Terrier, C. Arm, and J. P. Ansermet, Electrochem. *Solid State 8*, C43 (2005).

[22] K. Grünberg, E. Ch. Müller, A.Otto, R.Reszka, D. Linder, M. Kube, R. Reinhardt, and D. Schüler, *Appl. Eviron. Microbiol.* 70, 1040 (2004).

[23] D. Yamamoto, A. Taokac, T. Uchihashi, H. Sasakic, H. Watanabe, T. Ando, and Y. Fukumori, PNAS 107, 9382 (2010).

[24] Software version V4.0, user manual, *TexSEM Laboratories* (TSL), Draper, UT (2004).

[25] Koblischka-Veneva, F. Mücklich, M. R. Koblischka, N. Hari Babu, and D. A. Cardwell, *J. Am. Ceram. Soc.* 90, 2582 (2007).

[26] Koblischka-Veneva, M. R. Koblischka, N. Hari Babu, D. A. Cardwell, F. Mücklich, *Microscopy and Microanalyis 13* (Suppl. 3), 360 (2007).

[27] Koblischka-Veneva, M. R. Koblischka, F. Mücklich, S. Murphy, Y. Zhou, and I. V. Shvets, *IEEE Trans. Magn.* 42, 2873 (2006).

[28] Koblischka-Veneva, M. R. Koblischka, S. Murphy, S. K. Arora, F. Mücklich, U. Hartmann, and I. V. Shvets, *Mat. Sci. Eng. B* 144, 64 (2007).

[29] Koblischka-Veneva, M. R. Koblischka, C. L. Teng, M. P. Ryan, U. Hartmann, and F. Mücklich, *J. Magn. Magn. Mat.* 322, 1235 (2010).

In: Magnetic Thin Films
Editor: John P. Volkerts

ISBN: 978-1-61209-302-4

Chapter 14

THE LOW TEMPERATURE RESISTIVITY OF MN-NI FILMS

F. K. Ampong* and F. Boakye
Department of Physics, Kwame Nkrumah University of Science and Technology, Kumasi Ghana

ABSTRACT

Measurements on electrical resistivity of thermally evaporated $Mn_{100-x}Ni_x$ films (with x = 0.5, 1.5 and 2.5 at. %) have been carried out over the temperature range from 300 to 1.4 K using the van der Pauw four probe technique. The films were grown on a glass substrate held at a temperature of 300 K in an ambient pressure of 2 x 10^{-6} torr. All the films show the usual resistance minimum, a notable characteristic of α-Mn but their low temperature behavior reveal a tendency towards saturation of the resistivity as the temperature approaches zero. This saturation is reminiscent of the Kondo effect.

Keywords: Mn – Ni films, Electrical resistivity, Kondo effect.

INTRODUCTION

Manganese has four allotropic forms namely, α β γ and δ phases between the melting point and room temperature [1]. The α – phase is below 973 K and it undergoes a paramagnetic – antiferromagnetic phase transition at about 95 K. The α – phase has a complex crystal structure of lattice constant 8.91 Å, and with 29 atoms in a unit cell. These atoms are distributed over four crystallographically equivalent sites [2]. The magnetic moments at sites I, II, III and IV are 1.9, 1.7, 0.6 and 0.2 μ_B respectively at 4.4 K. Neutron diffuse scattering study by Shull and Wilkinson [3] has indicated an average moment of 0.5 μ_B in the paramagnetic state up to 500 K. This is consistent with the four different moments with a weighted average of 0.63 μ_B obtained from neutron diffraction experiments by Yamada

* E-mail: kampxx@yahoo.com , Tel: 0233-244-660301

et al [4]. Nickel on the other hand has an fcc structure at room temperature with a lattice constant of 3.52 Å. It has been reported [5] that it has an inter atomic spacing of 2.49 Å and an atomic weight of 28. The ferromagnetic phase of this metal exists below about 630 K [6]. It has been reported [7 - 12] that the saturation magnetization of the disordered Ni-Mn alloys increases linearly with increasing Mn content to about 6 at. % Mn, reaches a maximum at 10 at. % Mn, then decreases to zero near 25 at. % Mn. This behavior has been interpreted in terms of a nearest neighbor (NN) molecular – field model with ferromagnetic Ni – Ni and Ni – Mn antiferromagnetic Mn – Mn interactions [9, 13, 14]. Long wavelength neutron scattering experiments by Low et al [15, 16] have found that the moment disturbances about Fe and Mn impurities in Ni is effectively confined to the solute atom site differently from the case of Cr and V impurities. Moreover, polarized neutron diffuse scattering experiments by Cable et al [17, 18] and MNR measurements on Ni – rich and Ni – Mn alloys by Kitaoka et al [19] have shown that the magnetic state of Mn strongly depends on its NN environment. These suggests that the magnetic moment of isolated Mn atoms is parallel to the magnetization of the Ni – matrix, whereas the Mn atoms with three or more Mn neighbor, reduce or reverse its magnetic moment. The magnetic moments estimated by NMR experiments by Kitaoka et al [19] were found to be 3.0, 2.5 and -1.8 μ_B for isolated, reduced and reversed moment Mn atoms respectively. This existence of Mn atoms with reduced or reversed moments was shown by calculation of the electronic structure of Ni – Mn alloys in the coherent approximation [20]. The anomalous low temperature resistivity and critical scattering of conduction electrons around the Néel point of α – Mn films containing dilute concentration of Ni has been reported [21]. The results reveal that the Néel point of Mn – Ni alloys shift to upper values with concentration of Ni, and the low temperature resistivity of the 0.02 and 0.05 at. % Ni in Mn samples obeys a T^2 law indicating a spin fluctuation scattering mechanism in these samples. Accordingly, Mn – Ni alloy system is very complicated since its properties depend on the details of the microstructure, the distribution of Ni atoms in the alloy and consequently the way and how far Mn atoms interact with each other. Depending on the subtle changes in its chemical and topological environment, Mn may exhibit largely different spin states and exchange interaction with its neighbors. The result is that such alloys present a wealth of magnetic states ferromagnetic and antiferromagnetic glass properties. In this paper, we report on the low temperature anomalies of $Mn_{100-x}Ni_x$ thin films with x = 0.5, 1.5 and 2.5 at. %.

EXPERIMENTAL DETAILS

The starting material were 0.5, 1.5 and 2.5 at. % Ni in Mn in the form of flakes, all obtained from BDH chemicals Ltd, Poole England. The purity of the samples was quoted by the manufacturers as 99.98 %. The flakes were first cleaned in 5 % HCl in methanol to remove surface oxides and other contaminants and then ground to coarse powder before loaded in to a molybdenum boat as in our previous experiment [21]. An AUTO 306 coating plant from Edwards High Vacuum Ltd. UK was used in the coating process. A flash deposition rate of 350 Ås^{-1} was used for each specimen and the substrate temperature was kept at 300 K, by a radiant heater fitted to the vacuum plant. A quartz crystal monitor was used to monitor the film deposition rate. For accurate film thickness measurements, a Varian interferometer was used. This is discussed as follows: when two optical flats are brought into close proximity, interference fringes are observed. The fringes show a displacement as they pass over the film step edge. this displacement expressed as a fraction of λ/2 fringe spacing, gives the film thickness (λ being the wavelength of light being used). If x is the step height and y is the spacing between the fringes, the thickness t of the film is given by Tolansky [22] as

$$t = \frac{x}{y}.\frac{\lambda}{2}$$ where x and y are measured by the Varian interferometer.

The ambient pressure was kept at 2 x 10^{-6} Torr and to achieve this, the vacuum plant was left to pump for about 8 hours. The cryostat used in these experiments was a conventional design as described by Boakye et al [23]. Resistivity measurements were made by the van der Pauw [24] four probe technique. Temperatures between 300 and 60 K were measured with a copper resistance thermometer and below 60 K it was measured with a carbon resistance thermometer, since below 60 K copper resistance thermometer is insensitive. It took about nine hours for N_2 experimental run from 300 to 60 K and for He^4 experimental run, it took about five hours. Temperatures below 4.2 K were obtained by condensing liquid helium into the insert of the cryostat and pumping on it.

EXPERIMENTAL RESULTS AND DISCUSSION

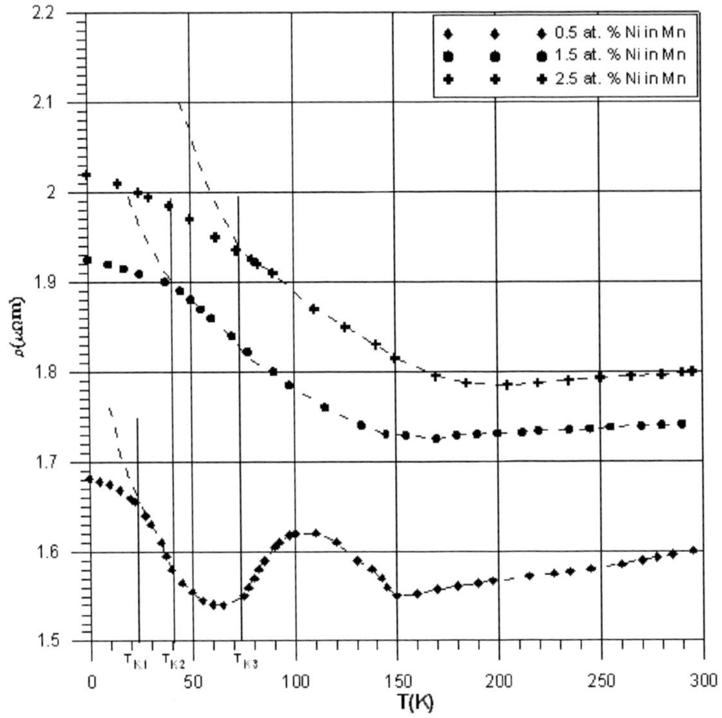

Figure 1. Temperature dependence of resistivity of 0.5, 1.5 and 2.5 at. % Ni in Mn. (Broken lines indicate the Kondo in ln T term).

Figure 1, illustrates the temperature dependence of resistivity of 0.5, 1.5 and 2.5 at. % Ni in Mn. It can be seen from the result of the 0.5 at. % Ni in Mn that the linear portion of the ρ – T curve drops to a resistivity minimum, (T_{min}) at 150 K. This is normally attributed to the phonon scattering of the conduction electrons. The resistivity goes through a 'hump', establishing a resistivity maximum at 100 K. The resistivity goes through a second minimum at about 70 K. The resistivity temperature curve then tends to a behavior revealing a tendency towards saturation of the resistivity as the temperature approaches zero. Comparing these results to those of Boakye et al [21], the resistivity at 300 K is higher and the residual resistivity ρ_o, is higher. This is not strange since the solute is ten times that of those reported by Boakye et al [21]. The present results have a shorter range of magnetic ordering. Craig and Goldberg [25] have defined the Néel point T_N of an antiferromagnet as the position of singularity of $\delta\rho/\delta T$ versus temperature curve. The line width of the transition corresponding to the difference in temperature of the first minimum to that corresponding to the second maximum is 50 K. This is larger than the previous [21] result. T_N could be anywhere between 90 and 95 K. Different features are displayed by the 1.5 at. % Ni and 2.5 at. % Ni in Mn samples. The resistivity minimum is seen at 170 K for the 1.5 at. % Ni in Mn, whilst it is seen at about 200 K for the 2.5 at. % Ni in Mn. The striking feature occurring in the two curves is that the resistivity maximum is completely collapsed. In both curves the resistivity results show a tendency towards saturation as the temperature approaches zero. In the previous results, the low temperature resistivity obeyed a T^2 law, indicating a spin fluctuation scattering mechanism in the specimens. It is important to note that the temperature T_{min} corresponding to the resistivity minimum, increases from 150 K for the 0.5 at. % Ni in Mn to about 200 K for the 2.5 at. % Ni in Mn. Bulk α – Mn has been shown [25, 26] to exhibit a sharp cusp in its resistivity – temperature curve around 95 K. This temperature has been taken by these authors to correspond to the temperature at which antiferromagnetism sets in, as reported in the neutron diffraction studies. Thus this minimum is associated with the ordering temperature T_N. Coles [27] has suggested that such changes in the resistivity curve can be explained by the presence of a large spin-disorder resistance above the Néel point. The temperature dependence of this resistivity anomaly depends on the amount and type of impurity. The thickness of the films measured in these experiments are in the 2000 – 3000 Å range. The mean free path of the conduction electrons for resistivities of 1.50 μΩm typical of the bulk material is to be estimated [28, 29] to be not more than 1.33 Å even if we assume that the five d-electrons per atom are of high effective mass. Size effects are not observed in these films and film thickness should not influence the resistivities of these samples. In view of the complex magnetic structure of α – Mn and also the fact that the magnetic moment is very dependent upon its environment, the low temperature resistivity reported by Boakye et al [21] might be the result of some disorder in Mn atom as Ni of dilute concentration is introduced. As pointed out earlier, the low temperature resistivity anomaly may depend upon the amount and type of impurities. Kondo [30] has given a theory providing valuable information in understanding the low temperature resistivity of some of these alloys. According to Kondo theory, the resistivity ρ can be given by the expression

$$\rho = \rho_{phon} + C\rho_A + C\rho_m + C\left(3ZJ\rho_m / E_F\right)\ln T$$

where ρ_{phon} is the phonon resistivity, C, the impurity concentration, ρ_A, the resistivity per unit concentration due to impurity potential, ρ_m, a temperature independent quantity occurring in the expression for the spin scattering resistivity, Z, the number of conduction electrons per atom, E_F, the Fermi energy and J denotes the exchange integral for the interaction between localized d-electrons and the conduction electrons. The value of J may be negative, and in this case the last term in the expression will increase as the temperature is reduced. The logarithmic Kondo term is dependent of the concentration of the solute and this might be responsible for the shift of T_{min} to upper values as the concentration is increased, since this term definitely affects the $\rho - T$ curves. The Kondo temperatures were established by doing a ln T curve fitting to match the experimental low temperature curve as illustrated in Hurd [31] for Mn dissolved in copper. The temperature corresponding to the point where the experimental curve departs from the ln T curve corresponds to the Kondo temperature T_K. These are established for 0.5 at. % Ni in Mn T_{K1}, as 23 K, 1.5 at. % Ni in Mn, T_{K2} as 40 K and 2.5 at. % Ni in Mn T_{K3} as 73 K respectively, indicating a shift to higher values as the concentration of the solute is increased. The logarithmic Kondo term therefore explains a tendency towards saturation of the resistivity as T approaches zero and it appears that the magnitude of C influences the shift of T_K to upper values. Within the context of the previous studies in Mn – Ni alloys, it appears that the present concentration of the solute are high to observe the maximum and the low temperature resistivity minimum of the 1.5 at. % Ni and 2.5 at. % Ni in Mn samples. The absence of these features appears to be a quantitative difference resulting from a different atomic size and the electronic structure of Ni.

CONCLUSION

The present results reveal that T_{min} a notable characteristic feature of α – Mn is found in all the Mn – Ni alloys. T_{min} increases as the concentration of Ni increases. The results show a tendency towards saturation of the resistivity as the temperature approaches zero, suggesting a Kondo scattering. The Kondo temperature T_K shifts to upper values as the concentration of Ni increases.

REFERENCES

[1] Sully, A. H., Manganese; Metallurgy of the rarer metals 3; Butterworth Scientific: London, 1955; pp. 135.

[2] Bradley, A.B.; Thewlis, *J. Proc. R. Soc.* 1927, A115, 456.

[3] Shull, C. G.; Wilkinson, M. K. *Rev. Mod. Phys.* 1953, 25, 106.

[4] Yamada, T.; Kunitomi, N.; Nakai, Y.; D. Cox, E.; Shirane, G. *J. Phys. Soc. Jpn.* 1970, 28, 615.

[5] Barrett, C. S.; Masalski, T. B. Structure of Metals; McGraw-Hill: New York, 1966; pp 629.

[6] Ashcroft, N. W.; Mermin, N. D. Solid State Physics; ISBN 0-03-083993-9; Holt Rinehart Winston: New York, 1975; pp 697.

[7] Kaya, S; Kussman, A. Z. *Phys.* 1931, 72, 293.

[8] Sadron, C. *Ann. Phys.* (France). 1932, 17, 371.
[9] Piercy, G. R.; Morgan, E. R. *Can. J. Phys.* 1953, 31, 529.
[10] Van Elst, H. C.; Lubach, B.; Van den Gerg, *G. J. Physica*. 1962, 28, 1297.
[11] Sidorov, S. K..; Doroshenko, A. V. *Phys. Status Solidi*. 1966, 16, 737.
[12] Tange, H.; Tokunaga, T.; Goto, *M. J. Phys. Soc. Jpn*. 1978, 45, 105.
[13] Carr Jr, W. J.; *Phys. Rev.* 1952, 85, 590.
[14] Sodorov, S. K.; Doroshenko, A. V. Fiz Metallou Metalloenie. 1964, 18, 811.
[15] Low, G. G.; Collins, M. F. *J. Appl. Phys.* 1963, 34, 1195.
[16] Collins, M. F.; Low, G. E. *Proc. Phys. Coc.* 1965, 86, 535.
[17] Cable, J. W.; Child, *H. R. Phys.* (France) 1971, 32, C1 – 67.
[18] Cable, J. W.; Child, H. R. *Phys. Rev.* 1974, B10, 4607.
[19] Kitaoka, Y.; Veno, K.; Asayama, *K. J. Phys. Soc. Jpn*. 1974, 44, 142.
[20] Jo, T. *J. Phys. Soc. Jpn*. 1976, 40, 715.
[21] Boakye, F.; Adanu, K. G.; Nkum, R. K. *Acter Mater*. 2001, 49, 2095.
[22] Tolansky, S. Multiple Beam Interferometry of Surface and Films; ASIN B0007IZ4YW; Oxford University Press: Fair Lawn, NJ, 1948; pp 70-73.
[23] Boakye, F.; Ampong, F. K.; Abavare, E. K. K. *Cryogenics*. 2007, 47, 153.
[24] van der Pauw, L. *J. Philips Res. Rep*. 1958, 13.
[25] Bellau, R. V.; Coles, B. R. *Proc. Phys. Soc*. 1963, 82, 121.
[26] Meaden, G. T. *Cryogenics* 1966, 6, 275.
[27] Coles, B. R. *Ad. Phys*. 1958, 7, 40.
[28] Meaden, G. T.; Peloux Gervais, P. *Cryogenics*. 1967, 7, 161.
[29] Guthrie, G. L.; Friedbergen, S. A.; Goldman, *J. E. Phys. Rev*. 4. 1965, section A, 1200.
[30] Kondo, *J. Prog. Therr. Phys*. 1964, 32, 37.
[31] Hurd, C. M. Electrons in metals; John Willey: New York, 1975; p 187.

INDEX

A

B

C

D

E

F

G

H

I

J

K

L

M

Q

R

S

X

Y

Z